VOLUME ONE HUNDRED AND THIRTY TWO

ADVANCES IN CANCER RESEARCH

Molecular and Cellular Basis of Metastasis: Road to Therapy

VOLUME ONE HUNDRED AND THIRTY TWO

ADVANCES IN CANCER RESEARCH

Molecular and Cellular Basis of Metastasis: Road to Therapy

Edited by

DANNY R. WELCH
The Kansas University Medical Center,
The University of Kansas Cancer Center,
Kansas City, KS, United States

PAUL B. FISHER
VCU Institute of Molecular Medicine,
School of Medicine,
Virginia Commonwealth University,
Richmond, VA, United States

ELSEVIER

AMSTERDAM • BOSTON • HEIDELBERG • LONDON
NEW YORK • OXFORD • PARIS • SAN DIEGO
SAN FRANCISCO • SINGAPORE • SYDNEY • TOKYO
Academic Press is an imprint of Elsevier

Academic Press is an imprint of Elsevier
50 Hampshire Street, 5th Floor, Cambridge, MA 02139, United States
525 B Street, Suite 1800, San Diego, CA 92101-4495, United States
The Boulevard, Langford Lane, Kidlington, Oxford OX5 1GB, United Kingdom
125 London Wall, London, EC2Y 5AS, United Kingdom

First edition 2016

ISBN: 978-0-12-804140-6
ISSN: 0065-230X

For information on all Academic Press publications
visit our website at https://www.elsevier.com/

Publisher: Zoe Kruze
Acquisition Editor: Zoe Kruze
Editorial Project Manager: Fenton Coulthurst
Production Project Manager: Surya Narayanan Jayachandran
Cover Designer: Maria Ines Cruz

Typeset by SPi Global, India

CONTENTS

Contributors ix
Preface xi

1. Detecting Tumor Metastases: The Road to Therapy Starts Here **1**

M.E. Menezes, S.K. Das, I. Minn, L. Emdad, X.-Y. Wang, D. Sarkar, M.G. Pomper, and P.B. Fisher

1. Introduction 2
2. Preclinical In Vitro and In Vivo Models of Metastasis 4
3. Approaches Currently Used to Detect Metastatic Lesions (on a Preclinical Level) 9
4. Approaches Used to Detect Metastatic Lesions (on a Clinical Level) 17
5. Metastasis Detection/Therapy: Combining Imaging with Therapy (Theranostics) 26
6. Challenges Faced in Detecting Metastatic Cells 29
7. Conclusions and Future Directions 30

Acknowledgments 30
References 30

2. The Relationship Between Dormant Cancer Cells and Their Microenvironment **45**

N. Linde, G. Fluegen, and J.A. Aguirre-Ghiso

1. Introduction 46
2. Models to Study Tumor Cell Dormancy 48
3. Intrinsic and Extrinsic Signals Converge to Induce Tumor Cell Dormancy 51
4. Cooperative Extrinsic Signals Induce Dormancy Within the Bone Microenvironment 53
5. Endosteal and Perivascular Niches Support Cancer Cell Dormancy 54
6. Reactivation from Dormancy 56
7. The Immune System and Dormancy 58
8. Summary and Outlook 65

References 66

3. Defining the Influence of Germline Variation on Metastasis Using Systems Genetics Approaches **73**

M. Lee and N.P.S. Crawford

1. Introduction 74

2. Hereditary Factors Associated with Cancer Aggressiveness and Metastasis in Human Populations 76
3. Systems Genetics Approaches to Identify Hereditary Modifiers of Metastasis 82
4. Conclusions 99
Acknowledgments 100
References 101

4. Breast Cancer Metastasis Suppressor 1 (BRMS1): Robust Biological and Pathological Data, But Still Enigmatic Mechanism of Action 111

D.R. Welch, C.A. Manton, and D.R. Hurst

Concluding Thoughts 129
Acknowledgments 129
References 129

5. Immune Regulation of the Metastatic Process: Implications for Therapy 139

A. de Mingo Pulido and B. Ruffell

1. Introduction 140
2. Key Players 141
3. Invasion and Intravasation 145
4. Survival and Extravasation 148
5. Ectopic Growth 151
6. Therapeutic Implications 154
Acknowledgments 156
References 156

6. MicroRNA and Metastasis 165

L. Ma

1. MicroRNAs 166
2. miRNAs in Cancer 176
3. miRNA Regulation of Metastasis 186
4. miRNAs as Therapeutic Targets 197
References 200

7. Plasticity of Cancer Cell Invasion—Mechanisms and Implications for Therapy 209

V. te Boekhorst and P. Friedl

1. Introduction 210
2. Cancer Types and Their Migration Modes 212

3. Cancer Cell Migration as a Plastic and Adaptive Process 219
4. Therapeutic Targeting of Cancer Cell Migration—Implications for Anticancer Therapy 229
5. Conclusions 240
6. Future Implications 241
Acknowledgments 241
References 241

8. Cytokine Regulation of Metastasis and Tumorigenicity 265
M. Yao, G. Brummer, D. Acevedo, and N. Cheng

1. Introduction 266
2. Interleukins 267
3. Chemokines 284
4. Chemokines in Therapy 307
5. Interferons 308
6. Tumor Necrosis Factor 321
7. Closing Remarks 333
References 334

Index 369

CONTRIBUTORS

D. Acevedo
University of Kansas Medical Center, Kansas City, KS, United States

J.A. Aguirre-Ghiso
Tisch Cancer Institute, Black Family Stem Cell Institute, Mount Sinai School of Medicine, New York, NY, United States

G. Brummer
University of Kansas Medical Center, Kansas City, KS, United States

N. Cheng
University of Kansas Medical Center, Kansas City, KS, United States

N.P.S. Crawford
Genetics and Molecular Biology Branch, National Human Genome Research Institute, NIH, Bethesda, MD, United States

S.K. Das
Virginia Commonwealth University; VCU Institute of Molecular Medicine, Virginia Commonwealth University; VCU Massey Cancer Center, Virginia Commonwealth University, School of Medicine, Richmond, VA, United States

A. de Mingo Pulido
H. Lee Moffitt Cancer Center and Research Institute, Tampa, FL, United States

L. Emdad
Virginia Commonwealth University; VCU Institute of Molecular Medicine, Virginia Commonwealth University; VCU Massey Cancer Center, Virginia Commonwealth University, School of Medicine, Richmond, VA, United States

P.B. Fisher
Virginia Commonwealth University; VCU Institute of Molecular Medicine, Virginia Commonwealth University; VCU Massey Cancer Center, Virginia Commonwealth University, School of Medicine, Richmond, VA, United States

G. Fluegen
Tisch Cancer Institute, Black Family Stem Cell Institute, Mount Sinai School of Medicine, New York, NY, United States

P. Friedl
David H. Koch Center for Applied Research of Genitourinary Cancers, The University of Texas MD Anderson Cancer Center, Houston, TX, United States; Radboud University Medical Centre, Nijmegen; Cancer Genomics Center (CGC.nl), Utrecht, The Netherlands

D.R. Hurst
University of Alabama at Birmingham, Birmingham, AL, United States

M. Lee
Genetics and Molecular Biology Branch, National Human Genome Research Institute, NIH, Bethesda, MD, United States

N. Linde
Tisch Cancer Institute, Black Family Stem Cell Institute, Mount Sinai School of Medicine, New York, NY, United States

L. Ma
The University of Texas MD Anderson Cancer Center, Houston, TX, United States

C.A. Manton
University of Kansas Medical Center; Kansas City, KS, United States

M.E. Menezes
Virginia Commonwealth University, School of Medicine, Richmond, VA, United States

I. Minn
The Johns Hopkins University School of Medicine, Baltimore, MD, United States

M.G. Pomper
The Johns Hopkins University School of Medicine, Baltimore, MD, United States

B. Ruffell
H. Lee Moffitt Cancer Center and Research Institute, Tampa, FL, United States

D. Sarkar
Virginia Commonwealth University; VCU Institute of Molecular Medicine, Virginia Commonwealth University; VCU Massey Cancer Center, Virginia Commonwealth University, School of Medicine, Richmond, VA, United States

V. te Boekhorst
David H. Koch Center for Applied Research of Genitourinary Cancers, The University of Texas MD Anderson Cancer Center, Houston, TX, United States

X.-Y. Wang
Virginia Commonwealth University; VCU Institute of Molecular Medicine, Virginia Commonwealth University; VCU Massey Cancer Center, Virginia Commonwealth University, School of Medicine, Richmond, VA, United States

D.R. Welch
University of Kansas Medical Center; University of Kansas Cancer Center, Kansas City, KS, United States

M. Yao
University of Kansas Medical Center, Kansas City, KS, United States

PREFACE

Cancer is a multifactorial and multistep process that is regulated by both genetic (ie, intrinsic to tumor cells) and epigenetic processes (ie, by the environment, including the microenvironment that comprises the tumor niche). The majority of cancer-related deaths in patients with solid cancers are attributed to the dissemination from a primary tumor site to discontinuous and distant areas in the body and colonization of those tissues, ie, metastasis. The process of metastasis is complex, involving a number of interrelated steps including local invasion, intravasation, survival in the circulation, adhesion, seeding, extravasation, establishing a new blood supply (angiogenesis) at a new site, and colonization of other tissues. Progress continues to be made in understanding the genetic and epigenetic factors that control metastasis. The role of the tumor microenvironment in facilitating cancer metastasis and recently the seminal role of the immune system and inflammation in regulating the end stage of the cancerous process are the subject of intense investigation.

This thematic issue of *Advances in Cancer Research* contains eight chapters written by experts in specific areas of cancer biology and metastasis and provides an up-to-date overview of "Molecular and Cellular Basis of Metastasis: Road to Therapy." This volume will be of use to both basic scientists and clinicians interested in an up-to-date overview of this pathological process that is a major determinant of cancer-patient death. Through understanding the fundamental processes involved in metastasis and the key players in this process, it may be possible in the future to develop improved approaches for both diagnosing and treating this important component of the cancerous process.

Advances in detecting metastases noninvasively using newer imaging approaches, which will provide a means of following and targeting antimetastatic therapies, and the ability to detect tumor (and potentially metastatic) cells or their DNA or miRNAs in the circulation (circulating tumor cells; liquid biopsy) offer promise of more effectively identifying and developing targeted therapies for metastasis. The first chapter by Dr. Menezes and colleagues, "Detecting Tumor Metastases: The Road to Therapy Starts Here," overviews various preclinical and clinical in vitro and in vivo assays developed to more efficiently detect tumor metastases, which provides a foundation for developing more effective therapies for this invariably fatal component of the cancerous process. Areas covered include preclinical in vitro and in vivo models of metastasis, approaches currently used to detect metastatic lesions (on a preclinical and clinical level), and challenges faced in detecting metastatic cells. This chapter introduces a recent approach that uses molecular-genetic imaging, employing cancer-selective promoters, to detect metastases in preclinical animal models and discusses the potential of using this strategy to develop "theranostics" which combines both imaging and therapy in the same genetic construct.

Chapter 2 by Dr. Linde and colleagues, "The Relationship Between Dormant Cancer Cells and Their Microenvironment," introduces the important concept of "tumor dormancy" in which disseminated tumor cells can survive in a dormant or quiescent state in a target organ and then reactivate resulting in metastases years or even decades after primary tumor diagnosis and treatment. A key to reactivation of dormant cancer cells may involve the microenvironment. This chapter reviews how the microenvironment regulates cancer dormancy and articulates new questions that may help move this field forward.

Chapter 3 by Dr. Lee and colleagues, "Defining the Influence of Germline Variation on Metastasis Using Systems Genetics Approaches," focuses on new methodologies that have allowed the identification of multiple hereditary metastasis susceptibility genes, with wide-ranging cellular functions including regulation of transcription, cell proliferation, and cell–cell adhesion. Approaches used to achieve these objectives include epidemiological studies of cohorts of cancer patients and systems genetics approaches in transgenic mouse models of human cancer. These strategies have facilitated the identification of hereditary metastasis modifiers, determination of the putative molecular functions of these metastasis-associated genes, and the implications of these findings to patient survival.

Chapter 4 by Dr. Welch and colleagues, "Breast Cancer Metastasis Suppressor 1 (BRMS1): Robust Biological and Pathological Data, but Still Enigmatic Mechanism of Action," focuses on the unique BRMS1 gene that has been shown by many groups to block specific steps in the metastatic cascade. This review discusses the data supporting the metastasis suppressing action of BRMS1 in multiple tumor types and focuses on the steps in metastasis that are inhibited. Defining the mechanistic data directly connecting molecular pathways with inhibition of specific steps in metastasis will provide further insights and highlight potential utility of this metastasis suppressing gene.

Chapter 5 by Drs. Pulido and Ruffell, "Immune Regulation of the Metastatic Process: Implications for Therapy," focuses on the role of the immune system and the microenvironment in regulating various components of the metastatic process. The authors discuss the key players of the immune system involved in regulating metastasis including monocytes, macrophages, neutrophils, T lymphocytes, and natural killer cells. The roles of these specific cell types in regulating invasion and intravasation, survival and extravasation, and ectopic growth of metastatic cells are discussed. Lastly, the potential therapeutic applications of targeting the key immune mediators of metastasis are overviewed.

Chapter 6 by Dr. Ma, "MicroRNA and Metastasis," focuses on the functions, mechanisms of action, and therapeutic potential of miRNAs, particularly "oncomirs" (miRNAs that function as oncogenes or tumor suppressor genes) and "metastamirs" (miRNAs that regulate molecular processes and pathways in malignant progression in either a tumor cell-autonomous or a cell-nonautonomous manner). Areas covered include a general discussion of miRNAs, roles of miRNAs in cancer, miRNAs as regulators of metastasis, and miRNAs as potential therapeutic targets.

Chapter 7 by Drs. te Boekhorst and Friedl, "Plasticity of Cancer Cell Invasion—Mechanisms and Implications for Therapy," focuses on basic mechanisms of invasion programs, both in vitro and in vivo, their plasticity, and relevance for metastatic dissemination. Key mechanisms, which were/are thought to underlie these invasion and metastasis programs have been taken forward to preclinical and clinical therapeutic intervention schemes, are discussed with emphasis on efficacy, adaptation, and resistance responses as well as gaps of knowledge between preclinical and clinical evidence. In conclusion, the authors discuss strategic choices to include aspects of metastasis research in clinical and coclinical routines, to better link mechanistic cancer research with clinical evidence.

Chapter 8 by Yao and colleagues, "Cytokine Regulation of Metastasis and Tumorigenicity," focuses on the structure/function of several cytokine families and reviews our current comprehension of the roles and mechanisms of cytokines in tumor progression. In addition, the authors discuss strategies for exploiting the expression and activity of cytokines in therapeutic intervention.

This compendium of state-of-the-art reviews in this important area of cancer biology, with significant import in the area of cancer pathogenesis and therapeutic interventions, will be relevant reading for students, postdoctoral scientists, early- and late-stage faculty members, and clinicians. The eight chapters have captured the current state of research in the area of metastasis, which remains a formidable challenge for the treatment of patients with cancer. We strongly believe: "To defeat the enemy, one must understand the enemy and define its weaknesses." Continued research will help us achieve this outcome.

Danny R. Welch
The Kansas University Medical Center,
The University of Kansas Cancer Center,
Kansas City, KS, United States

Paul B. Fisher
VCU Institute of Molecular Medicine, School of Medicine,
Virginia Commonwealth University,
Richmond, VA, United States

CHAPTER ONE

Detecting Tumor Metastases: The Road to Therapy Starts Here

M.E. Menezes[*,1], S.K. Das[*,†,‡,1], I. Minn[§], L. Emdad[*,†,‡], X.-Y. Wang[*,†,‡], D. Sarkar[*,†,‡], M.G. Pomper[§], P.B. Fisher[*,†,‡,2]

[*]Virginia Commonwealth University, School of Medicine, Richmond, VA, United States
[†]VCU Institute of Molecular Medicine, Virginia Commonwealth University, School of Medicine, Richmond, VA, United States
[‡]VCU Massey Cancer Center, Virginia Commonwealth University, School of Medicine, Richmond, VA, United States
[§]The Johns Hopkins University School of Medicine, Baltimore, MD, United States
[2]Corresponding author: e-mail address: paul.fisher@vcuhealth.org

Contents

1. Introduction 2
2. Preclinical In Vitro and In Vivo Models of Metastasis 4
 2.1 In Vitro Models 4
 2.2 In Vivo Models 6
3. Approaches Currently Used to Detect Metastatic Lesions (on a Preclinical Level) 9
 3.1 Lab Tests/Histopathology 9
 3.2 Noninvasive Blood Tests 9
 3.3 Small Animal Imaging 10
 3.4 Molecular-Genetic Imaging (Promoter-Based Protocols) 12
 3.5 Circulating Tumor Cells 16
4. Approaches Used to Detect Metastatic Lesions (on a Clinical Level) 17
 4.1 Biomarkers for Metastasis 17
 4.2 Imaging Procedures 20
 4.3 Circulating Tumor Cells 23
5. Metastasis Detection/Therapy: Combining Imaging with Therapy (Theranostics) 26
6. Challenges Faced in Detecting Metastatic Cells 29
7. Conclusions and Future Directions 30
Acknowledgments 30
References 30

Abstract

Metastasis is the complex process by which primary tumor cells migrate and establish secondary tumors in an adjacent or distant location in the body. Early detection of metastatic disease and effective therapeutic options for targeting these detected

[1] Contributed equally to this review: M.E.M. and S.K.D.

Advances in Cancer Research, Volume 132
ISSN 0065-230X
http://dx.doi.org/10.1016/bs.acr.2016.07.001

metastases remain impediments to effectively treating patients with advanced cancers. If metastatic lesions are identified early, patients might maximally benefit from effective early therapeutic interventions. Further, monitoring patients whose primary tumors are effectively treated for potential metastatic disease onset is also highly valuable. Finally, patients with metastatic disease can be monitored for efficacy of specific therapeutic interventions through effective metastatic detection techniques. Thus, being able to detect and visualize metastatic lesions is key and provides potential to greatly improve overall patient outcomes. In order to achieve these objectives, researchers have endeavored to mechanistically define the steps involved in the metastatic process as well as ways to effectively detect metastatic progression. We presently overview various preclinical and clinical in vitro and in vivo assays developed to more efficiently detect tumor metastases, which provides the foundation for developing more effective therapies for this invariably fatal component of the cancerous process.

1. INTRODUCTION

According to the American Cancer Society estimates for the year 2016, 1,685,210 new cases of cancer will be diagnosed while 595,690 people are estimated to die from the disease (American Cancer Society, 2016). With recent advances in therapeutic interventions for treating localized primary tumors, the main cause of death among cancer patients is metastasis. In fact, treating metastatic lesions remains the most challenging obstacle for effective therapy in cancer patients (Bacac & Stamenkovic, 2008). Further, early detection of metastatic lesions that might be latent or develop several years after removal of a primary tumor and when a patient is in remission is essential. Thus, developing novel methods for early detection, therapeutic intervention, monitoring, and prevention of metastasis is key to improved patient prognosis. Before we begin our discussion about detecting tumor metastasis, it is important to take a global look at the metastatic process (which is shown schematically in Fig. 1).

The ability of a primary tumor to metastasize to a secondary location in the body is one of the hallmarks of cancer (Hanahan & Weinberg, 2000, 2011). Metastasis from the primary tumor site to a secondary location involves a complex, multistep process (Fisher, 1983; Fisher & Weinstein, 1980; Sahai, 2007). The steps involved in the metastatic process include angiogenesis, epithelial mesenchymal transition (EMT), detachment, degradation of the basement membrane, invasion, migration, intravasation, survival in the circulation, extravasation, and proliferation (Bacac & Stamenkovic, 2008; Fidler, 2003; van Zijl, Krupitza, & Mikulits, 2011).

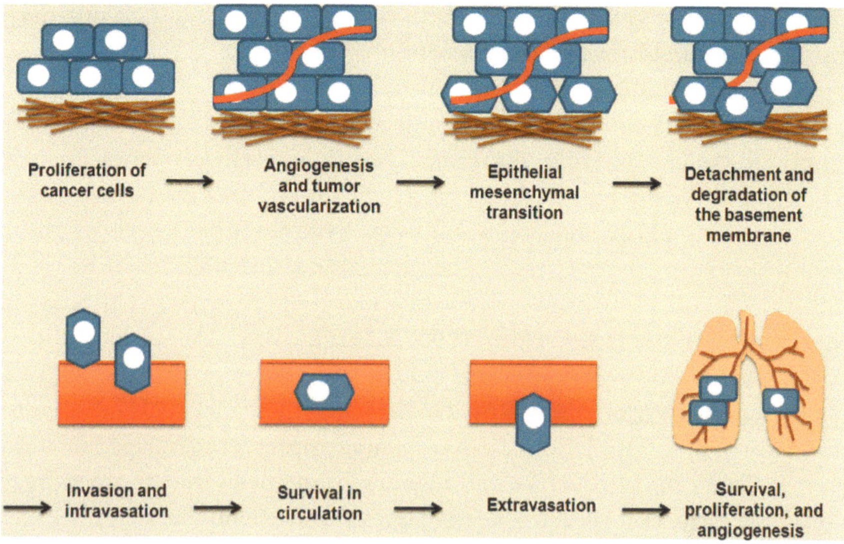

Fig. 1 Schematic diagram of the metastatic process.

In order for the primary tumor to grow and support its metabolic needs, once it exceeds 1–2 mm in diameter, tumor cells secrete various angiogenic factors to establish a capillary network from the surrounding host tissue, resulting in angiogenesis and tumor vascularization (Fidler, 2003). Epithelial, nonmotile tumor cells transform into mesenchymal, motile cells by the process of EMT, although this step remains controversial (Chui, 2013; Yang & Weinberg, 2008). Primary tumor cells that are transformed or have gained invasive/metastatic abilities must detach and degrade the basement membrane to either move through the basement membrane or between endothelial cells to gain access to the bloodstream/vasculature (Valastyan & Weinberg, 2011).

The tumor cells must then undergo the process of invasion and intravasation to enter the circulation. One of the most common routes for tumor cells to enter into the circulation is thin-walled venules, such as lymphatic channels, which tumor cells can easily penetrate. Tumor cells can migrate to distant locations via the blood and lymphatic circulatory channels. Once in the circulatory or lymphatic system, cancer cells must survive the hostile conditions in the circulation. Tumor cells must resist anoikis (programmed cell death associated with loss of cell–cell contact), they must evade recognition by the host immune system as well as successfully handle other physical stresses of being in the circulation (Joyce & Pollard, 2009). When circulating tumor cells (CTCs) arrest in capillary beds at a distant site, they can exit the circulation via extravasation or proliferate within the vessel. These tumor cells must survive in their new microenvironment, proliferate, and evade detection by the host immune system, as well as establish a blood supply for their growing nutritional needs as well as waste disposal (Hanahan & Weinberg, 2011). The newly formed metastases can create new blood vessels by angiogenesis, co-opt existing blood vessels or grow within an existing blood vessel. The metastatic lesion then needs to grow into a clinically relevant metastatic lesion (Chambers, Groom, & MacDonald, 2002; Fidler, 2002, 2003; Talmadge & Gabrilovich, 2013).

Having briefly overviewed the steps in the metastatic process, we will now discuss the various models used in detecting these metastatic cells. Early detection of metastatic lesions can have a huge impact on overall patient outcomes. Given the importance and urgency of the need to detect and treat metastasis early, as well as the need to better understand metastasis, several preclinical and clinical models of metastasis have been developed.

2. PRECLINICAL IN VITRO AND IN VIVO MODELS OF METASTASIS

To gain an enhanced, more complete understanding of the process and signaling mechanisms involved in metastasis, researchers have developed several in vitro and in vivo models that have aided preclinical development (Hulkower & Herber, 2011; Jung, 2014).

2.1 In Vitro Models

Several in vitro assays have been used to assess the various steps involved in the metastatic process. As a tumor grows, tumor cells support their growing needs by generating new blood vessels derived as extensions of the existing vasculature through the process of angiogenesis (Chambers et al., 2002). Tumor cells will also use the tumor vasculature to migrate to distant areas. The ability of tumor cells to generate new blood vessels (angiogenesis) can be assessed using the endothelial tube formation assay (Garrido, Riese, Aracil, & Perez-Aranda, 1995). In this assay, endothelial cells are introduced onto extracellular matrix along with conditioned media obtained from the tumor cells. Factors released into the conditioned media by the tumor cells will reprogram the endothelial cells to form tubes which correlate with angiogenic potential. Detailed instructions on performing the endothelial tube formation assay have been published elsewhere (Arnaoutova & Kleinman, 2010).

Cell–cell interactions as well as cell adhesion molecules play a central role in the metastatic process (Bendas & Borsig, 2012). The adhesion assay is utilized to determine changes in the ability of tumor cells to adhere to and interact with different extracellular matrices. During the metastatic process, tumor cells have to adhere to various extracellular matrices to intravasate and extravasate to successfully establish a distant metastasis. Extracellular matrix proteins such as fibronectin, collagen, and laminin are coated onto the bottom of culture dishes and the ability of tumor cells to adhere to the matrix is assessed in the adhesion assay. Several different combinations of extracellular matrix materials precoated onto cell culture plates are also commercially available. The ability of tumor cells to adhere to endothelium can also be assessed using commercially available reagents such as CytoSelect™ Tumor-Endothelium Adhesion assay by Cell Biolabs, Inc. This assay also provides a glimpse at the angiogenic capabilities of tumor cells.

Tumor cells must be capable of migrating from their primary site to a distant organ as well as within the circulation in order to metastasize. The scratch–wound assay is utilized to assess the motility and migration capabilities of tumor cells. Tumor cells are grown in a confluent monolayer. A scratch is made in the confluent layer and the ability of the cells to fill in the scratch or wound is measured over time. Detailed technical instructions on performing scratch–wound assays have been published elsewhere (Cory, 2011; Liang, Park, & Guan, 2007). Another method to determine migratory potential of tumor cells is by using a Transwell or modified Boyden chamber assay. Tumor cells are introduced into the upper chamber and their ability to migrate toward a chemoattractant in the lower chamber is assessed (Chen, 2005).

The ability of tumor cells to invade is another key attribute that ensures the successful development of metastatic lesions. This ability to invade can be assessed using the Boyden chamber invasion assay. The Boyden chamber contains a Transwell membrane with a gel of extracellular matrix proteins. One of the most commonly used Transwell membranes consists of Matrigel. The tumor cells are introduced into the upper chamber and their ability to invade through the Matrigel and toward a chemoattractant in the lower chamber is assessed. Detailed technical instructions on performing Transwell® Invasion assays (Marshall, 2011) and Boyden chamber assays have been published elsewhere (Falasca, Raimondi, & Maffucci, 2011).

Researchers have also developed three-dimensional (3D) culture assays to determine metastatic abilities of tumor cells in a more complex architectural context mimicking conditions in vivo. Here, tumor cells are allowed to grow within a 3D matrix and various attributes of the tumor spheroids that develop within the matrix are assessed (Debnath, Muthuswamy, & Brugge, 2003).

2.2 In Vivo Models

Preclinical in vivo models have been of significant benefit in evaluating innovative methods of detecting tumors and novel therapeutic intervention approaches for potential use in treating human cancers. Several different in vivo models of metastasis have been developed to model human metastatic disease (Khanna & Hunter, 2005; Menezes et al., 2014).

Experimental and spontaneous metastasis mouse models have been developed using both human and mouse cell lines (Price, 2001, 2014). In the experimental metastasis mouse model, human tumor cells with known or suspected metastatic capabilities are injected into immunodeficient mice. The injected tumor cells colonize organs depending on the intrinsic metastatic homing capabilities of the injected tumor cells as well as the site where tumor cells are introduced. Tumor cells that are injected into the tail vein mainly form lung metastases, tumor cells injected into the portal vein colonize the liver, intracardiac injection of tumor cells results in metastasis to a number of organs including bone (Fantozzi & Christofori, 2006), and intratibial injection of tumor cells results in bone metastasis (Park, Kim, McCauley, & Gallick, 2010). For example, when MDA-MB-231 human breast cancer cells are xenografted by injecting the tumor cells into the tail vein, mice develop lung metastasis (Yang, Zhang, & Huang, 2012). Another example is B16 melanoma cells that also form lung metastases when injected via the tail vein (Giavazzi & Decio, 2014). PC-3 prostate cancer cells when delivered via intratibial injection produce osteolytic metastatic lesions while LNCaP prostate cancer cells produce mixed osteoblastic and osteolytic lesions (Park et al., 2010). In these xenograft models, human tumor cells can safely be injected into mice without any rejection because the mice are immunodeficient. However, because these mice are immunodeficient, this model cannot be utilized to study the relevance of an intact immune system in metastatic progression. Other advantages of experimental metastasis models include the ability to control the number of tumor cells introduced into the mouse as well as the ability to target tumor cells to specific sites. Also experimental metastasis models generally develop tumors in a shorter duration of time as compared to spontaneous metastasis models. However, the experimental metastasis models use a more artificial route of tumor cell delivery to establish metastasis and limited steps within the metastatic process can be assessed using this model (Francia, Cruz-Munoz, Man, Xu, & Kerbel, 2011).

In the spontaneous tumor metastasis mouse model, tumor cells that have or are suspected to have metastatic capabilities are introduced into mice and metastasis is allowed to develop spontaneously (Fidler, 2006). For example, when 4T1 mouse mammary tumor cells are orthotopically injected into the mammary fat pad of syngeneic mice, the cells will spontaneously metastasize to the lungs, liver, bone, and brain (Fantozzi & Christofori, 2006; Yang et al., 2012). When PC-3MM2GL, a highly metastatic variant of PC-3 prostate cancer cells are orthotopically implanted intraprostatic, 100% lymph node metastases develop in 4 weeks (Park et al., 2010). Utilizing syngeneic mouse models are useful when evaluating the role of the immune system and the immune components in metastasis. Other advantages of using spontaneous metastasis models include the ability to follow and determine the mechanisms of metastatic spread via a more natural route that more closely resembles clinical disease as well as the ability to follow all of the steps involved in the metastatic process (Francia et al., 2011). The drawbacks of spontaneous metastasis models include prolonged time required for development of metastases and asynchronous development of metastatic lesions at multiple locations (Francia et al., 2011).

Researchers have also developed genetically engineered or transgenic mouse models that spontaneously develop tumors and metastases (Menezes et al., 2014; Smith & Muller, 2013). In these models, the genetic makeup of the mouse is altered to either inhibit the expression of a tumor suppressor gene or overexpress an oncogene or combinations of the two, so that mice will spontaneously develop tumors and metastases over their normal lifespan, sometimes as rapidly as a couple of months (Eklund, Bry, & Alitalo, 2013; Fantozzi & Christofori, 2006; Husemann & Klein, 2009). In addition, mice that develop tumors in a specific region have been developed by utilizing tissue-specific promoters so that the desired tumor suppressor is suppressed in the specific tissue while the desired oncogene is overexpressed in the specific tissue. These models have an intact immune system that facilitates studying the role of the immune system and immune components as well as the microenvironment in tumor metastasis. For example, the MMTV-PyMT transgenic mouse model has been used to model breast cancer progression and metastasis. In this model, the PyMT (polyoma virus middle T antigen) is expressed under the transcriptional control of the MMTV (mouse mammary tumor virus) promoter, which gives rise to multifocal mammary adenocarcinomas in all the mammary glands of female mice with 100% incidence and metastatic lesions develop in the lungs and lymph nodes by 3 months of age (Guy, Cardiff, & Muller, 1992). Similarly, female MMTV-c-myc transgenic mice that overexpress c-myc under the transcriptional control of the MMTV promoter in the mammary glands develop mammary adenocarcinomas in 5–6 months with a 100% incidence and develop metastatic tumors in the lungs (Stewart, Pattengale, & Leder, 1984). Another example is transgenic mice expressing $Kras^{G12D}$ with Ink4a/Arf deficiency specifically in the pancreas (Aguirre et al., 2003). These mice develop highly invasive and metastatic pancreatic cancers, with all mice succumbing to tumors by 11 weeks. The TRAMP (transgenic adenocarcinoma mouse prostate) mouse model expresses the SV40 T antigen under the transcriptional control of the rat probasin promoter to enhance prostate-specific expression (Parisotto & Metzger, 2013). These mice primarily develop lung and lymph node metastases but occasionally liver, kidney, and adrenal gland metastases are also observed. Transgenic mice expressing $BRaf^{V600E}$ and silenced for the tumor suppressor Pten develop melanoma with 100% penetrance and metastases develop in the lymph nodes and lungs (Dankort et al., 2009). Although we highlight a small list of interesting transgenic models used to study metastasis, research efforts continue to focus in multiple laboratories to develop genetically engineered mice that accurately recapitulate both primary tumor-specific development and metastases.

Finally, researchers have developed human patient-derived xenograft mouse models that might potentially lead to personalized medicine for cancer patients (Aparicio, Hidalgo, & Kung, 2015; Siolas & Hannon, 2013). In this model, a freshly harvested patient-derived tumor is digested and transplanted into an immunodeficient mouse. To maintain this model, cells are directly passaged from one mouse to another when the tumor burden becomes too high. Patient tumor can be transplanted into the subcutaneous flank of a mouse or orthotopically into the location where the tumor was derived from the patient. For example, human histologically-intact pancreatic cancer specimens have been orthotopically transplanted into athymic mouse pancreas and this model showed extensive local tumor growth, as well as metastases in the liver, regional lymph nodes, adrenal gland, diaphragm, and mediastinal lymph nodes (Fu, Guadagni, & Hoffman, 1992).

3. APPROACHES CURRENTLY USED TO DETECT METASTATIC LESIONS (ON A PRECLINICAL LEVEL)

3.1 Lab Tests/Histopathology

The presence of metastatic lesions in mice with tumors can be detected by sacrificing the mice at the appropriate time point and harvesting the desired organs to be tested for presence of metastases. These organs can be fixed in formalin, paraffin-embedded, sectioned, and placed on glass slides. These slides can then be stained using H&E (hematoxylin and eosin) staining and visually examined for the presence of metastatic cells by a trained pathologist. The slides can also be stained using other techniques to identify the presence of markers of the metastatic process. For example, immunohistochemical analysis (IHC) to identify the presence of CD31 and Factor VIII-related antigen are useful to detect and quantify tumor angiogenesis (Wang et al., 2008). Using human melanoma xenograft mouse model, lectin HPA and adhesion molecules CEACAM-1 (carcinoembryonic antigen-related cell adhesion molecule 1), and L1 expression were assessed by IHC and HPA, CEACAM-1 and L1 were shown to be markers of metastasis (Thies, Mauer, Fodstad, & Schumacher, 2007). Further, metastatic lesions can be probed for expression of protein with known metastatic capabilities to provide additional insight regarding the lesions. For example, melanoma differentiation associated gene-9 (MDA-9/syntenin), ie, frequently overexpressed in metastatic lesions can be assessed to identify metastases and can serve as a biomarker of metastasis (Boukerche et al., 2005).

3.2 Noninvasive Blood Tests

Research is also focused on developing noninvasive methods to detect and monitor tumor metastases using blood tests. miRNAs are small noncoding RNAs that play an important role in tumor progression and metastasis (Baranwal & Alahari, 2010; Zhang, Yang, & Wang, 2014). The presence of different miRNAs has been assessed from blood obtained from mice with tumors. As an example, whole blood was collected from transgenic mice with c-MYC-induced lymphoma, hepatocellular carcinoma, and osteosarcoma, and assessed for the expression of 20–30 miRNAs at different stages of the tumorigenic process (Fan et al., 2008). Specific miRNA expression profiles were identified based on the tumor type and stage. Interestingly, when the tumors regressed the expression of these miRNA returned to normal levels (Fan et al., 2008). Thus, detecting specific changes in miRNA expression from blood can aid in the detection of tumor progression as well as monitoring the efficacy of therapeutic interventions.

A blood-based tumor-activatable microcircle approach was devised for detection of tumors and metastases (Ronald, Chuang, Dragulescu-Andrasi, Hori, & Gambhir, 2015). Nonviral tumor-activatable minicircles encoding human SEAP (secreted embryonic alkaline phosphatase) under the transcriptional control of the tumor-specific survivin promoter along with a transfection agent was introduced systemically via the tail vein of the mouse. The reporter SEAP was only produced in tumor cells and was secreted into the blood stream of the mice. Blood samples collected from mice and plasma were screened for the presence of SEAP, which was indicative of the presence of tumors in the mouse (Ronald et al., 2015).

3.3 Small Animal Imaging

Visualizing metastases in vivo in mice provides an excellent model system to develop and evaluate approaches that may be translatable to visualization of metastases in human patients. The methods discussed earlier detect metastases from ex vivo samples but the methods discussed below detect metastasis in the mouse in vivo allowing for visualization and monitoring of metastasis onset and progression or regression over time without the need to sacrifice the mouse.

In order to directly follow and observe the metastatic process as it evolves over time, researchers have also utilized high-resolution in vivo videomicroscopy in living animals (Chambers, MacDonald, Schmidt, Morris, & Groom, 1998). Using a video camera attached to a light microscope, the movement of cancer cells within the circulation, escape from the circulation and events following escape in the surrounding tissue can be assessed (Chambers et al., 1995; MacDonald & Chambers, 2008).

Bioluminescence imaging (BLI) is a noninvasive, quantitative method of detecting metastases in mice in vivo (Badr, 2014; Contag et al., 1995; Minn et al., 2014). It is generally the first approach to image gene-tagged cells in vivo, when populations of cells rather than single cells are to be studied. Other, more complicated modalities are used for studies beyond small animals, namely clinical translation, as discussed later. Firefly luciferase is the most commonly used luciferase system for BLI. Tumor cells have to be stably transfected to express the firefly luciferase gene (or another detectable agent) and then introduced into the mice to assess experimental or spontaneous metastases. The substrate luciferin is introduced into mice intraperitoneally and metastases are visualized using a bioluminescent imager. Several other luciferase enzymes have been used for BLI including firefly luciferase, Renilla luciferase, Gaussia luciferase, Metridia luciferase, Vargula luciferase, and bacterial luciferase (Close, Xu, Sayler, & Ripp, 2011). Liver metastatic lesions were successfully visualized by injecting HCT-116 human colon carcinoma cells stably expressing firefly luciferase through the portal vein into the liver of athymic mice (Thalheimer et al., 2013). A slightly different approach is to deliver tumor-specific promoters either singly or within microbubbles via the tail vein of mice in order to detect the presence of tumor metastases. Using the tumor-specific PEG-3 (progression-elevated gene-3) promoter (Su et al., 2005), experimental metastases of human melanoma and breast cancer cells could be visualized in an athymic mouse using BLI (Bhang, Gabrielson, Laterra, Fisher, & Pomper, 2011). This is one instance of BLI being used to study inducible genetic systems, as further highlighted later (Section 3.4).

Magnetic resonance imaging (MRI) is another method for noninvasively detecting metastasis (Gauvain, Garbow, Song, Hirbe, & Weilbaecher, 2005). MRI is one of the standard imaging techniques used in the clinical setting for detection of tumors as well as metastasis (Nakanishi et al., 2007). MRI is performed using a magnetic resonance scanner and a small animal receiver coil. Using different xenograft mouse models of cancer, metastatic lesions in the liver, brain, adrenal glands, and lymph nodes were detected using MRI (Peldschus & Ittrich, 2014). Pancreatic tumors and liver metastases could be visualized using MRI in a orthotopic pancreatic cancer mouse model and liver metastasis mouse model (Partecke et al., 2011). Microcomputerized tomography (CT) has also successfully been used to detect metastases in mice. Using a hepatocyte-selective contrast agent and micro-CT, liver metastases established after injecting colon adenocarcinoma cells into the portal vein of mice could be detected (Kim et al., 2008). MR and CT are generally anatomic techniques for detection of metastases in vivo. Radionuclide-based techniques, such as positron emission tomography (PET) and single-photon emission computed tomography (SPECT) are frequently used for this purpose as well, when higher sensitivity is required at the expense of spatial resolution. Use of radionuclides for preclinical imaging of cancer and metastases has been extensively reviewed (Koba, Jelicks, & Fine, 2013; Vaquero & Kinahan, 2015; Weissleder & Nahrendorf, 2015) and is of course the mainstay of doing so clinically for molecular imaging with [^{18}F]fluorodeoxyglucose, when coupled with CT or MR, and an increasing number of newer, tumor-specific agents

(Elsinga & Dierckx, 2014; Spick, Herrmann, & Czernin, 2016). BLI and the radionuclide techniques are also frequently used to detect the presence and appearance of metastases using transcription-induced methods, as discussed later. One example is where MDA-MB-231 breast cancer cells stably expressing HSV1-tk (herpes simplex virus 1 thymidine kinase) was utilized with SPECT to detect bone metastasis in mice (Sanches et al., 2015). The precise location of the bone metastases could be detected by accumulation of the radiolabeled tracer promoted by HSV1-tk using SPECT.

3.4 Molecular-Genetic Imaging (Promoter-Based Protocols)

Molecular-genetic imaging approaches allow visualizing and quantifying biochemical processes at the cellular and molecular level (Bhang & Pomper, 2012; Minn et al., 2014; Pomper & Fisher, 2014). The basic components for molecular-genetic imaging are gene promoters, reporters, and gene delivery vehicles. Several gene promoters that are selectively active in cancer cells have been identified and utilized for molecular-genetic imaging approaches (Minn et al., 2014). Some of these promoters include the PEG-3 promoter (Su, Shi, & Fisher, 1997), the AEG-1 (astrocyte-elevated gene-1) promoter (Bhatnagar et al., 2014; Kang et al., 2005), the hTERT (human telomerase reverse transcriptase) promoter (Majumdar et al., 2001), a truncated tCCN1 (Cysteine-rich protein 61) promoter (Sarkar et al., 2015), and the survivin promoter (Chen et al., 2004). These promoters that are selectively active in tumor cells can be combined with different reporters and different gene delivery methods in order to visualize tumors and metastatic lesions in vivo. For example, the tumor- and metastasis-specific promoter PEG-3 was combined with the firefly luciferase reporter and HSV1-tk reporter, and delivered using in vivo-jetPEI® (Polyplus transfection) in order to detect micrometastatic disease in mouse models of human melanoma and breast cancer (Bhang et al., 2011) (Fig. 2). Similarly, the tumor-specific AEG-1 promoter was combined with the firefly luciferase reporter and HSV1-tk reporter and delivered using nanoparticles to detect metastases in soft tissues and bone in a mouse model of prostate cancer (Bhatnagar et al., 2014) (Fig. 2).

Use of tumor-specific promoter enables the molecular-genetic imaging applicable to all types of human cancer, whereas the application of conventional target-based imaging is usually limited to certain cancers that express specific targets. In addition, by adopting theranostic reporters, the molecular-genetic approach can provide therapies as well. Systemic

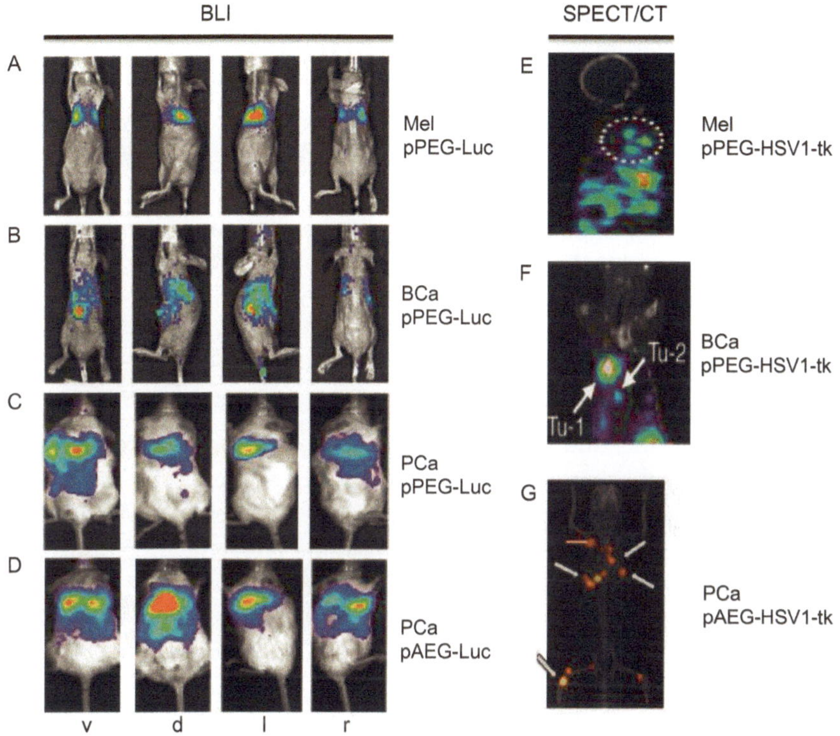

Fig. 2 Cancer-specific promoter-based imaging detects metastasis of human cancer in animal models. (A–D) Bioluminescent imaging (*BLI*). (E–G) Single-photon emission computed tomography (SPECT/CT). Melanoma (A and E), breast cancer (B and F), and PCa (PC3/ML) (C, D, and G) cells injected IV and developed as metastatic lesions in immunocompromised mice were imaged with reporter genes systemically delivered (IV) in an L-PEI nanoparticle under the control of tumor-specific promoters PEG-prom and AEG-prom. *v*, ventral; *d*, dorsal; *l*, left; and *r*, right views.

injection of the molecular-genetic vector formulated with in vivo delivery nanoparticle is suitable for targeting and imaging metastatic cancers with unknown locations in the body of patients. Although imaging large tumor lesions would be feasible with this technology, detection of micrometastatic lesions may be limited due to relatively lower promoter strength of majority of the tumor-specific promoters (Minn et al., 2014). In order to enhance the expression level of reporters, researchers have developed several molecular biology strategies.

Transcriptional enhancers have been added to expression vectors. Strong enhancers such as human cytomegalovirus (CMV) immediate-early enhancers (Penuelas et al., 2005) and simian virus 40 (SV40) enhancers

(Luke et al., 2011) can provide universal elevation of promoter strength. Other enhancers from target- or tissue-specific genes have also been tested. The examples include prostate-specific antigen (PSA) promoter/enhancer combination for targeting prostate cancer (Latham, Searle, Mautner, & James, 2000), prostate stem cell antigen enhancer with uroplakin II promoter for targeting bladder cancers (Wang et al., 2010), and endothelin-1 promoter/enhancer combination targeting tumor neovasculature (Dronadula et al., 2011).

Another successful approach to boost promoter strength was to adopt a research tool originally developed for firefly genetic research, GAL4/UAS system (Brand & Perrimon, 1993). The concept of amplifying transcriptional activity using the two step GAL4/UAS system was first tested by Nettelbeck, Jerome, and Muller (1998). The idea was to create a positive feedback loop of expression by constructing two expression cassettes in the expression vector. First, a week endothelial cell-specific von Willebrand factor promoter expresses a fusion protein of the DNA-binding domain of Lex A repressor and a strong VP16 activator of HSV1. Second, a reporter gene is under the control of multiple copies of Lex A binding sequences and minimal promoter. The study demonstrated 100-fold increase of tissue-specific expression of the reporter. This useful technique was further developed into a concept of two-step transcriptional amplification (TSTA) for molecular-genetic imaging (Iyer et al., 2001). Our preliminary in vivo study shows enhancement of sensitivity in detection of small metastatic lesions using the TSTA system (Fig. 3).

Size of the expression plasmid significantly affects the efficiency of transfection. A systemic study demonstrated that the expression level of a reporter was inversely proportional to the size of the expression vector (Yin, Xiang, & Li, 2005). The study also showed that plasmids larger than 5.1 kb exhibited severely decreased transfection efficiency, providing theoretical size limitation for in vivo expression vectors. A recent study using minicircle expression DNA demonstrated 5- to 10-fold enhanced expression of the reporter for cystic fibrosis gene therapy (Munye et al., 2016). This study also showed that the minicircle vector enabled prolonged expression of the reporter and conferred reduced inflammatory response. A practical use of minicircle expression for cancer detection used the survivin promoter upstream of a reporter protein, human SEAP (Ronald et al., 2015). Because SEAP is orthogonal to proteins normally expressed in adult tissue it could be used in a sensitive way to detect cancer in body fluids. Although molecular-genetic imaging was not used in this case, this example shows the versatility

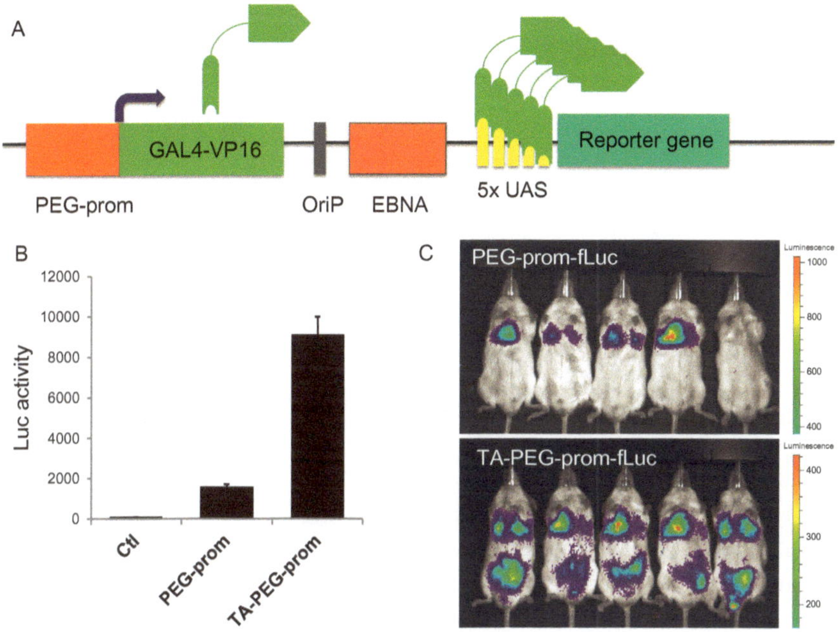

Fig. 3 Transcriptional amplification (TA) machinery. (A) Schematic diagram for the TA system. (B) In vitro enhancement of luciferase activity via the TA system. (C) In vivo sensitivity enhancement of TA vector. Note that TA system was able to detect small metastatic lesions (PC3/ML PCa) in liver and kidney (*lower images*), which were undetectable by the parental vector (*upper images*).

of such approaches, namely, that one may merely switch out the promoter, or imaging agent with other detectable substances or even gene-encoded therapeutics.

Promoters other than PEG-3 and survivin (Huang et al., 2011) have been used to good advantage for tumor-selective molecular-genetic imaging and therapy, including hTERT and AEG-1, as alluded to earlier. Because of the modular aspect to the plasmids the imaging reporter and even modality can be switched out. For example, hTERT has been used for tumor-selective expression of the human sodium–iodide symporter (hNIS) for therapy with ^{131}I, a β-particle emitter long used to treat thyroid cancer (Rajecki et al., 2012). In that study the ^{123}I was used as the imaging nuclide in conjunction with SPECT. Delivery of the transgene was through an oncolytic adenovirus. By using hTERT to drive the ferritin heavy chain MR imaging could be used to detect a variety of tumors in vivo, although at lower sensitivity than through methods employing radiotracers

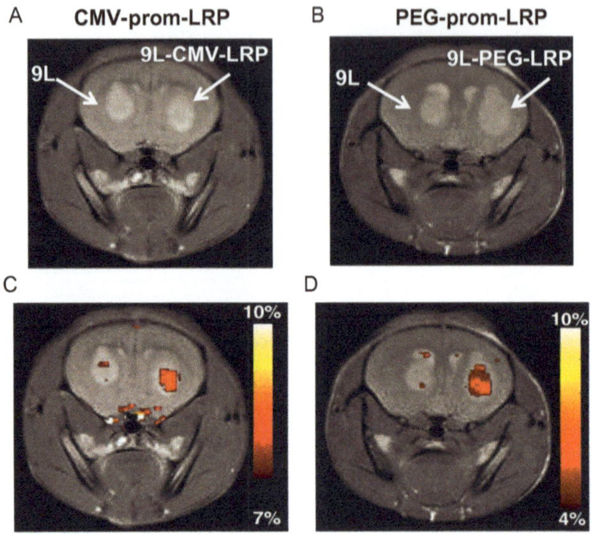

Fig. 4 In vivo CEST-MR imaging of glioma expressing LRP. Rat glioma cell line (9L) and 9L-expressing LRP were injected left and right side of a brain of mice, respectively. Both CMV-prom (A and C) and PEG-prom (B and D) successfully drive the expression of LRP, which gives CEST contrast (C and D).

(Yang et al., 2016). MR was also the modality of choice for imaging the lysine-rich protein, the expression of which was driven by PEG-3, using chemical exchange saturation transfer (Minn et al., 2015) (Fig. 4). A variety of other tumor-selective promoters, including those that are merely organ-specific, cell-type-specific, and more frankly cancer selective have recently been reviewed (Ahn, 2014; Bhang & Pomper, 2012). Use of promoters that display selective (or enhanced) expression in metastatic cells vs primary tumor cells, mesenchymal tumor cells during EMT, tumor vasculature, cancer stem cells, and hypoxic environments will also be of significant value in designing molecular genetic-based cancer imaging and therapeutic approaches (Talukdar, Emdad, Das, Sarkar, & Fisher, 2016).

3.5 Circulating Tumor Cells

Detecting CTCs in vivo is important and will greatly aid prognostic analysis as well as monitoring efficacy of therapy. Using fluorescently labeled tumor cells, researchers have been able to effectively monitor CTCs in mice (Hoffman, 2014). Using human PC-3 prostate cancer cells, Glinskii and colleagues showed that PC-3 cells growing orthotopically in athymic mice produced more viable CTCs that PC-3 cells growing ectopically (subcutaneously) (Glinskii et al., 2003). Further, using a dual-color orthotopic coimplantation model of human prostate cancer metastasis in athymic mice, Glinskii and colleagues were able to determine the metastatic potential of CTCs (Glinskii et al., 2003). An equal number of PC-3 GFP-expressing cells isolated from the circulation of mice with orthotopic tumors (CTCs) and parental PC-3 RFP-expressing cells were orthotopically transplanted into athymic mice and interestingly the metastatic lesions were almost exclusively GFP-expressing cells, indicating that cells isolated from the circulation (CTCs) had higher metastatic potential than parental cells (Glinskii et al., 2003). Tumors formed from CTCs isolated from orthotopic PC-3 GFP-expressing tumors in athymic mice could be imaged using the FluorVivo imaging system by INDEC Biosystems and the CTCs could be isolated using immunomagnetic beads coated with anti-EpCAM (epithelial cell adhesion molecule) and anti-PSMA (prostate-specific membrane antigen) within minutes (Kolostova, Pinterova, Hoffman, & Bobek, 2011).

4. APPROACHES USED TO DETECT METASTATIC LESIONS (ON A CLINICAL LEVEL)

In the preceding section, we discussed several approaches for detecting metastatic lesions on a preclinical level. Any diagnostic procedure is considered as perfect only when it displays 100% sensitivity and specificity (ie, everyone with cancer would have a positive test, while everyone without cancer would exhibit a negative test) (Mordente, Meucci, Martorana, & Silvestrini, 2015). Regrettably, not all approaches meet these criteria and have not successfully translated from "bench to bedside." Here, we provide a general overview of the different "conventional" and "state-of-the-art" processes that are routinely used for identifying metastatic lesions in different cancers.

4.1 Biomarkers for Metastasis

"Biomarker" is defined by US Food and Drug Administration (FDA) as a characteristic that is objectively measured and evaluated as an indicator of normal biologic processes, pathogenic processes, or pharmacologic responses to a therapeutic intervention (Taube, Jacobson, & Lively, 2005) that can be detected in circulation (whole-blood, serum, and plasma), secretions (stool, urine, and sputum), or organ biopsies (Kulasingam & Diamandis, 2008; Mordente et al., 2015). Until recently, FDA approved 19 protein-based biomarkers (Mordente et al., 2015). Clinically, although these biomarkers are routinely used for monitoring tumor progression, staging, and in some contexts screening purposes, their correlation with tumor metastasis are not fully reliable and often need confirmation using a second approach, eg, imaging.

PSA is a well-described biomarker for prostate cancer and its utility as a diagnostic tool is established (Wilt, Scardino, Carlsson, & Basch, 2014). In 1994, Vijayakumar et al. retrospectively evaluated 90 patients with prostate cancer, on the basis of initial serum-PSA level and bone scans, and found that patients with PSA more than 10 ng/mL had evidence of bone metastasis (Vijayakumar, Vijayakumar, Quadri, & Blend, 1994). Similar results were obtained in another retrospective study conducted by Kamaleshwaran et al. where the cut-off value for PSA was set at 20 ng/mL or greater for predicting bone metastases (Kamaleshwaran et al., 2012). In subsequent studies, to determine the correlation between PSA levels and bone metastasis risk, Moreira et al. (2015) developed a bone metastasis predictive table using serum PSA level that was further validated by Freedland et al. (2016). According to these two recent studies the cut-off value for PSA was 5 ng/mL and below that level the incidence of skeletal metastasis was very rare (Freedland et al. (2016). In another pooled study using 8644 patients (from 23 studies), the investigators observed that the likelihood of a positive bone scan increases markedly in patients who exhibit a PSA level ≥20 ng/mL, locally advanced disease, or a Gleason score ≥8 (Abuzallouf, Dayes, & Lukka, 2004; Briganti et al., 2014). Overall, all of these retrospective studies argue that PSA levels are a valid predictor of bone metastasis, however, which cut-off value most accurately predicts risk is unclear.

Mucins are a family of high-molecular weight glycoproteins expressed/produced by epithelial cells (Nicolini, Ferrari, & Rossi, 2015). Members of this family are involved in breast cancer development and altered expression is associated with cancer progression (Nicolini et al., 2015). Carbohydrate antigen 15.3 (CA15.3) is the most common member in this family and approved by FDA as a biomarker for breast cancer and some other malignancies such as ovarian cancer, endometrial carcinoma, and nonsmall-cell lung cancer (Molina et al., 2008; Moore et al., 2008; Nicolini et al., 2015). Geraghty et al. demonstrated that the serum CA15.3 level was elevated in 50–80% of breast cancer patients with metastasis (Geraghty, Coveney, Sherry, O'Higgins, & Duffy, 1992). In a study conducted in 2007, Keshaviah et al. investigated the relevance of this biomarker with development of recurrence and found that the risk of recurrence increased by 30% in breast cancer patients with abnormal levels of CA15.3 (Keshaviah et al., 2007). In another study, 88% (23 out of 33) patients showed a positive correlation between serum CA15.3 levels and breast cancer bone metastasis (O'Brien et al., 1992). In a prospective study (Kokko, Holli, & Hakama, 2002), 243 patients with localized disease were followed for relapse and metastasis development after primary treatment. Fifty-nine relapses were noticed within 5 years and 36% of these had an elevated level of CA15.3. More than 50% of bone metastasis in these subjects demonstrated a higher level of CA15.3, supporting the use of CA15.3 as an alternative to conventional bone scintigraphy (Byrne, Horgan, England, Callaghan, & Given, 1992). Begić et al. compared bone scintigraphy with serum CA15.3 levels and found a weak correlation between the number of metastases and CA15.3 levels (Begić et al., 2005). However, a significant difference was observed in CA15.3 levels when comparing patients with metastases to patients without metastases. This particular study also compared CEA (carcinoembryonic antigen, another serum biomarker approved by the FDA) levels with bone metastasis development and observed a positive correlation between CEA levels and the number of metastatic lesions (Begić et al., 2006). CA19.9, another member of the mucin family is being considered as a potent diagnostic factor for pancreatic cancer with overall sensitivity of 81% and specificity of 90% (Duffy et al., 2010), although the utility of this factor to screen pancreatic cancer is questionable. Kim et al. analyzed the serum levels of CA19.9 among 84 pancreatic cancer patients who had undergone curative resection and found a positive association between CA19.9 levels in about 69% of the patients that developed distant metastasis within 6 months (Kim et al., 2011). Additionally, this study confirmed that patients with higher preoperative levels of CA19.9 also had a higher tendency to develop distant metastasis (Kim et al., 2011).

Epidermal growth factor receptor (EGFR) and v-erb-b2 erythroblastic leukemia viral oncogene homolog 2 (HER2-neu) are two additional biomarkers detected in tissue specimens and the correlation of these biomarkers with disease progression are well studied although their diagnostic value in metastasis remain to be confirmed. In a recent study, Wang and Wang systematically performed a meta-analysis to define EGFR mutations in primary and matching metastatic nonsmall cell lung cancer (NSCLC) and found that EGFR mutations are present both in primary and metastatic NSCLC lesion, and therefore routine analysis of EGFR is not recommended in primary and metastatic tumors (Wang & Wang, 2015). In another study, Westood and colleagues analyzed 12 databases and verified EGFR mutation status in NSCLCs (Westwood et al., 2014). Consistent with Wang and Wang's

(2015) observation they also did not find any greater accuracy for EGFR mutation as a diagnostic measure. However, some positive correlations of EGFR with breast cancer metastasis are evident. Gaedcke et al. demonstrated that EGFR expression was increased by 40% in brain metastases compared to primary tumors, which showed only 16% EGFR expression (Gaedcke et al., 2007). However, it is important to also note that 75–85% of primary and brain metastatic tumors were shown to have constant EGFR expression (Gaedcke et al., 2007; Grupka, Lear-Kaul, Kleinschmidt-DeMasters, & Singh, 2004). In agreement with breast cancer data, Deng et al. analyzed and concluded that higher expression of EGFR in metastatic lymph nodes may be more accurate in predicting survival than in primary or metastatic tissues (Deng et al., 2009).

A study of elevated HER2 levels, a protein-based biomarker, detected in tissue specimens of primary tumors and axillary lymph node or distant metastases has been shown to correlate in several clinical studies (Brufsky et al., 2011; Kuba et al., 2014; Shao et al., 2011). In contrast, multiple studies indicate that HER2-positive metastases with negative primary tumors are more frequent (Dieci et al., 2013; Jensen, Knoop, Ewertz, & Laenkholm, 2012; Strien, Leidenius, von Smitten, & Heikkila, 2010; Xiao, Gong, Han, Gonzalez-Angulo, & Sneige, 2011). Thus, Rossi et al. concluded that this phenomenon could be correlated with enhanced tumor aggressiveness or with an underestimation of HER2 protein overexpression in the primary tumor by the pathologist (Rossi et al., 2012). Although several protein-based biomarkers have been used to detect or predict the propensity of metastatic lesion development, until now clinical oncologists rely heavily on various imaging modalities for higher accuracy and sensitivity.

4.2 Imaging Procedures

X-ray, radiographs, computed tomography scan (CT scan), nuclear imaging including PET and SPECT, MRI are few examples of currently available approaches in the clinical arena (Minn et al., 2014; Pomper & Fisher, 2014). In this particular section, we will focus on the clinical applications of these approaches, particularly in the context of bone metastasis, which is very difficult to detect using conventional protein-based biomarkers.

Bone is the third most common site for cancer metastasis and a major reason for mortality for prostate and breast cancer (Bussard, Gay, & Mastro, 2008; Yu, Tsai, & Hoffe, 2012). Vertebrae, pelvis, ribs, and the ends of long bones are preferred destinations, whereas mandible, patella, and distal extremities are less common for metastases (Bussard et al., 2008; Roberts et al., 2010; Saha, Burke, Desai, Vijayanathan, & Gnanasegaran, 2013). Detection of bone lesions or metastasis are always challenging in comparison with lesions in soft tissue or solid organs, such as the lungs or liver. Skeletal scintigraphy (SS), radiography, PET, and MRI are the most common approaches that are currently being used to detect bone metastases. Plain radiographs are recommended to assess the risk of pathological fracture; however, this approach is not of adequate sensitivity to routinely screen for asymptomatic metastasis (Costelloe et al., 2009; Roberts et al., 2010). Additionally, this approach is not suitable for monitoring treatment response (Bussard et al., 2008; Vassiliou et al., 2011). Computed tomography (CT) has advantages over radiography based on resolution, and sensitivity/specificity (O'Sullivan, Carty, & Cronin, 2015), and evaluating treatment responses (Vassiliou et al., 2011). It is an excellent approach to detect bone metastasis in bone marrow before initiation of bone destruction—thus pertinent to early diagnosis (Bauerle & Semmler, 2009). PET is another imaging modality that can also detect skeletal metastasis and it is superior in terms of spatial resolution (O'Sullivan et al., 2015). The sensitivity of PET is dependent on the type of radiotracers employed. ^{18}F-FDG and ^{18}F Sodium fluoride (NaF) are two common radiopharmaceuticals most frequently employed to detect skeletal metastasis. Existing literature suggests that ^{18}F NaF-PET is both sensitive and specific with improved resolution and better discrimination capability to distinct normal and abnormal bone (Langsteger, Heinisch, & Fogelman, 2006). Other tracers, such as ^{18}F-choline, ^{11}C-choline (half-life of 20 min) are also used in staging bone disease in prostate cancer. ^{11}C-choline PET may have other advantages over ^{18}F-FDG PET for detection of pelvic disease and bone metastases (Messiou, Cook, & deSouza, 2009). PET is a functional rather than anatomic imaging approach such as CT and depends on the uptake of radiotracers. In patients with primary osteoscleorotic metastasis from prostate cancer, ^{18}F-FDG PET has shown less sensitivity due to potential uptake problems. The other disadvantage of PET is inability to assess the treatment response in patients who underwent hormone therapy, a phenomenon known as "flare phenomenon" characterized by enhanced uptake of radionucleotides resulting in false-positive findings (Lecouvet et al., 2014; Vassiliou et al., 2011). In breast cancer patients, PET scans are not recommended in some contexts (Khan et al., 2007). In a retrospective study conducted in University of Kansas only 2% of patients were confirmed as having metastasis although 18% were primarily diagnosed with cancer by PET scanning (Khan et al., 2007). However, it is more reliable to determine locally advanced breast cancer and to detect extra-axial nodal disease (Bellon et al., 2004; Fuster et al., 2008; Mahner et al., 2008). Therefore, this approach is optional to detect breast cancer metastasis.

As per guidelines established by the National Comprehensive Cancer Network, MRI is the first line imaging modality for cancer of the head and neck, central nervous system, prostate, and hepatobiliary system (Spick et al., 2016). With respect to sensitivity and specificity issues, MRI is comparatively better than PET in detecting metastasis spreading in the marrow cavity and extension of tumors toward surrounding tissues (Costelloe et al., 2009; O'Sullivan et al., 2015). It is extremely beneficial to detect early tumor cells seeding into the hematopoietic compartment (Tombal & Lecouvet, 2012). In one study, MRI was able to detect over 37.5% of positive cases, which were primarily considered as indecisive by other approaches such as bone scan and plain X-ray (Lecouvet et al., 2007; Messiou et al., 2009). The development of whole-body MRI advances our capabilities to survey the entire body for detecting any marrow abnormalities, which was further improved using perfusion (DCE) and diffusion-weighted imaging (DWI) that refine the assessment of lesions during the (early) phases of therapy, providing tools to evaluate the efficacy of treatments targeting bone lesions (Attariwala & Picker, 2013; Dutoit, Vanderkerken, & Verstraete, 2013; Essig et al., 2013). It should be noted that DWI approach is extremely valuable to detect metastasis in ribs, which are very difficult through conventional MRI (Lecouvet et al., 2010; Venkitaraman et al., 2009). A meta-analysis was conducted by Yang and colleagues to compare the different approaches (eg, ^{18}FDG PET, CT, MRI, and bone scintigraphy) used for detection of bone metastasis in clinical settings and concluded that PET and MRI are equivalent and both significantly more accurate than bone scan and CT (Yang, Liu, Wang, Xu, & Deng, 2011) to detect bone metastasis. This conclusion is based on data from 67 studies published during 1995 to January 2010 in MEDLINE and EMBASE database.

Diagnostic accuracy can be significantly enhanced by combining two approaches, introduced as "hybrid technology" in the filed of cancer imaging (Cherry, 2009; Yoo, Lee, & Lee, 2015). PET/CT, an example of this hybrid technology which utilizes the metabolic information of PET with the anatomic detail of CT, which overcomes the inherent barriers of the individual approach. In a further refinement of the combined approach, PET/MRI was recently introduced with excellent soft tissue resolution

(Partovi et al., 2014). The superiority of PET/MRI over PET/CT in cancer diagnosis were confirmed in multiple recent studies including oncolytic bone lesions (Beiderwellen et al., 2014), liver metastasis (Beiderwellen, Geraldo, et al., 2015), detecting malignant/benign lesions in recurrent breast cancer patients (Sawicki et al., 2016), recurrent female malignancies such as ovarian cancer (Beiderwellen, Grueneisen, et al., 2015), thyroid cancer (Nagarajah et al., 2011), pancreatic (Nagamachi et al., 2013), and head and neck (Queiroz & Huellner, 2015) cancers. In pancreatic and head and neck cancer studies, although PET/MRI is sensitive over PET/CT, the differences were not statistically significant. A very recent review compared these two approaches in different malignancies and concluded that both function equally well for cancer assessment (Spick et al., 2016). PET/MRI has advantages over PET/CT in clinical management, detecting bone metastasis and locating intraprostatic sites of disease (Spick et al., 2016). On the other hand, PET/CT approach is advantageous in detecting pulmonary metastasis in some contexts (Spick et al., 2016). However, it is too early to reach a firm conclusion due to the limited numbers of comparative studies.

Despite technical advantages of imaging modalities, clinical oncologists often face several obstacles in bone metastasis detection that includes age-associated benign pathologies that might mimic the signal from metastatic cells, "flare phenomenon" (increased uptake of radiotracer) after hormone therapy, and the difficulty in detecting individual lesions when they are closely spaced (termed as superscan pattern frequently observed with elderly patients, late stage of disease, and individuals who are predisposed to bone metabolic disorders) (Lecouvet et al., 2014). Other issues such as scan duration (^{99}mTc-bisphosphonates must be imaged for several hours, whereas ^{18}F-NaF by 10 min) (Win & Aparici, 2014), longer on-camera acquisition, soft tissue uptake, extraosseous uptake of ^{18}F-NaF due to hypocalcemia, calcified soft tissues might also impose some technical challenges to accurately detect the lesions (Lecouvet et al., 2014).

4.3 Circulating Tumor Cells

Presence of circulating tumor cells in blood was first reported almost 140 years ago and, very recently CTC research expanded rapidly to emphasize its potential as both a diagnostic and prognostic marker of cancer. As of March 2016, there were approximately 17,571 publications and more than 767 clinical trials under the search term "circulating tumor cells" in

PubMed/Clinicaltrail.gov database reflecting the logarithmic expansion of this emerging technology. Extremely low numbers of CTCs in the circulation (as low as 1 CTC in 10^6–10^7 leukocytes of peripheral blood; Hong & Zu, 2013), lack of universal CTC detection antibodies, absence of appropriate sensitive methods for rapid molecular characterization of CTCs, and the need for sensitive technical equipment have proven to be challenges in adapting this approach for routine screening (Hong & Zu, 2013). Additionally, various methodologies practiced by different laboratories, lack of reference samples, selection biases, use of diverse capture antibodies from different sources, and oversimplification of cytopathology processes significantly impact on the validity of outcomes (Hong & Zu, 2013). Despite these limitations, a substantial number of clinical studies have shown the power of the CTC technology in patients with multiple cancer indications. The seminal work from Cristofanilli and colleagues revolutionized the clinical applications of CTCs and established the correlation of CTCs with progression-free survival and overall survival in patients with metastatic breast cancer (Cristofanilli et al., 2004). Following this study, a plethora of supportive studies have been published by multiple independent laboratories (Giordano & Cristofanilli, 2012; Giordano et al., 2013; Lianidou, Strati, & Markou, 2014; Pierga et al., 2012). In addition to tumor progression, the CTC approach has also been successfully applied to follow treatment responses including adjuvant chemotherapy in a phase II trial, after surgery (Ignatiadis et al., 2007; Pierga et al., 2008). The clinical relevance of CTCs has also been studied in prostate cancer at both early and advanced stages. In a pilot study, Lowes et al. demonstrated that CTCs can be detected at early stages of prostate cancer and may be pertinent to follow therapeutic responses (Lowes et al., 2012). Castration resistant prostate cancer (CRPC) is an advanced stage disease and both localized (Helo et al., 2009) and metastatic stages (Danila et al., 2007; de Bono et al., 2008; Scher et al., 2009) of CRPC correlated with CTCs in these independent studies. Interesting phenomenon was observed by Armstrong et al. (2011) that CTCs isolated from patients expressed both mesenchymal and epithelial markers, and the numbers of CTCs served as surrogate marker. Molecular characterization of CTCs isolated from melanomas showed some inconsistencies when BRAF mutation (Sakaizawa et al., 2012), a signature for 81% cases of melanoma status (Kitago et al., 2009), was analyzed although the clinical significance of CTCs in patient's blood was correlated with disease-free survival and overall survival (Hoshimoto et al., 2012). Presence of CTCs in pancreatic cancers correlated with an unfavorable prognosis (Tjensvoll, Nordgard, &

Smaaland, 2014). In 2013, Han et al. published a meta study with total 623 pancreatic cancer patients and established a positive correlation with CTCs and disease outcome (Han, Chen, & Zhao, 2014). This particular study also compared the survival among CTC-positive and CTC-negative patients, and the overall survival was worse in the latter group further supporting the prognostic potential of CTCs in the context of pancreatic cancer (Han et al., 2014). The prognostic value of CTCs in the contexts of resectable colorectal liver metastasis and metastatic colorectal cancer were studied in 11 independent publications. As per this summary report (Huang et al., 2015), liver metastasis are more prominent in CTC-positive patients and the presence of reduced CTCs are associated with overall progression-free survival. The study also suggested that the presence of CTCs could act as an indicator for treatment response (Huang et al., 2015). Apart from these studies, promising correlations of the presence of CTCs with head and neck (Wu et al., 2016), bladder cancers (Gazzaniga et al., 2014) were also reported by different groups. Although lung cancer mortality is a major clinical concern, however, the correlation of CTCs with lung cancer detection was less predictive at least with current sets of methods, which mostly rely on use of epithelial marker to identify CTCs. In lung cancer, CTCs often exhibit non-epithelial characteristics (Wu et al., 2015; Zhang, Ramnath, & Nagrath, 2015). In this context, developing appropriate reagents to detect both epithelial and mesenchymal markers will be beneficial. Despite this challenge, attempts were taken and positive correlations between CTCs with lung cancer were reported (Lecharpentier et al., 2011; Zhang et al., 2015; Zhu et al., 2014). Regarding treatment response, a meta-analysis considering 12 relevant studies demonstrated that prior to treatment CTCs correlated with lymph node status, distant metastasis, and disease staging, however, post-treatment CTCs only correlated with staging (Ma et al., 2012). Including the earlier mentioned meta-analysis, a number of meta-analyses were conducted in different cancers such as gastric cancer (Wang, Wei, Zou, Qian, & Liu, 2016), metastatic breast cancer (Lv et al., 2016), ovarian cancer (Cui, Kwong, & Wang, 2015; Zhou et al., 2015), head and neck (Wang, Cui, Xue, Tong, & Li, 2015), colorectal cancer (Huang et al., 2015), liver cancer (Jin, Peng, & Wu, 2013), prostate cancer (Wang et al., 2011), bladder and urothelial cancer (Wang et al., 2011), and positive correlations with CTCs were established.

Sample volume and sampling number are the major determining factors in quantifying CTCs. In current settings, a small amount of blood (around 7.5 mL, 0.15% of our total volume of blood) is used to detect CTCs. It can be argued that this amount of sample may not be statistically adequate to represent the whole circulation. A study compared the average CTCs in 7.5 vs 30 mL from 15 patients and the results suggested that larger volume is preferable for detecting CTCs (Lalmahomed et al., 2010). As per Allan and Keeney's mathematical model, at least 20 mL of whole blood sample needs to be analyzed to detect the lower frequency (1 CTC in 10^7 leukocytes) (Allan & Keeney, 2010). Multistep and complex sample processing represent other issues that need to be optimized to circumvent inconsistency. Additionally, operator variability and data interpretation could also impact the results of CTCs in a clinical setting (Hong & Zu, 2013). Regardless of the primary success of this approach that has been reported in various studies, monitoring CTCs is not well accepted as a diagnostic tool by different oncology associations such as the American Society of Clinical Oncology, the National Academy of Clinical Biochemistry (reviewed by Hong & Zu, 2013; Sturgeon et al., 2008; Zhang et al., 2015). Turnaround time for typical *Point of Care* (normally within 1 h, but 30 min is preferred) (Kilgore, Steindel, & Smith, 1998; Louie, Tang, Shelby, & Kost, 2000) is another limitation for practical implication of CTC approaches (Hong & Zu, 2013). However, considering all the published results from several clinical studies CTC biology might have significant future potential in the clinical arena. At this stage, major advances are needed to improve the tools and methodologies for monitoring CTCs, which if achieved might significantly enhance the power and accuracy of this approach for monitoring patient cancers noninvasively.

5. METASTASIS DETECTION/THERAPY: COMBINING IMAGING WITH THERAPY (THERANOSTICS)

"Theranostics" integrates the processes of diagnosis and therapy (Baum & Kulkarni, 2012), and represents a rapidly emerging area in cancer. For efficient disease management, being able to diagnose, target, and monitor therapeutic responses are essential. In the clinical arena over the past few decades', significant achievements have been made individually in these three aspects of patient care. The "theranostic" concept has merged these three independent aims into a single platform and examples of successful applications have been documented in preclinical animal and in limited clinical studies establishing this approach as a promising therapeutic strategy.

Theranostic approaches are relatively new and studies are currently conducted to provide evidences for developing effective treatment regimes in cancer biology including gene therapy, chemotherapy, and radiation therapy. Gene therapy is an exciting area in cancer research with additional applications in other inherited genetic diseases such as cystic fibrosis where a faulty gene is being replaced with a correct gene (Das et al., 2015). Delivery of a "gene of interest" specifically in targeted cells is always challenging. Various approaches including both viral- and lipid-based molecules are being optimized for efficient delivery of nucleotides (Das et al., 2015). Recently, strategies have been developed where both vehicles and nucleotides are labeled with organic dyes, which can be tracked by the imaging techniques (Hong, Yang, Zhang, & Cai, 2010). Quantum dots are an alternative to organic dyes and have advantages over fluorescent dyes for nucleic acid trafficking (Walling, Novak, & Shepard, 2009). However, these latter approaches are still predominantly at an experimental stage in the laboratory and no information is available relating to the successful application of these strategies in the clinic.

Microbubbles, gas filled spheres currently used as FDA-approved ultrasound contrast agents (Castle et al., 2013), have been evaluated as drug and therapeutic virus carriers in numerous studies and have potential as theranostic tools (Kiessling, Fokong, Koczera, Lederle, & Lammers, 2012). Application of ultrasound in conjunction with microbubbles can enhance the acoustic forces that facilitate drug/therapeutic delivery (Azab et al., 2012; Dash et al., 2011; Sarkar et al., 2015) across various biological barriers including the blood–brain barrier (Meairs, 2015). Although it is beyond the scope of this review, but it is worth noting that utility of microbubbles as a therapeutic in thrombolysis in addition to imaging has been validated clinically (Molina et al., 2006). Theranostic property of microbubbles was initially observed by Leong-Poi et al. in a study where microbubbles containing a VEGF expression plasmid (permitting VEGF expression) were injected into the rat to treat arteriogenesis (Leong-Poi et al., 2007). In the same year, Rapoport and colleagues provided more confirmative evidence from in vivo studies showing the success of this approach (Rapoport, Gao, & Kennedy, 2007). Doxorubicin-loaded nanobubbles were injected which extravasated into the tumor and coalesced into microbubbles at physiologic temperatures (Rapoport et al., 2007). Ultrasound was applied to visualize the tumor site as well as trigger the destruction and delivery of drugs that eventually killed the tumor. Recently, research has focused on developing target-specific microbubbles by decorating their surface, which can enhance and expand the utility of this approach in the clinic.

Nanoparticles, which are commonly used to deliver drug(s) (Mudshinge, Deore, Patil, & Bhalgat, 2011), also have proven amenable as imaging and therapeutic modalities (Kievit & Zhang, 2011). Indeed, although nanoparticles as theranostics are still in their early stage of development, a substantial number of studies support their potential (Xie, Lee, & Chen, 2010). Iron oxide nanoparticles (IONPs), also known as magnetic nanoparticles due to their magnetic properties are used as contrast agents in MRI (Bu et al., 2012). For example, several dextran-based IONPs are approved and currently used for detection of liver and spleen lesions by AMAG pharmaceuticals (Xie et al., 2010). Various research groups verified IONPs as a carrier for chemotherapeutics (eg, methotrexate, paclitaxel, doxorubicin) and nucleotides (Hwu et al., 2009; Xie et al., 2010). Besides their roles as contrast agents and therapeutic carriers, their magnetic properties have also been utilized for therapy of tumors (Wadajkar et al., 2013). Application of external alternating magnetic fields can convert electromagnetic energy into heat, which raises the temperature of the tumor above 43°C resulting in thermal killing of tumor cells (Shen et al., 2015). Gold nanoparticles, due to stability, biosafety, and ability to be modified for better delivery and targeting are routinely used to deliver a wide variety of therapeutics (Khan, Rashid, Murtaza, & Zahra, 2014). Utility of gold nanoparticles as imaging or diagnostic agents is related to particle absorption spectrum and upon application of laser irradiation, gold nanoparticles serve as energy transducers and induce photothermal killing as shown by different studies including breast (Au et al., 2008; Jin, Hong, Kakar, & Kang, 2008), glioblastoma (Jin et al., 2008), oral (El-Sayed, Huang, & El-Sayed, 2006), and urothelial cancer (Chen, Wu, & Chen, 2015). Other nanoparticles such as carbon nanotubes, silica-based nanoparticles have also demonstrated theranostic properties in specific situations (Xie et al., 2010).

Molecular-genetic approach is suitable for theranostics of metastatic tumors. One can take advantage of a single reporter that can serve as both an imaging and therapeutic gene. One of the most widely utilized theranostic genes is HSV1-tk. Many HSV1-tk-specific substrates have been developed for imaging and therapeutic applications and these include pyrimidine nucleoside derivatives and acycloguanosine derivatives. 2′fluoro-2′-deoxy-1-β-D-arabinofuranosyl-5-iodouracil (FIAU), 2′fluoro-2′-deoxy-5-methyl-1-β-D-arabinofuranosyl-5-iodouracil (FMAU), and 2′fluoro-2′-deoxy-5-ethyl-1-β-D-arabinofuranosyl-5-iodouracil (FEAU) belongs to the pyrimidine nucleoside derivatives. Acyclovir, ganciclovir, penciclovir, and 9-(4-fluoro-3-hydroxymethylbytyl)guanine (FHBG).

FIAU, FMAU, FEAU, and FHBG labeled with ^{18}F can serve as PET tracers. FIAU labeled with a therapeutic radioisotope such as ^{131}I can be utilized as radiopharmaceutical therapy (Yaghoubi et al., 2005). Finally, acyclovir, ganciclovir, and penciclovir are potent prodrugs that require HSV1-tk to be converted to active ingredients. Clinical application for theranostic applications using HSV1-tk have been tested in human glioma (Jacobs et al., 2001), liver cancer (Penuelas et al., 2005), and relapsed allogenic stem cell transplant patients (Eissenberg et al., 2015). Somatosatin respect 2 (SSTr2) has been tested as a theranostic gene. Radiolabeled octreotides such as [^{123}I]Tyr3-octreotide (Krenning et al., 1989), [^{111}In]DTPA-D-Phe-octretide (Bakker et al., 1991), and [^{94m}Tc]Tyr3-octreotate (Rogers et al., 2005) have been tested for nuclear imaging probes targeting SSTr2. Y-90 DOTA-Phe1-Tyr3-octreotide has been used to treat neuroendocrine tumors (Bushnell et al., 2004). NIS can also be used as a theranostic gene by using [^{99m}Tc]pertechnetate and [^{124}I]NaI as SPECT and PET tracer, respectively (Chung, 2002), and [^{131}I]NaI (Dadachova & Carrasco, 2004) or [188Re] (Kang et al., 2004) for therapeutic radionuclides.

6. CHALLENGES FACED IN DETECTING METASTATIC CELLS

Even with our ever-increasing understanding of the mechanism(s) underlying cancer progression and metastatic spread, detecting metastatic cells still remains challenging. Detecting metastases at their inception would greatly aid in overall patient outcome as targeted therapies could be administered at the onset of this progressive process. Further inhibiting the growth of metastases at a distant location will greatly benefit overall patient outcomes. Although researchers have determined multiple molecular changes, immunologic, and genetic factors that support the metastatic process, as well as several methods that aid in detection of metastases, these findings have not been easy to translate into the clinical setting. Conventional methods of detecting metastatic lesions such as MRI and CT are not sensitive enough to detect micrometastases.

CTCs can play a crucial role in the development of metastasis. Several groups have explored techniques to detect CTCs (Hong & Zu, 2013; Lurje, Schiesser, Claudius, & Schneider, 2010). However, CTCs are present at extremely low concentrations—as low as one CTC in 10^6–10^7 leukocytes in peripheral blood of cancer patients (Hong & Zu, 2013). Several devices/instruments have been developed to detect CTCs, including the only US

FDA-cleared CellSearch® system by Veridex which captures and enumerates CTCs in cancer patients with metastatic breast, prostate, and colorectal cancer (Andree, van Dalum, & Terstappen, 2016). However, challenges in detecting CTCs still remain and have been described in detail elsewhere (Andree et al., 2016; Hong & Zu, 2013; Pantel & Alix-Panabieres, 2010).

7. CONCLUSIONS AND FUTURE DIRECTIONS

As discussed earlier, several in vitro and in vivo preclinical and clinical methods have been developed to gain a better understanding of the metastatic process. However, while several research groups have attempted to dissect the various steps of metastasis as well as develop methods to detect metastatic lesions; it is clear that several challenges still remain and a lot still needs to be done before these techniques become commonplace in the clinic for detecting metastatic lesions in patients. As newer information becomes available, we will hopefully be able to develop highly specific methods to detect metastasis in patients which will greatly benefit overall patient outcome.

ACKNOWLEDGMENTS

Support for our laboratories was provided in part by National Institutes of Health Grants R01 CA097318 (P.B.F.), R01 CA168517 (Maurizio Pellecchia and P.B.F.), and P50 CA058326 (M.G.P. and P.B.F.); the Samuel Waxman Cancer Research Foundation (P.B.F. and D.S.); National Foundation for Cancer Research (P.B.F.); NCI Cancer Center Support Grant to VCU Massey Cancer Center P30 CA016059 (P.B.F.); and VCU Massey Cancer Center developmental funds (P.B.F.). P.B.F. and D.S. are SWCRF investigators. P.B.F. holds the Thelma Newmeyer Corman Chair in Cancer Research in the VCU Massey Cancer Center.

Conflict of interest: Drs. M.G.P. and P.B.F. are cofounders of, serve as consultants to and have ownership interest in CTS, Inc. Dr. M.G.P. is a member of the board of directors of CTS, Inc., Johns Hopkins University, Virginia Commonwealth University, and Columbia University have ownership interest in CTS, Inc.

REFERENCES

Abuzallouf, S., Dayes, I., & Lukka, H. (2004). Baseline staging of newly diagnosed prostate cancer: A summary of the literature. *The Journal of Urology*, *171*, 2122–2127.

Aguirre, A. J., Bardeesy, N., Sinha, M., Lopez, L., Tuveson, D. A., Horner, J., et al. (2003). Activated Kras and Ink4a/Arf deficiency cooperate to produce metastatic pancreatic ductal adenocarcinoma. *Genes Development*, *17*, 3112–3126.

Ahn, B. C. (2014). Requisites for successful theranostics with radionuclide-based reporter gene imaging. *Journal of Drug Targeting*, *22*, 295–303.

Allan, A. L., & Keeney, M. (2010). Circulating tumor cell analysis: Technical and statistical considerations for application to the clinic. *Journal of Oncology*, *2010*, 426218.

American Cancer Society (2016). Cancer facts & figures 2016. Atlanta: American Cancer Society.
Andree, K. C., van Dalum, G., & Terstappen, L. W. (2016). Challenges in circulating tumor cell detection by the cell search system. *Molecular Oncology, 10*, 395–407.
Aparicio, S., Hidalgo, M., & Kung, A. L. (2015). Examining the utility of patient-derived xenograft mouse models. *Nature Reviews Cancer, 15*, 311–316.
Armstrong, A. J., Marengo, M. S., Oltean, S., Kemeny, G., Bitting, R. L., Turnbull, J. D., et al. (2011). Circulating tumor cells from patients with advanced prostate and breast cancer display both epithelial and mesenchymal markers. *Molecular Cancer Research, 9*, 997–1007.
Arnaoutova, I., & Kleinman, H. K. (2010). In vitro angiogenesis: Endothelial cell tube formation on gelled basement membrane extract. *Nature Protocols, 5*, 628–635.
Attariwala, R., & Picker, W. (2013). Whole body MRI: Improved lesion detection and characterization with diffusion weighted techniques. *Journal of Magnetic Resonance Imaging, 38*, 253–268.
Au, L., Zheng, D., Zhou, F., Li, Z. Y., Li, X., & Xia, Y. (2008). A quantitative study on the photothermal effect of immuno gold nanocages targeted to breast cancer cells. *ACS Nano, 2*, 1645–1652.
Azab, B., Dash, R., Das, S. K., Bhutia, S. K., Shen, X. N., Quinn, B. A., et al. (2012). Enhanced delivery of mda-7/IL-24 using a serotype chimeric adenovirus (Ad.5/3) in combination with the Apogossypol derivative BI-97C1 (Sabutoclax) improves therapeutic efficacy in low CAR colorectal cancer cells. *Journal of Cellular Physiology, 227*, 2145–2153.
Bacac, M., & Stamenkovic, I. (2008). Metastatic cancer cell. *Annual Review of Pathology, 3*, 221–247.
Badr, C. E. (2014). Bioluminescence imaging: Basics and practical limitations. *Methods in Molecular Biology, 1098*, 1–18.
Bakker, W. H., Albert, R., Bruns, C., Breeman, W. A., Hofland, L. J., Marbach, P., et al. (1991). [111In-DTPA-D-Phe1]-octreotide, a potential radiopharmaceutical for imaging of somatostatin receptor-positive tumors: Synthesis, radiolabeling and in vitro validation. *Life Sciences, 49*, 1583–1591.
Baranwal, S., & Alahari, S. K. (2010). miRNA control of tumor cell invasion and metastasis. *International Journal of Cancer, 126*, 1283–1290.
Bauerle, T., & Semmler, W. (2009). Imaging response to systemic therapy for bone metastases. *European Radiology, 19*, 2495–2507.
Baum, R. P., & Kulkarni, H. R. (2012). THERANOSTICS: From molecular imaging using Ga-68 labeled tracers and PET/CT to personalized radionuclide therapy—The Bad Berka experience. *Theranostics, 2*, 437–447.
Begić, A., Kucukalic-Selimovic, E., Obralic, N., Duric, O., Lacevic, N., Begovic, S., et al. (2005). Role of bone scintigraphy and tumor marker-Ca 15-3 in detection of bone metastases in patients with breast cancer. *Bosnian Journal of Basic Medical Sciences/Udruzenje basicnih mediciniskih znanosti Association of Basic Medical Sciences, 5*, 23–26.
Begić, A., Kucukalic-Selimovic, E., Obralić, N., Durić, O., Lacević, N., & Skopljak, A. (2006). Correlation between bone scintigraphy and tumor markers in patients with breast carcinoma. *Bosnian Journal of Basic Medical Sciences, 6*(1), 7.
Beiderwellen, K., Geraldo, L., Ruhlmann, V., Heusch, P., Gomez, B., Nensa, F., et al. (2015). Accuracy of [18 F]FDG PET/MRI for the detection of liver metastases. *Plos One, 10*, e0137285.
Beiderwellen, K., Grueneisen, J., Ruhlmann, V., Buderath, P., Aktas, B., Heusch, P., et al. (2015). [(18)F]FDG PET/MRI vs. PET/CT for whole-body staging in patients with recurrent malignancies of the female pelvis: Initial results. *European Journal of Nuclear Medicine and Molecular Imaging, 42*, 56–65.

Beiderwellen, K., Huebner, M., Heusch, P., Grueneisen, J., Ruhlmann, V., Nensa, F., et al. (2014). Whole-body [(1)(8)F]FDG PET/MRI vs PET/CT in the assessment of bone lesions in oncological patients: Initial results. *European radiology, 24*, 2023–2030.

Bellon, J. R., Livingston, R. B., Eubank, W. B., Gralow, J. R., Ellis, G. K., Dunnwald, L. K., et al. (2004). Evaluation of the internal mammary lymph nodes by FDG-PET in locally advanced breast cancer (LABC). *American Journal of Clinical Oncology, 27*, 407–410.

Bendas, G., & Borsig, L. (2012). Cancer cell adhesion and metastasis: Selectins, integrins, and the inhibitory potential of heparins. *International Journal of Cell Biology, 2012*, 676–731.

Bhang, H. E., Gabrielson, K. L., Laterra, J., Fisher, P. B., & Pomper, M. G. (2011). Tumor-specific imaging through progression elevated gene-3 promoter-driven gene expression. *Nature Medicine, 17*, 123–129.

Bhang, H. E., & Pomper, M. G. (2012). Cancer imaging: Gene transcription-based imaging and therapeutic systems. *The International Journal of Biochemistry & Cell Biology, 44*, 684–689.

Bhatnagar, A., Wang, Y., Mease, R. C., Gabrielson, M., Sysa, P., Minn, I., et al. (2014). AEG-1 promoter-mediated imaging of prostate cancer. *Cancer Research, 74*, 5772–5781.

Boukerche, H., Su, Z. Z., Emdad, L., Baril, P., Balme, B., Thomas, L., et al. (2005). mda-9/Syntenin: A positive regulator of melanoma metastasis. *Cancer Research, 65*, 10901–10911.

Brand, A. H., & Perrimon, N. (1993). Targeted gene expression as a means of altering cell fates and generating dominant phenotypes. *Development, 118*, 401–415.

Briganti, A., Suardi, N., Gallina, A., Abdollah, F., Novara, G., Ficarra, V., et al. (2014). Predicting the risk of bone metastasis in prostate cancer. *Cancer Treatment Reviews, 40*, 3–11.

Brufsky, A. M., Mayer, M., Rugo, H. S., Kaufman, P. A., Tan-Chiu, E., Tripathy, D., et al. (2011). Central nervous system metastases in patients with HER2-positive metastatic breast cancer: Incidence, treatment, and survival in patients from registHER. *Clinical Cancer Research, 17*, 4834–4843.

Bu, L. H., Xie, J., Chen, K., Huang, J., Aguilar, Z. P., Wang, A., et al. (2012). Assessment and comparison of magnetic nanoparticles as MRI contrast agents in a rodent model of human hepatocellular carcinoma. *Contrast Media & Molecular Imaging, 7*, 363–372.

Bushnell, D., Menda, Y., O'Dorisio, T., Madsen, M., Miller, S., Carlisle, T., et al. (2004). Effects of intravenous amino acid administration with Y-90 DOTA-Phe1-Tyr3-Octreotide (SMT487[OctreoTher) treatment. *Cancer Biotherapy & Radiopharmaceuticals, 19*, 35–41.

Bussard, K. M., Gay, C. V., & Mastro, A. M. (2008). The bone microenvironment in metastasis; what is special about bone? *Cancer Metastasis Reviews, 27*, 41–55.

Byrne, J., Horgan, P. G., England, S., Callaghan, J., & Given, H. F. (1992). An evaluation of the usefulness of primary tumour expression of MCA and CA15-3 as prognostic indicators in breast carcinoma. *European Journal of Surgical Oncology, 18*, 230–234.

Castle, J., Butts, M., Healey, A., Kent, K., Marino, M., & Feinstein, S. B. (2013). Ultrasound-mediated targeted drug delivery: Recent success and remaining challenges. *American Journal of Physiology Heart and Circulatory Physiology, 304*, H350–H357.

Chambers, A. F., Groom, A. C., & MacDonald, I. C. (2002). Dissemination and growth of cancer cells in metastatic sites. *Nature Reviews Cancer, 2*, 563–572.

Chambers, A. F., MacDonald, I. C., Schmidt, E. E., Koop, S., Morris, V. L., Khokha, R., et al. (1995). Steps in tumor metastasis: New concepts from intravital videomicroscopy. *Cancer Metastasis Reviews, 14*, 279–301.

Chambers, A. F., MacDonald, I. C., Schmidt, E. E., Morris, V. L., & Groom, A. C. (1998). Preclinical assessment of anti-cancer therapeutic strategies using in vivo videomicroscopy. *Cancer Metastasis Reviews, 17*, 263–269.

Chen, H. C. (2005). Boyden chamber assay. *Methods in Molecular Biology, 294*, 15–22.

Chen, J. S., Liu, J. C., Shen, L., Rau, K. M., Kuo, H. P., Li, Y. M., et al. (2004). Cancer-specific activation of the survivin promoter and its potential use in gene therapy. *Cancer Gene Therapy, 11*, 740–747.

Chen, C. H., Wu, Y. J., & Chen, J. J. (2015). Gold nanotheranostics: Photothermal therapy and imaging of Mucin 7 conjugated antibody nanoparticles for urothelial cancer. *BioMed Research International, 2015*, 813632.

Cherry, S. R. (2009). Multimodality imaging: Beyond PET/CT and SPECT/CT. *Seminars in Nuclear Medicine, 39*(5), 348–353.

Chui, M. H. (2013). Insights into cancer metastasis from a clinicopathologic perspective: Epithelial-mesenchymal transition is not a necessary step. *International Journal of Cancer, 132*, 1487–1495.

Chung, J. K. (2002). Sodium iodide symporter: Its role in nuclear medicine. *Journal of Nuclear Medicine, 43*, 1188–1200.

Close, D. M., Xu, T., Sayler, G. S., & Ripp, S. (2011). In vivo bioluminescent imaging (BLI): Noninvasive visualization and interrogation of biological processes in living animals. *Sensors (Basel), 11*, 180–206.

Contag, C. H., Contag, P. R., Mullins, J. I., Spilman, S. D., Stevenson, D. K., & Benaron, D. A. (1995). Photonic detection of bacterial pathogens in living hosts. *Molecular Microbiology, 18*, 593–603.

Cory, G. (2011). Scratch-wound assay. *Methods in Molecular Biology, 769*, 25–30.

Costelloe, C. M., Rohren, E. M., Madewell, J. E., Hamaoka, T., Theriault, R. L., Yu, T. K., et al. (2009). Imaging bone metastases in breast cancer: Techniques and recommendations for diagnosis. *The Lancet Oncology, 10*, 606–614.

Cristofanilli, M., Budd, G. T., Ellis, M. J., Stopeck, A., Matera, J., Miller, M. C., et al. (2004). Circulating tumor cells, disease progression, and survival in metastatic breast cancer. *The New England Journal of Medicine, 351*, 781–791.

Cui, L., Kwong, J., & Wang, C. C. (2015). Prognostic value of circulating tumor cells and disseminated tumor cells in patients with ovarian cancer: A systematic review and meta-analysis. *Journal of Ovarian Research, 8*, 38.

Dadachova, E., & Carrasco, N. (2004). The Na/I symporter (NIS): Imaging and therapeutic applications. *Seminars in Nuclear Medicine, 34*, 23–31.

Danila, D. C., Heller, G., Gignac, G. A., Gonzalez-Espinoza, R., Anand, A., Tanaka, E., et al. (2007). Circulating tumor cell number and prognosis in progressive castration-resistant prostate cancer. *Clinical Cancer Research, 13*, 7053–7058.

Dankort, D., Curley, D. P., Cartlidge, R. A., Nelson, B., Karnezis, A. N., Damsky, W. E., et al. (2009). Braf(V600E) cooperates with Pten loss to induce metastatic melanoma. *Nature Genetics, 41*, 544–552.

Das, S. K., Menezes, M. E., Bhatia, S., Wang, X. Y., Emdad, L., Sarkar, D., et al. (2015). Gene therapies for cancer: Strategies, challenges and successes. *Journal of Cellular Physiology, 230*, 259–271.

Dash, R., Azab, B., Quinn, B. A., Shen, X. N., Wang, X. Y., Das, S. K., et al. (2011). Apogossypol derivative BI-97C1 (Sabutoclax) targeting Mcl-1 sensitizes prostate cancer cells to mda-7/IL-24-mediated toxicity. *Proceedings of the National Academy of Sciences of theUnited States of America, 108*, 8785–8790.

de Bono, J. S., Scher, H. I., Montgomery, R. B., Parker, C., Miller, M. C., Tissing, H., et al. (2008). Circulating tumor cells predict survival benefit from treatment in metastatic castration-resistant prostate cancer. *Clinical Cancer Research, 14*, 6302–6309.

Debnath, J., Muthuswamy, S. K., & Brugge, J. S. (2003). Morphogenesis and oncogenesis of MCF-10A mammary epithelial acini grown in three-dimensional basement membrane cultures. *Methods, 30*, 256–268.

Deng, Y. H., Kurland, B. F., Wang, J. P., Bi, J., Li, W., Rao, S., et al. (2009). High epidermal growth factor receptor expression in metastatic colorectal cancer lymph nodes may me more prognostic of poor survival than in primary tumor. *American Journal of Clinical Oncology*, *32*, 245–252.

Dieci, M. V., Barbieri, E., Piacentini, F., Ficarra, G., Bettelli, S., Dominici, M., et al. (2013). Discordance in receptor status between primary and recurrent breast cancer has a prognostic impact: A single-institution analysis. *Annals of Oncology*, *24*, 101–108.

Dronadula, N., Du, L., Flynn, R., Buckler, J., Kho, J., Jiang, Z., et al. (2011). Construction of a novel expression cassette for increasing transgene expression in vivo in endothelial cells of large blood vessels. *Gene Therapy*, *18*, 501–508.

Duffy, M. J., Sturgeon, C., Lamerz, R., Haglund, C., Holubec, V. L., Klapdor, R., et al. (2010). Tumor markers in pancreatic cancer: A European group on tumor markers (EGTM) status report. *Annals of Oncology*, *21*, 441–447.

Dutoit, J. C., Vanderkerken, M. A., & Verstraete, K. L. (2013). Value of whole body MRI and dynamic contrast enhanced MRI in the diagnosis, follow-up and evaluation of disease activity and extent in multiple myeloma. *European Journal of Radiology*, *82*, 1444–1452.

Eissenberg, L. G., Rettig, M. P., Ritchey, J. K., Prior, J. L., Schwarz, S. W., Frye, J., et al. (2015). [(18)F]FHBG PET/CT imaging of CD34-TK75 transduced donor T cells in relapsed allogeneic stem cell transplant patients: Safety and feasibility. *Molecular Therapy*, *23*, 1110–1122.

Eklund, L., Bry, M., & Alitalo, K. (2013). Mouse models for studying angiogenesis and lymphangiogenesis in cancer. *Molecular Oncology*, 7, 259–282.

El-Sayed, I. H., Huang, X. H., & El-Sayed, M. A. (2006). Selective laser photo-thermal therapy of epithelial carcinoma using anti-EGFR antibody conjugated gold nanoparticles. *Cancer Letters*, *239*, 129–135.

Elsinga, P. H., & Dierckx, R. A. (2014). Small molecule PET-radiopharmaceuticals. *Current Pharmaceutical Design*, *20*, 2268–2274.

Essig, M., Shiroishi, M. S., Nguyen, T. B., Saake, M., Provenzale, J. M., Enterline, D., et al. (2013). Perfusion MRI: The five most frequently asked technical questions. *American Journal of Roentgenology*, *200*, 24–34.

Falasca, M., Raimondi, C., & Maffucci, T. (2011). Boyden chamber. *Methods in Molecular Biology*, *769*, 87–95.

Fan, A. C., Goldrick, M. M., Ho, J., Liang, Y., Bachireddy, P., & Felsher, D. W. (2008). A quantitative PCR method to detect blood microRNAs associated with tumorigenesis in transgenic mice. *Molecular Cancer*, 7, 74.

Fantozzi, A., & Christofori, G. (2006). Mouse models of breast cancer metastasis. *Breast Cancer Research*, *8*, 212.

Fidler, I. J. (2002). Critical determinants of metastasis. *Seminars in Cancer Biology*, *12*, 89–96.

Fidler, I. J. (2003). The pathogenesis of cancer metastasis: The 'seed and soil' hypothesis revisited. *Nature Reviews Cancer*, *3*, 453–458.

Fidler, I. J. (2006). Models for spontaneous metastasis. *Cancer Research*, *66*, 9787.

Fisher, P. B. (1983). Chemical-viral interactions in cell transformation. *Cancer Investigation*, *1*, 495–509.

Fisher, P. B., & Weinstein, I. B. (1980). Chemical-viral interactions and multistep aspects of cell transformation. *The International Agency for Research on Cancer*, *27*, 113–131.

Francia, G., Cruz-Munoz, W., Man, S., Xu, P., & Kerbel, R. S. (2011). Mouse models of advanced spontaneous metastasis for experimental therapeutics. *Nature Reviews Cancer*, *11*, 135–141.

Freedland, S. J., Howard, L. E., Hanyok, B. T., Kadiyala, V. K., Kuang, J. Y., Whitney, C. A., et al. (2016). Validation of a bone scan positivity risk table in nonmetastatic castration-resistant prostate cancer. *BJU International*. Epub Jan 13, 2016.

Fu, X., Guadagni, F., & Hoffman, R. M. (1992). A metastatic nude-mouse model of human pancreatic cancer constructed orthotopically with histologically intact patient specimens. *Proceedings of the National Academy of Sciences of the United States of America, 89*, 5645–5649.

Fuster, D., Duch, J., Paredes, P., Velasco, M., Munoz, M., Santamaria, G., et al. (2008). Preoperative staging of large primary breast cancer with [18 F]fluorodeoxyglucose positron emission tomography/computed tomography compared with conventional imaging procedures. *Journal of Clinical Oncology, 26*, 4746–4751.

Gaedcke, J., Traub, F., Milde, S., Wilkens, L., Stan, A., Ostertag, H., et al. (2007). Predominance of the basal type and HER-2/neu type in brain metastasis from breast cancer. *Modern Pathology, 20*, 864–870.

Garrido, T., Riese, H. H., Aracil, M., & Perez-Aranda, A. (1995). Endothelial cell differentiation into capillary-like structures in response to tumour cell conditioned medium: A modified chemotaxis chamber assay. *British Journal of Cancer, 71*, 770–775.

Gauvain, K. M., Garbow, J. R., Song, S. K., Hirbe, A. C., & Weilbaecher, K. (2005). MRI detection of early bone metastases in b16 mouse melanoma models. *Clinical and Experimental Metastasis, 22*, 403–411.

Gazzaniga, P., de Berardinis, E., Raimondi, C., Gradilone, A., Busetto, G. M., De Falco, E., et al. (2014). Circulating tumor cells detection has independent prognostic impact in high-risk non-muscle invasive bladder cancer. *International Journal of Cancer, 135*, 1978–1982.

Geraghty, J. G., Coveney, E. C., Sherry, F., O'Higgins, N. J., & Duffy, M. J. (1992). CA 15-3 in patients with locoregional and metastatic breast carcinoma. *Cancer, 70*, 2831–2834.

Giavazzi, R., & Decio, A. (2014). Syngeneic murine metastasis models: B16 melanoma. *Methods in Molecular Biology, 1070*, 131–140.

Giordano, A., & Cristofanilli, M. (2012). CTCs in metastatic breast cancer. Recent results in cancer research. *Fortschritte der Krebsforschung. Progres dans les recherches sur le cancer, 195*, 193–201.

Giordano, A., Egleston, B. L., Hajage, D., Bland, J., Hortobagyi, G. N., Reuben, J. M., et al. (2013). Establishment and validation of circulating tumor cell-based prognostic nomograms in first-line metastatic breast cancer patients. *Clinical Cancer Research, 19*, 1596–1602.

Glinskii, A. B., Smith, B. A., Jiang, P., Li, X. M., Yang, M., Hoffman, R. M., et al. (2003). Viable circulating metastatic cells produced in orthotopic but not ectopic prostate cancer models. *Cancer Research, 63*, 4239–4243.

Grupka, N. L., Lear-Kaul, K. C., Kleinschmidt-DeMasters, B. K., & Singh, M. (2004). Epidermal growth factor receptor status in breast cancer metastases to the central nervous system. Comparison with HER-2/neu status. *Archives of Pathology & Laboratory Medicine, 128*, 974–979.

Guy, C. T., Cardiff, R. D., & Muller, W. J. (1992). Induction of mammary tumors by expression of polyomavirus middle T oncogene: A transgenic mouse model for metastatic disease. *Molecular and Cellular Biology, 12*, 954–961.

Han, L., Chen, W., & Zhao, Q. (2014). Prognostic value of circulating tumor cells in patients with pancreatic cancer: A meta-analysis. *Tumour Biology, 35*, 2473–2480.

Hanahan, D., & Weinberg, R. A. (2000). The hallmarks of cancer. *Cell, 100*, 57–70.

Hanahan, D., & Weinberg, R. A. (2011). Hallmarks of cancer: The next generation. *Cell, 144*, 646–674.

Helo, P., Cronin, A. M., Danila, D. C., Wenske, S., Gonzalez-Espinoza, R., Anand, A., et al. (2009). Circulating prostate tumor cells detected by reverse transcription-PCR in men with localized or castration-refractory prostate cancer: Concordance with cell search assay and association with bone metastases and with survival. *Clinical Chemistry, 55*, 765–773.

Hoffman, R. M. (2014). Orthotopic mouse models of tumor metastasis expressing fluorescent reporters produce imageable circulating tumor cells. *Cancer Microenvironment*, *7*, 133–138.

Hong, H., Yang, Y., Zhang, Y., & Cai, W. (2010). Non-invasive cell tracking in cancer and cancer therapy. *Current Topics in Medicinal Chemistry*, *10*, 1237–1248.

Hong, B., & Zu, Y. (2013). Detecting circulating tumor cells: Current challenges and new trends. *Theranostics*, *3*, 377–394.

Hoshimoto, S., Faries, M. B., Morton, D. L., Shingai, T., Kuo, C., Wang, H. J., et al. (2012). Assessment of prognostic circulating tumor cells in a phase III trial of adjuvant immunotherapy after complete resection of stage IV melanoma. *Annals of Surgery*, *255*, 357–362.

Huang, X., Gao, P., Song, Y., Sun, J., Chen, X., Zhao, J., et al. (2015). Meta-analysis of the prognostic value of circulating tumor cells detected with the cell search system in colorectal cancer. *BMC Cancer*, *15*, 202–208.

Huang, R., Zhao, Z., Ma, X., Li, S., Gong, R., & Kuang, A. (2011). Targeting of tumor radioiodine therapy by expression of the sodium iodide symporter under control of the survivin promoter. *Cancer Gene Therapy*, *18*, 144–152.

Hulkower, K. I., & Herber, R. L. (2011). Cell migration and invasion assays as tools for drug discovery. *Pharmaceutics*, *3*, 107–124.

Husemann, Y., & Klein, C. A. (2009). The analysis of metastasis in transgenic mouse models. *Transgenic Research*, *18*, 1–5.

Hwu, J. R., Lin, Y. S., Josephrajan, T., Hsu, M. H., Cheng, F. Y., Yeh, C. S., et al. (2009). Targeted paclitaxel by conjugation to iron oxide and gold nanoparticles. *Journal of the American Chemical Society*, *131*, 66–68.

Ignatiadis, M., Xenidis, N., Perraki, M., Apostolaki, S., Politaki, E., Kafousi, M., et al. (2007). Different prognostic value of cytokeratin-19 mRNA positive circulating tumor cells according to estrogen receptor and HER2 status in early-stage breast cancer. *Journal of Clinical Oncology*, *25*, 5194–5202.

Iyer, M., Wu, L., Carey, M., Wang, Y., Smallwood, A., & Gambhir, S. S. (2001). Two-step transcriptional amplification as a method for imaging reporter gene expression using weak promoters. *Proceedings of the National Academy of Sciences of the United States of America*, *98*, 14595–14600.

Jacobs, A., Voges, J., Reszka, R., Lercher, M., Gossmann, A., Kracht, L., et al. (2001). Positron-emission tomography of vector-mediated gene expression in gene therapy for gliomas. *Lancet*, *358*, 727–729.

Jensen, J. D., Knoop, A., Ewertz, M., & Laenkholm, A. V. (2012). ER, HER2, and TOP2A expression in primary tumor, synchronous axillary nodes, and asynchronous metastases in breast cancer. *Breast Cancer Research and Treatment*, *132*, 511–521.

Jin, H., Hong, B., Kakar, S. S., & Kang, K. A. (2008). Tumor-specific nano-entities for optical detection and hyperthermic treatment of breast cancer. *Advances in Experimental Medicine and Biology*, *614*, 275–284.

Jin, T., Peng, H., & Wu, H. (2013). Clinical value of circulating liver cancer cells for the diagnosis of hepatocellular carcinoma: A meta-analysis. *Biomedical Reports*, *1*, 731–736.

Joyce, J. A., & Pollard, J. W. (2009). Microenvironmental regulation of metastasis. *Nature Reviews Cancer*, *9*, 239–252.

Jung, J. (2014). Human tumor xenograft models for preclinical assessment of anticancer drug development. *Toxicology Research*, *30*, 1–5.

Kamaleshwaran, K. K., Mittal, B. R., Harisankar, C. N. B., Bhattacharya, A., Singh, S. K., & Mandal, A. K. (2012). Predictive value of serum prostate specific antigen in detecting bone metastasis in prostate cancer patients using bone scintigraphy. *Indian Journal of Nuclear Medicine*, *27*(2), 81–84.

Kang, J. H., Chung, J. K., Lee, Y. J., Shin, J. H., Jeong, J. M., Lee, D. S., et al. (2004). Establishment of a human hepatocellular carcinoma cell line highly expressing sodium iodide symporter for radionuclide gene therapy. *Journal of Nuclear Medicine*, *45*, 1571–1576.

Kang, D. C., Su, Z. Z., Sarkar, D., Emdad, L., Volsky, D. J., & Fisher, P. B. (2005). Cloning and characterization of HIV-1-inducible astrocyte elevated gene-1, AEG-1. *Gene*, *353*, 8–15.

Keshaviah, A., Dellapasqua, S., Rotmensz, N., Lindtner, J., Crivellari, D., Collins, J., et al. (2007). CA15-3 and alkaline phosphatase as predictors for breast cancer recurrence: A combined analysis of seven International Breast Cancer Study Group trials. *Annals of Oncology*, *18*, 701–708.

Khan, Q. J., O'Dea, A. P., Dusing, R., Sharma, P., Roling, S. M., Fabian, C. J., et al. (2007). Integrated FDG-PET/CT for initial staging of breast cancer. *Journal Clinical Oncology*, *25*(18 Suppl.). Abstract 558.

Khan, A. K., Rashid, R., Murtaza, G., & Zahra, A. (2014). Gold nanoparticles: Synthesis and applications in drug delivery. *Tropical Journal of Pharmaceutical Research*, *13*, 1169–1177.

Khanna, C., & Hunter, K. (2005). Modeling metastasis in vivo. *Carcinogenesis*, *26*, 513–523.

Kiessling, F., Fokong, S., Koczera, P., Lederle, W., & Lammers, T. (2012). Ultrasound microbubbles for molecular diagnosis, therapy, and theranostics. *Journal of Nuclear Medicine*, *53*, 345–348.

Kievit, F. M., & Zhang, M. (2011). Cancer nanotheranostics: Improving imaging and therapy by targeted delivery across biological barriers. *Advanced Materials*, *23*, H217–H247.

Kilgore, M. L., Steindel, S. J., & Smith, J. A. (1998). Evaluating stat testing options in an academic health center: Therapeutic turnaround time and staff satisfaction. *Clinical Chemistry*, *44*, 1597–1603.

Kim, H. W., Cai, Q. Y., Jun, H. Y., Chon, K. S., Park, S. H., Byun, S. J., et al. (2008). Micro-CT imaging with a hepatocyte-selective contrast agent for detecting liver metastasis in living mice. *Academic Radiology*, *15*, 1282–1290.

Kim, T. H., Han, S. S., Park, S. J., Lee, W. J., Woo, S. M., Yoo, T., et al. (2011). Ca 19-9 level as indicator of early distant metastasis and therapeutic selection in resected pancreatic cancer. *International Journal of Radiation Oncology*, *81*, E743–E748.

Kitago, M., Koyanagi, K., Nakamura, T., Goto, Y., Faries, M., O'Day, S. J., et al. (2009). mRNA expression and BRAF mutation in circulating melanoma cells isolated from peripheral blood with high molecular weight melanoma-associated antigen-specific monoclonal antibody beads. *Clinical Chemistry*, *55*, 757–764.

Koba, W., Jelicks, L. A., & Fine, E. J. (2013). MicroPET/SPECT/CT imaging of small animal models of disease. *The American Journal of Pathology*, *182*, 319–324.

Kokko, R., Holli, K., & Hakama, M. (2002). Ca 15-3 in the follow-up of localised breast cancer: A prospective study. *European Journal of Cancer*, *38*, 1189–1193.

Kolostova, K., Pinterova, D., Hoffman, R. M., & Bobek, V. (2011). Circulating human prostate cancer cells from an orthotopic mouse model rapidly captured by immunomagnetic beads and imaged by GFP expression. *Anticancer Research*, *31*, 1535–1539.

Krenning, E. P., Bakker, W. H., Breeman, W. A., Koper, J. W., Kooij, P. P., Ausema, L., et al. (1989). Localisation of endocrine-related tumours with radioiodinated analogue of somatostatin. *Lancet*, *1*, 242–244.

Kuba, S., Ishida, M., Nakamura, Y., Yamanouchi, K., Minami, S., Taguchi, K., et al. (2014). Treatment and prognosis of breast cancer patients with brain metastases according to intrinsic subtype. *Japanese Journal of Clinical Oncology*, *44*, 1025–1031.

Kulasingam, V., & Diamandis, E. P. (2008). Strategies for discovering novel cancer biomarkers through utilization of emerging technologies. *Nature Clinical Practice Oncology*, *5*, 588–599.

Lalmahomed, Z. S., Kraan, J., Gratama, J. W., Mostert, B., Sleijfer, S., & Verhoef, C. (2010). Circulating tumor cells and sample size: The more, the better. *Journal of Clinical Oncology*, *28*, e288–e289.

Langsteger, W., Heinisch, M., & Fogelman, I. (2006). The role of fluorodeoxyglucose, 18 F-dihydroxyphenylalanine, 18 F-choline, and 18 F-fluoride in bone imaging with emphasis on prostate and breast. *Seminars in Nuclear Medicine*, *36*, 73–92.

Latham, J. P., Searle, P. F., Mautner, V., & James, N. D. (2000). Prostate-specific antigen promoter/enhancer driven gene therapy for prostate cancer: Construction and testing of a tissue-specific adenovirus vector. *Cancer Research*, *60*, 334–341.

Lecharpentier, A., Vielh, P., Perez-Moreno, P., Planchard, D., Soria, J. C., & Farace, F. (2011). Detection of circulating tumour cells with a hybrid (epithelial/mesenchymal) phenotype in patients with metastatic non-small cell lung cancer. *British Journal of Cancer*, *105*, 1338–1341.

Lecouvet, F. E., Geukens, D., Stainier, A., Jamar, F., Jamart, J., d'Othee, B. J., et al. (2007). Magnetic resonance imaging of the axial skeleton for detecting bone metastases in patients with high-risk prostate cancer: Diagnostic and cost-effectiveness and comparison with current detection strategies. *Journal of Clinical Oncology*, *25*, 3281–3287.

Lecouvet, F. E., Simon, M., Tombal, B., Jamart, J., Vande Berg, B. C., & Simoni, P. (2010). Whole-body MRI (WB-MRI) versus axial skeleton MRI (AS-MRI) to detect and measure bone metastases in prostate cancer (PCa). *European Radiology*, *20*, 2973–2982.

Lecouvet, F. E., Talbot, J. N., Messiou, C., Bourguet, P., Liu, Y., de Souza, N. M., et al. (2014). Monitoring the response of bone metastases to treatment with Magnetic Resonance Imaging and nuclear medicine techniques: A review and position statement by the European Organisation for Research and Treatment of Cancer imaging group. *European Journal of Cancer*, *50*, 2519–2531.

Leong-Poi, H., Kuliszewski, M. A., Lekas, M., Sibbald, M., Teichert-Kuliszewska, K., Klibanov, A. L., et al. (2007). Therapeutic arteriogenesis by ultrasound-mediated VEGF165 plasmid gene delivery to chronically ischemic skeletal muscle. *Circulation Research*, *101*, 295–303.

Liang, C. C., Park, A. Y., & Guan, J. L. (2007). In vitro scratch assay: A convenient and inexpensive method for analysis of cell migration in vitro. *Nature Protocols*, *2*, 329–333.

Lianidou, E. S., Strati, A., & Markou, A. (2014). Circulating tumor cells as promising novel biomarkers in solid cancers. *Critical Reviews in Clinical Laboratory Sciences*, *51*, 160–171.

Louie, R. F., Tang, Z. P., Shelby, D. G., & Kost, G. J. (2000). Point-of-care testing: Millennium technology for critical care. *Laboratory Medicine*, *31*, 402–408.

Lowes, L. E., Lock, M., Rodrigues, G., D'Souza, D., Bauman, G., Ahmad, B., et al. (2012). Circulating tumour cells in prostate cancer patients receiving salvage radiotherapy. *Clinical & Translational Oncology*, *14*, 150–156.

Luke, J. M., Vincent, J. M., Du, S. X., Gerdemann, U., Leen, A. M., Whalen, R. G., et al. (2011). Improved antibiotic-free plasmid vector design by incorporation of transient expression enhancers. *Gene Therapy*, *18*, 334–343.

Lurje, G., Schiesser, M., Claudius, A., & Schneider, P. M. (2010). Circulating tumor cells in gastrointestinal malignancies: Current techniques and clinical implications. *Journal of Oncology*, *2010*, 392652.

Lv, Q., Gong, L., Zhang, T., Ye, J., Chai, L., Ni, C., et al. (2016). Prognostic value of circulating tumor cells in metastatic breast cancer: A systemic review and meta-analysis. *Clinical & Translational Oncology*, *18*, 322–330.

Ma, X. L., Xiao, Z. L., Liu, L., Liu, X. X., Nie, W., Li, P., et al. (2012). Meta-analysis of circulating tumor cells as a prognostic marker in lung cancer. *Asian Pacific Journal of Cancer Prevention: APJCP*, *13*, 1137–1144.

MacDonald, I. C., & Chambers, A. F. (2008). Chapter 9. Intravital videomicroscopy in angiogenesis research. *Methods in Enzymology*, *444*, 201–230.

Mahner, S., Schirrmacher, S., Brenner, W., Jenicke, L., Habermann, C. R., Avril, N., et al. (2008). Comparison between positron emission tomography using 2-[fluorine-18] fluoro-2-deoxy-D-glucose, conventional imaging and computed tomography for staging of breast cancer. *Annals of Oncology*, *19*, 1249–1254.

Majumdar, A. S., Hughes, D. E., Lichtsteiner, S. P., Wang, Z., Lebkowski, J. S., & Vasserot, A. P. (2001). The telomerase reverse transcriptase promoter drives efficacious tumor suicide gene therapy while preventing hepatotoxicity encountered with constitutive promoters. *Gene Therapy*, *8*, 568–578.

Marshall, J. (2011). Transwell((R)) invasion assays. *Methods in Molecular Biology*, *769*, 97–110.

Meairs, S. (2015). Facilitation of drug transport across the blood-brain barrier with ultrasound and microbubbles. *Pharmaceutics*, 7, 275–293.

Menezes, M. E., Das, S. K., Emdad, L., Windle, J. J., Wang, X. Y., Sarkar, D., et al. (2014). Genetically engineered mice as experimental tools to dissect the critical events in breast cancer. *Advances in Cancer Research*, *121*, 331–382.

Messiou, C., Cook, G., & deSouza, N. M. (2009). Imaging metastatic bone disease from carcinoma of the prostate. *British Journal of Cancer*, *101*, 1225–1232.

Minn, I., Bar-Shir, A., Yarlagadda, K., Bulte, J. W., Fisher, P. B., Wang, H., et al. (2015). Tumor-specific expression and detection of a CEST reporter gene. *Magnetic Resonance in Medicine*, *74*, 544–549.

Minn, I., Menezes, M. E., Sarkar, S., Yarlagadda, K., Das, S. K., Emdad, L., et al. (2014). Molecular-genetic imaging of cancer. *Advances in Cancer Research*, *124*, 131–169.

Molina, R., Auge, J. M., Escudero, J. M., Marrades, R., Vinolas, N., Carcereny, E., et al. (2008). Mucins CA 125, CA 19.9, CA 15.3 and TAG-72.3 as tumor markers in patients with lung cancer: Comparison with CYFRA 21-1, CEA, SCC and NSE. *Tumor Biology*, *29*, 371–380.

Molina, C. A., Ribo, M., Rubiera, M., Montaner, J., Santamarina, E., Delgado-Mederos, R., et al. (2006). Microbubble administration accelerates clot lysis during continuous 2-MHz ultrasound monitoring in stroke patients treated with intravenous tissue plasminogen activator. *Stroke*, *37*, 425–429.

Moore, R. G., Brown, A. K., Miller, M. C., Skates, S., Allard, W. J., Verch, T., et al. (2008). The use of multiple novel tumor biomarkers for the detection of ovarian carcinoma in patients with a pelvic mass. *Gynecological Oncology*, *108*, 402–408.

Mordente, A., Meucci, E., Martorana, G. E., & Silvestrini, A. (2015). Cancer biomarkers discovery and validation: State of the art, problems and future perspectives. *Advances in Experimental Medicine and Biology*, *867*, 9–26.

Moreira, D. M., Howard, L. E., Sourbeer, K. N., Amarasekara, H. S., Chow, L. C., Cockrell, D. C., et al. (2015). Predicting bone scan positivity in non-metastatic castration-resistant prostate cancer. *Prostate Cancer and Prostatic Diseases*, *18*, 333–337.

Mudshinge, S. R., Deore, A. B., Patil, S., & Bhalgat, C. M. (2011). Nanoparticles: Emerging carriers for drug delivery. *The Saudi Pharmaceutical Journal*, *19*, 129–141.

Munye, M. M., Tagalakis, A. D., Barnes, J. L., Brown, R. E., McAnulty, R. J., Howe, S. J., et al. (2016). Minicircle DNA provides enhanced and prolonged transgene expression following airway gene transfer. *Scientific Reports*, *6*, 23125.

Nagamachi, S., Nishii, R., Wakamatsu, H., Mizutani, Y., Kiyohara, S., Fujita, S., et al. (2013). The usefulness of (18)F-FDG PET/MRI fusion image in diagnosing pancreatic tumor: Comparison with (18)F-FDG PET/CT. *Annals of Nuclear Medicine*, *27*, 554–563.

Nagarajah, J., Jentzen, W., Hartung, V., Rosenbaum-Krumme, S., Mikat, C., Heusner, T. A., et al. (2011). Diagnosis and dosimetry in differentiated thyroid carcinoma using 124I PET: Comparison of PET/MRI vs PET/CT of the neck. *European Journal of Nuclear Medicine and Molecular Imaging*, *38*, 1862–1868.

Nakanishi, K., Kobayashi, M., Nakaguchi, K., Kyakuno, M., Hashimoto, N., Onishi, H., et al. (2007). Whole-body MRI for detecting metastatic bone tumor: Diagnostic value of diffusion-weighted images. *Magnetic Resonance in Medical Sciences*, *6*, 147–155.

Nettelbeck, D. M., Jerome, V., & Muller, R. (1998). A strategy for enhancing the transcriptional activity of weak cell type-specific promoters. *Gene Therapy*, *5*, 1656–1664.

Nicolini, A., Ferrari, P., & Rossi, G. (2015). Mucins and cytokeratins as serum tumor markers in breast cancer. *Advances in Experimental Medicine and Biology*, *867*, 197–225.

O'Brien, D. P., Horgan, P. G., Gough, D. B., Skehill, R., Grimes, H., & Given, H. F. (1992). CA15-3: A reliable indicator of metastatic bone disease in breast cancer patients. *Annals of the Royal College of Surgeons of England*, *74*, 9–11.

O'Sullivan, G. J., Carty, F. L., & Cronin, C. G. (2015). Imaging of bone metastasis: An update. *World Journal of Radiology*, *7*, 202–211.

Pantel, K., & Alix-Panabieres, C. (2010). Circulating tumour cells in cancer patients: Challenges and perspectives. *Trends in Molecular Medicine*, *16*, 398–406.

Parisotto, M., & Metzger, D. (2013). Genetically engineered mouse models of prostate cancer. *Molecular Oncology*, *7*, 190–205.

Park, S. I., Kim, S. J., McCauley, L. K., & Gallick, G. E. (2010). Pre-clinical mouse models of human prostate cancer and their utility in drug discovery. *Current Protocols in Pharmacology*, *51*, Chapter 14, Unit 14.15.1–14.15.27.

Partecke, I. L., Kaeding, A., Sendler, M., Albers, N., Kuhn, J. P., Speerforck, S., et al. (2011). In vivo imaging of pancreatic tumours and liver metastases using 7 Tesla MRI in a murine orthotopic pancreatic cancer model and a liver metastases model. *BMC Cancer*, *11*, 40.

Partovi, S., Kohan, A., Rubbert, C., Vercher-Conejero, J. L., Gaeta, C., Yuh, R., et al. (2014). Clinical oncologic applications of PET/MRI: A new horizon. *American Journal of Nuclear Medicine & Molecular Imaging*, *4*, 202–212.

Peldschus, K., & Ittrich, H. (2014). Magnetic resonance imaging of metastases in xenograft mouse models of cancer. *Methods in Molecular Biology*, *1070*, 213–222.

Penuelas, I., Mazzolini, G., Boan, J. F., Sangro, B., Marti-Climent, J., Ruiz, M., et al. (2005). Positron emission tomography imaging of adenoviral-mediated transgene expression in liver cancer patients. *Gastroenterology*, *128*, 1787–1795.

Pierga, J. Y., Bidard, F. C., Mathiot, C., Brain, E., Delaloge, S., Giachetti, S., et al. (2008). Circulating tumor cell detection predicts early metastatic relapse after neoadjuvant chemotherapy in large operable and locally advanced breast cancer in a phase II randomized trial. *Clinical Cancer Research*, *14*, 7004–7010.

Pierga, J. Y., Hajage, D., Bachelot, T., Delaloge, S., Brain, E., Campone, M., et al. (2012). High independent prognostic and predictive value of circulating tumor cells compared with serum tumor markers in a large prospective trial in first-line chemotherapy for metastatic breast cancer patients. *Annals of Oncology*, *23*, 618–624.

Pomper, M. G., & Fisher, P. B. (2014). Emerging applications of molecular imaging to oncology. *Advances in Cancer Research*, *124*, 1–402.

Price, J. E. (2001). Xenograft models in immunodeficient animals: I. Nude mice: Spontaneous and experimental metastasis models. *Methods in Molecular Medicine*, *58*, 205–213.

Price, J. E. (2014). Spontaneous and experimental metastasis models: Nude mice. *Methods in Molecular Biology*, *1070*, 223–233.

Queiroz, M. A., & Huellner, M. W. (2015). PET/MR in cancers of the head and neck. *Seminars in Nuclear Medicine*, *45*, 248–265.

Rajecki, M., Sarparanta, M., Hakkarainen, T., Tenhunen, M., Diaconu, I., Kuhmonen, V., et al. (2012). SPECT/CT imaging of hNIS-expression after intravenous delivery of an oncolytic adenovirus and 131I. *PloS One*, *7*, e32871.

Rapoport, N., Gao, Z., & Kennedy, A. (2007). Multifunctional nanoparticles for combining ultrasonic tumor imaging and targeted chemotherapy. *Journal of the National Cancer Institute, 99*, 1095–1106.

Roberts, C. C., Daffner, R. H., Weissman, B. N., Bancroft, L., Bennett, D. L., Blebea, J. S., et al. (2010). ACR appropriateness criteria on metastatic bone disease. *Journal of the American College of Radiology*, 7, 400–409.

Rogers, B. E., Parry, J. J., Andrews, R., Cordopatis, P., Nock, B. A., & Maina, T. (2005). MicroPET imaging of gene transfer with a somatostatin receptor-based reporter gene and (94 m)Tc-Demotate 1. *Journal of Nuclear Medicine, 46*, 1889–1897.

Ronald, J. A., Chuang, H. Y., Dragulescu-Andrasi, A., Hori, S. S., & Gambhir, S. S. (2015). Detecting cancers through tumor-activatable minicircles that lead to a detectable blood biomarker. *Proceedings of the National Academy of Sciences of the United States of America, 112*, 3068–3073.

Rossi, V., Sarotto, I., Maggiorotto, F., Berchialla, P., Kubatzki, F., Tomasi, N., et al. (2012). Moderate immunohistochemical expression of HER-2 (2+) without HER-2 gene amplification is a negative prognostic factor in early breast cancer. *The Oncologist, 17*, 1418–1425.

Saha, S., Burke, C., Desai, A., Vijayanathan, S., & Gnanasegaran, G. (2013). SPECT-CT: Applications in musculoskeletal radiology. *The British Journal of Radiology, 86*, 20120519.

Sahai, E. (2007). Illuminating the metastatic process. *Nature Reviews Cancer*, 7, 737–749.

Sakaizawa, K., Goto, Y., Kiniwa, Y., Uchiyama, A., Harada, K., Shimada, S., et al. (2012). Mutation analysis of BRAF and KIT in circulating melanoma cells at the single cell level. *British Journal of Cancer, 106*, 939–946.

Sanches, P. G., Peters, S., Rossin, R., Kaijzel, E. L., Que, I., Lowik, C. W., et al. (2015). Bone metastasis imaging with SPECT/CT/MRI: A preclinical toolbox for therapy studies. *Bone, 75*, 62–71.

Sarkar, S., Quinn, B. A., Shen, X. N., Dash, R., Das, S. K., Emdad, L., et al. (2015). Therapy of prostate cancer using a novel cancer terminator virus and a small molecule BH-3 mimetic. *Oncotarget, 6*, 10712–10727.

Sawicki, L. M., Grueneisen, J., Schaarschmidt, B. M., Buchbender, C., Nagarajah, J., Umutlu, L., et al. (2016). Evaluation of (18)F-FDG PET/MRI, (18)F-FDG PET/CT, MRI, and CT in whole-body staging of recurrent breast cancer. *European Journal of Radiology, 85*, 459–465.

Scher, H. I., Jia, X. Y., de Bono, J. S., Fleisher, M., Pienta, K. J., Raghavan, D., et al. (2009). Circulating tumour cells as prognostic markers in progressive, castration-resistant prostate cancer: A reanalysis of IMMC38 trial data. *Lancet Oncology, 10*, 233–239.

Shao, M. M., Liu, J., Vong, J. S., Niu, Y., Germin, B., Tang, P., et al. (2011). A subset of breast cancer predisposes to brain metastasis. *Medical Molecular Morphology, 44*, 15–20.

Shen, S., Wang, S., Zheng, R., Zhu, X., Jiang, X., Fu, D., et al. (2015). Magnetic nanoparticle clusters for photothermal therapy with near-infrared irradiation. *Biomaterials, 39*, 67–74.

Siolas, D., & Hannon, G. J. (2013). Patient-derived tumor xenografts: Transforming clinical samples into mouse models. *Cancer Research, 73*, 5315–5319.

Smith, H. W., & Muller, W. J. (2013). Transgenic mouse models—A seminal breakthrough in oncogene research. *Cold Spring Harbor Protocols, 2013*, 1099–1108.

Spick, C., Herrmann, K., & Czernin, J. (2016). 18F-FDG PET/CT and PET/MRI perform equally well in cancer: Evidence from studies on more than 2,300 patients. *Journal of Nuclear Medicine, 57*, 420–430.

Stewart, T. A., Pattengale, P. K., & Leder, P. (1984). Spontaneous mammary adenocarcinomas in transgenic mice that carry and express MTV/myc fusion genes. *Cell, 38*, 627–637.

Strien, L., Leidenius, M., von Smitten, K., & Heikkila, P. (2010). Concordance between HER-2 and steroid hormone receptor expression between primary breast cancer, sentinel node metastases, and isolated tumor cells. *Pathology, Research and Practice*, *206*, 253–258.

Sturgeon, C. M., Duffy, M. J., Stenman, U. H., Lilja, H., Brunner, N., Chan, D. W., et al. (2008). National Academy of Clinical Biochemistry laboratory medicine practice guidelines for use of tumor markers in testicular, prostate, colorectal, breast, and ovarian cancers. *Clinical Chemistry*, *54*, e11–e79.

Su, Z. Z., Sarkar, D., Emdad, L., Duigou, G. J., Young, C. S., Ware, J., et al. (2005). Targeting gene expression selectively in cancer cells by using the progression-elevated gene-3 promoter. *Proceedings of the National Academy of Sciences of the United States of America*, *102*, 1059–1064.

Su, Z. Z., Shi, Y., & Fisher, P. B. (1997). Subtraction hybridization identifies a transformation progression-associated gene PEG-3 with sequence homology to a growth arrest and DNA damage-inducible gene. *Proceedings of the National Academy of Sciences of the United States of America*, *94*, 9125–9130.

Talmadge, J. E., & Gabrilovich, D. I. (2013). History of myeloid-derived suppressor cells. *Nature Reviews Cancer*, *13*, 739–752.

Talukdar, S., Emdad, L., Das, S. K., Sarkar, D., & Fisher, P. B. (2016). Evolving strategies for therapeutically targeting cancer stem cells (CSCs). *Advances in Cancer Research*, *131*, 159–191.

Taube, S. E., Jacobson, J. W., & Lively, T. G. (2005). Cancer diagnostics: Decision criteria for marker utilization in the clinic. *American Journal of Pharmacogenomics*, *5*, 357–364.

Thalheimer, A., Korb, D., Bonicke, L., Wiegering, A., Muhling, B., Schneider, M., et al. (2013). Noninvasive visualization of tumor growth in a human colorectal liver metastases xenograft model using bioluminescence in vivo imaging. *The Journal of Surgical Research*, *185*, 143–151.

Thies, A., Mauer, S., Fodstad, O., & Schumacher, U. (2007). Clinically proven markers of metastasis predict metastatic spread of human melanoma cells engrafted in scid mice. *British Journal of Cancer*, *96*, 609–616.

Tjensvoll, K., Nordgard, O., & Smaaland, R. (2014). Circulating tumor cells in pancreatic cancer patients: Methods of detection and clinical implications. *International Journal of Cancer*, *134*, 1–8.

Tombal, B., & Lecouvet, F. (2012). Modern detection of prostate cancer's bone metastasis: Is the bone scan era over? *Advances in Urology*, *2012*, 893193.

Valastyan, S., & Weinberg, R. A. (2011). Tumor metastasis: Molecular insights and evolving paradigms. *Cell*, *147*, 275–292.

van Zijl, F., Krupitza, G., & Mikulits, W. (2011). Initial steps of metastasis: Cell invasion and endothelial transmigration. *Mutation Research*, *728*, 23–34.

Vaquero, J. J., & Kinahan, P. (2015). Positron emission tomography: Current challenges and opportunities for technological advances in clinical and preclinical imaging systems. *Annual Review of Biomedical Engineering*, *17*, 385–414.

Vassiliou, V., Andreopoulos, D., Frangos, S., Tselis, N., Giannopoulou, E., & Lutz, S. (2011). Bone metastases: Assessment of therapeutic response through radiological and nuclear medicine imaging modalities. *Clinical oncology (Royal College of Radiologists (Great Britain))*, *23*, 632–645.

Venkitaraman, R., Cook, G. J., Dearnaley, D. P., Parker, C. C., Khoo, V., Eeles, R., et al. (2009). Whole-body magnetic resonance imaging in the detection of skeletal metastases in patients with prostate cancer. *Journal of Medical Imaging and Radiation Oncology*, *53*, 241–247.

Vijayakumar, V., Vijayakumar, S., Quadri, S. F., & Blend, M. J. (1994). Can prostate-specific antigen levels predict bone-scan evidence of metastases in newly-diagnosed prostate-cancer. *American Journal of Clinical Oncology-Cancer*, *17*, 432–436.

Wadajkar, A. S., Menon, J. U., Kadapure, T., Tran, R. T., Yang, J., & Nguyen, K. T. (2013). Design and application of magnetic-based theranostic nanoparticle systems. *Recent Patents on Biomedical Engineering*, *6*, 47–57.

Walling, M. A., Novak, J. A., & Shepard, J. R. (2009). Quantum dots for live cell and in vivo imaging. *International Journal of Molecular Sciences*, *10*, 441–491.

Wang, Z., Cui, K., Xue, Y., Tong, F., & Li, S. (2015). Prognostic value of circulating tumor cells in patients with squamous cell carcinoma of the head and neck: A systematic review and meta-analysis. *Medical Oncology*, *32*, 164.

Wang, D., Stockard, C. R., Harkins, L., Lott, P., Salih, C., Yuan, K., et al. (2008). Immunohistochemistry in the evaluation of neovascularization in tumor xenografts. *Biotechnic & Histochemistry*, *83*, 179–189.

Wang, S., & Wang, Z. (2015). Meta-analysis of epidermal growth factor receptor and kras gene status between primary and corresponding metastatic tumours of non-small cell lung cancer. *Clinical oncology (Royal College of Radiologists (Great Britain))*, *27*, 30–39.

Wang, D., Wang, Z., Tian, J., He, X., Chowdhury, W. H., Zhang, X., et al. (2010). Prostate stem cell antigen enhancer and uroplakin II promoter based bladder cancer targeted tissue-specific vector. *Urologic Oncology*, *28*, 164–169.

Wang, H. Y., Wei, J., Zou, Z. Y., Qian, X. P., & Liu, B. R. (2016). Circulating tumour cells predict survival in gastric cancer patients: A meta-analysis. *Contemporary Oncology*, *19*, 451–457.

Wang, F. B., Yang, X. Q., Yang, S., Wang, B. C., Feng, M. H., & Tu, J. C. (2011). A higher number of circulating tumor cells (CTC) in peripheral blood indicates poor prognosis in prostate cancer patients—A meta-analysis. *Asian Pacific Journal of Cancer Prevention*, *12*, 2629–2635.

Weissleder, R., & Nahrendorf, M. (2015). Advancing biomedical imaging. *Proceedings of the National Academy of Sciences of the United States of America*, *112*, 14424–14428.

Westwood, M., Joore, M., Whiting, P., van Asselt, T., Ramaekers, B., Armstrong, N., et al. (2014). Epidermal growth factor receptor tyrosine kinase (EGFR-TK) mutation testing in adults with locally advanced or metastatic non-small cell lung cancer: A systematic review and cost-effectiveness analysis. *Health Technology Assessment*, *18*, 1–166.

Wilt, T. J., Scardino, P. T., Carlsson, S. V., & Basch, E. (2014). Prostate-specific antigen screening in prostate cancer: Perspectives on the evidence. *Journal of the National Cancer Institute*, *106*(3), dju010.

Win, A. Z., & Aparici, C. M. (2014). Factors affecting uptake of NaF-18 by the normal skeleton. *Journal of Clinical Medicine Research*, *6*, 435–442.

Wu, S., Liu, S., Liu, Z., Huang, J., Pu, X., Li, J., et al. (2015). Classification of circulating tumor cells by epithelial-mesenchymal transition markers. *PloS One*, *10*, e0123976.

Wu, X. L., Tu, Q., Faure, G., Gallet, P., Kohler, C., & Bittencourt Mde, C. (2016). Diagnostic and prognostic value of circulating tumor cells in head and neck squamous cell carcinoma: A systematic review and meta-analysis. *Scientific Reports*, *6*, 20210.

Xiao, C., Gong, Y., Han, E. Y., Gonzalez-Angulo, A. M., & Sneige, N. (2011). Stability of HER2-positive status in breast carcinoma: A comparison between primary and paired metastatic tumors with regard to the possible impact of intervening trastuzumab treatment. *Annals of Oncology*, *22*, 1547–1553.

Xie, J., Lee, S., & Chen, X. (2010). Nanoparticle-based theranostic agents. *Advanced Drug Delivery Reviews*, *62*, 1064–1079.

Yaghoubi, S. S., Barrio, J. R., Namavari, M., Satyamurthy, N., Phelps, M. E., Herschman, H. R., et al. (2005). Imaging progress of herpes simplex virus type 1 thymidine kinase suicide gene therapy in living subjects with positron emission tomography. *Cancer Gene Therapy*, *12*, 329–339.

Yang, Y., Gong, M. F., Yang, H., Zhang, S., Wang, G. X., Su, T. S., et al. (2016). MR molecular imaging of tumours using ferritin heavy chain reporter gene expression mediated by the hTERT promoter. *European Radiology*.

Yang, H. L., Liu, T., Wang, X. M., Xu, Y., & Deng, S. M. (2011). Diagnosis of bone metastases: A meta-analysis comparing (1)(8)FDG PET, CT, MRI and bone scintigraphy. *European Radiology*, *21*, 2604–2617.

Yang, J., & Weinberg, R. A. (2008). Epithelial-mesenchymal transition: At the crossroads of development and tumor metastasis. *Developmental Cell*, *14*, 818–829.

Yang, S., Zhang, J. J., & Huang, X. Y. (2012). Mouse models for tumor metastasis. *Methods in Molecular Biology*, *928*, 221–228.

Yin, W., Xiang, P., & Li, Q. (2005). Investigations of the effect of DNA size in transient transfection assay using dual luciferase system. *Analytical Biochemistry*, *346*, 289–294.

Yoo, H. J., Lee, J. S., & Lee, J. M. (2015). Integrated whole body MR/PET: Where are we? *Korean Journal of Radiology*, *16*, 32–49.

Yu, H. H., Tsai, Y. Y., & Hoffe, S. E. (2012). Overview of diagnosis and management of metastatic disease to bone. *Cancer Control*, *19*, 84–91.

Zhang, Z., Ramnath, N., & Nagrath, S. (2015). Current status of CTCs as liquid biopsy in lung cancer and future directions. *Frontiers in Oncology*, *5*, 209.

Zhang, Y., Yang, P., & Wang, X. F. (2014). Microenvironmental regulation of cancer metastasis by miRNAs. *Trends in Cell Biology*, *24*, 153–160.

Zhou, Y., Bian, B., Yuan, X., Xie, G., Ma, Y., & Shen, L. (2015). Prognostic value of circulating tumor cells in ovarian cancer: A meta-analysis. *PloS One*, *10*, e0130873.

Zhu, W. F., Li, J., Yu, L. C., Wu, Y., Tang, X. P., Hu, Y. M., et al. (2014). Prognostic value of EpCAM/MUC1 mRNA-positive cells in non-small cell lung cancer patients. *Tumour Biology*, *35*, 1211–1219.

CHAPTER TWO

The Relationship Between Dormant Cancer Cells and Their Microenvironment

N. Linde[1,2], G. Fluegen[1,2,3], J.A. Aguirre-Ghiso[1]
Tisch Cancer Institute, Black Family Stem Cell Institute, Mount Sinai School of Medicine, New York, NY, United States
[1]Corresponding authors: e-mail address: nina.linde@mssm.edu; georg.fluegen@med.uni-duesseldorf.de; julio.aguirre-ghiso@mssm.edu

Contents

1. Introduction 46
2. Models to Study Tumor Cell Dormancy 48
3. Intrinsic and Extrinsic Signals Converge to Induce Tumor Cell Dormancy 51
4. Cooperative Extrinsic Signals Induce Dormancy Within the Bone Microenvironment 53
5. Endosteal and Perivascular Niches Support Cancer Cell Dormancy 54
6. Reactivation from Dormancy 56
7. The Immune System and Dormancy 58
8. Summary and Outlook 65
References 66

Abstract

The majority of cancer deaths are due to metastases that can occur years or decades after primary tumor diagnosis and treatment. Disseminated tumor cells (DTCs) surviving in a dormant state in target organs appear to explain the timing of this phenomenon. Knowledge on this process is important as it might provide a window of opportunity to prevent recurrences by eradicating dormant DTCs and/or by maintaining DTCs in a dormant state. Importantly, this research might offer markers of dormancy for early monitoring of metastatic relapse. However, our understanding of the mechanisms underlying the regulation of entry into and exit from dormancy is still limited and crippling any therapeutic opportunity. While cancer cell-intrinsic signaling pathways have been linked to dormancy regulation, it is likely that these pathways and the switch controlling reactivation from dormancy are regulated by microenvironmental cues. Here we review and discuss recent findings on how the microenvironment regulates cancer dormancy and raise new questions that may help advance the field.

[2] Equal contribution.
[3] Current address: General-, Visceral- and Pediatric Surgery, University Hospital Duesseldorf, Duesseldorf, Germany.

Advances in Cancer Research, Volume 132
ISSN 0065-230X
http://dx.doi.org/10.1016/bs.acr.2016.07.002

1. INTRODUCTION

Metastasis formation is responsible for the majority of cancer deaths and is caused by cancer cells disseminated from primary tumors that persist in the host after primary tumor removal. Metastasis formation consists of several steps: local invasion from the primary tumor and intravasation, survival in circulation, extravasation, and proliferation in a target organ microenvironment. Importantly, after extravasation and before proliferation into detectable metastasis, years or even decades can pass. Long time periods where patients present with no evidence of disease (NED) followed by late recurrences are explained by the survival of disseminated tumor cells (DTCs) in a dormant state. The mechanisms that determine the amount of time that can pass between the extravasation of DTCs and their proliferation into metastatic masses are one of the most important questions in cancer biology.

From a cell biology perspective the asymptomatic phase that precedes the reactivation of DTCs to form detectable metastases can be explained by *cellular dormancy*, where single DTCs survive in a quiescent, reversibly growth-arrested state for long stretches of time, and *population-based dormancy*, where micrometastases are limited in their growth due to the lack of vascularization and increased cell death that compensates for proliferation or due to immune-mediated responses, which results in a balance between proliferation and DTC death (Aguirre-Ghiso, 2007). While these models of dormancy are not mutually exclusive and might even coexist in the same patient, we will focus on cellular dormancy that is best supported by clinical evidence (Banys, Hartkopf, et al., 2012; Morrissey, Vessella, Lange, & Lam, 2015). Several studies showed that DTCs are frequently found in cancer patients with NED (Braun et al., 2005; Chery et al., 2014; Klein, 2009; Schardt et al., 2005), and the detection of DTCs in the bone marrow is associated with worse clinical outcome in many solid cancers (Braun et al., 2000; Hartkopf et al., 2014, 2015; Thorban, Rosenberg, Busch, & Roder, 2000; Wollenberg et al., 2004). These studies demonstrate that cancer patients with dormant DTCs are at higher risk for metastatic relapses and thus underline the clinical importance of dormancy. The time of NED prior to the reemergence of DTCs from dormancy could be used to for therapeutic intervention but is currently not used to treat patients. Being able to predict if a patient is at risk of recurrence, or not, based on DTC-associated biology would allow more specific therapeutic strategies and avoid overtreatment with antiproliferative therapies that might not address the residual disease biology. Also, dormant DTCs could either be awakened and eradicated or maintained in a dormant state during this time (Aguirre-Ghiso, 2007; Ghajar, 2015; Sosa, Bragado, & Aguirre-Ghiso, 2014). Yet, this window of opportunity is missed, mostly because our understanding of the mechanisms that characterize the dormant state and how DTCs can reemerge from it is cripplingly limited (Aguirre-Ghiso, Bragado, & Sosa, 2013).

Several cancer cell-intrinsic signal pathways that lead to cellular dormancy have been described. Historically, the balance between activated extracellular regulated kinase (ERK1/2) and activated p38α/β was the first signaling mechanism that has been connected reproducibly to DTC dormancy (Adam et al., 2009; Aguirre Ghiso, Kovalski, & Ossowski, 1999; Aguirre-Ghiso, Estrada, Liu, & Ossowski, 2003; Aguirre-Ghiso, Liu, Mignatti, Kovalski, & Ossowski, 2001; Bragado et al., 2013; Kobayashi et al., 2011; Najmi, Korah, Chandra, Abdellatif, & Wieder, 2005; Ruppender et al., 2015; Sosa, Avivar-Valderas, Bragado, Wen, & Aguirre-Ghiso, 2011). Interestingly, early on, a link to the microenvironment was identified in these studies, as the balance between ERK1/2 and p38 signaling was regulated by fibronectin and uPA signaling via the uPA receptor and specific integrins (Aguirre Ghiso et al., 1999; Aguirre-Ghiso et al., 2001). Phosphorylation of p38 leads to the activation of the unfolded protein response pathway, which promotes cell survival and dormancy through ATF6/Rheb/mTOR signaling (Schewe & Aguirre-Ghiso, 2008) and leads to the induction of the dormancy-associated transcription factors DEC2/Sharp1, $p27^{Kip1}$, p21, and NR2F1 (Bragado et al., 2013; Sosa et al., 2015). These mechanisms are integrated (Fig. 1A) to coordinate a deep but still reversible growth arrest and robust survival pathways. For more in-depth information we would like to refer to reviews previously published by our group (Aguirre-Ghiso, 2007; Sosa et al., 2014). These mechanisms explain *how* tumor cells regulate specific signal transducers to enter a state of cellular dormancy (G0–G1 arrest). Yet, the fact that tumor cells, which have disseminated from proliferating tumor masses, enter quiescence and stop proliferating but yet maintain reactivating capacity is puzzling. One likely explanation could be the microenvironment partially controlling the switch between DTC proliferation and dormancy.

The tumor microenvironment is usually defined as the sum of all cellular and extracellular components surrounding cancer cells. In the context of a healthy epithelial tissue, the microenvironment will maintain tissue integrity and is in turn regulated by stromal cells such as fibroblasts and myeloid cells. Several studies support that changes that subvert the tumor

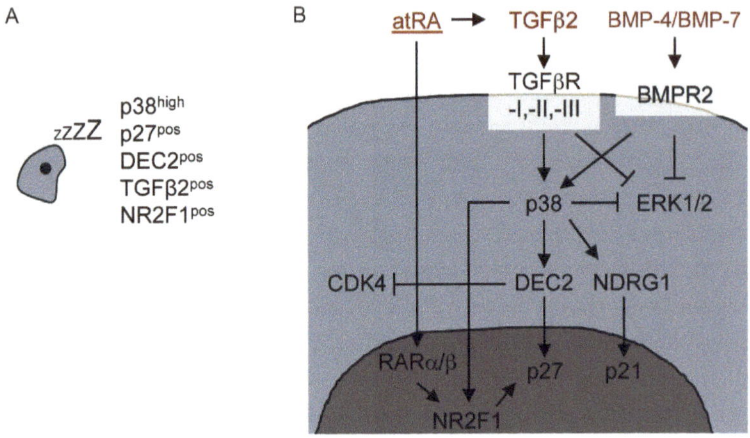

Fig. 1 Overview of dormancy-inducing signaling pathways. (A) Overview of dormancy marker expression in DTCs based on known dormancy-signaling pathways. (B) Microenvironment-derived atRA, TGFβ2, and BMP-4 and -7 cooperate to induce a dormant state in DTCs characterized by activating p38 and NR2F1 and inhibiting ERK1/2 signaling. p38 and NR2F1 induce the cell cycle inhibitors p27 and p21, which results in cell cycle arrest (Bragado et al., 2013; Kobayashi et al., 2011; Sosa et al., 2015).

microenvironment are required for malignant cells to grow into tumors (Hanahan & Coussens, 2012; Mueller & Fusenig, 2004). Thus, since all adult tissues encode mechanisms to essentially prevent uncontrolled ectopic growth, it is reasonable to hypothesize that a tumor-naïve target organ microenvironment may encode regulatory mechanisms to prevent the expansion of DTCs and this may result in dormancy onset. Similarly, one could propose that changes in the target organ microenvironment might awaken dormant DTCs and allow them to proliferate and thus induce late recurrences. In this chapter we will focus on reviewing recent findings that analyzed the influence of microenvironmental cues and cellular components on dormancy and hypothesize about their influence on dormancy induction and exit from dormancy. The goal is to develop potential answers to persistent questions that need to be addressed to find a solution to the urgent clinical problem of dormancy.

2. MODELS TO STUDY TUMOR CELL DORMANCY

One of the challenges in studying dormancy is that by definition it is undetectable using conventional whole-body imaging tools and takes place over long time periods.

This provides a challenge to drug development, as clinical trials are usually performed with far-progressed patient cohorts. Testing drugs in a metastasis prevention setting with adjuvant therapies would be a radical shift in the standard of clinical trials and requires better insight into dormant disease.

One of the main obstacles to studying dormancy, cited by basic researchers repeatedly, is the lack of model systems. Most basic research relies on fast-growing cancer cell lines and fast transgenic oncogene models. It is also common to use aggressively growing metastasis models where metastases develop without any latency. Moreover, most metastasis assays focus on macrometastases as an end point and rely on the use of clones selected for aggressiveness. Commonly, the presence of solitary DTCs or micrometastases is not investigated and the absence of macrometastases is interpreted as the inability of cancer cells to disseminate without investigating which step of the metastatic cascade was not completed. Therefore, many studies simply miss solitary DTC biology.

Yet, the notion that there is a lack of good models to study dormancy is not correct. Table 1 provides examples of some of the model systems used to study dormancy that will be briefly discussed here.

The in vivo metastasis assay used most commonly is the experimental metastasis assay where cell lines are injected into circulation, either through the tail vein or into the left cardiac ventricle or the iliac artery (Box & Eccles, 2011; Rosol, Tannehill-Gregg, Corn, Schneider, & McCauley, 2004; Wang et al., 2015). Alternatively, cancer cell lines are grown orthotopically or subcutaneously and metastases derived from spontaneously DTCs are monitored. This system allows monitoring multiorgan dormancy by tracking tagged DTCs to monitor solitary cell biology and/or label retention at single-cell resolution, as used in cancer models and studies of hematopoietic stem cell dormancy (Lawson, McDonald, et al., 2015). Even pathway biosensors can be used to monitor dormancy pathways in DTCs to understand their regulation (Aguirre-Ghiso, Ossowski, & Rosenbaum, 2004). Additionally, dormant DTCs can be identified in different target organs by detection of proliferation and dormancy markers (Aguirre-Ghiso et al., 2004; Bragado et al., 2013; Sosa et al., 2015), summarized in Fig. 1A.

Instead of using highly aggressive cancer cell lines (i.e. MDA-MB-231), cell lines that give rise predominantly to dormant DTCs can be used in experimental assays or in in vitro assays (Barkan & Green, 2011). Examples are the dormant breast cancer cell line variants 4T07 variant (Aslakson & Miller, 1992) or D2O.R cells (Morris et al., 1993). Moreover, complementary to the common approach to select aggressively growing clonal cell line

Table 1 Models to Study Cancer Dormancy

Model System	Description	Examples
In vitro models	Several dormant cancer cells retain quiescence when grown as in 3D in vitro models	Ghajar et al. (2013), Barkan and Green (2011), and Marlow and Dontu (2015)
Isolation of latent cancer cell lines	Latent clones can be isolated from spontaneous mouse tumors	4T07 (Aslakson & Miller, 1992; Gao et al., 2012); D2O.R (Morris, Tuck, Wilson, Percy, & Chambers, 1993)
Clonal cell line variants	Cell line variants are established through in vivo selection of clones derived from indolent cell lines that show metastatic growth (e.g., D2.1) or selection of indolent clones from metastatic lines (e.g., MDA-MB231-SCP6)	MDA-MB231-SCP6 (Lu et al., 2011); H2087-LCC; HCC1954-LCC (Malladi et al., 2016)
Dormant DTC analysis in PDX tumors	Dormant tumor cells derived from PDX tumors can be analyzed in vivo	Aguirre-Ghiso et al. (2001), Lawson, Bhakta, et al. (2015), and Ossowski and Reich (1980)
Spontaneous mouse cancer models	Dormant DTCs and latent metastases can be detected using spontaneous mouse cancer models. Inducible transgenic mouse cancer models allow for investigation of residual dormant DTCs after oncogene withdrawal	Dormant DTCs: MMTV-HER2 (Husemann et al., 2008); indolent lung metastases: MMTV-Wnt1 (Li, Hively, & Varmus, 2000); inducible transgenic mouse models (Abravanel et al., 2015; Felsher & Bishop, 1999; Gunther et al., 2002)
Human sample validation	DTCs and CTCs expressing dormancy markers can be isolated from patients	DTCs (Chery et al., 2014; Sosa et al., 2015); CTCs (Vishnoi et al., 2015)

variants, several studies have described the selection of latent clonal variants that may be informative (Lu et al., 2011; Malladi et al., 2016). A caveat is that the mechanisms identified may be biased by the clonal selection and not encompass important additional mechanisms of microenvironmental and epigenetic regulation of dormancy regulation. Additionally, dormant DTCs can be identified in murine patient-derived xenograft (PDX) experiments, for example, in head and neck squamous cell carcinoma (Aguirre-Ghiso et al., 2001, 2004; Bragado et al., 2013; Sosa et al., 2015) and breast cancer (Lawson, Bhakta, et al., 2015).

Transgenic models also provide an opportunity to study dormancy if the right microenvironment and stage of progression are considered. The group of Dr. Klein has shown that in the MMTV-HER2 and MMTV-PyMT breast cancer models, dissemination occurs in parallel to primary tumor development and that these early disseminating tumor cells enter a dormancy period from which they can reemerge (Husemann et al., 2008). This model allows studying how DTCs evolve ectopically and how the tissue microenvironment might shape genetic and epigenetic changes. Thus, using MMTV-HER2 and MMTV-PyMT in early stages allows for analysis of dormant DTCs. The MMTV-WNT1 breast cancer model that is usually considered to be nonmetastatic frequently develops metastasis after prolonged period of indolence that can only be reached when primary tumors are surgically removed (Li et al., 2000). Our lab has described that some target organ microenvironments such as the bone and liver microenvironment foster dormancy in DTCs (Bragado et al., 2013), a finding that we could reproduce in MMTV-HER2 mice (unpublished data). Transgenic models where the oncogene is under the control of an inducible promoter are another way to model dormancy. Several labs have used these models to mimic therapy-induced dormancy (Abravanel et al., 2015; Moody et al., 2002) and how loss of oncogene expression can induce a residual population that allowed studying both intrinsic and microenvironmental regulation of dormancy induced by oncogene inactivation (Felsher, 2008, 2010; Felsher & Bishop, 1999; Giuriato et al., 2006; Rakhra et al., 2010).

Finally, while Kaplan–Meier plots provide some information on how primary tumor signatures affect metastasis-free intervals, they do not inform on the actual gene signatures and mechanisms at play in the residual disease. Thus, it is essential to model dormancy by studying cancer patients beyond the analysis of Kaplan–Meier plots and by isolating CTCs and DTCs and performing single-cell analysis to investigate dormancy markers (Chery et al., 2014; Sosa et al., 2015; Vishnoi et al., 2015). This summary indicates that there are multiplicities of models to extract information related to dormancy that can be validated in human samples.

3. INTRINSIC AND EXTRINSIC SIGNALS CONVERGE TO INDUCE TUMOR CELL DORMANCY

Several studies showed how cues that regulate adult stem cell biology and the interplay between mitogenic and stress-signaling pathways regulate dormancy onset.

These mechanisms are summarized briefly in Fig. 1B and have been reviewed elsewhere recently (Aguirre-Ghiso, 2007; Sosa et al., 2014). However, less work is available on how dormancy and pluripotency pathways, which are commonly associated with aggressive cancer behavior, are integrated. Insight into this problem came from recent work from our lab showing that dormancy of HNSCC DTCs is dependent on NR2F1 signaling, an orphan receptor of the retinoic acid-signaling pathway (Sosa et al., 2015). This study revealed that NR2F1 upregulated both quiescence and pluripotency genes SOX9, OCT4, SOX2, and NANOG. Thus, dormant DTCs coordinate a profound growth arrest with the upregulation of self-renewal genes. This suggests that dormant cancer cells share characteristics of embryonic and adult stem cells. Interestingly, Scognamiglio et al. (2016) showed recently that upon withdrawal of *myc* signaling, murine embryonic stem cells enter a reversible dormant state where they are quiescent while retaining the expression of NANOG and OCT4. Upon reintroduction of *myc* signaling, the cells continue to form normal embryos. This *myc*-dependent dormancy is recapitulated in many mammals, where blastocysts enter a phase of dormancy, called diapause, induced by microenvironmental signals from the uterus. Interestingly, dormancy induced by p38 and NR2F1 signaling is associated with a downregulation of *myc* transcriptional activity and NR2F1-antagonized *myc* signaling for proliferation (Sosa et al., 2015). Cancer cell dormancy might therefore resemble an epigenetic state in which cells remain quiescent while retaining their self-renewing capacities. This tumor cell-intrinsic network can be induced by external factors. All-*trans* retinoic acid (atRA) is abundant in the bone marrow and regulates hematopoietic stem cell renewal (Ghiaur et al., 2013; Purton et al., 2006). Treatment of T-HEp3 cells with atRA induced NR2F1 and TGFβ2 expression in HNSCC DTCs (Sosa et al., 2015). TGFβ2 is a member of the TGFβ family also abundantly present in the bone and has been shown by our group to induce DTC dormancy through p38-dependent signaling, which leads to induction of the dormancy-associated proteins DEC2/Sharp1 and $p27^{Kip1}$ (Bragado et al., 2013). These findings indicate that in the bone the microenvironment-derived signals TGFβ2 and atRA might cooperate with tumor-intrinsic signals to allow DTCs to enter a dormant state characterized by growth arrest, survival, and pluripotency gene expression. This might lead DTCs to produce an autocrine loop to maintain dormancy but still retain reactivation plasticity. The clinical relevance of these findings was confirmed by a recent study from the group of Dr. Morrisey. They found that DTCs are present in the bone marrow of both prostate cancer patients with late recurrence or with NED, years after treatment of the primary tumor (Chery et al., 2014). While DTCs in the NED group mostly had a dormant and NR2F1/p38-high transcriptome, both dormant and proliferating DTCs were detected in recurring patients (Chery et al., 2014). Additionally, a transcriptome analysis showed considerable heterogeneity in both groups, demonstrating that there may be more than one form of dormancy even within the same patient and organ, and that dormant and slow cycling/proliferating DTCs can be present at the same time within the same organ.

4. COOPERATIVE EXTRINSIC SIGNALS INDUCE DORMANCY WITHIN THE BONE MICROENVIRONMENT

The bone marrow is home to hematopoietic stem cells that undergo tightly controlled steps of differentiation and proliferation during hematopoiesis (Eaves, 2015). DTCs are frequently found in the bone marrow of cancer patients, even in patients with no evidence of metastases (Braun et al., 2005; Pierga et al., 2003). Many of these patients never develop bone metastases and when they do, they usually have long periods of NED before, while many patients with bone DTCs never develop metastases (Sherry, Greco, Johnson, & Hainsworth, 1986). All this indicates that the bone marrow microenvironment therefore seems to be a growth-inhibitory microenvironment and therefore most DTCs enter dormancy. This might seem counterintuitive to the fact that the bone marrow is also the site of hematopoiesis that includes proliferative steps. However, hematopoietic cell expansion is immediately followed by differentiation (not proliferation) and dormancy of adult stem cells coexists with events of proliferation. Therefore, growth events in the bone microenvironment need to be tightly controlled. One way to resolve the paradox of the coexistence of proliferation and quiescence is the formation of niches within the bone microenvironment. Quiescent hematopoietic stem cells are localized in the hematopoietic stem cell niche (Shiozawa et al., 2011). Several studies indicate that DTCs enter the hematopoietic stem cell niche, where they remain dormant. For a detailed summary of these studies, please refer to a recent review by Ghajar (2015). In addition to these hematopoietic stem cell niche-specific effects, atRA and TGFβ2 are both abundant in the bone marrow. Additionally, Kobayashi et al. found that BMP-7, another TGFβ family member secreted by bone stromal cells, was able to induce a reversible dormancy in intra-tibially injected prostate cancer cells through induction of p38 signaling and upregulation of the metastasis suppressor gene NDRG1

(Kobayashi et al., 2011). The selective bone metastatic suppressive function of recombinant BMP-7 in a metastatic prostate cancer xenograft system was previously described by Buijs et al. (2007), who reported a synergistic mechanism in which TGFβ1, amply present in the bone microenvironment, contributes to this BMP-7 function. Thus, in the bone microenvironment, atRA, TGFβ2, and BMP-7 might provide redundant dormancy-inducing cues that might explain why dormancy can be such a stable state, lasting for years or decades (Fig. 1B). Which stromal cells are secreting atRA, TGFβ2, and BMP-7 and whether their expression is uniformly high or enriched in niches within target organs need to be elucidated.

5. ENDOSTEAL AND PERIVASCULAR NICHES SUPPORT CANCER CELL DORMANCY

In addition to cooperative microenvironment-derived dormancy-inducing cues, the tendency of DTCs to enter dormancy depends on their specific localization within the organ. A recent review by Ghajar summarizes most of the available data on the dormant niche (Ghajar, 2015). Below, we will therefore only briefly cover the most recent publications and raise questions associated with this topic. Lawson et al. found that dormant myeloma bone marrow DTCs were engaging with osteoblasts in the endosteum, whereas proliferating DTCs were not (Lawson, McDonald, et al., 2015). Remarkably, proliferating cells introduced to an experimental endosteal niche entered dormancy. Dormant DTCs can be released from the endosteum and activated through enhanced osteoclast activity induced by sRANKL (Lawson, McDonald, et al., 2015). This shows that dormancy is a reversible state crucially regulated by the microenvironment. It further indicates that while osteoblasts might be involved in inducing dormancy, osteoclast activity might be involved in the escape from dormancy by triggering a vicious cycle that characterizes osteolytic bone metastases (Fig. 2A). The group of Dr. Taichman accordingly showed the dormancy-inducing effect of osteoblasts, showing that osteoblast-derived growth arrest-specific 6 (GAS6) protein signaling through its receptor Axl induces dormancy in prostate cancer DTCs (Shiozawa et al., 2010; Taichman et al., 2013). Interestingly, under hypoxic conditions a negative feedback loop of GAS6/Axl is inhibited, leading to increased Axl expression and possibly maintenance of dormancy in hypoxic microenvironments (Mishra et al., 2012). Additionally, another study had shown that osteoclasts might mediate escape from dormancy (Lu et al., 2011). They found that osteoclast progenitors are

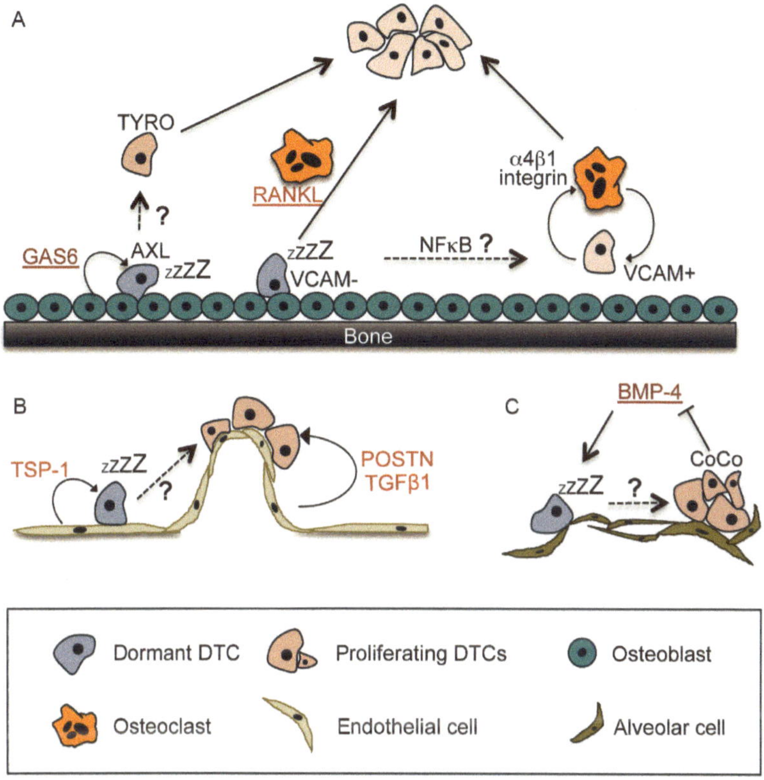

Fig. 2 Extrinsic signals inducing dormancy in disseminated tumor cells (DTCs) in different microenvironments. (A) Dormancy induction in the bone microenvironment is in part mediated by osteoblasts through GAS6/AXL signaling (Shiozawa et al., 2010; Taichman et al., 2013). DTCs can escape osteoblast-induced dormancy through activation of TYRO or through activation of osteoclasts. Osteoclasts can be activated through RANKL (Lawson, McDonald, et al., 2015) or through recruitment of osteoclast progenitors via VCAM1/integrin α4β1 signaling (Lu et al., 2011). How these dormancy escape mechanisms occur spontaneously in patients and whether they resemble alternative dormancy pathways or cooperate remain to be established. (B) DTCs in the perivascular niche frequently enter dormancy due to thrombospondin (TSP-1) signaling (Ghajar et al., 2013). Activated endothelium on the other hand releases periostin (POSTN) and TGFβ1 that induce DTC proliferation and possibly escape from dormancy. (C) BMP-4, a TGFβ family member enriched in the lung microenvironment, induces dormancy (Gao et al., 2012). Overexpression of CoCo, a BMP inhibitor, allows DTCs to escape dormancy.

recruited to VCAM1+ DTCs through integrin α4β1 signaling. This triggered osteoclast formation in vicinity of DTCs and entry into the vicious cycle (Lu et al., 2011). In contrast, in absence of VCAM1 expression, DTCs failed to recruit osteoclast progenitors and entered prolonged periods of dormancy before the formation of osteolytic macrometastases. While this indicates that osteoclasts drive metastasis formation, the mechanisms of dormancy and whether DTCs entered cellular or population-based dormancy were not addressed (Lu et al., 2011). Additionally, this study compared cell lines with different bone metastasis properties, suggesting that whether DTCs enter dormancy depends on their expression of VCAM1 upon arrival in the bone (Fig. 2B). However, clinical evidence implies that DTCs in the bone commonly go through a phase of dormancy from which they can reemerge spontaneously (Braun et al., 2005; Chery et al., 2014; Pierga et al., 2003). Thus, whether the spontaneous exit from dormancy might be due to spontaneous upregulation of VCAM1 or due to different mechanisms leading to the recruitment of osteoclasts remains to be elucidated.

In addition to the endosteal niche, Ghajar et al. found a link between the proliferative state of endothelial cells and the dormancy of DTCs associated with the vessels (Fig. 2B). They identified that resting endothelium produces thrombospondin-1 (TSP-1) that induces quiescence in DTCs (Ghajar et al., 2013). In contrast, DTCs in contact with neovascular tips showed enhanced proliferation and this was mediated by secretion of TGFβ1 and periostin (POSTN) by endothelial tip cells. Since all DTCs by default have to pass through the perivascular niche upon dissemination, the question arises whether dormancy might be the default program of extravasating DTCs and only DTCs entering through activated endothelial tip cell niches are the ones who immediately start to proliferate.

Overall, these studies indicate that DTC dormancy within one target organ microenvironment might be regulated by the specific localization of DTCs in the endosteal or the perivascular niche. It will be interesting to understand whether DTCs might move incidentally into a niche where they enter dormancy or whether they are actively recruited. Furthermore, it remains to be elucidated whether the dormant niche is identical with the adult stem cell niche.

6. REACTIVATION FROM DORMANCY

The studies discussed above show that the induction and maintenance of dormancy could be influenced by microenvironment-derived factors such as atRA, TGFβ2, BMP-7, and TSP-1 and by interaction with resting endothelial cells and osteoblasts. Yet, it is a hallmark of dormancy that it is a reversible growth arrest and the exit from dormancy is the clinically most relevant phase. Several studies describe how DTCs escape dormancy, but not how these dormancy escape mechanisms are triggered spontaneously.

However, dormancy can last for years and decades, indicating that dormancy-inducing cues can be very stable. This implies that relapses include the switch of DTCs from a dormant into a proliferative state, most likely due to changes in their microenvironment that disrupted the stable dormant state. However, most of the dormancy escape mechanisms described to this date are dormancy exit signals, but do not address how these exit signals are activated spontaneously.

For example, as mentioned above, engagement with osteoclasts in the bone environment can allow exit of DTCs from dormancy (Lawson, McDonald, et al., 2015; Lu et al., 2011). What leads to the spontaneous recruitment of osteoclasts toward DTCs? Lu et al. showed that NFkB signaling leads to activation of VCAM1 expression (Lu et al., 2011). Which factors derived from the microenvironment might be able to activate NFkB and subsequent VCAM1 upregulation in dormant DTCs? Alternatively, can osteoclasts be activated systemically? Could comorbidities such as bone injury or osteoporosis be triggering osteoclast activation? Can dormancy escape be a matter of chance, induced by coincidental colocalization of osteoclasts and dormant DTCs? Does aging affect the expression of dormancy inducers allowing for a slow reactivation process?

Ghajar et al. showed that activation of the endothelium can lead to reactivation of dormant DTCs residing in the perivascular niche (Ghajar et al., 2013). Thus, in the case of perivascular dormancy, is the reemergence of DTCs a matter of chance when the endothelium in their vicinity becomes activated, as shown by Ghajar et al. in vitro (Ghajar et al., 2013)? Alternatively, could DTCs be activated by interaction with stromal cells such as macrophages and therefore move out of the perivascular niche that maintained their dormancy?

Investigating the mechanisms of breast cancer dormancy in the lung, the group of Dr. Giancotti found that the secreted BMP-4 inhibitor protein CoCo reactivates dormant breast cancer cells in the lung through blockade of BMP signaling (Gao et al., 2012). CoCo is secreted by cancer cells and accumulates specifically in the pericellular matrix, where it can neutralize BMP-4 secreted by the stroma, and thus enable DTCs to exit dormancy. The blockade of BMP/p-Smad-driven inhibition of self-renewal led to proliferation of single breast cancer DTCs in the lungs of mice. In contrast, breast cancer cells with low expression of CoCo remained dormant (Fig. 2C). This effect was specific to BMP-rich target organs, as CoCo expression predicted for lung, a BMP-rich organ, but not bone or brain metastasis in a large patient cohort (Gao et al., 2012). However, Gao et al. used two clonal variants that either entered dormancy due to low expression of CoCo or were able to escape due to high CoCo expression (4T1/4T07 breast cancer cell lines). While these identified CoCo as a mechanism to evade dormancy, it remains to be elucidated whether dormant DTCs can upregulate CoCo spontaneously.

A recent preclinical study by Sansone et al. (2016) found that chronic hormone therapy of mice with ER-positive luminal breast cancer led to induction of a dormant, $CD133^{high}/ER^{low}/mitochondria^{low}$ population of cells, which produced IL-6 in an autocrine manner. These cells showed an increase in pluripotency genes and were able to exit dormancy by utilizing IL-6/Stat3/Notch3-driven, ER-independent self-renewal and mitochondria reactivation. The use of an anti-IL6R antibody restored ER expression in HT-resistant cells and double treatment with HT/anti-IL6R antibodies in vivo was effective even in HT-resistant tumors (Sansone et al., 2016). These results demonstrate that dormancy may be induced by therapy and that microenvironment-derived cytokines such as IL-6 may be able to reawaken dormant cells. Yet, the cellular source of IL-6 and how its secretion is regulated were not addressed.

In summary, while the phase of dormancy exit is clinically relevant, our understanding how spontaneous exit from dormancy occurs is still very limited. Part of the reason for this is that this process is hard to study. Many commonly used metastasis models use highly aggressive clonal cell line variants that do not undergo dormancy. Only a few cell lines enter dormancy in vivo and while these models do allow identification and study of dormant cells, following the spontaneous escape of dormant cells from dormancy in situ remains a technical challenge. Possibly, recent advances in live imaging might allow to follow the fate of dormant tumor cells and thereby allow us to gain insight into the microenvironmental changes that facilitate dormancy escape. This could be achieved by the use of fluorescently tagged spontaneous mouse models where the fate of single DTCs and their switch from single cells into proliferating metastases in target organs could be followed by live imaging. Such studies could be combined with systemic effects on DTC behavior such as therapy or activation of bone marrow progenitor cells.

7. THE IMMUNE SYSTEM AND DORMANCY

The ability of the immune system to affect tumor progression has been established over the last decades. Interestingly, the immune system can have both tumor-promoting and tumor-inhibitory functions. Inflammatory reactions and association of tumor cells with myeloid cells have been implicated with tumor-promoting functions (for reviews, see Baxter & Hodgkin, 2002; Grivennikov, Greten, & Karin, 2010; Mantovani, Allavena, Sica, & Balkwill, 2008). On the other hand, it has been shown that the adaptive immune system is able to recognize tumor-specific antigens and inhibit tumor growth (Baxter & Hodgkin, 2002; Gajewski, Schreiber, & Fu, 2013).

Metastasis-promoting effects of myeloid cells have recently been reviewed elsewhere (Quail & Joyce, 2013). Yet, the majority of these studies focus on mechanisms how myeloid cells facilitate macrometastatic growth but do not address the interaction of solitary DTCs with myeloid cells and whether this might provide a dormancy/growth switch. Work by Pollard and colleagues revealed that inflammatory Ly6C+ monocytes are required for breast cancer cell extravasation in the lung, a process mediated by chemokine C-C-motif ligand 2 (CCL2) and vascular endothelial growth factor (VEGF) (Qian et al., 2009, 2011; Qian & Pollard, 2010) (Fig. 3A). In addition, a study from the lab of Dr. Massague showed that macrophages are involved in the survival of breast cancer cells in the lung after their extravasation (Chen, Zhang, & Massague, 2011). This is mediated by binding of VCAM1 expressed on cancer cells to β-1-integrin-positive macrophages which leads to induction of Akt signaling in DTCs and allows them to escape TRAIL-induced apoptosis (Chen et al., 2011). These studies imply that monocytes and monocyte-derived macrophages are crucial for extravasation and survival of DTCs.

Work from the group of Dr. Pollard also showed that in the lung tissue, monocytes differentiate into metastasis-associated macrophages (MAMs) where they support metastasis growth (Kitamura et al., 2015; Qian et al., 2015) (Fig. 3A). Blocking of MAM reduced metastasis burden even when metastases had established already. These studies were performed in highly aggressive mouse models where metastases form rapidly after extravasation. In these models, monocytes are required for extravasation, thereafter differentiate into MAMs, and as such mediate metastatic growth. However, these models do not show a dormancy phase, precluding the formation of metastases. They therefore do not contribute to our understanding how the fate of solitary DTCs that enter a dormant state prior to their entry into proliferation is affected by MAMs or other immune cells. Clinical evidence shows that in the majority of patients, there is a significant lag between extravasation (that likely already occurred at the time of surgery) and proliferation into detectable metastasis (Hartkopf et al., 2014; Lilleby,

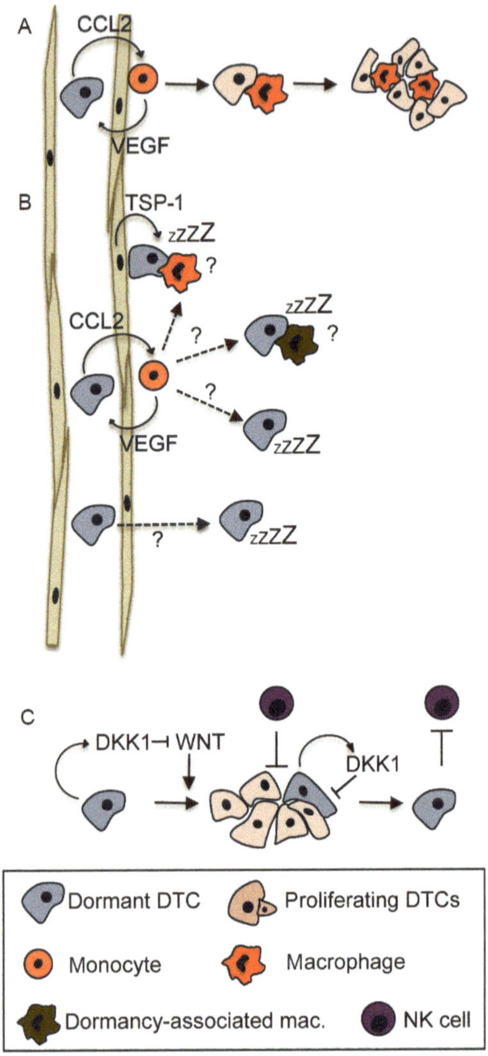

Fig. 3 Effects of macrophages and NK cells on dormancy. (A) Monocytes are recruited to DTCs by CCL2 and assist their extravasation in the lung through VEGF secretion (Qian et al., 2009, 2011). In the lung tissue, these monocytes differentiate into metastasis-associated macrophages that promote DTC proliferation (Kitamura et al., 2015; Qian et al., 2015). (B) The role of monocytes and macrophages in dormancy has not been established. If DTCs require monocytes to extravasate, why do some DTCs enter dormancy instead of entering a cycle of macrophage-assisted proliferation? Can niche-derived factors such as TSP-1 in the endothelial niche override the growth-promoting effect of macrophages? Are there macrophage subtypes that induce dormancy? Do some DTCs fail to retain growth-promoting macrophages? Can DTCs extravasate independent of monocytes and therefore enter dormancy? (C) Activation of WNT signaling allows DTCs to enter a proliferative state in which they are more susceptible to cytotoxic signals from NK cells. Once cells enter a dormant state through activation of DKK1 and subsequent WNT inhibition, they are able to escape NK cytotoxic signals (Malladi et al., 2016).

Stensvold, Mills, & Nesland, 2013). Therefore, it seems that extravasation and activation of proliferation can be significantly uncoupled processes. This raises several questions (depicted in Fig. 3B): Can other microenvironment-derived signals, such as TSP-1 in the perivascular niche, induce dormancy even when DTCs are associated with macrophages? Can monocytes differentiate into a growth-suppressive macrophage phenotype? Do some DTCs fail to retain macrophages and therefore enter dormancy? Could it be a matter of stoichiometry where monocytes supporting extravasation interact with several DTCs so that there are not enough macrophages to support subsequent growth? Is there a macrophage-independent extravasation step in some cases and some DTCs may thus not interact with macrophages at all? Can DTCs be irresponsive to MAM-derived growth signals? Live imaging analysis might provide the answers to the exact stoichiometry of the macrophage–DTC interaction during extravasation and subsequent growth promotion. However, all will depend on what models are used to test this hypothesis and it is likely that the use of clonal variants selected for aggressive growth will reveal the same mechanisms.

While macrophages are mostly associated with tumor-promoting features, a recent study by the lab of Dr. Massague showed that NK cells are involved in dormancy of DTCs after reactivation (Malladi et al., 2016). Using a model of latency competent cancer cells where dormant clones were selected from an experimental metastasis assay with a HER2+ breast cancer and a lung adenocarcinoma cell line, the authors could confirm previous findings that dormant cancer cells activate the p38 and self-renewal pathways through Sox2 and Sox9 (Sosa et al., 2015). Malladi et al. found that Sox2 also induced a growth arrest by inhibition of Wnt signaling and downstream proliferative pathways through activation of DKK1 (Fig. 3C). Once DKK1 induced a dormant state, DTCs were able to avoid NK cell-mediated cell death, whereas DKK1 low-proliferating DTCs were susceptible to NK cell cytotoxicity. This suggests that immune evasion is a result of dormancy induced by different microenvironmental signals rather than dormancy being a result of NK cells on DTCs.

Adaptive immune cells have also been shown to negatively affect tumor growth. This has recently been confirmed in clinical practice when a new generation of immune therapies targeting immune checkpoint molecules such as CTLA4, PD-1, and PDL-1 has led to great success (Pardoll, 2012). Yet again, our understanding of the role of adaptive immunity in dormancy is surprisingly limited. We have known for a long time that immune suppression after organ transplants is linked to an enhanced risk to develop cancer (Ross, 2007), indicating that the immune system can suppress tumor growth. Cases where melanoma was transferred through a kidney transplant from organ donors with NED for up to 16 years (MacKie, Reid, & Junor, 2003) indicate that dormancy of kidney DTCs can be mediated through adaptive immunity and that these DTCs exit dormancy upon immune suppression after organ transplantation. However, whether this phenomenon is functionally linked to a loss of CD8 cell cytotoxic activity or NK cell activity is still unclear.

Only few studies investigated the role of CD4 and CD8 T-cells in maintaining dormancy. It has been reported that dormant tumor cells seem to be less susceptible to adaptive immune cell responses and show reduced tumor antigen expression (Matsuzawa, Takeda, Narita, & Ozawa, 1991; Weinhold, Miller, & Wheelock, 1979). Additionally, it has been shown that dormant leukemia cells express PDL1, which allows them to inhibit T-cells (Saudemont & Quesnel, 2004). These studies indicate tumor cells escape immune-mediated cell killing and are therefore able to survive in a dormant state (Fig. 4A). Other studies indicate that T-cell-derived factors might induce dormancy in DTCs (Fig. 4B). For example, CD8 T-cell-derived interferon γ (IFNγ) has been shown to induce dormancy in a murine lymphoma model (Farrar et al., 1999). In the Rip-Tag2 mouse model for pancreatic cancer, CD4 T-cells induced angiogenic population-based dormancy through IFNγ and tumor necrosis factor α (TNFα) signaling (Muller-Hermelink et al., 2008). It remains to be elucidated whether evasion of T-cell-induced cell death and T-cell-induced dormancy are exclusive processes or whether they might be different mechanisms, depending on

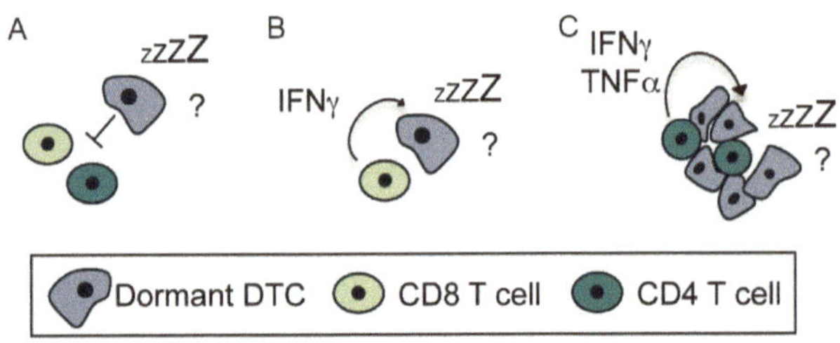

Fig. 4 Effects of T-cells on dormancy. (A) Some studies suggest that dormant DTCs survive by escaping adaptive immune responses (Matsuzawa et al., 1991; Saudemont & Quesnel, 2004; Weinhold et al., 1979). (B) Other studies indicate that T-cell-derived IFNγ and TNFR signaling induce dormancy (*bottom*) (Farrar et al., 1999; Muller-Hermelink et al., 2008). (C) Whether immune-mediated dormancy resembles cellular dormancy or population-based dormancy, where tumor cell killing and proliferation are in equilibrium, is not clear either.

the tumor type and the specific immune response mechanisms. Furthermore, it remains to be elucidated whether immune-mediated dormancy always resembles population-based dormancy or whether T-cell-derived factors such as IFNγ might also be able to induce cellular dormancy in DTCs (Fig. 4C).

According to the immune editing hypothesis, tumors go through a phase of elimination, followed by a phase of equilibrium and a phase of escape where the evolutionary pressure of the immune system has resulted in immune escape mechanisms, through intrinsic changes such as reduced tumor antigen expression, through extrinsic changes by creating an immune-suppressive microenvironment, or through a combination of both (Schreiber, Old, & Smyth, 2011). The immune editing hypothesis refers to phases during primary tumor development. However, we now have strong evidence that primary tumors and metastases develop in parallel (Klein, 2009). This is based on large cohort patient studies (Banys, Gruber, et al., 2012; Braun et al., 2005; Sanger et al., 2011; Schardt et al., 2005; Turajlic & Swanton, 2016) and studies with spontaneous murine tumor models for breast cancer (Husemann et al., 2008) and pancreatic cancer (Rhim et al., 2012), showing that tumor cells can disseminate during premalignant and preinvasive asymptomatic disease stages. Thus, while primary tumors undergo phases of immune editing, tumor cells already disseminate and enter tumor-naïve target organs. This raises several interesting questions posed below.

The immune editing hypothesis further postulates an equilibrium phase characterized by a balance between tumor cell killing through immune cells and tumor cell proliferation. However, there is only little tumor cell apoptosis and tumor cell proliferation during this phase (Koebel et al., 2007). This implies that limited cancer growth during the equilibrium phase might rather be due to growth-inhibitory effects of the immune system with some contribution of tumor cell killing. This is supported by a recent study showing that in a murine pancreatic cancer model, inhibition of focal adhesion kinase led to disease stabilization and reduced proliferation and augmented immunity (Jiang et al., 2016). Further, if DTCs disseminate during the equilibrium phase, could they be more susceptible to enter immune-induced dormancy in target organs? Another interesting question is how DTCs manage the immune escape phase in target organs. Are those DTCs that have retained high tumor antigen expression and are derived from primary tumors with immune-suppressive microenvironments more prone to be eradicated by the immune system when they enter a tumor-naïve organ microenvironment or do they enter immune-mediated dormancy? Interestingly, a recent study by Payne et al. found that therapy-induced dormant mammary carcinoma cells were unable to use immune editing to escape immunotherapy in a mouse model (Payne et al., 2016). Even slowly cycling, Ki-67^{low} cells were able to escape immunotherapy, indicating that cellular but not immune or vascular dormancy is the driving mechanism behind this biology. This observation might lead to combination therapies where immunotherapy targets dormant cells, while established antiproliferative therapies target the proliferating or slow cycling populations.

Another interesting unanswered question is whether DTCs are able to escape the immune system in target organs by entering immune-protected sites. As mentioned above and reviewed elsewhere (Ghajar, 2015), DTCs can become dormant by entering stem cell niches. Adult stem cell niches have been shown to be immune-protected sites. The hematopoietic stem cell niche, for example, is rich in CD4+CD25+ regulatory T-cells (Zou et al., 2004) that mediate immune protection of hematopoietic stem cells (Fujisaki et al., 2011). It will be interesting to address which DTC niches provide immune-privileged sites that protect DTCs from eradication throughout the immune system and how these DTCs can exit their dormant niche and form metastases without being eradicated by the immune system. Would systemic immune suppression allow DTCs to exit immune-protected sites and form metastases—a phenomenon observed in immune-suppressed organ transplant recipients?

Novel immune therapies against CTLA-4 and PD-1 achieved great immediate success in certain patient populations (Pardoll, 2012). However, it is unknown whether the success of these immune therapies is based on eradication of cancer cells or whether they also contribute to a switch into immune-mediated dormancy. Data about long-term disease progression are not available yet since these new therapies have only been applied in the clinic for less than 5 years. It therefore remains to be elucidated whether novel immune therapies might in fact contribute to dormant minimal residual disease and whether late relapses might occur.

In addition to a limited understanding of long-term effects of immune therapies on minimal residual disease and late relapses due to dormant DTCs, we also do not fully understand whether there might be an association of dormancy and nonresponders to immune therapy. A recent study has shown that the total load of nonsynonymous mutations seems to correlate with a patient's response to CTLA4 therapy (Van Allen et al., 2015). Thus, the more mutation a tumor has acquired, the better the response to immune therapy, possibly because the developed escape mechanisms rely on inhibition of costimulatory factors such as CTLA4 and PD-1. However, not all DTCs are necessarily derived from a highly proliferative invasive tumor with high mutator phenotypes. As mentioned above, it is now accepted that dissemination occurs in parallel to primary tumor development and results in increased heterogeneity and discordance of mutations in primary tumors and metastases (Klein, 2009; Turajlic & Swanton, 2016). DTCs that disseminated early will have a lower mutational burden than the primary tumor at the time of diagnosis. Could those early DTCs therefore be resistant to immune therapy?

In summary, it is important to highlight that our understanding of immune-mediated dormancy currently limits our understanding of long-term effects of immune therapies. More insight into this subject might hold the key to more targeted immune therapies and possibly combined immune and dormancy therapies.

8. SUMMARY AND OUTLOOK

Two alternative strategies to use the therapeutic window of dormancy have been discussed. Since dormancy provides a mechanism by which DTCs evade current therapies, one could devise ways to reawaken dormant DTCs prior to therapy. Forcing DTCs into proliferation and using currently available antiproliferative therapies might kill the majority of the cells. Yet, from the treatment of primary cancers we know that this might also induce the selection of resistant clones (Pisco et al., 2013), leading to a potentially worsened outcome. The other alternative would be to induce a perpetual dormancy. By evaluating which signals keep DTCs dormant, one could artificially keep the cells in a dormant state, thus preventing overt metastasis. Both possibilities have been reviewed in depth elsewhere (Aguirre-Ghiso et al., 2013; Ghajar, 2015; Sosa et al., 2011, 2014). For the latter strategy, we need to find out whether dormancy therapies need to be administered continuously or can be given in pulses which would be more tolerable for NED patients. Can we identify markers that indicate a dormant vs a reawakening phenotype of DTCs (Aguirre-Ghiso et al., 2013)? One possibility might be to measure levels of dormancy-inducing factors in the blood. A few studies indicate that this might be possible. High TGFβ2 levels in the blood seem to correlate with therapy success (Kopp, Jonat, Schmahl, & Knabbe, 1995; Lucia, Sporn, Roberts, Stewart, & Danielpour, 1998). In addition, analysis of the dormancy status of DTCs isolated from bone marrow of patients with NED might allow predicting relapses. In fact, recent work (Chery et al., 2014; Sosa et al., 2015) indicated for the first time that analysis of bone marrow DTCs can be predictive of late recurrences and that markers derived from understanding the basic mechanisms of dormancy are expressed in DTCs from NED patients arguing they could serve as dormancy markers. These studies indicate that understanding the mechanism of dormancy might allow to follow patients with NED more closely in order to predict their metastatic relapse before it occurs. Development of dormancy therapies might allow treating those patients with reduced levels of dormancy markers. This could change the life expectancy of patients with disseminated cancer drastically since current therapies administered to patients with overt metastases often fail. Moreover, once we better understand the mechanisms of dormancy and immune responses, combined dormancy and immune therapy might be an exciting new avenue in cancer therapy.

REFERENCES

Abravanel, D. L., Belka, G. K., Pan, T. C., Pant, D. K., Collins, M. A., Sterner, C. J., et al. (2015). Notch promotes recurrence of dormant tumor cells following HER2/neu-targeted therapy. *The Journal of Clinical Investigation*, *125*(6), 2484–2496.

Adam, A. P., George, A., Schewe, D., Bragado, P., Iglesias, B. V., Ranganathan, A. C., et al. (2009). Computational identification of a p38SAPK-regulated transcription factor network required for tumor cell quiescence. *Cancer Research*, *69*(14), 5664–5672.

Aguirre Ghiso, J. A., Kovalski, K., & Ossowski, L. (1999). Tumor dormancy induced by downregulation of urokinase receptor in human carcinoma involves integrin and MAPK signaling. *The Journal of Cell Biology*, *147*(1), 89–104.

Aguirre-Ghiso, J. A. (2007). Models, mechanisms and clinical evidence for cancer dormancy. *Nature Reviews Cancer*, 7(11), 834–846.

Aguirre-Ghiso, J. A., Bragado, P., & Sosa, M. S. (2013). Metastasis awakening: Targeting dormant cancer. *Nature Medicine*, *19*(3), 276–277.

Aguirre-Ghiso, J. A., Estrada, Y., Liu, D., & Ossowski, L. (2003). ERK(MAPK) activity as a determinant of tumor growth and dormancy; regulation by p38(SAPK). *Cancer Research*, *63*(7), 1684–1695.

Aguirre-Ghiso, J. A., Liu, D., Mignatti, A., Kovalski, K., & Ossowski, L. (2001). Urokinase receptor and fibronectin regulate the ERK(MAPK) to p38(MAPK) activity ratios that determine carcinoma cell proliferation or dormancy in vivo. *Molecular Biology of the Cell*, *12*(4), 863–879.

Aguirre-Ghiso, J. A., Ossowski, L., & Rosenbaum, S. K. (2004). Green fluorescent protein tagging of extracellular signal-regulated kinase and p38 pathways reveals novel dynamics of pathway activation during primary and metastatic growth. *Cancer Research*, *64*(20), 7336–7345.

Aslakson, C. J., & Miller, F. R. (1992). Selective events in the metastatic process defined by analysis of the sequential dissemination of subpopulations of a mouse mammary tumor. *Cancer Research*, *52*(6), 1399–1405.

Banys, M., Gruber, I., Krawczyk, N., Becker, S., Kurth, R., Wallwiener, D., et al. (2012). Hematogenous and lymphatic tumor cell dissemination may be detected in patients diagnosed with ductal carcinoma in situ of the breast. *Breast Cancer Research and Treatment*, *131*(3), 801–808.

Banys, M., Hartkopf, A. D., Krawczyk, N., Kaiser, T., Meier-Stiegen, F., Fehm, T., et al. (2012). Dormancy in breast cancer. *Breast Cancer (Dove Medical Press)*, *4*, 183–191.

Barkan, D., & Green, J. E. (2011). An in vitro system to study tumor dormancy and the switch to metastatic growth. *Journal of Visualized Experiments*, *54*.

Baxter, A. G., & Hodgkin, P. D. (2002). Activation rules: The two-signal theories of immune activation. *Nature Reviews Immunology*, *2*(6), 439–446.

Box, G. M., & Eccles, S. A. (2011). Simple experimental and spontaneous metastasis assays in mice. *Methods in Molecular Biology*, *769*, 311–329.

Bragado, P., Estrada, Y., Parikh, F., Krause, S., Capobianco, C., Farina, H. G., et al. (2013). TGF-beta2 dictates disseminated tumour cell fate in target organs through TGF-beta-RIII and p38alpha/beta signalling. *Nature Cell Biology*, *15*(11), 1351–1361.

Braun, S., Pantel, K., Muller, P., Janni, W., Hepp, F., Kentenich, C. R., et al. (2000). Cytokeratin-positive cells in the bone marrow and survival of patients with stage I, II, or III breast cancer. *The New England Journal of Medicine*, *342*(8), 525–533.

Braun, S., Vogl, F. D., Naume, B., Janni, W., Osborne, M. P., Coombes, R. C., et al. (2005). A pooled analysis of bone marrow micrometastasis in breast cancer. *The New England Journal of Medicine*, *353*(8), 793–802.

Buijs, J. T., Rentsch, C. A., van der Horst, G., van Overveld, P. G., Wetterwald, A., Schwaninger, R., et al. (2007). BMP7, a putative regulator of epithelial homeostasis in the human prostate, is a potent inhibitor of prostate cancer bone metastasis in vivo. *The American Journal of Pathology*, *171*(3), 1047–1057.

Chen, Q., Zhang, X. H., & Massague, J. (2011). Macrophage binding to receptor VCAM-1 transmits survival signals in breast cancer cells that invade the lungs. *Cancer Cell*, *20*(4), 538–549.

Chery, L., Lam, H. M., Coleman, I., Lakely, B., Coleman, R., Larson, S., et al. (2014). Characterization of single disseminated prostate cancer cells reveals tumor cell heterogeneity and identifies dormancy associated pathways. *Oncotarget*, *5*(20), 9939–9951.

Eaves, C. J. (2015). Hematopoietic stem cells: Concepts, definitions, and the new reality. *Blood*, *125*(17), 2605–2613.

Farrar, J. D., Katz, K. H., Windsor, J., Thrush, G., Scheuermann, R. H., Uhr, J. W., et al. (1999). Cancer dormancy. VII. A regulatory role for CD8+ T cells and IFN-gamma in establishing and maintaining the tumor-dormant state. *Journal of Immunology*, *162*(5), 2842–2849.

Felsher, D. W. (2008). Oncogene addiction versus oncogene amnesia: Perhaps more than just a bad habit? *Cancer Research*, *68*(9), 3081–3086. discussion 3086.

Felsher, D. W. (2010). MYC inactivation elicits oncogene addiction through both tumor cell-intrinsic and host-dependent mechanisms. *Genes & Cancer*, *1*(6), 597–604.

Felsher, D. W., & Bishop, J. M. (1999). Reversible tumorigenesis by MYC in hematopoietic lineages. *Molecular Cell*, *4*(2), 199–207.

Fujisaki, J., Wu, J., Carlson, A. L., Silberstein, L., Putheti, P., Larocca, R., et al. (2011). In vivo imaging of Treg cells providing immune privilege to the haematopoietic stem-cell niche. *Nature*, *474*(7350), 216–219.

Gajewski, T. F., Schreiber, H., & Fu, Y. X. (2013). Innate and adaptive immune cells in the tumor microenvironment. *Nature Immunology*, *14*(10), 1014–1022.

Gao, H., Chakraborty, G., Lee-Lim, A. P., Mo, Q., Decker, M., Vonica, A., et al. (2012). The BMP inhibitor Coco reactivates breast cancer cells at lung metastatic sites. *Cell*, *150*(4), 764–779.

Ghajar, C. M. (2015). Metastasis prevention by targeting the dormant niche. *Nature Reviews. Cancer*, *15*(4), 238–247.

Ghajar, C. M., Peinado, H., Mori, H., Matei, I. R., Evason, K. J., Brazier, H., et al. (2013). The perivascular niche regulates breast tumour dormancy. *Nature Cell Biology*, *15*(7), 807–817.

Ghiaur, G., Yegnasubramanian, S., Perkins, B., Gucwa, J. L., Gerber, J. M., & Jones, R. J. (2013). Regulation of human hematopoietic stem cell self-renewal by the microenvironment's control of retinoic acid signaling. *Proceedings of the National Academy of Sciences of the United States of America*, *110*(40), 16121–16126.

Giuriato, S., Ryeom, S., Fan, A. C., Bachireddy, P., Lynch, R. C., Rioth, M. J., et al. (2006). Sustained regression of tumors upon MYC inactivation requires p53 or thrombospondin-1 to reverse the angiogenic switch. *Proceedings of the National Academy of Sciences of the United States of America*, *103*(44), 16266–16271.

Grivennikov, S. I., Greten, F. R., & Karin, M. (2010). Immunity, inflammation, and cancer. *Cell*, *140*(6), 883–899.

Gunther, E. J., Belka, G. K., Wertheim, G. B., Wang, J., Hartman, J. L., Boxer, R. B., et al. (2002). A novel doxycycline-inducible system for the transgenic analysis of mammary gland biology. *FASEB J*, *16*(3), 283–292.

Hanahan, D., & Coussens, L. M. (2012). Accessories to the crime: Functions of cells recruited to the tumor microenvironment. *Cancer Cell*, *21*(3), 309–322.

Hartkopf, A. D., Taran, F. A., Wallwiener, M., Hahn, M., Becker, S., Solomayer, E. F., et al. (2014). Prognostic relevance of disseminated tumour cells from the bone marrow of early stage breast cancer patients—Results from a large single-centre analysis. *European Journal of Cancer*, *50*(15), 2550–2559.

Hartkopf, A. D., Wallwiener, M., Fehm, T. N., Hahn, M., Walter, C. B., Gruber, I., et al. (2015). Disseminated tumor cells from the bone marrow of patients with nonmetastatic primary breast cancer are predictive of locoregional relapse. *Annals of Oncology*, *26*(6), 1155–1160.

Husemann, Y., Geigl, J. B., Schubert, F., Musiani, P., Meyer, M., Burghart, E., et al. (2008). Systemic spread is an early step in breast cancer. *Cancer Cell*, *13*(1), 58–68.

Jiang, H., Hegde, S., Knolhoff, B. L., Zhu, Y., Herndon, J. M., Meyer, M. A., et al. (2016). Targeting focal adhesion kinase renders pancreatic cancers responsive to checkpoint immunotherapy. *Nature Medicine*, *22*(8), 851–860.

Kitamura, T., Qian, B. Z., Soong, D., Cassetta, L., Noy, R., Sugano, G., et al. (2015). CCL2-induced chemokine cascade promotes breast cancer metastasis by enhancing retention of metastasis-associated macrophages. *The Journal of Experimental Medicine*, *212*(7), 1043–1059.

Klein, C. A. (2009). Parallel progression of primary tumours and metastases. *Nature Reviews Cancer*, *9*(4), 302–312.

Kobayashi, A., Okuda, H., Xing, F., Pandey, P. R., Watabe, M., Hirota, S., et al. (2011). Bone morphogenetic protein 7 in dormancy and metastasis of prostate cancer stem-like cells in bone. *The Journal of Experimental Medicine*, *208*(13), 2641–2655.

Koebel, C. M., Vermi, W., Swann, J. B., Zerafa, N., Rodig, S. J., Old, L. J., et al. (2007). Adaptive immunity maintains occult cancer in an equilibrium state. *Nature*, *450*(7171), 903–907.

Kopp, A., Jonat, W., Schmahl, M., & Knabbe, C. (1995). Transforming growth factor beta 2 (TGF-beta 2) levels in plasma of patients with metastatic breast cancer treated with tamoxifen. *Cancer Research*, *55*(20), 4512–4515.

Lawson, D. A., Bhakta, N. R., Kessenbrock, K., Prummel, K. D., Yu, Y., Takai, K., et al. (2015). Single-cell analysis reveals a stem-cell program in human metastatic breast cancer cells. *Nature*, *526*(7571), 131–135.

Lawson, M. A., McDonald, M. M., Kovacic, N., Hua Khoo, W., Terry, R. L., Down, J., et al. (2015). Osteoclasts control reactivation of dormant myeloma cells by remodelling the endosteal niche. *Nature Communications*, *6*, 8983.

Li, Y., Hively, W. P., & Varmus, H. E. (2000). Use of MMTV-Wnt-1 transgenic mice for studying the genetic basis of breast cancer. *Oncogene, 19*(8), 1002–1009.
Lilleby, W., Stensvold, A., Mills, I. G., & Nesland, J. M. (2013). Disseminated tumor cells and their prognostic significance in nonmetastatic prostate cancer patients. *International Journal of Cancer, 133*(1), 149–155.
Lu, X., Mu, E., Wei, Y., Riethdorf, S., Yang, Q., Yuan, M., et al. (2011). VCAM-1 promotes osteolytic expansion of indolent bone micrometastasis of breast cancer by engaging alpha4beta1-positive osteoclast progenitors. *Cancer Cell, 20*(6), 701–714.
Lucia, M. S., Sporn, M. B., Roberts, A. B., Stewart, L. V., & Danielpour, D. (1998). The role of transforming growth factor-beta1, −beta2, and -beta3 in androgen-responsive growth of NRP-152 rat prostatic epithelial cells. *Journal of Cellular Physiology, 175*(2), 184–192.
MacKie, R. M., Reid, R., & Junor, B. (2003). Fatal melanoma transferred in a donated kidney 16 years after melanoma surgery. *The New England Journal of Medicine, 348*(6), 567–568.
Malladi, S., Macalinao, D. G., Jin, X., He, L., Basnet, H., Zou, Y., et al. (2016). Metastatic latency and immune evasion through autocrine inhibition of WNT. *Cell, 165*(1), 45–60.
Mantovani, A., Allavena, P., Sica, A., & Balkwill, F. (2008). Cancer-related inflammation. *Nature, 454*(7203), 436–444.
Marlow, R., & Dontu, G. (2015). Modeling the breast cancer bone metastatic niche in complex three-dimensional cocultures. *Methods in Molecular Biology, 1293*, 213–220.
Matsuzawa, A., Takeda, Y., Narita, M., & Ozawa, H. (1991). Survival of leukemic cells in a dormant state following cyclophosphamide-induced cure of strongly immunogenic mouse leukemia (DL811). *International Journal of Cancer, 49*(2), 303–309.
Mishra, A., Wang, J., Shiozawa, Y., McGee, S., Kim, J., Jung, Y., et al. (2012). Hypoxia stabilizes GAS6/Axl signaling in metastatic prostate cancer. *Molecular Cancer Research, 10*(6), 703–712.
Moody, S. E., Sarkisian, C. J., Hahn, K. T., Gunther, E. J., Pickup, S., Dugan, K. D., et al. (2002). Conditional activation of Neu in the mammary epithelium of transgenic mice results in reversible pulmonary metastasis. *Cancer Cell, 2*(6), 451–461.
Morris, V. L., Tuck, A. B., Wilson, S. M., Percy, D., & Chambers, A. F. (1993). Tumor progression and metastasis in murine D2 hyperplastic alveolar nodule mammary tumor cell lines. *Clinical & Experimental Metastasis, 11*(1), 103–112.
Morrissey, C., Vessella, R. L., Lange, P. H., & Lam, H. M. (2015). The biology and clinical implications of prostate cancer dormancy and metastasis. *Journal of Molecular Medicine (Berlin, Germany), 94*, 259–265.
Mueller, M. M., & Fusenig, N. E. (2004). Friends or foes—Bipolar effects of the tumour stroma in cancer. *Nature Reviews Cancer, 4*(11), 839–849.
Muller-Hermelink, N., Braumuller, H., Pichler, B., Wieder, T., Mailhammer, R., Schaak, K., et al. (2008). TNFR1 signaling and IFN-gamma signaling determine whether T cells induce tumor dormancy or promote multistage carcinogenesis. *Cancer Cell, 13*(6), 507–518.
Najmi, S., Korah, R., Chandra, R., Abdellatif, M., & Wieder, R. (2005). Flavopiridol blocks integrin-mediated survival in dormant breast cancer cells. *Clinical Cancer Research, 11*(5), 2038–2046.
Ossowski, L., & Reich, E. (1980). Experimental model for quantitative study of metastasis. *Cancer Research, 40*(7), 2300–2309.
Pardoll, D. M. (2012). The blockade of immune checkpoints in cancer immunotherapy. *Nature Reviews. Cancer, 12*(4), 252–264.
Payne, K. K., Keim, R. C., Graham, L., Idowu, M. O., Wan, W., Wang, X. Y., et al. (2016). Tumor-reactive immune cells protect against metastatic tumor and induce immunoediting of indolent but not quiescent tumor cells. *Journal of Leukocyte Biology, 100*.

Pierga, J. Y., Bonneton, C., Magdelenat, H., Vincent-Salomon, A., Nos, C., Pouillart, P., et al. (2003). Clinical significance of proliferative potential of occult metastatic cells in bone marrow of patients with breast cancer. *British Journal of Cancer, 89*(3), 539–545.

Pisco, A. O., Brock, A., Zhou, J., Moor, A., Mojtahedi, M., Jackson, D., et al. (2013). Non-Darwinian dynamics in therapy-induced cancer drug resistance. *Nature Communications, 4*, 2467.

Purton, L. E., Dworkin, S., Olsen, G. H., Walkley, C. R., Fabb, S. A., Collins, S. J., et al. (2006). RARgamma is critical for maintaining a balance between hematopoietic stem cell self-renewal and differentiation. *The Journal of Experimental Medicine, 203*(5), 1283–1293.

Qian, B., Deng, Y., Im, J. H., Muschel, R. J., Zou, Y., Li, J., et al. (2009). A distinct macrophage population mediates metastatic breast cancer cell extravasation, establishment and growth. *PLoS One, 4*(8), e6562.

Qian, B. Z., Li, J., Zhang, H., Kitamura, T., Zhang, J., Campion, L. R., et al. (2011). CCL2 recruits inflammatory monocytes to facilitate breast-tumour metastasis. *Nature, 475*(7355), 222–225.

Qian, B. Z., & Pollard, J. W. (2010). Macrophage diversity enhances tumor progression and metastasis. *Cell, 141*(1), 39–51.

Qian, B. Z., Zhang, H., Li, J., He, T., Yeo, E. J., Soong, D. Y., et al. (2015). FLT1 signaling in metastasis-associated macrophages activates an inflammatory signature that promotes breast cancer metastasis. *The Journal of Experimental Medicine, 212*(9), 1433–1448.

Quail, D. F., & Joyce, J. A. (2013). Microenvironmental regulation of tumor progression and metastasis. *Nature Medicine, 19*(11), 1423–1437.

Rakhra, K., Bachireddy, P., Zabuawala, T., Zeiser, R., Xu, L., Kopelman, A., et al. (2010). CD4(+) T cells contribute to the remodeling of the microenvironment required for sustained tumor regression upon oncogene inactivation. *Cancer Cell, 18*(5), 485–498.

Rhim, A. D., Mirek, E. T., Aiello, N. M., Maitra, A., Bailey, J. M., McAllister, F., et al. (2012). EMT and dissemination precede pancreatic tumor formation. *Cell, 148*(1–2), 349–361.

Rosol, T. J., Tannehill-Gregg, S. H., Corn, S., Schneider, A., & McCauley, L. K. (2004). Animal models of bone metastasis. *Cancer Treatment and Research, 118*, 47–81.

Ross, K. (2007). For organ transplant recipients, cancer threatens long-term survival. *Journal of the National Cancer Institute, 99*(6), 421–422.

Ruppender, N., Larson, S., Lakely, B., Kollath, L., Brown, L., Coleman, I., et al. (2015). Cellular adhesion promotes prostate cancer cells escape from dormancy. *PLoS One, 10*(6), e0130565.

Sanger, N., Effenberger, K. E., Riethdorf, S., Van Haasteren, V., Gauwerky, J., Wiegratz, I., et al. (2011). Disseminated tumor cells in the bone marrow of patients with ductal carcinoma in situ. *International Journal of Cancer, 129*(10), 2522–2526.

Sansone, P., Ceccarelli, C., Berishaj, M., Chang, Q., Rajasekhar, V. K., Perna, F., et al. (2016). Self-renewal of CD133(hi) cells by IL6/Notch3 signalling regulates endocrine resistance in metastatic breast cancer. *Nature Communications*, 7, 10442.

Saudemont, A., & Quesnel, B. (2004). In a model of tumor dormancy, long-term persistent leukemic cells have increased B7-H1 and B7.1 expression and resist CTL-mediated lysis. *Blood, 104*(7), 2124–2133.

Schardt, J. A., Meyer, M., Hartmann, C. H., Schubert, F., Schmidt-Kittler, O., Fuhrmann, C., et al. (2005). Genomic analysis of single cytokeratin-positive cells from bone marrow reveals early mutational events in breast cancer. *Cancer Cell, 8*(3), 227–239.

Schewe, D. M., & Aguirre-Ghiso, J. A. (2008). ATF6alpha-Rheb-mTOR signaling promotes survival of dormant tumor cells in vivo. *Proceedings of the National Academy of Sciences of the United States of America, 105*(30), 10519–10524.

Schreiber, R. D., Old, L. J., & Smyth, M. J. (2011). Cancer immunoediting: Integrating immunity's roles in cancer suppression and promotion. *Science*, *331*(6024), 1565–1570.
Scognamiglio, R., Cabezas-Wallscheid, N., Thier, M. C., Altamura, S., Reyes, A., Prendergast, A. M., et al. (2016). Myc depletion induces a pluripotent dormant state mimicking diapause. *Cell*, *164*(4), 668–680.
Sherry, M. M., Greco, F. A., Johnson, D. H., & Hainsworth, J. D. (1986). Metastatic breast cancer confined to the skeletal system. An indolent disease. *The American Journal of Medicine*, *81*(3), 381–386.
Shiozawa, Y., Pedersen, E. A., Havens, A. M., Jung, Y., Mishra, A., Joseph, J., et al. (2011). Human prostate cancer metastases target the hematopoietic stem cell niche to establish footholds in mouse bone marrow. *The Journal of Clinical Investigation*, *121*(4), 1298–1312.
Shiozawa, Y., Pedersen, E. A., Patel, L. R., Ziegler, A. M., Havens, A. M., Jung, Y., et al. (2010). GAS6/AXL axis regulates prostate cancer invasion, proliferation, and survival in the bone marrow niche. *Neoplasia*, *12*(2), 116–127.
Sosa, M. S., Avivar-Valderas, A., Bragado, P., Wen, H. C., & Aguirre-Ghiso, J. A. (2011). ERK1/2 and p38alpha/beta signaling in tumor cell quiescence: Opportunities to control dormant residual disease. *Clinical Cancer Research*, *17*(18), 5850–5857.
Sosa, M. S., Bragado, P., & Aguirre-Ghiso, J. A. (2014). Mechanisms of disseminated cancer cell dormancy: An awakening field. *Nature Reviews. Cancer*, *14*(9), 611–622.
Sosa, M. S., Parikh, F., Maia, A. G., Estrada, Y., Bosch, A., Bragado, P., et al. (2015). NR2F1 controls tumour cell dormancy via SOX9- and RARbeta-driven quiescence programmes. *Nature Communications*, *6*, 6170.
Taichman, R. S., Patel, L. R., Bedenis, R., Wang, J., Weidner, S., Schumann, T., et al. (2013). GAS6 receptor status is associated with dormancy and bone metastatic tumor formation. *PLoS One*, *8*(4), e61873.
Thorban, S., Rosenberg, R., Busch, R., & Roder, R. J. (2000). Epithelial cells in bone marrow of oesophageal cancer patients: A significant prognostic factor in multivariate analysis. *British Journal of Cancer*, *83*(1), 35–39.
Turajlic, S., & Swanton, C. (2016). Metastasis as an evolutionary process. *Science*, *352*(6282), 169–175.
Van Allen, E. M., Miao, D., Schilling, B., Shukla, S. A., Blank, C., Zimmer, L., et al. (2015). Genomic correlates of response to CTLA-4 blockade in metastatic melanoma. *Science*, *350*(6257), 207–211.
Vishnoi, M., Peddibhotla, S., Yin, W., Scamardo, A. T., George, G. C., Hong, D. S., et al. (2015). The isolation and characterization of CTC subsets related to breast cancer dormancy. *Scientific Reports*, *5*, 17533.
Wang, H., Yu, C., Gao, X., Welte, T., Muscarella, A. M., Tian, L., et al. (2015). The osteogenic niche promotes early-stage bone colonization of disseminated breast cancer cells. *Cancer Cell*, *27*(2), 193–210.
Weinhold, K. J., Miller, D. A., & Wheelock, E. F. (1979). The tumor dormant state. Comparison of L5178Y cells used to establish dormancy with those that emerge after its termination. *The Journal of Experimental Medicine*, *149*(3), 745–757.
Wollenberg, B., Walz, A., Kolbow, K., Pauli, C., Chaubal, S., & Andratschke, M. (2004). Clinical relevance of circulating tumour cells in the bone marrow of patients with SCCHN. *Onkologie*, *27*(4), 358–362.
Zou, L., Barnett, B., Safah, H., Larussa, V. F., Evdemon-Hogan, M., Mottram, P., et al. (2004). Bone marrow is a reservoir for CD4+CD25+ regulatory T cells that traffic through CXCL12/CXCR4 signals. *Cancer Research*, *64*(22), 8451–8455.

CHAPTER THREE

Defining the Influence of Germline Variation on Metastasis Using Systems Genetics Approaches

M. Lee, N.P.S. Crawford[1]

Genetics and Molecular Biology Branch, National Human Genome Research Institute, NIH, Bethesda, MD, United States

[1]Corresponding author: e-mail address: crawforn@mail.nih.gov

Contents

1. Introduction 74
2. Hereditary Factors Associated with Cancer Aggressiveness and Metastasis in Human Populations 76
   2.1 Epidemiological Analysis 76
   2.2 Genetic Analysis Human Cohorts 80
3. Systems Genetics Approaches to Identify Hereditary Modifiers of Metastasis 82
   3.1 Breast Cancer 82
   3.2 Prostate Cancer 93
   3.3 Melanoma 96
4. Conclusions 99
Acknowledgments 100
References 101

Abstract

Cancer is estimated to be responsible for 8 million deaths worldwide and over half a million deaths every year in the United States. The majority of cancer-related deaths in solid tumors is directly associated with the effects of metastasis. While the influence of germline factors on cancer risk and development has long been recognized, the contribution of hereditary variation to tumor progression and metastasis has only gained acceptance more recently. A variety of approaches have been used to define how hereditary variation influences tumor progression and metastasis. One approach that garnered much early attention was epidemiological studies of cohorts of cancer patients, which demonstrated that specific loci within the human genome are associated with a differential propensity for aggressive tumor development. However, a powerful, and somewhat underutilized approach has been the use of systems genetics approaches in transgenic mouse models of human cancer. Such approaches are typically multifaceted, and involve integration of multiple lines of evidence derived, for example, from genetic and transcriptomic screens of genetically diverse mouse models of cancer, coupled with bioinformatics analysis of human cancer datasets, and

Advances in Cancer Research, Volume 132
ISSN 0065-230X
http://dx.doi.org/10.1016/bs.acr.2016.07.003

functional analysis of candidate genes. These methodologies have allowed for the identification of multiple hereditary metastasis susceptibility genes, with wide-ranging cellular functions including regulation of gene transcription, cell proliferation, and cell–cell adhesion. In this chapter, we review how each of these approaches have facilitated the identification of these hereditary metastasis modifiers, the molecular functions of these metastasis-associated genes, and the implications of these findings upon patient survival.

1. INTRODUCTION

In 2016, it is estimated that nearly 1.7 million new cases of cancer will be diagnosed, and approximately 600,000 patients will succumb to cancer-related causes in the United States alone (Siegel, Miller, & Jemal, 2016). This is a 1.6% increase in new cases and 1.1% increase in deaths compared to 2015 estimations. Metastasis, which is defined as the dissemination of tumor cells from primary lesions to distant organs to form new secondary tumors, has been estimated to be directly associated with over 90% of deaths in solid tumors (Hunter, 2015). Metastasis is a multistep, intricate process that requires the cooperation of tumor cells and microenvironmental factors at both the primary and secondary sites. Due to its complex nature and the host factors involved in this process, our understanding of the molecular determinants of metastasis is incomplete despite recent advances in the field.

One of the most widely accepted models that aim to explain the underlying mechanisms of metastasis is the "progression model," where it is postulated that populations of cells within the primary tumor acquire different somatic mutations during cell division as the tumor progresses. As these somatic mutations accumulate, select cells acquire mutations that allow them to gain a more metastatic phenotype, which facilitates local invasion and intravasation, survival in the circulatory system, and finally extravasation and proliferation at a secondary site. The progression model is supported by many observations, including the presence of metastasis-specific mutations (Haffner et al., 2013; Huang et al., 2014; Xie et al., 2014; Zhang et al., 2013), and that metastases appear to be clonal derivatives of a subset of cells originating from the primary tumor (Fidler & Kripke, 1977; Fidler & Talmadge, 1986; Haffner et al., 2013; Talmadge, Wolman, & Fidler, 1982). However, several observations remain that cannot be explained by the progression model. For instance, if the spontaneous somatic mutations acquired by the metastatic cells are sufficient to confer metastatic potential, then these cells should be able to metastasize when transplanted. This, however, is not the case (Giavazzi, Alessandri, Spreafico, Garattini, & Mantovani, 1980; Mantovani, Giavazzi, Alessandri, Spreafico, & Garattini, 1981), implying that other factors may be required for successful tumor cell dissemination and proliferation at secondary sites. In addition, early studies profiling global gene expression within primary tumor samples demonstrate that distinct transcriptomic patterns are associated with tumor progression and patient survival (Ramaswamy, Ross, Lander, & Golub, 2003; van de Vijver et al., 2002; van't Veer et al., 2002; Yang et al., 2005). Most significantly, these metastatic gene expression signatures arise at an early time point in tumorigenesis, prior to the occurrence of clinically detectable metastasis, which is an observation at odds with the progression model. Specifically, the progression model postulates that metastasis-associated somatic events should occur later in the evolution of the primary lesion, and thus tumor cells exhibiting metastasis-associated gene expression signatures should not be sufficiently represented at an early time point. Furthermore, metastatic tumors with unknown primary lesions have been observed in approximately 5% of patients (Ribelles, Santonja, Pajares, Llacer, & Alba, 2014; Riethmuller & Klein, 2001), which leaves open the question as to how these presumably small primary tumors are able to metastasize given it is statistically unlikely that a tumor composed of a relatively small number of cells will have evolved sufficiently to acquire the necessary, random metastasis-permissive somatic mutations. Thus, these inconsistencies of the conventional progression model have necessitated the development of alternative models that can explain these phenomena associated with metastasis.

Germline polymorphisms are naturally occurring hereditary variations in the genetic sequence and are important determinants of morphological traits such as eye color and height, as well as other traits such as disease susceptibility and metabolism (Welter et al., 2014). Hereditary variation has long been recognized as a driving factor in the development of most forms of cancer (Goldgar, Easton, Cannon-Albright, & Skolnick, 1994; Hemminki & Vaittinen, 1998), with first-degree relatives of cancer sufferers being at higher risk of disease development for most tumor types. Yet, the influence of such hereditary components in respect to tumor progression and patient survival has only been established in more recent times (Hemminki, Ji, Forsti, Sundquist, & Lenner, 2008a,2008b). While genome-wide association studies (GWAS) have identified hundreds of genomic regions or "loci" for overall disease susceptibility for multiple tumor types (eg, prostate cancer; Al Olama et al., 2014; Berndt et al., 2015; Hazelett et al., 2014), breast cancer (Kainu et al., 2000), the search for hereditary modifiers of aggressive tumorigenesis and metastasis has proven more elusive. The likely cause of this is that disease modifiers that act later in a disease course (as would be the case with modifiers of metastasis) are likely to act in a more "context-dependent" manner (Bjorkegren, Kovacic, Dudley, & Schadt, 2015) and require exposure to a specific activating factor (ie, the influence of the primary tumor upon the host) to become active as a disease modifier. GWAS are also significantly impacted by disease treatment, which alters the natural course of disease in many patients, thereby reducing the likelihood that genes associated with metastasis formation will be detected. It should be noted that this issue does not likely apply to more generalized measures of outcome such as disease specific or overall survival, which measure not only metastasis formation but response to treatment as well. Finally, replication of GWAS findings in additional independent cohorts has proven to be a recurrent problem (Pirie et al., 2015; Schumacher et al., 2011). Thus, to more comprehensively define the effects of the germline upon metastasis and other forms of aggressive tumorigenesis, more integrative "systems genetics" approaches are required, which span the fields of genetics, genomics, bioinformatics, and molecular biology (Bjorkegren et al., 2015; Civelek & Lusis, 2014; Pirie et al., 2015). Here, we review the evidence demonstrating the influence of the host upon tumor progression and metastasis and discuss how systems genetics approaches have yielded fresh insights into the role of hereditary variation in the modulation of metastasis susceptibility in various types of cancer.

2. HEREDITARY FACTORS ASSOCIATED WITH CANCER AGGRESSIVENESS AND METASTASIS IN HUMAN POPULATIONS

2.1 Epidemiological Analysis

The significant role of family history as a risk factor in cancer development has long been recognized (Amin Al Olama et al., 2015; Amos & Struewing, 1993; Eeles et al., 2014; Hemminki, Li, & Czene, 2004; Hemminki & Vaittinen, 1998). A schematic of how hereditary variation influences susceptibility to cancer development as well as susceptibility to aggressive disease forms is presented in Fig. 1. Epidemiological studies have proven to be a powerful tool in defining how germline variation influences cancer development. For example, a study using a cohort of 9512 pairs of twins found significant association for development of sporadic prostate, colorectal, and breast cancers for twins with the same type of cancer, with greater

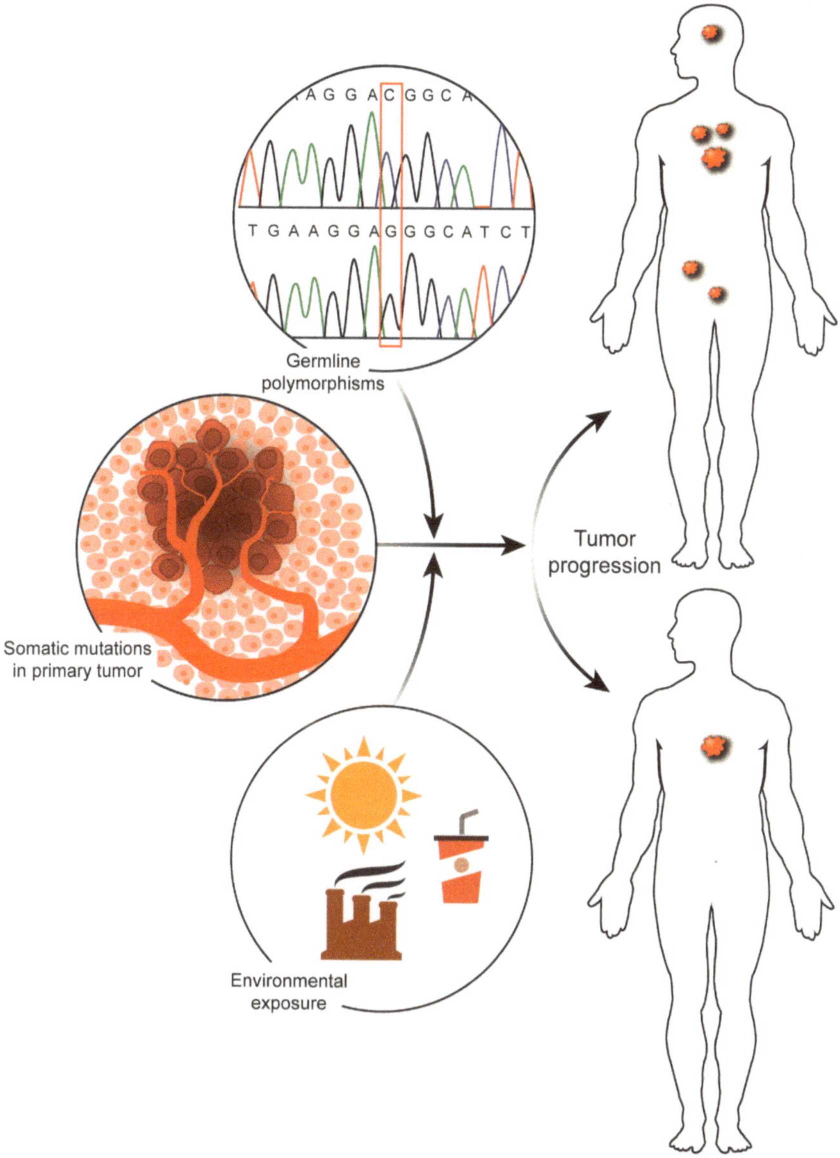

Fig. 1 Factors that contribute to tumor progression and metastasis susceptibility. While the accumulation of somatic mutations in the primary tumor act as the primary driving force of tumor progression, additional factors such as genetic variation present in the host genome (ie, germline polymorphisms) and environmental factors (ie, diet) influence how the tumor progresses in different individuals. These factors collectively determine differences between individuals in tumor progression and metastasis.

association for monozygotic compared to dizygotic twins (Lichtenstein et al., 2000). Furthermore, men with first-degree relatives with prostate cancer are more than twice as likely to be diagnosed with prostate cancer compared to the control population (relative risk [RR]; 2.48, 95% confidence interval [CI]; 2.25–2.74) (Goldgar et al., 1994; Kicinski, Vangronsveld, & Nawrot, 2011). Similarly, first-degree relatives of women with breast cancer are also at higher risk of approximately 1.5–1.8-fold of developing breast cancer compared to others (Hemminki & Vaittinen, 1998; Peto, Easton, Matthews, Ford, & Swerdlow, 1996). In daughters of women diagnosed before the age of 40, the risk of developing breast cancer under 40 years old increases four-fold compared to matched population controls (Hemminki & Vaittinen, 1998), suggesting that genetic factors influence both age of onset and disease aggressiveness. Finally, ethnic differences in cancer aggressiveness also suggest a modifying role for hereditary variation. For example, the disproportionately high risk and mortality rates of prostate cancer in African Americans compared to men of other ethnicities and races also suggests that inherited genetic factors mediate prostate cancer development and progression (Ferlay et al., 2015; Howlader et al., 2014; Tan, Mok, & Rebbeck, 2016).

In the past decade, it has become apparent that the factors influencing cancer progression and survival are different from those factors determining cancer risk (Hemminki et al., 2008a,2008b; Lindstrom et al., 2007). To identify whether familial components also influence tumor progression and metastasis, survival studies were performed using the Swedish Family Cancer Database in various cancer types. Familial correlations with survival were demonstrated in breast cancer using 1277 mother–daughter patient pairs (Hemminki et al., 2008b). Familial breast cancer was defined in cases of mother–daughter pairs when daughters were diagnosed with breast cancer between 1990 and 1999 and mothers were affected with breast cancer before 1990. Sporadic cases of breast cancer were excluded. Daughters who had mothers with more than 120 months survival exhibited a significantly better cause-specific survival compared to daughters who had mothers with less than 36 months survival (hazard ratio [HR] = 0.65, 95% CI 0.46–0.92). Reversing the analyses by using the daughters as a reference, which eliminates the effect of disease awareness on survival, produced similar results, with better survival being observed in mothers whose daughters had survived more than 120 months compared to mothers whose daughters had poorer survival of less than 36 months (HR = 0.68, 95% CI 0.50–0.93). Interestingly, cause-specific survival was equal in sporadic and familial breast cancer in mothers but not daughters, suggesting that early diagnosis and other behavioral factors may influence survival in later generations. In a similar study, familial correlation in survival in breast cancer was confirmed in 2162 parent–child pairs; children of a parent with poor survival had worse outcomes compared to those whose parents were living 10 years postdiagnosis (HR = 0.91, 95% CI 0.88–0.93 vs HR = 0.86, 95% CI 0.82–0.89) (Lindstrom et al., 2007). In addition to the likelihood of shared environmental risk factors, these results suggest the existence of genetic risk factors that modulate survival in breast cancer. Given that, as mentioned earlier, the key determinant of survival is the presence or absence of metastasis, it undoubtedly holds true that host genetics also have a significant modifying effect upon metastasis.

A similar study was performed in prostate cancer using 609 father–son pairs from the Swedish Family Cancer Database (Hemminki et al., 2008a). In a manner similar to that seen in breast cancer, favorable survival was also observed among family members with prostate cancer; in sons with fathers whose prostate cancer-specific survival was greater than 60 months, cause-specific survival was significantly better than those with fathers whose survival was less than 24 months (HR = 0.62, 95% CI 0.41–0.94). Conversely, fathers whose sons had survival of over 60 months had better outcomes compared to those whose sons had poorer survival of less than 60 months (HR = 0.78, 95% CI 0.62–1.00). Although enrollment of eligible father–son pairs was more challenging for prostate cancer due to the later median age of onset, the results from this study illustrate the familial influence in prostate cancer survival. Familial influence in prostate cancer survival was also apparent in an earlier study of 1304 father–son pairs, in which sons whose fathers had survived more than 10 years had better survival than those whose parents died within 10 years (HR = 0.91, 95% CI 0.78–0.96 vs HR = 0.87, 95% CI 0.82–0.91) (Lindstrom et al., 2007).

In addition to these findings, survival has also been demonstrated to be influenced by familial factors for numerous other cancer types. In bladder or renal cancer, children who have parents with cancer of the same site were demonstrated to have a decreased risk for cause-specific death if their affected parents had survival periods of 5 years or more (bladder cancer; HR = 0.27, 95% CI 0.07–1.00; renal cell cancer; HR = 0.35, 95% CI 0.13–0.94) (Ji, Forsti, Sundquist, Lenner, & Hemminki, 2008a). In pancreatic cancer, which is associated with a particularly low overall survival, both parents and children with familial cancer have a significantly poorer survival than those with sporadic cancer (children; HR = 1.44, 95% CI 1.13–1.84;

parents; HR = 1.37, 95% 1.09–1.72) (Ji, Forsti, Sundquist, Lenner, & Hemminki, 2008b). In a population-based Swedish family database, an increase in risk for cancer-related death was observed in children with parents with poor survival compared to those with parents with good survival in colorectal cancer (HR = 1.44, 95% CI 1.01–2.01) and lung cancer (HR = 1.39, 95% CI 1.00–1.94) (Lindstrom et al., 2007).

2.2 Genetic Analysis Human Cohorts

A number of approaches to investigate the genetic origins of the differential survival of cancer patients described earlier have been utilized in human cohorts. Family based linkage studies were among some of the first forms of genetic analyses to identify specific regions of the genome associated with susceptibility to metastasis and other forms of aggressive disease, especially prostate cancer. These studies are typically performed in "high-risk" cancer families, where multiple family members are affected with the same tumor type, and aim to correlate patterns of germline allelic variation among multiple patient families with specific disease traits. This approach has been successful in multiple cancer types and has led to the identification of multiple loci associated with aggressive disease. For instance, a genome-wide linkage analysis of 244 men with aggressive prostate cancer, where one of the clinical variables used to define aggressiveness was distant metastasis, identified regions on chromosome Xq27–28 and Xq22 as being associated with an increased incidence of aggressive disease (Chang et al., 2005). Additionally, a family based linkage scan of 513 brothers affected with prostate cancer identified chromosome 7q32 to be associated with aggressive prostate cancer (Witte et al., 2000). Tumor aggressiveness was measured by Gleason score, a parameter that reflects primary tumor grade. Linkage of this region was confirmed using genotype data from 216 men diagnosed with prostate cancer from 100 German families, in which chromosome 7q31–33 was found to be associated with tumor grade and late onset prostate cancer (Paiss et al., 2003). However, although these linkage studies have proven useful in defining regions of the genome associated with aggressive cancer phenotypes, it is typically difficult to achieve mapping resolutions that readily facilitate the identification of candidate modifier genes, even in cases where hundreds of families are available (Boehnke, 1994; Janer et al., 2003; Stanford et al., 2009). Thus, in spite of the aforementioned successes, it has proven challenging to identify causative genes associated with susceptibility to metastasis and other forms of aggressive disease using these approaches.

GWAS typically involve genotyping millions of single nucleotide polymorphisms (SNPs) in unrelated individuals, followed by comparison of allele frequencies in cases and controls to identify variants associated with tumor dissemination and other aggressive traits. This approach, which requires large cohorts composed of many thousands of cases and controls, has enabled the detection of hundreds of causative loci for complex trait diseases (Bjorkegren et al., 2015), and has been used extensively to study the genetic origins of cancer development and aggressiveness. For prostate cancer, GWAS have led to the identification of over 100 loci associated with disease development, which are estimated to account for approximately 30% of the heritable risk (Al Olama et al., 2014; Eeles et al., 2014). Yet, identification of hereditary factors associated with tumor progression and metastasis, which differentiate aggressive from nonaggressive prostate cancer, has proven to be more challenging (Al Olama et al., 2014; Amin Al Olama et al., 2013; Teerlink et al., 2014). Although identification of aggressive disease susceptibility loci is clinically important and could potentially aid in the stratification of patient risk in aggressive forms of cancer (Amin Al Olama et al., 2013; Farber, 2013; Teerlink et al., 2014), detection of these loci in GWAS is hindered by variables including small cohort sizes, the confounding effects of disease treatment, and the stringent conditions required to correct for multiple testing. A recent GWAS of 12,518 prostate cancer cases reports two novel loci, rs35148638 at chromosome 5q14.3 ($P=6.49\times10^{-9}$) and rs78943174 at chromosome 3q26.31 ($P=4.18\times10^{-8}$), associated with aggressive prostate cancer which was measured by the Gleason score (Berndt et al., 2015). In breast cancer, a recent metaanalysis that examined SNPs previously identified as germline polymorphisms associated with breast cancer survival confirms the need for these associated studies to be validated in larger cohorts (Pirie et al., 2015). In a pooled cohort of 37,954 breast cancer patients, 56 variants previously reported to be associated with breast cancer survival were evaluated. This metaanalysis demonstrated that none of these variants achieved genome-wide significance ($P<5\times10^{-8}$). However, five variants in five different genes (*FGFR2*; rs2981582, *NQO1*; rs1800566, *LOC100506172*; rs9934948, *TGF*; rs1800470, *SULT1E1*; rs3775775) reached nominal significance ($P<0.05$), and seven additional variants in four different genes (*CYP19A1*; rs700519, rs12900137, rs1902586, rs28566535, *VDR*; rs731236, *PPP2R2B*; rs10477313, *MPO*; rs2333227) reached significance ($P<0.05$) in estrogen receptor (ER)-positive cases (Pirie et al., 2015). These findings highlight the requirement for extremely large cohort sizes in GWAS metaanalyses in order to perform adequately powered studies that will facilitate the detection of associations at the very stringent genome-wide level of statistical significance.

In conclusion, the analysis of human cohorts has proved to be a powerful tool for identifying genetic factors associated with overall susceptibility to cancer development as well as susceptibility to some forms of aggressive disease. However, although population-based epidemiological studies demonstrate a hereditary component to patient survival, which serves as a proxy for metastasis for the reasons stated earlier, it has proven somewhat challenging to identify genes associated with metastasis using linkage analysis and GWAS. This has therefore led to recent calls to reevaluate human GWAS data using systems-based approaches, which aim to integrate genome-wide analyses of DNA, RNA, and protein levels (Bjorkegren et al., 2015; Civelek & Lusis, 2014). It is likely that these systems genetics approaches will yield in the identification of "context-dependent" susceptibility loci that affect later stages of disease progression. In the latter part of this review, we will focus on an alternative and very much complementary approach to human GWAS and other types of genetic analysis, which centers on systems genetics studies using mouse models of tumorigenesis. These studies, which allow for the control of confounding variables such as genetic and environmental variation, have proven to be a particularly powerful means of identifying hereditary metastasis susceptibility genes (Fig. 2).

3. SYSTEMS GENETICS APPROACHES TO IDENTIFY HEREDITARY MODIFIERS OF METASTASIS

3.1 Breast Cancer

Breast cancer is the most common malignancy in women in almost all countries, including the United States (Cardoso et al., 2012; Siegel et al., 2016). The American Cancer Society estimates that the lifetime likelihood of women in the United States developing invasive breast cancer is 12.3% (Siegel et al., 2016). It is estimated that over 40,000 women in the United States will succumb to breast cancer in 2016 (Siegel et al., 2016), and that metastasis will be the direct cause of death in over 90% of cases (Gupta & Massague, 2006). Despite advances in research and treatment, 5–10% of breast cancers are metastatic at the time of diagnosis, and it is estimated that 20–30% of all breast cancer patients will develop metastatic disease with approximately 20% of these surviving 5 years (Cardoso et al., 2012; Chung & Carlson, 2003).

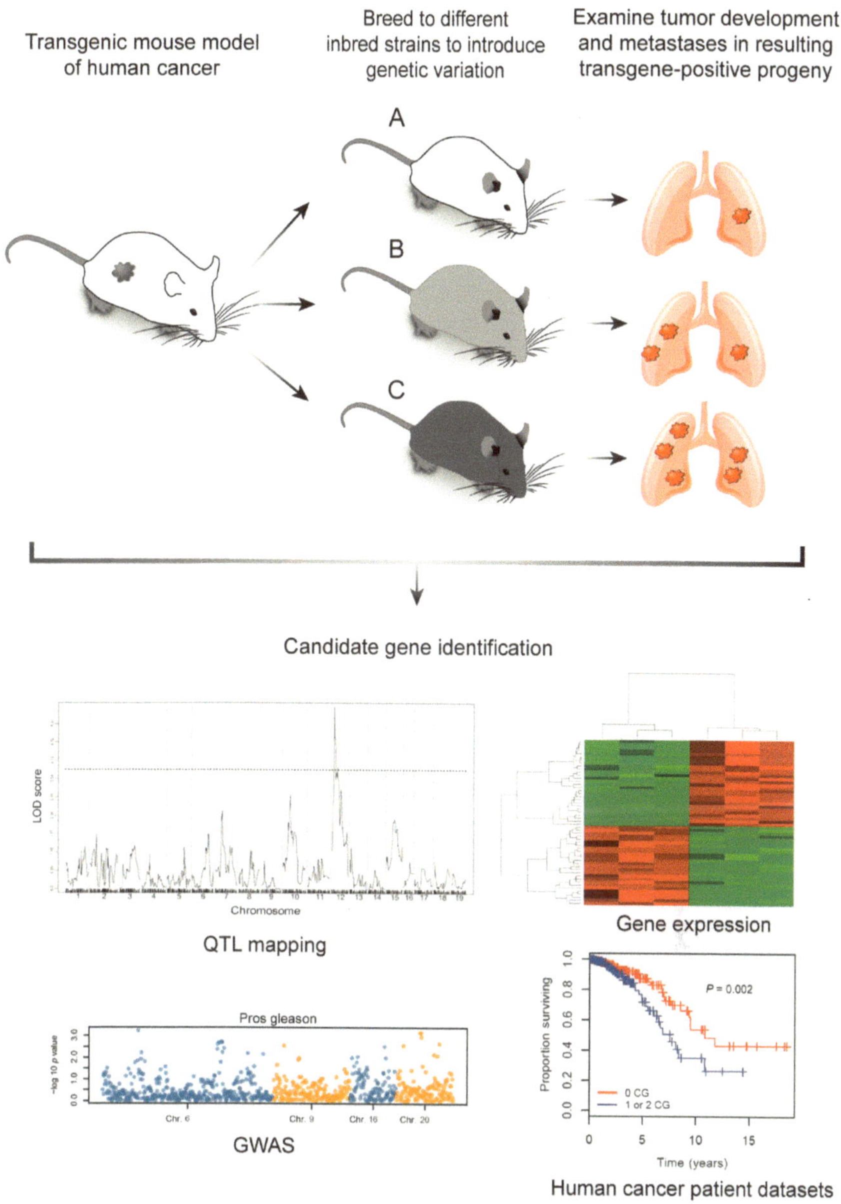

Fig. 2 Breeding scheme with transgenic mouse models of human cancers for QTL to identify metastasis susceptibility loci. Transgenic mouse models of various human cancers have been used to identify hereditary factors influencing tumor progression and metastasis. Genetic variation is introduced into these transgenic mouse models by breeding, and tumor development and metastases are quantified in the transgene-positive progeny of these crosses. This is followed by quantification of disease aggressiveness in these genetically diverse mouse populations to determine whether there is significant variation in the phenotype depending on the paternal or maternal genetic background. These polymorphic mouse populations form the backbone of systems genetics approaches, which are performed in an integrative fashion and typically utilize an array of genetic, genomic, and functional assays coupled with bioinformatics analyses to identify candidate metastasis modifier genes.

The FVB/N-Tg(MMTV-PyVT)634Mul/J (PyMT) transgenic mouse model of mammary tumorigenesis (Guy, Cardiff, & Muller, 1992) was the first model in which systems genetics approaches were used to identify novel germline metastasis susceptibility genes. These mice, which have been most extensively studied using such systems genetics methodologies, express the polyoma virus middle T antigen under the control of the mouse mammary tumor virus (MMTV) promoter. All female mice develop palpable mammary tumors by 5 weeks of age, which progress through stages similar to those seen in human breast cancer, and culminates in pulmonary metastasis in over 90% of mice by 12 weeks of age (Guy et al., 1992; Lin et al., 2003). In the work of Lifsted et al. (1998), which was the first to prove that genetic background affects metastasis susceptibility, germline variation was introduced by breeding male PyMT mice to female mice of 27 different inbred strains. Pulmonary metastases and tumor burden were quantified in transgene-positive female F1 progeny after 40 days. Metastatic efficiency varied substantially depending on the maternal strain used to generate F1 mice, ranging from approximately a 3-fold increase to a 10-fold decrease compared to that of wild-type PyMT mice. The majority of strains tested were observed to have a metastasis suppressive effect, with 13/27F1 strains exhibiting a significant reduction in pulmonary metastases compared to the wild-type FVB/NJ PyMT mouse. Given that levels and timing of PyMT expression was equal across each of the different F1 strains, this study conclusively demonstrated that germline variation has a powerful modifying effect on mammary tumor metastasis.

The results from this study established the foundation for future work that led to the identification of multiple breast cancer metastasis modifier genes (Table 1). The primary approach used in these studies is known as "quantitative trait locus" (QTL) mapping, which is a form of genetic linkage analysis used to identify genomic loci associated with traits of interest. The first of these studies to identify modifiers of mammary tumor metastasis was performed by crossing the PyMT mouse model to either AKXD recombinant inbred strains (Mucenski, Taylor, Jenkins, & Copeland, 1986) or a variety of inbred strains with low-metastatic capacities. By correlating patterns of aggressive disease development with patterns of hereditary variation, a number of loci associated with metastasis were identified, the most notable of which was a locus on mouse chromosome 19 that was significantly associated with metastatic suppression (Hunter et al., 2001). Haplotype analysis in two metastasis suppressive strains of mice (DBA/2J and NZB/B1NJ)

Table 1 Hereditary Modifiers of Metastasis in Breast Cancer

Gene	Description	Mouse Chromosome	Method of Identification	Human Ortholog	Location	Cellular Function of Protein	OMIM
Sipa1	Signal-induced proliferation-associated 1	19	(AKXD × PyMT) QTL analysis	*SIPA1*	11q13	Mitogen-induced GTPase-activating protein	602180
Rrp1b	Ribosomal RNA processing 1 homolog B	17	(AKXD × PyMT) QTL analysis	*RRP1B*	21q22.3	Regulation of gene expression by modulating chromatin modification and mRNA splicing	610654
Brd4	Bromodomain containing 4	17	(AKXD × PyMT) QTL analysis	*BRD4*	19p13.1	Chromosome association	608749
Ndn	Necdin, MAGE family member	7	(AKXD × PyMT) QTL analysis	*NDN*	15q11.2	Transcription factor	602117
Arid4b	AT-rich interaction domain 4B	13	(AKXD × PyMT) QTL analysis	*ARID4B*	1q42.3	Subunit of the histone deacetylase-dependent SIN3A transcriptional corepressor complex	609696
Cadm1	Cell adhesion molecule 1	9	FVB × (NZB × PyMT) QTL analysis	*CADM1*	11q23.2	Immunoglobulin superfamily; cell adhesion	605686
Cnot2	CCR4-NOT transcription complex subunit 2	10	(AKXD × PyMT) microarray data analysis	*CNOT2*	12q15	Regulation of mRNA synthesis, splicing, transport, localization, and degradation	604909
Mir216/217	*MicroRNA 216/217* cluster	11	(AKXD × PyMT) microarray data analysis	*MIR216/217*	2p16.1	Tumor formation and progression	610944, 615096

and two metastasis permissive strains (FVB/NJ and AKR/J) facilitated the nomination of 23 high-priority candidate genes within this locus on chromosome 19 (Park, Clifford, Buetow, & Hunter, 2003). Further analyses of sequence variation among these strains identified *Signal-Induced Proliferation-Associated gene 1* (*Sipa1*) as the first hereditary breast cancer metastasis modifier gene (Park et al., 2005). *Sipa1* encodes for a GTPase-activating protein (GAP) specific for Rap1 and Rap2 GTPases (Kurachi et al., 1997) and functions in pathways involved in cell polarity (Shimonaka et al., 2003), cell–cell contact (Hogan et al., 2004; Price et al., 2004), and cell proliferation (Altschuler & Ribeiro-Neto, 1998). Mouse *Sipa1* harbors a nonsynonymous polymorphism that encodes an alanine to threonine substitution, which distinguishes the low- and high-metastatic efficiency mouse strains. This polymorphism is located in a PDZ protein–protein interaction domain and was shown to affect the binding ability of SIPA1 to AQP2, a binding partner of SIPA1. In cells expressing the *Sipa1* allele from low-metastatic strains, overexpression of AQP2 increased levels of GTP-Rap1, thus inhibiting the ability of SIPA1 to regulate Rap-GAP activity (Park et al., 2005). Furthermore, to examine the role of SIPA1 in tumor progression and metastasis, shRNA-mediated knockdown of *Sipa1* was performed in the highly metastatic mammary mouse tumor cell line Mvt-1. Implantation of these cells into the mammary fat pad of FVB/NJ female mice caused an average 20-fold decrease in pulmonary metastases compared to control cells, demonstrating that *Sipa1* promotes metastasis. To confirm these results, Mvt-1 cells stably overexpressing the metastatic FVB allele of *Sipa1* were subcutaneously implanted into FVB female mice. As expected, overexpression of *Sipa1* resulted in a significant increase of over two-fold in pulmonary metastases after 4 weeks compared to mice implanted with control cells (Park et al., 2005).

To confirm the relevance of these findings to metastasis susceptibility in human populations, the relative expression of *SIPA1* was examined in metastatic and nonmetastatic tumors of various cancer types. These analyses revealed that higher levels of *Sipa1* within the primary tumor is correlated with metastatic progression of human prostate cancer ($P < 0.001$) (Park et al., 2005), implying that metastasis susceptibility genes may be of importance in multiple cancer types. To further investigate the relevance of *SIPA1* to human disease, a follow-up study was performed where the frequencies of three SNPs within the promoter and coding regions of *SIPA1* [−313 G > A (rs931127), 545 C > T (rs3741378), 2760 G > A (rs746429)] were quantified in a group of 300 women (154 cases of regional or metastatic breast cancer and 146 cases of randomly selected localized disease). Each of these hereditary variants were found to be strongly associated with aggressive breast cancer (Crawford et al., 2006), which not only validates the relevance of mouse-based systems genetics approaches to human disease, but also demonstrates the importance of *SIPA1* as a hereditary modifier of metastasis in human breast cancer.

A subsequent study utilized primary tumor microarray data from a previous study that crossed the PyMT model to 18 strains of the AKXD recombinant inbred panel (Hunter et al., 2001; Yang et al., 2005), with the latter being generated through over 20 generations of inbreeding of the metastasis permissive strain AKR/J with the metastasis suppressive DBA/2J strain. Expression QTL (eQTL) mapping using microarray data derived from (AKXD × PyMT) F1 primary tumors identified three loci on mouse chromosomes 7, 17, and 18 as being associated with extracellular matrix (ECM) gene expression (Crawford et al., 2007). This was particularly interesting since ECM genes are a common component of metastasis-predictive gene expression signatures in breast cancer (Yang et al., 2004). More significantly, two of these three ECM eQTL loci, on chromosomes 7 and 17, overlapped with previously described metastasis efficiency loci, which adds credence to the hypothesis that ECM gene expression and metastasis are influenced by the same germline polymorphisms. Correlation analysis was performed to identify candidate genes within these regions, which allowed for the identification of *Ribosomal RNA Processing 1 Homolog B* (*Rrp1b*) as both an ECM gene expression and metastasis modifier. The role of RRP1B as a germline modifier of metastasis was further strengthened by concurrent findings from yeast two-hybrid screening and immunoprecipitation assays, which demonstrated that RRP1B is an interactor of SIPA1 (Crawford et al., 2007). Binding of RRP1B to SIPA1 inhibited the Rap-GAP activity of SIPA1 similar to the binding effect observed with AQP2. Overexpression of *Rrp1b* in Mvt-1 cells produced a gene expression signature that predicts overall survival of human breast cancer patients (Crawford et al., 2007). A significant decrease in tumor burden and pulmonary metastases was observed compared to the control with subcutaneous implantation of Mvt-1 cells stably overexpressing *Rrp1b* into syngeneic FVB/NJ female mice. Similarly, knockdown of RRP1B expression in the highly metastatic mouse mammary tumor cell lines Mvt-1 and 4T1 caused an increase in in vivo pulmonary metastases and in vitro tumor cell invasiveness (Lee, Dworkin, Gildea, et al., 2014). The functionality of RRP1B appears to be very much multifaceted, with it being demonstrated to orchestrate these metastasis-associated changes in gene expression at both the transcriptional level by modulating global chromatin H3K9 trimethylation levels (Crawford, Yang, Mattaini, & Hunter, 2009; Lee, Dworkin, Lichtenberg, et al., 2014) as well as at the post-transcriptional level by interacting with mRNA splicing factors (Lee, Dworkin, Gildea et al., 2014). Finally, in terms of its relevance to human disease, a nonsynonymous polymorphism within *RRP1B* (1421 G>A, rs9306160) that encodes a proline with a leucine substitution was found to be associated with better outcome in two independent breast cancer cohorts (one cohort of 139 cases with regional or metastatic disease and 130 cases localized disease, and another cohort of 248 patients). The variant A allele was found at a higher frequency in patients with ER-positive tumors and low grade tumors in both cohorts (Crawford et al., 2007). Furthermore, a later study demonstrated that this SNP in *RRP1B* acted synergistically with the *SIPA1* SNP rs2448490, with a combination of *RRP1B* and *SIPA1* SNPs being associated with metastasis-free survival in a cohort of ER-positive/lymph node-negative breast cancer patients. Specifically, carriers of the T variant allele (CT + TT) for *RRP1B* SNP rs9306160 that were homozygous for the variant A allele for *SIPA1* SNP rs2448490 had significantly lower risk of developing metastases compared to other genotypes (HR = 0.40, 95% CI 0.24–0.68, $P = 0.001$) (Hsieh, Look, Sieuwerts, Foekens, & Hunter, 2009).

Bromodomain containing 4 (BRD4), another interactor of SIPA1 (Farina et al., 2004), was identified through the same (AKXD × PyMT) F1 ECM eQTL mapping experiment as a hereditary modifier of breast cancer metastasis (Crawford, Alsarraj, et al., 2008). *Brd4* is located on mouse chromosome 17, in proximity to *Rrp1b*, and in the same region identified as being associated with the expression of metastasis-predictive ECM genes. Like RRP1B, dysregulation of *Brd4* expression in the highly metastatic Mvt-1 cell line has similar effects upon metastasis and ECM gene expression (Crawford, Alsarraj, et al., 2008; Crawford, Walker, et al., 2008). Orthotopic implantation of Mvt-1 cells stably overexpressing *Brd4* significantly reduced tumor burden and pulmonary metastases in syngeneic FVB/NJ mice compared to control cells. As was the case with *Rrp1b*, overexpression of *Brd4* generated a gene expression signature that predicts overall survival in breast cancer patients in multiple datasets, with its effects apparently limited to lymph node-negative and ER-positive breast cancer patients (Crawford, Alsarraj, et al., 2008).

Brd4 and *Rrp1b* were subsequently identified as central components in a transcriptional network, termed the "Diasporin Pathway," in which all core components are involved in regulating ECM gene expression, and tumor progression and metastasis (Crawford, Walker, et al., 2008). This pathway is composed of seven core ECM eQTL/metastasis modifier candidate genes identified through analysis of (AKXD × PyMT) F1 primary tumors. One of these seven genes is *Necdin* (*Ndn*), which is located on mouse chromosome 7 and is known to encode a transcription factor. Overexpression of *Ndn* in the Mvt-1 cell line was found to regulate the expression of metastasis-predictive ECM genes, as well as suppressing both tumor growth and metastasis (Crawford, Walker, et al., 2008). In humans, *NDN* is located in a region on chromosome 15 that normally undergoes maternal imprinting and is completely nonfunctional in Prader–Willi syndrome (Jay et al., 1997), a neurogenetic disorder. Overexpression of *Ndn* in Mvt-1 cells generates a gene expression signature that predicts survival in human breast cancer patients (Crawford, Walker, et al., 2008). Additionally, haplotype analysis of *NDN* in a cohort of 466 breast cancer patients derived from the Cancer Genome Atlas (TCGA) repository identified a haplotype 35 kb upstream of NDN (rs850815–rs850814 AA) that was more common in aggressive breast cancer subtypes, such as ER-negative cases (frequency; ER-positive, 0.459; ER-negative, 0.716; $P = 3.57 \times 10^{-4}$; FDR = 0.020). In addition, two 2-marker haplotypes in linkage disequilibrium with *NDN*, (rs11632341–rs824195 AT, rs1717831–rs4267267 CG), were associated with tumor progression and overall patient outcome in this cohort.

In mice, a nonsynonymous coding polymorphism in *Ndn* (50 T > C; rs261911330) was found to differentiate the metastasis permissive AKR/J strain from the metastasis suppressive DBA/2J strain (Lee, Beggs, et al., 2015), which are the two founder strains of the AKXD recombinant inbred panel. Microarray analysis using cells stably overexpressing either *Ndn* 50T or *Ndn* 50C identified a 71-gene expression signature that distinguished the two allelic variants. Stable overexpression of *Ndn* 50T, but not *Ndn* 50C, in the highly metastatic mouse mammary tumor cell lines Mvt-1 and 4T1 caused a significant decrease in pulmonary metastases compared to control when subcutaneously implanted into the mammary fat pad of NU/J or BALB/cJ female mice, respectively. Chromatin immunoprecipitation assays were used to define differences in chromatin binding activity of the two allelic variants and revealed that NDN interacts with regulatory regions of c-*Myc* to regulate its expression (Lee, Beggs, et al., 2015). This is of particular interest since MYC has been reported to promote tumorigenesis but suppress metastasis (Liu et al., 2012), in a similar manner as observed with subcutaneous implantation of cells overexpressing the *Ndn* 50T allelic variant.

In a subsequent study that utilized (AKXD × PyMT) F1 primary tumor microarray data to identify differentially expressed genes associated with metastasis, *Arid4b*, which resides on proximal chromosome 13, was implicated as a germline metastasis susceptibility gene (Winter, Lukes, Walker, Welch, & Hunter, 2012). Specifically, differential expression of *Arid4b* was found to be correlated with both primary tumor burden and pulmonary metastases. Similar to *Ndn*, *Arid4b* harbors polymorphisms that distinguish the metastasis permissive AKR/J strain from the metastasis suppressive DBA/2J strain, with the DBA/2J polymorphisms being associated with higher expression levels of *Arid4b*. Paradoxically, however, overexpression of the DBA allelic variant of *Arid4b* in the mouse mammary tumor cell line Met-1 increased in vitro tumor cell migration and invasion. Stable knockdown of *Arid4b* in the highly metastatic mouse mammary tumor cell line 6DT1 inhibited in vivo pulmonary metastases. Interestingly, for both *Arid4b* and *Ndn*, the allelic variant derived from the metastasis permissive AKR/J strain was associated with metastasis suppressive phenotypes, and the allelic variant from the poorly metastatic DBA/2J strain-promoted metastasis. This paradox is likely due to the fact that other metastasis-promoting factors present in the genetic background of AKR/J, with a stronger overall effect, mask the metastasis-suppressive qualities of *Arid4b* and *Ndn*. Conversely, DBA/2J will harbor additional metastasis-suppressive alleles, the sum of which out-weigh the metastasis-promoting alleles such as those described for *Ndn* and *Arid4b*. This evidence further strengthens the role that the intrinsic genetic makeup of the host plays in tumor progression and metastasis, and consequently, survival.

In a further genetic screen performed by breeding PyMT transgenic mice to either NZB/B1NJ or C58/J, with both of the latter strains being classed as metastasis suppressive, a region of chromosome 9 was identified as a metastasis susceptibility locus (Faraji et al., 2012; Hunter et al., 2001). Candidate genes were nominated by transcriptomic analysis of polymorphic mouse tumors and identified *Cell Adhesion Molecule 1* (*Cadm1*) as a candidate germline metastasis susceptibility gene by virtue of its differential expression in tumors derived from metastasis permissive and suppressive strains (Faraji et al., 2012). *Cadm1* has been reported to function as a tumor suppressor gene (Kuramochi et al., 2001), with downregulation of this gene being associated with poor tumor prognosis in lung, ovarian, and skin cancers (Kikuchi et al., 2006; Yang et al., 2011; You et al., 2010). Overexpression of *Cadm1* in Mvt-1 and 6DT1 cell lines, followed by either subcutaneous implantation into the mammary fat pad of syngeneic mice or intravenous injection into the tail vein caused a significant decrease in pulmonary metastases, indicating that *Cadm1* most likely affects metastasis downstream of local invasion and intravasation. Given that CADM1 interacts with CRTAM (class-I restricted T cell adhesion molecule), a receptor present on the surface of activated $CD8^+$ cells (Galibert et al., 2005), orthotopic implantation of Mvt-1 cells overexpressing *Cadm1* into the mammary fat pad of mice treated with either control rat IgG or rat monoclonal anti-CD8 IgG was performed. In the mice depleted of $CD8^+$ lymphocytes, overexpression of *Cadm1* failed to reduce pulmonary metastases compared to control, demonstrating that the metastasis suppressive properties of *Cadm1* are dependent upon T cell-mediated immunity (Faraji et al., 2012). Finally, to confirm the relevance of these observations to human breast cancer, it was demonstrated that high levels of *CADM1* expression correlated with better survival in ER-positive breast cancer patients (Faraji et al., 2012).

Gene expression analysis of (AKXD × PyMT) F1 primary tumors focusing on defining networks of coexpressed genes identified *Cnot2* (*CCR4-NOT Transcription Complex, Subunit 2*) as a key regulator of a transcriptomic network correlated to distant metastasis-free survival (DMFS) (Faraji et al., 2014). The gene expression signature associated with *Cnot2* distinguished high-risk DMFS breast cancer patients from low-risk DMFS patients, and *CNOT2* expression levels were found to be lower in invasive breast carcinoma compared to normal breast tissue samples. Stable shRNA-mediated knockdown of *Cnot2* in Mvt-1 and 6DT1 cells increased pulmonary metastases when orthotopically implanted into syngeneic FVB/NJ female mice. Interestingly, another gene that encodes for a component of the CCR4-NOT RNA deadenylase complex, *Cnot7*, was also found to be associated with metastasis, which highlights the importance of this complex in the modulation of metastasis at the hereditary level (Faraji et al., 2016). Orthotopic implantation of Mvt-1 or 6DT1 cells stably overexpressing wild-type CNOT7, but not of cells stably overexpressing deadenylase-inactive CNOT7, caused an increase in pulmonary metastases.

As reviewed earlier, regulators of the transcriptome have proven to play a significant role in modulating metastasis. Another important regulator of metastasis-related gene expression are microRNAs (miRNAs), which are noncoding transcripts that act to induce translational repression and promote mRNA deadenylation. As is the case with protein coding transcripts, miRNAs have been reported to harbor allelic variations that modulate metastasis susceptibility (Goldberger, Walker, Kim, Winter, & Hunter, 2013). In particular, the gene expression signature induced by the predicted targets of mir216/217 has been shown to be associated with metastasis susceptibility in ER-negative tumors. Furthermore, 6DT1 and 4T1 cells stably overexpressing *Mir216a*, *Mir216b*, *Mir217*, which are located in the *Mir216/217* cluster, suppressed pulmonary metastases when orthotopically implanted into an ER-negative mouse mammary tumor model (Faraji et al., 2014), which confirms that these miRNAs act as suppressors of tumor progression and metastasis.

The means of identification of these hereditary metastasis modifiers, in which females of other inbred strains were crossed with male transgenic mice, did not account for one important source of maternal DNA: the mitochondria. Previous studies have implicated mitochondrial DNA (mtDNA) mutations in cancer development and metastasis (Ishikawa et al., 2008; Petros et al., 2005). To investigate whether mtDNA polymorphism influences tumor progression in breast cancer, nuclear transfer was used to generate mice that have FVB/NJ nuclei and cytoplasm—and thus mitochondria—of either C57BL/6J (FVB^{n}:$C57^{mt}$) or BALB/cJ (FVB^{n}:$BALBc^{mt}$) (Fetterman et al., 2013), and crossed with hemizygous FVB/NJ PyMT male mice (Feeley et al., 2015). The rationale for using these strains is that it had previously been observed that C57BL/6J and BALB/cJ produce F1 progeny that were either low- or high-metastatic capacity, respectively (Lifsted et al., 1998). The mitochondrial genome alone, without any alteration of the nuclear genome, was sufficient to influence both primary tumor latency and pulmonary metastasis, mirroring the results reported in the previous study by Lifsted and colleagues (Lifsted et al., 1998). Early formation of tumors in PyMT-positive F1 females was suppressed by the transmission of C57BL/6J mtDNA but stimulated by BALB/cJ mtDNA compared to FVB/NJ control mice, which have FVB nuclei and cytoplasm (Feeley et al., 2015). Additionally, although the median number of metastases was no different between all three groups, inheritance of C57BL/6J mtDNA suppressed the overall average size of the metastatic lesions whereas BALB/cJ mtDNA caused an increase compared to that of the FVB/NJ control mice. Since mitochondria play a key role in cell metabolism and consequently tumor growth (Bell, Emerling, Ricoult, & Guarente, 2011; Petros et al., 2005; van Ginkel et al., 2007), mammary tumor epithelial cells from each PyMT-positive F1 female were isolated to measure metabolic capacities including oxygen consumption rate (OCR) and average baseline extracellular acidification (ECAR) (Feeley et al., 2015). Baseline OCR and ECAR levels were significantly elevated in the FVB^{n}:$C57^{mt}$ cells but not the FVB^{n}:$BALBc^{mt}$ cells compared to that of the control. Reserve capacity, which is equal to the difference between the maximum OCR and baseline OCR and reflects the ability of cells to respond to stress, (eg, metastasis) (Dranka, Hill, & Darley-Usmar, 2010; Liu et al., 2014), was the lowest in the FVB^n:$C57^{mt}$ cells, which again suggests a linear relationship between cellular metabolism and metastatic capacity as previously reported (Liu et al., 2014). The lack of difference in tumor burden and overall number of pulmonary metastases suggests that these traits are mediated by hereditary variation in the nuclear DNA. Yet, the significant differences in tumor latency, size of pulmonary metastases, and metabolism demonstrate the importance of host mitochondrial genetics upon these processes (Feeley et al., 2015). Given that metastasis is a highly energy-intensive process that results in an extreme amount of cellular stress, it is not surprising that mitochondrial genetics plays a prominent role in modulating metastasis.

3.2 Prostate Cancer

Prostate cancer is the most common malignancy in men and is only surpassed by lung cancer in terms of cancer mortality in men in the United States (Siegel et al., 2016). While most men develop an indolent disease and succumb to unrelated causes, the vast majority of fatalities directly attributable to prostate cancer are due to the clinical effects of metastasis. Assays used to assess prognosis at the time of clinical presentation, such as PSA screening, are quite inaccurate, which leads to significant challenges in early identification of high-risk patients of metastatic prostate cancer. As a result, prostate cancer is frequently overtreated, which presents both a significant public health issue and a prominent cause of patient morbidity. The mainstay of treatment for locally advanced and recurrent prostate cancer is the use of androgen deprivation therapy (ADT), which exploits the androgen dependency of prostate tumor cells by inhibiting androgen receptor signaling and consequently suppressing cell growth (Karantanos, Corn, & Thompson, 2013; Patel et al., 2014; Terry & Beltran, 2014). However, continued use of ADT invariably leads to recurrent or metastatic tumors, which arise from preexisting prostate adenocarcinomas. Importantly, approximately 25% of the recurrent and ultimately fatal prostate tumors histologically display neuroendocrine (NE) characteristics and are consequently termed "treatment-associated" NE prostate cancer (Epstein et al., 2014; Wang et al., 2014). This is of significance since de novo NE prostate cancer is rare and comprises 0.3–1.0% of all prostate malignancies at the time of diagnosis (Humphrey, 2012). However, unlike the far more commonly diagnosed prostate adenocarcinoma, which constitutes 90–95% of all cases at the time of diagnosis, NE prostate cancer patients typically present visceral as opposed to bony metastases and have a significantly poorer overall survival (median survival = 10.0 months vs 125.0 months for adenocarcinoma) (Marcus, Goodman, Jani, Osunkoya, & Rossi, 2012). Like de novo NE prostate cancer, treatment-associated NE prostate cancer is associated with visceral metastasis and a poor prognosis (Priftakis, Kritikos, Stavrinides, Kleanthous, & Baziotis, 2015). The precise mechanisms of how a proportion of fatal forms of adenocarcinoma gain these NE characteristics are, for the most part, unclear.

The influence of germline polymorphisms in prostate cancer has been examined using the C57BL/6-Tg(TRAMP)8247Ng/J (TRAMP) mouse model (Gingrich, Barrios, Foster, & Greenberg, 1999; Gingrich et al., 1997). These mice express the SV40 virus T-antigen specifically in the prostate epithelium and develop invasive NE prostate tumors (Chiaverotti et al., 2008) that metastasize to local lymph nodes and distant, and primarily visceral, organs by 28 weeks of age (Berman-Booty & Knudsen, 2015; Gingrich et al., 1999; Greenberg et al., 1995). Germline polymorphisms were introduced by breeding these mice to the eight progenitor strains of the Collaborative Cross (CC) recombinant inbred panel; C57BL/6J, A/J, 129S1/SvImJ, NOD/ShiLtJ, NZO/HlLtJ, CAST/EiJ, PWK/PhJ, and WSB/EiJ. The CC is a panel of recombinant inbred strains derived from eight founders, which will eventually be composed of hundreds of strains (Threadgill & Churchill, 2012). The tremendous amount of allelic variation represented across CC strains will ultimately represent over 90% of genetic diversity in mouse species, and will make it a powerful tool for mapping genes associated with complex traits. In the F1 progeny from the crosses between the TRAMP mouse and CC progenitor strains, tumor burden and local metastasis to lymph nodes as well as distant metastasis were quantified. The introduction of genetic variation by breeding was found to modulate tumor-associated mortality, tumor burden, and both local and distant metastasis in the various TRAMP F1 mice, even though the temporal expression and levels of the SV40 virus T-antigen did not vary between different F1 strains (Patel, Molinolo, Gutkind, & Crawford, 2013). The most aggressive form of disease was seen in (TRAMP × NOD/ShiLtJ) F1 males, and of the 42F1 mice examined, a six-fold increase in average primary tumor burden was observed compared to wild-type TRAMP mice. Additionally, 75% developed pulmonary metastases and 95% developed local lymph node metastases, which represented a significant increase compared to wild-type

TRAMP mice. In contrast, the least aggressive disease was seen in (TRAMP × PWK/PhJ) F1 males where only 1 of the 37 mice developed macroscopic tumors with no evidence of either local or distant metastasis at the experimental endpoint. This work, which facilitated several follow-up studies that led to the identification of multiple novel prostate cancer metastasis susceptibility genes, demonstrates the influence of germline variation on both tumor progression and metastasis in prostate cancer.

Based on these striking results, NOD/ShiLtJ and PWK/PhJ strains were selected for further investigation. QTL mapping identified 11 loci in transgene-positive (TRAMP × NOD/ShiLtJ) F2 male mice associated with tumor progression and metastasis; DMFS; chromosomes 1 and 11, lymph node metastasis burden; chromosome 13, liver surface metastasis count; chromosome 11, prostate tumor burden; chromosome 13, seminal vesicle tumor burden; chromosomes 2, 4, 8, and 17, and age of death; and chromosomes 7 and 8 (Williams et al., 2014). To nominate metastasis susceptibility candidate genes within these loci, transcriptomic analysis of (TRAMP × NOD/ShiLtJ) F2 tumors was performed using microarrays. High confidence aggressive prostate cancer susceptibility genes were identified through trait correlation and eQTL analyses. Through these approaches, a total of 35 candidate genes were identified, of which 29 had identifiable human orthologs. It was subsequently hypothesized that if any of these human orthologs were a true aggressive disease modifier then they should display similar characteristics as their murine counterparts (ie, both having an expression level associated with aggressive prostate cancer and being in linkage disequilibrium with aggressive disease-associated germline SNPs). To test this hypothesis, the human orthologs of these genes were further examined by integrating expression and survival data from publicly available human prostate cancer patient cohorts and the Cancer Genetic Markers of Susceptibility (CGEMS) GWAS of patients from the Prostate, Lung, Colorectal, and Ovarian (PLCO) cancer screening trial (Gohagan et al., 2000; O'Brien et al., 2000). Three genes, *CXCL14*, *ITGAX*, and *LPCAT2*, were found to have expression levels associated with a poorer disease-free survival in TCGA ($P=0.025$) and Prostate Oncogenome Project (GSE21032) ($P=0.010$) (Taylor et al., 2010) and also harbor SNPs associated with aggressive prostate cancer traits. Particularly, SNPs in *CXCL14* and *LPCAT2* were found to be associated with metastasis; rs801564 (OR = 1.05, 95% CI 1.01–1.09, permutation $P=0.011$) and rs2289119 (OR = 1.06, 95% CI 1.01–1.10, permutation $P=0.009$), respectively. Additionally, all three aggressive disease susceptibility genes harbored SNPs associated with Gleason score; *CXCL14*, rs10515473 (OR = 0.72, 95% CI 0.59–0.88, permutation *P* = 0.001); *ITGAX*, rs8047538 (OR = 1.33, 95% CI 1.08–1.62, permutation *P* = 0.007); and *LPCAT2*, rs289707 (OR = 1.22, 95% CI 1.07–1.38, permutation *P* = 0.002) and rs17369578 (OR = 1.41, 95% CI 1.13–1.77, permutation *P* = 0.003).

Similarly, in transgene-positive (TRAMP × PWK/PhJ) F2 male mice, two loci were identified (lymph node metastasis burden; chromosome 12 and DMFS; chromosome 14) as being associated with prostate cancer metastasis (Lee, Williams, et al., 2015). For both of these loci, a total of 35 candidate genes with an identifiable human ortholog were nominated based on having an expression level correlated with metastasis. Five of these genes were found to have expression levels associated with poor disease-free survival in TCGA patient cohort (*P* = 0.032). Four of these genes, *GNL3*, *MATA1A*, *SKA3*, and *ZMYM5*, were in linkage disequilibrium with haplotypes or SNPs associated with markers of aggressive prostate cancer development in the CGEMS GWAS analysis. In addition, functional assays using human prostate cancer PC-3 cell lines stably overexpressing *GNL3* and *SKA3* demonstrated that overexpression of *GNL3* and *SKA3* in caused a significant decrease in in vitro tumor cell migration and invasiveness compared to control cell lines (Lee, Williams, et al., 2015).

In conclusion, these studies, which have sought to integrate multiple lines of evidence from a mouse model of prostate cancer that has been bred to be genetically diverse and various forms of publicly available human prostate cancer datasets, have identified *ITGAX*, *GNL3*, *SKA3*, *MAT1A*, and *ZMYM5* as metastasis susceptibility candidate genes in prostate cancer (Table 2).

3.3 Melanoma

Melanoma, the most aggressive type of skin cancer, is a tumor originating from melanocytes, which are pigment-producing cells located in the skin, hair follicles, and the eye. Most often, they arise from the melanocytes in the basal layer of the epidermis, with dermal invasion being associated with tumor progression. In 2016 in the United States, it is estimated that melanoma will account for over 76,000 new cases of skin cancer (excluding squamous cell and basal cell skin cancers), and that over 10,000 individuals will succumb to this disease (Siegel et al., 2016). The median survival for metastatic melanoma is particularly poor, with patients surviving approximately 8–18 months following the diagnosis of disseminated disease.

Table 2 Hereditary Modifiers of Metastasis in Prostate Cancer

Gene	Description	Mouse Chromosome	Method of Identification	QTL Analysis	Human Ortholog	Human Location	Cellular Function of Protein	Associated Aggressive Disease Traits		OMIM
								Logistic regression in human gene expression datasets	GWAS associations	
Cxcl14	Chemokine (C–X–C motif) ligand 14	13	(TRAMP × NOD) F2 QTL analysis	Primary tumor burden	*CXCL14*	5q31.1	Homeostasis of monocyte-derived macrophages	Disease-free survival; pathological stage	Nodal metastasis; Gleason score	604186
Itgax	Integrin, alpha × (complement component three receptors four subunits)	7	(TRAMP × NOD) F2 QTL analysis	Age of death	*ITGAX*	16p11.2	Cell–cell adhesion	Gleason score	Gleason score	151510
Lpcat2	Lysophosphatidylcholine acyltransferase 2	8	(TRAMP × NOD) F2 QTL analysis	Seminal vesicle tumor burden	*LPCAT2*	16q12.2	Membrane biogenesis; production of platelet-activating factor in inflammatory cells	Pathological stage	Nodal metastasis; Gleason score	612040
Gnl3	Guanine nucleotide-binding protein-like 3	14	(TRAMP × PWK) F2 QTL analysis	Distant metastasis-free survival	*GNL3*	3p21.1	Cell cycle regulation	Gleason score	Gleason score	608011
Mat1a	Methionine adenosyltransferase 1 alpha	14	(TRAMP × PWK) F2 QTL analysis	Distant metastasis-free survival	*MAT1A*	10q22.3	S-adenosylmethionine formation	Gleason score	Distant metastasis	610550

Continued

Table 2 Hereditary Modifiers of Metastasis in Prostate Cancer—cont'd

Gene	Description	Mouse Chromosome	Method of Identification	QTL Analysis	Human Ortholog	Human Location	Cellular Function of Protein	Associated Aggressive Disease Traits		OMIM
								Logistic regression in human gene expression datasets	GWAS associations	
Ska3	Spindle and kinetochore-associated complex subunit 3	14	(TRAMP × PWK) F2 QTL analysis	Distant metastasis-free survival	*SKA3*	13q12.11	Kinetochore-microtubule attachment during mitosis	Disease-free survival; pathological stage	Clnical stage; Gleason score	None
Zmym5	Zinc finger, MYM-type 5	14	(TRAMP × PWK) F2 QTL analysis	Distant metastasis-free survival	*ZMYM5*	13q12	Transcriptional regulation	Disease-free survival; pathological stage	Gleason score	616443

The influence of host genetic variation upon tumor progression in malignant melanoma was explored in a recent study that utilized the CC recombinant inbred strains for the first time in cancer research (Ferguson et al., 2015). To identify hereditary modifiers of melanoma, more than 50 CC strains were bred to the Cdk^{R24C}::*Tyr-NRAS*^{Q61K} (*Cdk4::NRAS*) transgenic mouse model of malignant melanoma. These mice, which express two mutated oncogenes implicated in human melanoma pathogenesis, *Cdk4* and *NRAS*, on an FVB/NJ background, develop numerous spontaneous, benign nevi that occasionally progress to malignant melanoma (Ferguson et al., 2015). CC strains were bred to these mice and traits associated with aggressive tumorigenesis were observed in both the male and female F1 progeny that was heterozygous for Cdk^{R24C} and harbored $NRAS^{Q61K}$. Genetic variation was observed to affect many aspects of tumor progression, such as age of onset of nevoid plaques, number of nevoid plaques at death, melanoma age of onset, number of melanomas at death, and average time for transformation of nevoid plaques to melanoma. Depending on the genetic background, the age of onset of nevoid plaques ranged from 110 and 355 days and the age of onset of melanoma ranged from 145 to 426 days. Interestingly, morphological differences were not observed with genetic variation. Although distant metastases rarely develop in FVB/NJ *Cdk4::NRAS* transgenic mice, significant differences in local lymph node metastasis were observed in the F1 progeny of several CC parental strains. Surprisingly, comparisons between the different phenotypic traits did not reveal any significant correlation to onset of malignant melanoma, which confirms the complexity of tumor progression, and the various factors involved in the process.

4. CONCLUSIONS

Metastasis is an intricate process that requires the orchestration of many different cellular processes and signaling pathways at various steps. Due to this complex nature, it has proven to be challenging to identify factors responsible, and the molecular determinants of metastasis remain, for the most part, unclear. Somatic mutations are known to initiate both primary tumor formation and metastasis (Fidler & Kripke, 1977; Greenman et al., 2007; Knudson, 2001; Pleasance et al., 2010), and while some population-based studies still argue that hereditary factors do not influence cancer survival (Verkooijen et al., 2012; Whittemore et al., 2009), the evidence reviewed here clearly demonstrates the influence of genetic variation on tumor progression and survival, and how the propensity for metastasis is modified by genetic inheritance. Human population-based studies have proved somewhat successful in defining factors associated with metastasis and other forms of aggressive tumorigenesis. However, their implementation is challenging and is surrounded with complicating issues such as the necessarily high statistical threshold for defining statistical significance due to problems surrounding compounding of type I error, population substructure, and controlling for environmental or socioeconomic factors. Therefore, we have attempted to demonstrate how systems genetics approaches in transgenic mouse models of human diseases can be successfully employed in a manner complementary to human studies, to overcome some of these difficulties. Complex genetic screens using mouse models of breast, prostate, and skin cancers have demonstrated that hereditary factors strongly influence tumor progression and metastasis (Ferguson et al., 2015; Lifsted et al., 1998; Patel et al., 2013). These studies have led to the identification of multiple loci and genes that encode factors involved in a myriad of cellular functions, ranging from transcriptional regulation, cell division, and cell migration (Faraji et al., 2012; Lee, Williams, et al., 2015). Based on the evidence presented in this review, we postulate that hereditary factors most likely modulate the different steps of metastasis by influencing a network of coregulated genes that, to some extent, interact with each other, as opposed to a single gene or loci being primarily responsible for the observed effects. These dynamic and fluid interactions add to the complexity of tumor progression and metastasis. Yet, germline factors that have an influence across different types of cancer, including risk factors, such as *BRCA1* (Ford, Easton, Bishop, Narod, & Goldgar, 1994), as well as metastasis susceptibility genes that have been identified for multiple cancer types, such as *SIPA1* that is associated with metastatic progression in both breast and prostate cancer patients (Crawford et al., 2006; Park et al., 2005), demonstrate that common regulatory factors do exist throughout the various cancer types, providing some consistency. Considering that risk loci identified through GWAS only account for a fraction of genetic variance in complex diseases, systems genetics will likely prove a powerful tool that will facilitate the identification of hereditary factors associated with the most deadly sequelae of cancer.

ACKNOWLEDGMENTS

We thank Drs. Kent Hunter and Jean Winter for the careful review of this manuscript. This research was supported [in part] by the Intramural Research Program of the National Human Genome Research Institute, National Institutes of Health.

REFERENCES

Al Olama, A. A., Kote-Jarai, Z., Berndt, S. I., Conti, D. V., Schumacher, F., Han, Y., et al. (2014). A meta-analysis of 87,040 individuals identifies 23 new susceptibility loci for prostate cancer. *Nature Genetics*, *46*(10), 1103–1109. http://dx.doi.org/10.1038/ng.3094.

Altschuler, D. L., & Ribeiro-Neto, F. (1998). Mitogenic and oncogenic properties of the small G protein Rap1b. *Proceedings of the National Academy of Sciences of the United States of America*, *95*(13), 7475–7479.

Amin Al Olama, A., Benlloch, S., Antoniou, A. C., Giles, G. G., Severi, G., Neal, D. E., et al. (2015). Risk analysis of prostate cancer in PRACTICAL, a multinational consortium, using 25 known prostate cancer susceptibility loci. *Cancer Epidemiology, Biomarkers & Prevention: A Publication of the American Association for Cancer Research, Cosponsored by the American Society of Preventive Oncology*, *24*(7), 1121–1129. http://dx.doi.org/10.1158/1055-9965.EPI-14-0317.

Amin Al Olama, A., Kote-Jarai, Z., Schumacher, F. R., Wiklund, F., Berndt, S. I., Benlloch, S., et al. (2013). A meta-analysis of genome-wide association studies to identify prostate cancer susceptibility loci associated with aggressive and non-aggressive disease. *Human Molecular Genetics*, *22*(2), 408–415. http://dx.doi.org/10.1093/hmg/dds425.

Amos, C. I., & Struewing, J. P. (1993). Genetic epidemiology of epithelial ovarian cancer. *Cancer*, *71*(2 Suppl.), 566–572.

Bell, E. L., Emerling, B. M., Ricoult, S. J., & Guarente, L. (2011). SirT3 suppresses hypoxia inducible factor 1alpha and tumor growth by inhibiting mitochondrial ROS production. *Oncogene*, *30*(26), 2986–2996. http://dx.doi.org/10.1038/onc.2011.37.

Berman-Booty, L. D., & Knudsen, K. (2015). Models of neuroendocrine prostate cancer. *Endocrine-Related Cancer*, *22*, R33–R49. http://dx.doi.org/10.1530/ERC-14-0393.

Berndt, S. I., Wang, Z., Yeager, M., Alavanja, M. C., Albanes, D., Amundadottir, L., et al. (2015). Two susceptibility loci identified for prostate cancer aggressiveness. *Nature Communications*, *6*, 6889. http://dx.doi.org/10.1038/ncomms7889.

Bjorkegren, J. L., Kovacic, J. C., Dudley, J. T., & Schadt, E. E. (2015). Genome-wide significant loci: How important are they? Systems genetics to understand heritability of coronary artery disease and other common complex disorders. *Journal of the American College of Cardiology*, *65*(8), 830–845. http://dx.doi.org/10.1016/j.jacc.2014.12.033.

Boehnke, M. (1994). Limits of resolution of genetic linkage studies: Implications for the positional cloning of human disease genes. *American Journal of Human Genetics*, *55*(2), 379–390.

Cardoso, F., Harbeck, N., Fallowfield, L., Kyriakides, S., Senkus, E., & Group, E. G. W. (2012). Locally recurrent or metastatic breast cancer: ESMO Clinical Practice Guidelines for diagnosis, treatment and follow-up. *Annals of Oncology: Official Journal of the European Society for Medical Oncology/ESMO*, *23*(Suppl. 7), vii11–vii19. http://dx.doi.org/10.1093/annonc/mds232.

Chang, B. L., Isaacs, S. D., Wiley, K. E., Gillanders, E. M., Zheng, S. L., Meyers, D. A., et al. (2005). Genome-wide screen for prostate cancer susceptibility genes in men with clinically significant disease. *The Prostate*, *64*(4), 356–361. http://dx.doi.org/10.1002/pros.20249.

Chiaverotti, T., Couto, S. S., Donjacour, A., Mao, J. H., Nagase, H., Cardiff, R. D., et al. (2008). Dissociation of epithelial and neuroendocrine carcinoma lineages in the transgenic adenocarcinoma of mouse prostate model of prostate cancer. *The American Journal of Pathology*, *172*(1), 236–246. http://dx.doi.org/10.2353/ajpath.2008.070602.

Chung, C. T., & Carlson, R. W. (2003). Goals and objectives in the management of metastatic breast cancer. *The Oncologist*, *8*(6), 514–520.

Civelek, M., & Lusis, A. J. (2014). Systems genetics approaches to understand complex traits. *Nature Reviews Genetics*, *15*(1), 34–48. http://dx.doi.org/10.1038/nrg3575.

Crawford, N. P., Alsarraj, J., Lukes, L., Walker, R. C., Officewala, J. S., Yang, H. H., et al. (2008). Bromodomain 4 activation predicts breast cancer survival. *Proceedings of the*

National Academy of Sciences of the United States of America, 105(17), 6380–6385. http://dx.doi.org/10.1073/pnas.0710331105.

Crawford, N. P., Qian, X., Ziogas, A., Papageorge, A. G., Boersma, B. J., Walker, R. C., et al. (2007). Rrp1b, a new candidate susceptibility gene for breast cancer progression and metastasis. *PLoS Genetics, 3*(11), e214. http://dx.doi.org/10.1371/journal.pgen.0030214.

Crawford, N. P., Walker, R. C., Lukes, L., Officewala, J. S., Williams, R. W., & Hunter, K. W. (2008). The Diasporin pathway: A tumor progression-related transcriptional network that predicts breast cancer survival. *Clinical & Experimental Metastasis, 25*(4), 357–369. http://dx.doi.org/10.1007/s10585-008-9146-6.

Crawford, N. P., Yang, H., Mattaini, K. R., & Hunter, K. W. (2009). The metastasis efficiency modifier ribosomal RNA processing 1 homolog B (RRP1B) is a chromatin-associated factor. *The Journal of Biological Chemistry, 284*(42), 28660–28673. http://dx.doi.org/10.1074/jbc.M109.023457.

Crawford, N. P., Ziogas, A., Peel, D. J., Hess, J., Anton-Culver, H., & Hunter, K. W. (2006). Germline polymorphisms in SIPA1 are associated with metastasis and other indicators of poor prognosis in breast cancer. *Breast Cancer Research, 8*(2), R16. http://dx.doi.org/10.1186/bcr1389.

Dranka, B. P., Hill, B. G., & Darley-Usmar, V. M. (2010). Mitochondrial reserve capacity in endothelial cells: The impact of nitric oxide and reactive oxygen species. *Free Radical Biology & Medicine, 48*(7), 905–914. http://dx.doi.org/10.1016/j.freeradbiomed.2010.01.015.

Eeles, R., Goh, C., Castro, E., Bancroft, E., Guy, M., Al Olama, A. A., et al. (2014). The genetic epidemiology of prostate cancer and its clinical implications. *Nature Reviews. Urology, 11*(1), 18–31. http://dx.doi.org/10.1038/nrurol.2013.266.

Epstein, J. I., Amin, M. B., Beltran, H., Lotan, T. L., Mosquera, J. M., Reuter, V. E., et al. (2014). Proposed morphologic classification of prostate cancer with neuroendocrine differentiation. *The American Journal of Surgical Pathology, 38*(6), 756–767. http://dx.doi.org/10.1097/PAS.0000000000000208.

Faraji, F., Hu, Y., Wu, G., Goldberger, N. E., Walker, R. C., Zhang, J., et al. (2014). An integrated systems genetics screen reveals the transcriptional structure of inherited predisposition to metastatic disease. *Genome Research, 24*(2), 227–240. http://dx.doi.org/10.1101/gr.166223.113.

Faraji, F., Hu, Y., Yang, H. H., Lee, M. P., Winkler, G. S., Hafner, M., et al. (2016). Post-transcriptional control of tumor cell autonomous metastatic potential by CCR4-NOT deadenylase CNOT7. *PLoS Genetics, 12*(1), e1005820. http://dx.doi.org/10.1371/journal.pgen.1005820.

Faraji, F., Pang, Y., Walker, R. C., Nieves Borges, R., Yang, L., & Hunter, K. W. (2012). Cadm1 is a metastasis susceptibility gene that suppresses metastasis by modifying tumor interaction with the cell-mediated immunity. *PLoS Genetics, 8*(9), e1002926. http://dx.doi.org/10.1371/journal.pgen.1002926.

Farber, C. R. (2013). Systems-level analysis of genome-wide association data. *G3, 3*(1), 119–129. http://dx.doi.org/10.1534/g3.112.004788.

Farina, A., Hattori, M., Qin, J., Nakatani, Y., Minato, N., & Ozato, K. (2004). Bromodomain protein Brd4 binds to GTPase-activating SPA-1, modulating its activity and subcellular localization. *Molecular and Cellular Biology, 24*(20), 9059–9069. http://dx.doi.org/10.1128/MCB.24.20.9059-9069.2004.

Feeley, K. P., Bray, A. W., Westbrook, D. G., Johnson, L. W., Kesterson, R. A., Ballinger, S. W., et al. (2015). Mitochondrial genetics regulate breast cancer tumorigenicity and metastatic potential. *Cancer Research, 75*(20), 4429–4436. http://dx.doi.org/10.1158/0008-5472.CAN-15-0074.

Ferguson, B., Ram, R., Handoko, H. Y., Mukhopadhyay, P., Muller, H. K., Soyer, H. P., et al. (2015). Melanoma susceptibility as a complex trait: Genetic variation controls all stages of tumor progression. *Oncogene*, *34*(22), 2879–2886. http://dx.doi.org/10.1038/onc.2014.227.

Ferlay, J., Soerjomataram, I., Dikshit, R., Eser, S., Mathers, C., Rebelo, M., et al. (2015). Cancer incidence and mortality worldwide: Sources, methods and major patterns in GLOBOCAN 2012. *International Journal of Cancer*, *136*(5), E359–E386. http://dx.doi.org/10.1002/ijc.29210.

Fetterman, J. L., Zelickson, B. R., Johnson, L. W., Moellering, D. R., Westbrook, D. G., Pompilius, M., et al. (2013). Mitochondrial genetic background modulates bioenergetics and susceptibility to acute cardiac volume overload. *The Biochemical Journal*, *455*(2), 157–167. http://dx.doi.org/10.1042/BJ20130029.

Fidler, I. J., & Kripke, M. L. (1977). Metastasis results from preexisting variant cells within a malignant tumor. *Science*, *197*(4306), 893–895.

Fidler, I. J., & Talmadge, J. E. (1986). Evidence that intravenously derived murine pulmonary melanoma metastases can originate from the expansion of a single tumor cell. *Cancer Research*, *46*(10), 5167–5171.

Ford, D., Easton, D. F., Bishop, D. T., Narod, S. A., & Goldgar, D. E. (1994). Risks of cancer in BRCA1-mutation carriers. Breast Cancer Linkage Consortium. *Lancet*, *343*(8899), 692–695.

Galibert, L., Diemer, G. S., Liu, Z., Johnson, R. S., Smith, J. L., Walzer, T., et al. (2005). Nectin-like protein 2 defines a subset of T-cell zone dendritic cells and is a ligand for class-I-restricted T-cell-associated molecule. *The Journal of Biological Chemistry*, *280*(23), 21955–21964. http://dx.doi.org/10.1074/jbc.M502095200.

Giavazzi, R., Alessandri, G., Spreafico, F., Garattini, S., & Mantovani, A. (1980). Metastasizing capacity of tumour cells from spontaneous metastases of transplanted murine tumours. *British Journal of Cancer*, *42*(3), 462–472.

Gingrich, J. R., Barrios, R. J., Foster, B. A., & Greenberg, N. M. (1999). Pathologic progression of autochthonous prostate cancer in the TRAMP model. *Prostate Cancer and Prostatic Diseases*, *2*(2), 70–75. http://dx.doi.org/10.1038/sj.pcan.4500296.

Gingrich, J. R., Barrios, R. J., Kattan, M. W., Nahm, H. S., Finegold, M. J., & Greenberg, N. M. (1997). Androgen-independent prostate cancer progression in the TRAMP model. *Cancer Research*, *57*(21), 4687–4691.

Gohagan, J. K., Prorok, P. C., Hayes, R. B., Kramer, B. S., & Prostate, Lung, Colorectal and Ovarian Cancer Screening Trial Project Team. (2000). The Prostate, Lung, Colorectal and Ovarian (PLCO) Cancer Screening Trial of the National Cancer Institute: History, organization, and status. *Controlled Clinical Trials*, *21*(6 Suppl.), 251S–272S.

Goldberger, N., Walker, R. C., Kim, C. H., Winter, S., & Hunter, K. W. (2013). Inherited variation in miR-290 expression suppresses breast cancer progression by targeting the metastasis susceptibility gene Arid4b. *Cancer Research*, *73*(8), 2671–2681. http://dx.doi.org/10.1158/0008-5472.CAN-12-3513.

Goldgar, D. E., Easton, D. F., Cannon-Albright, L. A., & Skolnick, M. H. (1994). Systematic population-based assessment of cancer risk in first-degree relatives of cancer probands. *Journal of the National Cancer Institute*, *86*(21), 1600–1608.

Greenberg, N. M., DeMayo, F., Finegold, M. J., Medina, D., Tilley, W. D., Aspinall, J. O., et al. (1995). Prostate cancer in a transgenic mouse. *Proceedings of the National Academy of Sciences of the United States of America*, *92*(8), 3439–3443.

Greenman, C., Stephens, P., Smith, R., Dalgliesh, G. L., Hunter, C., Bignell, G., et al. (2007). Patterns of somatic mutation in human cancer genomes. *Nature*, *446*(7132), 153–158. http://dx.doi.org/10.1038/nature05610.

Gupta, G. P., & Massague, J. (2006). Cancer metastasis: Building a framework. *Cell*, *127*(4), 679–695. http://dx.doi.org/10.1016/j.cell.2006.11.001.

Guy, C. T., Cardiff, R. D., & Muller, W. J. (1992). Induction of mammary tumors by expression of polyomavirus middle T oncogene: A transgenic mouse model for metastatic disease. *Molecular and Cellular Biology, 12*(3), 954–961.

Haffner, M. C., Mosbruger, T., Esopi, D. M., Fedor, H., Heaphy, C. M., Walker, D. A., et al. (2013). Tracking the clonal origin of lethal prostate cancer. *The Journal of Clinical Investigation, 123*(11), 4918–4922. http://dx.doi.org/10.1172/JCI70354.

Hazelett, D. J., Rhie, S. K., Gaddis, M., Yan, C., Lakeland, D. L., Coetzee, S. G., et al. (2014). Comprehensive functional annotation of 77 prostate cancer risk loci. *PLoS Genetics, 10*(1), e1004102. http://dx.doi.org/10.1371/journal.pgen.1004102.

Hemminki, K., Ji, J., Forsti, A., Sundquist, J., & Lenner, P. (2008a). Concordance of survival in family members with prostate cancer. *Journal of Clinical Oncology, 26*(10), 1705–1709. http://dx.doi.org/10.1200/JCO.2007.13.3355.

Hemminki, K., Ji, J., Forsti, A., Sundquist, J., & Lenner, P. (2008b). Survival in breast cancer is familial. *Breast Cancer Research and Treatment, 110*(1), 177–182. http://dx.doi.org/10.1007/s10549-007-9692-7.

Hemminki, K., Li, X., & Czene, K. (2004). Familial risk of cancer: Data for clinical counseling and cancer genetics. *International Journal of Cancer, 108*(1), 109–114. http://dx.doi.org/10.1002/ijc.11478.

Hemminki, K., & Vaittinen, P. (1998). Familial breast cancer in the family-cancer database. *International Journal of Cancer, 77*(3), 386–391.

Hogan, C., Serpente, N., Cogram, P., Hosking, C. R., Bialucha, C. U., Feller, S. M., et al. (2004). Rap1 regulates the formation of E-cadherin-based cell–cell contacts. *Molecular and Cellular Biology, 24*(15), 6690–6700. http://dx.doi.org/10.1128/MCB.24.15.6690-6700.2004.

Howlader, N., Chen, V. W., Ries, L. A., Loch, M. M., Lee, R., DeSantis, C., et al. (2014). Overview of breast cancer collaborative stage data items—Their definitions, quality, usage, and clinical implications: A review of SEER data for 2004–2010. *Cancer, 120*(Suppl. 23), 3771–3780. http://dx.doi.org/10.1002/cncr.29059.

Hsieh, S. M., Look, M. P., Sieuwerts, A. M., Foekens, J. A., & Hunter, K. W. (2009). Distinct inherited metastasis susceptibility exists for different breast cancer subtypes: A prognosis study. *Breast Cancer Research, 11*(5), R75. http://dx.doi.org/10.1186/bcr2412.

Huang, Y., Gao, S., Wu, S., Song, P., Sun, X., Hu, X., et al. (2014). Multilayered molecular profiling supported the monoclonal origin of metastatic renal cell carcinoma. *International Journal of Cancer, 135*(1), 78–87. http://dx.doi.org/10.1002/ijc.28654.

Humphrey, P. A. (2012). Histological variants of prostatic carcinoma and their significance. *Histopathology, 60*(1), 59–74. http://dx.doi.org/10.1111/j.1365-2559.2011.04039.x.

Hunter, K. (2015). The role of individual inheritance in tumor progression and metastasis. *Journal of Molecular Medicine, 93*(7), 719–725. http://dx.doi.org/10.1007/s00109-015-1299-6.

Hunter, K. W., Broman, K. W., Voyer, T. L., Lukes, L., Cozma, D., Debies, M. T., et al. (2001). Predisposition to efficient mammary tumor metastatic progression is linked to the breast cancer metastasis suppressor gene Brms1. *Cancer Research, 61*(24), 8866–8872.

Ishikawa, K., Takenaga, K., Akimoto, M., Koshikawa, N., Yamaguchi, A., Imanishi, H., et al. (2008). ROS-generating mitochondrial DNA mutations can regulate tumor cell metastasis. *Science, 320*(5876), 661–664. http://dx.doi.org/10.1126/science.1156906.

Janer, M., Friedrichsen, D. M., Stanford, J. L., Badzioch, M. D., Kolb, S., Deutsch, K., et al. (2003). Genomic scan of 254 hereditary prostate cancer families. *The Prostate, 57*(4), 309–319. http://dx.doi.org/10.1002/pros.10305.

Jay, P., Rougeulle, C., Massacrier, A., Moncla, A., Mattei, M. G., Malzac, P., et al. (1997). The human necdin gene, NDN, is maternally imprinted and located in the Prader–Willi syndrome chromosomal region. *Nature Genetics*, *17*(3), 357–361. http://dx.doi.org/10.1038/ng1197-357.

Ji, J., Forsti, A., Sundquist, J., Lenner, P., & Hemminki, K. (2008a). Survival in bladder and renal cell cancers is familial. *Journal of the American Society of Nephrology*, *19*, 985–991. ASN.2007070818 [pii].

Ji, J., Forsti, A., Sundquist, J., Lenner, P., & Hemminki, K. (2008b). Survival in familial pancreatic cancer. *Pancreatology*, *8*(3), 252–256. http://dx.doi.org/10.1159/000134272. 000134272 [pii].

Kainu, T., Juo, S. H., Desper, R., Schaffer, A. A., Gillanders, E., Rozenblum, E., et al. (2000). Somatic deletions in hereditary breast cancers implicate 13q21 as a putative novel breast cancer susceptibility locus. *Proceedings of the National Academy of Sciences of the United States of America*, *97*(17), 9603–9608.

Karantanos, T., Corn, P. G., & Thompson, T. C. (2013). Prostate cancer progression after androgen deprivation therapy: Mechanisms of castrate resistance and novel therapeutic approaches. *Oncogene*, *32*(49), 5501–5511. http://dx.doi.org/10.1038/onc.2013.206.

Kicinski, M., Vangronsveld, J., & Nawrot, T. S. (2011). An epidemiological reappraisal of the familial aggregation of prostate cancer: A meta-analysis. *PLoS One*, *6*(10), e27130. http://dx.doi.org/10.1371/journal.pone.0027130.

Kikuchi, S., Yamada, D., Fukami, T., Maruyama, T., Ito, A., Asamura, H., et al. (2006). Hypermethylation of the TSLC1/IGSF4 promoter is associated with tobacco smoking and a poor prognosis in primary nonsmall cell lung carcinoma. *Cancer*, *106*(8), 1751–1758. http://dx.doi.org/10.1002/cncr.21800.

Knudson, A. G. (2001). Two genetic hits (more or less) to cancer. *Nature Reviews. Cancer*, *1*(2), 157–162. http://dx.doi.org/10.1038/35101031.

Kurachi, H., Wada, Y., Tsukamoto, N., Maeda, M., Kubota, H., Hattori, M., et al. (1997). Human SPA-1 gene product selectively expressed in lymphoid tissues is a specific GTPase-activating protein for Rap1 and Rap2. Segregate expression profiles from a rap1GAP gene product. *The Journal of Biological Chemistry*, *272*(44), 28081–28088.

Kuramochi, M., Fukuhara, H., Nobukuni, T., Kanbe, T., Maruyama, T., Ghosh, H. P., et al. (2001). TSLC1 is a tumor–suppressor gene in human non-small-cell lung cancer. *Nature Genetics*, *27*(4), 427–430. http://dx.doi.org/10.1038/86934.

Lee, M., Beggs, S. M., Gildea, D., Bupp, S., Lichtenberg, J., Trivedi, N. S., et al. (2015). Necdin is a breast cancer metastasis suppressor that regulates the transcription of c-Myc. *Oncotarget*, *6*(31), 31557–31568. http://dx.doi.org/10.18632/oncotarget.5230.

Lee, M., Dworkin, A. M., Gildea, D., Trivedi, N. S., Program, N. C. S., Moorhead, G. B., et al. (2014a). RRP1B is a metastasis modifier that regulates the expression of alternative mRNA isoforms through interactions with SRSF1. *Oncogene*, *33*(14), 1818–1827. http://dx.doi.org/10.1038/onc.2013.133.

Lee, M., Dworkin, A. M., Lichtenberg, J., Patel, S. J., Trivedi, N. S., Gildea, D., et al. (2014b). Metastasis-associated protein ribosomal RNA processing 1 homolog B (RRP1B) modulates metastasis through regulation of histone methylation. *Molecular Cancer Research: MCR*, *12*(12), 1818–1828. http://dx.doi.org/10.1158/1541-7786.MCR-14-0167.

Lee, M., Williams, K. A., Hu, Y., Andreas, J., Patel, S. J., Zhang, S., et al. (2015). GNL3 and SKA3 are novel prostate cancer metastasis susceptibility genes. *Clinical & Experimental Metastasis*, *32*(8), 769–782. http://dx.doi.org/10.1007/s10585-015-9745-y.

Lichtenstein, P., Holm, N. V., Verkasalo, P. K., Iliadou, A., Kaprio, J., Koskenvuo, M., et al. (2000). Environmental and heritable factors in the causation of cancer—Analyses of cohorts of twins from Sweden, Denmark, and Finland. *The New England Journal of Medicine*, *343*(2), 78–85. http://dx.doi.org/10.1056/NEJM200007133430201.

Lifsted, T., Le Voyer, T., Williams, M., Muller, W., Klein-Szanto, A., Buetow, K. H., et al. (1998). Identification of inbred mouse strains harboring genetic modifiers of mammary tumor age of onset and metastatic progression. *International Journal of Cancer, 77*(4), 640–644.

Lin, E. Y., Jones, J. G., Li, P., Zhu, L., Whitney, K. D., Muller, W. J., et al. (2003). Progression to malignancy in the polyoma middle T oncoprotein mouse breast cancer model provides a reliable model for human diseases. *The American Journal of Pathology, 163*(5), 2113–2126. http://dx.doi.org/10.1016/S0002-9440(10)63568-7.

Lindstrom, L. S., Hall, P., Hartman, M., Wiklund, F., Gronberg, H., & Czene, K. (2007). Familial concordance in cancer survival: A Swedish population-based study. *The Lancet Oncology, 8*(11), 1001–1006. http://dx.doi.org/10.1016/S1470-2045(07)70282-6.

Liu, W., Beck, B. H., Vaidya, K. S., Nash, K. T., Feeley, K. P., Ballinger, S. W., et al. (2014). Metastasis suppressor KISS1 seems to reverse the Warburg effect by enhancing mitochondrial biogenesis. *Cancer Research, 74*(3), 954–963. http://dx.doi.org/10.1158/0008-5472.CAN-13-1183.

Liu, H., Radisky, D. C., Yang, D., Xu, R., Radisky, E. S., Bissell, M. J., et al. (2012). MYC suppresses cancer metastasis by direct transcriptional silencing of alphav and beta3 integrin subunits. *Nature Cell Biology, 14*(6), 567–574. http://dx.doi.org/10.1038/ncb2491.

Mantovani, A., Giavazzi, R., Alessandri, G., Spreafico, F., & Garattini, S. (1981). Characterization of tumor lines derived from spontaneous metastases of a transplanted murine sarcoma. *European Journal of Cancer, 17*(1), 71–76.

Marcus, D. M., Goodman, M., Jani, A. B., Osunkoya, A. O., & Rossi, P. J. (2012). A comprehensive review of incidence and survival in patients with rare histological variants of prostate cancer in the United States from 1973 to 2008. *Prostate Cancer and Prostatic Diseases, 15*(3), 283–288. http://dx.doi.org/10.1038/pcan.2012.4.

Mucenski, M. L., Taylor, B. A., Jenkins, N. A., & Copeland, N. G. (1986). AKXD recombinant inbred strains: Models for studying the molecular genetic basis of murine lymphomas. *Molecular and Cellular Biology, 6*(12), 4236–4243.

O'Brien, B., Nichaman, L., Browne, J. E., Levin, D. L., Prorok, P. C., Gohagan, J. K., et al. (2000). Coordination and management of a large multicenter screening trial: The Prostate, Lung, Colorectal and Ovarian (PLCO) cancer screening trial. *Controlled Clinical Trials, 21*(6 Suppl.), 310S–328S.

Paiss, T., Worner, S., Kurtz, F., Haeussler, J., Hautmann, R. E., Gschwend, J. E., et al. (2003). Linkage of aggressive prostate cancer to chromosome 7q31-33 in German prostate cancer families. *European Journal of Human Genetics: EJHG, 11*(1), 17–22. http://dx.doi.org/10.1038/sj.ejhg.5200898.

Park, Y. G., Clifford, R., Buetow, K. H., & Hunter, K. W. (2003). Multiple cross and inbred strain haplotype mapping of complex-trait candidate genes. *Genome Research, 13*(1), 118–121. http://dx.doi.org/10.1101/gr.786403.

Park, Y. G., Zhao, X., Lesueur, F., Lowy, D. R., Lancaster, M., Pharoah, P., et al. (2005). Sipa1 is a candidate for underlying the metastasis efficiency modifier locus Mtes1. *Nature Genetics, 37*(10), 1055–1062. http://dx.doi.org/10.1038/ng1635.

Patel, J. D., Krilov, L., Adams, S., Aghajanian, C., Basch, E., Brose, M. S., et al. (2014). Clinical Cancer Advances 2013: Annual Report on Progress Against Cancer from the American Society of Clinical Oncology. *Journal of Clinical Oncology, 32*(2), 129–160. http://dx.doi.org/10.1200/JCO.2013.53.7076.

Patel, S. J., Molinolo, A. A., Gutkind, S., & Crawford, N. P. (2013). Germline genetic variation modulates tumor progression and metastasis in a mouse model of neuroendocrine prostate carcinoma. *PLoS One, 8*(4), e61848. http://dx.doi.org/10.1371/journal.pone.0061848.

Peto, J., Easton, D. F., Matthews, F. E., Ford, D., & Swerdlow, A. J. (1996). Cancer mortality in relatives of women with breast cancer: The OPCS Study. Office of Population

Censuses and Surveys. *International Journal of Cancer, 65*(3), 275–283. http://dx.doi.org/10.1002/(SICI)1097-0215(19960126)65:3<275::AID-IJC1>3.0.CO;2-X.

Petros, J. A., Baumann, A. K., Ruiz-Pesini, E., Amin, M. B., Sun, C. Q., Hall, J., et al. (2005). mtDNA mutations increase tumorigenicity in prostate cancer. *Proceedings of the National Academy of Sciences of the United States of America, 102*(3), 719–724. http://dx.doi.org/10.1073/pnas.0408894102.

Pirie, A., Guo, Q., Kraft, P., Canisius, S., Eccles, D. M., Rahman, N., et al. (2015). Common germline polymorphisms associated with breast cancer-specific survival. *Breast Cancer Research, 17*, 58. http://dx.doi.org/10.1186/s13058-015-0570-7.

Pleasance, E. D., Cheetham, R. K., Stephens, P. J., McBride, D. J., Humphray, S. J., Greenman, C. D., et al. (2010). A comprehensive catalogue of somatic mutations from a human cancer genome. *Nature, 463*(7278), 191–196. http://dx.doi.org/10.1038/nature08658.

Price, L. S., Hajdo-Milasinovic, A., Zhao, J., Zwartkruis, F. J., Collard, J. G., & Bos, J. L. (2004). Rap1 regulates E-cadherin-mediated cell–cell adhesion. *The Journal of Biological Chemistry, 279*(34), 35127–35132. http://dx.doi.org/10.1074/jbc.M404917200.

Priftakis, D., Kritikos, N., Stavrinides, S., Kleanthous, S., & Baziotis, N. (2015). Neuroendocrine differentiation in castration-resistant prostate cancer: A case report. *Molecular and Clinical Oncology, 3*(6), 1392–1394. http://dx.doi.org/10.3892/mco.2015.645.

Ramaswamy, S., Ross, K. N., Lander, E. S., & Golub, T. R. (2003). A molecular signature of metastasis in primary solid tumors. *Nature Genetics, 33*(1), 49–54. http://dx.doi.org/10.1038/ng1060.

Ribelles, N., Santonja, A., Pajares, B., Llacer, C., & Alba, E. (2014). The seed and soil hypothesis revisited: Current state of knowledge of inherited genes on prognosis in breast cancer. *Cancer Treatment Reviews, 40*(2), 293–299. http://dx.doi.org/10.1016/j.ctrv.2013.09.010.

Riethmuller, G., & Klein, C. A. (2001). Early cancer cell dissemination and late metastatic relapse: Clinical reflections and biological approaches to the dormancy problem in patients. *Seminars in Cancer Biology, 11*(4), 307–311. http://dx.doi.org/10.1006/scbi.2001.0386.

Schumacher, F. R., Berndt, S. I., Siddiq, A., Jacobs, K. B., Wang, Z., Lindstrom, S., et al. (2011). Genome-wide association study identifies new prostate cancer susceptibility loci. *Human Molecular Genetics, 20*(19), 3867–3875. http://dx.doi.org/10.1093/hmg/ddr295.

Shimonaka, M., Katagiri, K., Nakayama, T., Fujita, N., Tsuruo, T., Yoshie, O., et al. (2003). Rap1 translates chemokine signals to integrin activation, cell polarization, and motility across vascular endothelium under flow. *The Journal of Cell Biology, 161*(2), 417–427. http://dx.doi.org/10.1083/jcb.200301133.

Siegel, R. L., Miller, K. D., & Jemal, A. (2016). Cancer statistics, 2016. *CA: A Cancer Journal for Clinicians, 66*, 7–30. http://dx.doi.org/10.3322/caac.21332.

Stanford, J. L., FitzGerald, L. M., McDonnell, S. K., Carlson, E. E., McIntosh, L. M., Deutsch, K., et al. (2009). Dense genome-wide SNP linkage scan in 301 hereditary prostate cancer families identifies multiple regions with suggestive evidence for linkage. *Human Molecular Genetics, 18*(10), 1839–1848. http://dx.doi.org/10.1093/hmg/ddp100.

Talmadge, J. E., Wolman, S. R., & Fidler, I. J. (1982). Evidence for the clonal origin of spontaneous metastases. *Science, 217*(4557), 361–363.

Tan, D. S., Mok, T. S., & Rebbeck, T. R. (2016). Cancer genomics: Diversity and disparity across ethnicity and geography. *Journal of Clinical Oncology, 34*(1), 91–101. http://dx.doi.org/10.1200/JCO.2015.62.0096.

Taylor, B. S., Schultz, N., Hieronymus, H., Gopalan, A., Xiao, Y., Carver, B. S., et al. (2010). Integrative genomic profiling of human prostate cancer. *Cancer Cell, 18*(1), 11–22. http://dx.doi.org/10.1016/j.ccr.2010.05.026.

Teerlink, C. C., Thibodeau, S. N., McDonnell, S. K., Schaid, D. J., Rinckleb, A., Maier, C., et al. (2014). Association analysis of 9,560 prostate cancer cases from the International Consortium of Prostate Cancer Genetics confirms the role of reported prostate cancer associated SNPs for familial disease. *Human Genetics*, *133*(3), 347–356. http://dx.doi.org/10.1007/s00439-013-1384-2.

Terry, S., & Beltran, H. (2014). The many faces of neuroendocrine differentiation in prostate cancer progression. *Frontiers in Oncology*, *4*, 60. http://dx.doi.org/10.3389/fonc.2014.00060.

Threadgill, D. W., & Churchill, G. A. (2012). Ten years of the collaborative cross. *Genetics*, *190*(2), 291–294. http://dx.doi.org/10.1534/genetics.111.138032.

van de Vijver, M. J., He, Y. D., van't Veer, L. J., Dai, H., Hart, A. A., Voskuil, D. W., et al. (2002). A gene-expression signature as a predictor of survival in breast cancer. *The New England Journal of Medicine*, *347*(25), 1999–2009. http://dx.doi.org/10.1056/NEJMoa021967.

van Ginkel, P. R., Sareen, D., Subramanian, L., Walker, Q., Darjatmoko, S. R., Lindstrom, M. J., et al. (2007). Resveratrol inhibits tumor growth of human neuroblastoma and mediates apoptosis by directly targeting mitochondria. *Clinical Cancer Research*, *13*(17), 5162–5169. http://dx.doi.org/10.1158/1078-0432.CCR-07-0347.

van't Veer, L. J., Dai, H., van de Vijver, M. J., He, Y. D., Hart, A. A., Mao, M., et al. (2002). Gene expression profiling predicts clinical outcome of breast cancer. *Nature*, *415*(6871), 530–536. http://dx.doi.org/10.1038/415530a.

Verkooijen, H. M., Hartman, M., Usel, M., Benhamou, S., Neyroud-Caspar, I., Czene, K., et al. (2012). Breast cancer prognosis is inherited independently of patient, tumour, and treatment characteristics. *International Journal of Cancer*, *130*, 2103–2110. http://dx.doi.org/10.1002/ijc.26206.

Wang, H. T., Yao, Y. H., Li, B. G., Tang, Y., Chang, J. W., & Zhang, J. (2014). Neuroendocrine Prostate Cancer (NEPC) progressing from conventional prostatic adenocarcinoma: Factors associated with time to development of NEPC and survival from NEPC diagnosis—A systematic review and pooled analysis. *Journal of Clinical Oncology*, *32*(30), 3383–3390. http://dx.doi.org/10.1200/JCO.2013.54.3553.

Welter, D., MacArthur, J., Morales, J., Burdett, T., Hall, P., Junkins, H., et al. (2014). The NHGRI GWAS catalog, a curated resource of SNP-trait associations. *Nucleic Acids Research*, *42*(Database issue), D1001–D1006. http://dx.doi.org/10.1093/nar/gkt1229.

Whittemore, A. S., Stearman, B., Venne, V., Halpern, J., Felberg, A., McGuire, V., et al. (2009). No evidence of familial correlation in breast cancer metastasis. *Breast Cancer Research and Treatment*, *118*(3), 575–581. http://dx.doi.org/10.1007/s10549-009-0368-3.

Williams, K. A., Lee, M., Hu, Y., Andreas, J., Patel, S. J., Zhang, S., et al. (2014). A systems genetics approach identifies CXCL14, ITGAX, and LPCAT2 as novel aggressive prostate cancer susceptibility genes. *PLoS Genetics*, *10*(11), e1004809. http://dx.doi.org/10.1371/journal.pgen.1004809.

Winter, S. F., Lukes, L., Walker, R. C., Welch, D. R., & Hunter, K. W. (2012). Allelic variation and differential expression of the mSIN3A histone deacetylase complex gene Arid4b promote mammary tumor growth and metastasis. *PLoS Genetics*, *8*(5), e1002735. http://dx.doi.org/10.1371/journal.pgen.1002735.

Witte, J. S., Goddard, K. A., Conti, D. V., Elston, R. C., Lin, J., Suarez, B. K., et al. (2000). Genomewide scan for prostate cancer-aggressiveness loci. *American Journal of Human Genetics*, *67*(1), 92–99. http://dx.doi.org/10.1086/302960.

Xie, T., Cho, Y. B., Wang, K., Huang, D., Hong, H. K., Choi, Y. L., et al. (2014). Patterns of somatic alterations between matched primary and metastatic colorectal tumors characterized by whole-genome sequencing. *Genomics*, *104*(4), 234–241. http://dx.doi.org/10.1016/j.ygeno.2014.07.012.

Yang, H., Crawford, N., Lukes, L., Finney, R., Lancaster, M., & Hunter, K. W. (2005). Metastasis predictive signature profiles pre-exist in normal tissues. *Clinical & Experimental Metastasis*, *22*(7), 593–603. http://dx.doi.org/10.1007/s10585-005-6244-6.

Yang, G., He, W., Cai, M., Luo, F., Kung, H., Guan, X., et al. (2011). Loss/Down-regulation of tumor suppressor in lung cancer 1 expression is associated with tumor progression and is a biomarker of poor prognosis in ovarian carcinoma. *International Journal of Gynecological Cancer: Official Journal of the International Gynecological Cancer Society*, *21*(3), 486–493. http://dx.doi.org/10.1097/IGC.0b013e31820fa168.

Yang, H., Rouse, J., Lukes, L., Lancaster, M., Veenstra, T., Zhou, M., et al. (2004). Caffeine suppresses metastasis in a transgenic mouse model: A prototype molecule for prophylaxis of metastasis. *Clinical & Experimental Metastasis*, *21*(8), 719–735.

You, Y., Ma, L., You, M., Li, X., Wang, S., Li, H., et al. (2010). TSLC1 gene silencing in cutaneous melanoma. *Melanoma Research*, *20*(3), 179–183. http://dx.doi.org/10.1097/CMR.0b013e32833413c0.

Zhang, X. C., Xu, C., Mitchell, R. M., Zhang, B., Zhao, D., Li, Y., et al. (2013). Tumor evolution and intratumor heterogeneity of an oropharyngeal squamous cell carcinoma revealed by whole-genome sequencing. *Neoplasia*, *15*(12), 1371–1378.

CHAPTER FOUR

Breast Cancer Metastasis Suppressor 1 (BRMS1): Robust Biological and Pathological Data, But Still Enigmatic Mechanism of Action

D.R. Welch[*,†,1], **C.A. Manton**[*], **D.R. Hurst**[‡,1]

*University of Kansas Medical Center; Kansas City, KS, United States
†University of Kansas Cancer Center, Kansas City, KS, United States
‡University of Alabama at Birmingham, Birmingham, AL, United States
[1]Corresponding authors: e-mail address: dwelch@kumc.edu; dhurst@uab.edu

Contents

Concluding Thoughts 129
Acknowledgments 129
References 129

Abstract

Metastasis requires coordinated expression of multiple genetic cassettes, often via epigenetic regulation of gene transcription. BRMS1 blocks metastasis, but not orthotopic tumor growth in multiple tumor types, presumably via SIN3 chromatin remodeling complexes. Although there is an abundance of strong data supporting BRMS1 as a metastasis suppressor, the mechanistic data directly connecting molecular pathways with inhibition of particular steps in metastasis are not well defined. In this review, the data for BRMS1-mediated metastasis suppression in multiple tumor types are discussed along with the steps in metastasis that are inhibited.

Cancer is a relatively curable disease as long as it is diagnosed early. However, if neoplastic cells have disseminated to vital organs and established colonies at those sites, prognosis is usually extremely poor. It is anticipated that deeper understanding of the metastatic process and the molecules that contribute to its success will lead to improved rates of cure. Similarly, identifying molecules that block metastasis, whether intrinsic or pharmacologic, may represent a new avenue toward cancer control (Eccles & Welch, 2007; Khanna et al., 2014; Steeg et al., 2009).

Advances in Cancer Research, Volume 132
ISSN 0065-230X
http://dx.doi.org/10.1016/bs.acr.2016.05.003

Multiple steps are required for invasive cells to establish discontiguous secondary foci. The process of metastasis is highly inefficient and complex and it requires the coordinate expression of genes at every step to confer survival advantages by interacting with the ever-changing microenvironment (Bragado et al., 2013; Fidler, Kim, & Langley, 2007; Hurst & Welch, 2011a; Langley & Fidler, 2011; Liu et al., 2014; Obenauf & Massague, 2015; Sleeman, Nazarenko, & Thiele, 2011). These metastasis-associated genes are still in the early stages of characterization but can be categorized based on their involvement in metastasis initiation (including invasion, angiogenesis, bone marrow mobilization, and circulation), metastasis progression (including extravasation, survival, and reinitiation), and metastasis virulence (organ-specific colonization) (Nguyen, Bos, & Massague, 2009). Most of the genes listed in these somewhat arbitrary classes do not regulate/control only one step; rather, they typically function in multiple processes involved in the continuum of the metastatic cascade.

Metastasis-associated genes are regulated by multiple mechanisms (Bohl, Harihar, Denning, Sharma, & Welch, 2013; Liu et al., 2014; Sethi & Kang, 2011), but are not typically mutated. Rather, the development of metastasis more frequently involves changes in gene expression or posttranslational modifications of gene products or pathways (Jones & Baylin, 2002). Epigenetic regulation is important for cells to rapidly respond to changes in their environment. This is accomplished by chemical modifications to DNA and/or chromatin-associated histones (Rodriguez-Paredes & Esteller, 2011). Histone (de)acetylation regulates gene transcription by altering the three-dimensional structure of DNA. Histone acetyltransferases (HAT) add acetyl moieties to histone lysine side chains, which results in a more open conformation of DNA (euchromatin) that is generally more accessible to transcription factors. Histone deacetylases (HDAC) remove those acetyl moieties, compacting DNA (heterochromatin), typically leading to transcriptional repression. HAT and HDAC complexes are large and are often linked with DNA methyltransferases and other protein or DNA modifying enzymes. The enzymatic activity and overall function of these complexes are intricately regulated by the specific protein composition of each complex.

There are many HDAC-containing complexes that promote and suppress tumorigenicity and metastasis. Nucleosome remodeling and histone deacetylase (NuRD) has been extensively studied as a metastasis promoter after metastasis-associated protein 1 (MTA1) was determined to be a critical component of NuRD complexes (Li, Pakala, Nair, Eswaran, & Kumar, 2012; Xue et al., 1998). Some long noncoding RNAs that alter chromatin state have been implicated in metastasis promotion (eg, HOTAIR and MALAT, reviewed in Crea, Clermont, Parolia, Wang, & Helgason, 2013; Gupta et al., 2010; Gutschner et al., 2013; Kogo et al., 2011; Li & Kang, 2014; Zheng et al., 2015) while chromatin remodeling proteins LSD1 (Wang et al., 2009) and RRP1B (Crawford et al., 2007; Crawford, Yang, Mattaini, & Hunter, 2009) have been linked to suppression. Although such links between chromatin remodeling and metastasis regulation have been made, these are still relatively new areas of investigation.

Chromatin remodeling is typically accomplished via large, multiprotein complexes. Scaffolding proteins recruit histone-binding proteins, transcription factors, and other protein modifying enzymes. SIN3 (switch-independent 3) is one of the most abundant scaffolding proteins and was initially identified as a global regulator of gene transcription (Grzenda, Lomberk, Zhang, & Urrutia, 2009; Kadamb, Mittal, Bansal, Batra, & Saluja, 2013; McDonel, Costello, & Hendrich, 2008; Silverstein & Ekwall, 2005). The composition of these complexes has been extensively characterized in several yeast species; however, the functional complexity of SIN3 is amplified in mammalian cells with two isoforms that arise from distinct genes. Mammalian SIN3 complexes are quite dynamic and have yet to be fully determined. BRMS1, which does not exist in yeast, has been repeatedly associated with both SIN3 isoforms (SIN3A and SIN3B). The relevance of these associations will be discussed in more detail later.

The epigenetic changes alluded to above occur selectively on both metastasis-promoting and metastasis-inhibiting genes. Discovery of metastasis-promoting genes is complicated by recognition that successful cells have completed every step of the complex metastatic cascade (primary tumor growth, invasion of surrounding host tissue with eventual entry into a circulatory compartment, arrest and extravasation at nearby or distant sites, and colonization of secondary tissues). The inefficiency of the metastatic process is due, at least in part, to deficiencies of individual cells to complete any one of those steps. However, individual cells can be incapable of completing more than one step. Therefore, establishing a gene as metastasis-promoting necessarily means testing it experimentally in cells that are competent for all other steps in the metastatic cascade. Unfortunately, it is impossible to know the status of all other capabilities a priori. As a result, experimental assessment of metastasis promoters is fraught with a higher probability of a false negative result. Therefore, identification of a metastasis suppressor is more experimentally straightforward, because one need only impede any single step of the process.

In 1986, Patricia Steeg compared gene expression in paired samples of K1735 melanoma and identified the first metastasis suppressor gene, Nm23 (NME1). Interestingly, Nm23 blocked establishment of secondary foci but failed to block growth of tumor cells inoculated orthotopically. These characteristics defined a new class of molecules and established, definitively, that metastasis and primary tumor formation are genetically distinct phenotypes. Since that time, we and others have discovered more than 30 metastasis suppressors which function at different stages of the metastatic cascade and in different cancer histotypes (Bohl et al., 2013; Liu et al., 2014).

BRMS1 (breast cancer metastasis suppressor 1) was discovered using differential display by Jabed Seraj and colleagues in 2000 as a molecule that suppresses metastasis without preventing primary tumor growth (Seraj, Samant, Verderame, & Welch, 2000). The discovery of *BRMS1* was the culmination of a series of experiments that began with clinical karyotypic correlations establishing that both the long and short arms of chromosome 11 are frequent sites of amplification/deletion and are associated with the progression of breast cancer (Phillips et al., 1996). This led Karen Phillips, Bernard "Buddy" Weissman and colleagues to hypothesize that a metastasis suppressor might exist on chromosome 11. To test that hypothesis, they transferred an intact copy of neomycin-tagged wild-type chromosome 11 into metastatic breast cancer cells (MDA-MB-435; designated neo11/435 with the introduced chromosome 11) using microcell-mediated chromosome transfer. They observed a significant reduction in the formation of spontaneous metastases in neo11/435 cells compared to parental cells or transfer with another chromosome. There were no significant changes in orthotopic tumor incidence (Phillips et al., 1996). When differential display RT–PCR was used to compare MDA-MB-435 and neo11/435 cell clones, transcripts with higher expression in neo11/435 cells were identified. Among those transcripts were BRMS1 which, when reexpressed in MDA-MB-435 and MDA-MB-231 metastatic breast cancer cells decreased the incidence and number of metastases to lung and lymph nodes in both experimental and spontaneous metastasis assays. Orthotopic tumor growth rates were similar, but there was a lag in growth in the 435-BRMS1 cells. Thus, these results demonstrated that BRMS1 suppresses metastasis without blocking tumorigenicity, satisfying the functional definition of a metastasis suppressor gene.

Subsequently, several labs utilized different model systems to confirm functionally that BRMS1 is a metastasis suppressor. Metastasis suppression has been demonstrated in melanoma, breast, ovarian, rectal, gall bladder,

Table 1 BRMS1 Suppresses Metastatic Potential in Multiple Cancer Types

Cancer Type	Article	Key Findings
Breast (TNBC)	Seraj et al. (2000)	Identification of BRMS1, encoded at 11q13, and characterization of its ability to suppress metastasis to lungs and lymph nodes after injection IV or orthotopically (mammary fat pad)
	Hedley, Vaidya, et al. (2008)	BRMS1 expression decreased metastatic burden to liver (mesenteric vein injection), bone, brain (intracardiac left ventricle injections), and lung (lateral tail vein injection)
Melanoma	Shevde et al. (2002)	BRMS1 expression suppressed metastasis to lung after IV injection or orthotopic (intradermal) injection
Ovarian	Zhang, Huang, et al. (2006)	BRMS1 expression reduced formation of lung metastases of IV injected cells and resulted in fewer metastases to organs in the peritoneal cavity for orthotopically implanted cells
Nonsmall cell lung cancer	Smith et al. (2009)	BRMS1 expression reduced pulmonary and hepatic metastases from NSCLC cells injected into the flank
Rectal	Zhang, Guan, et al. (2014)	BRMS1 expression inhibited metastasis to lung and liver of rectal cancer cells injected IV
Bladder[a]	Seraj, Harding, Gildea, Welch, and Theodorescu (2001)	Higher expression of BRMS1 mRNA observed in T24 bladder cancer cells compared to the more metastatic variant line T24T
	Ponnusamy et al. (2012)	BRMS1 expression inversely associated with tumor grade. MB49 experimental metastasis was decreased in BRMS1-overexpressing cells

[a]Although experimental data exist, the rigor of defining BRMS1 as a metastasis suppressor is not fulfilled due to lack of presented data on orthotopic tumor growth.

and nonsmall cell lung carcinomas. The key papers are summarized in Table 1. Suggestive data in bladder carcinoma were also described—*BRMS1* mRNA is reduced in the metastatic variant cell line T24T compared to the less metastatic T24 (Seraj et al., 2001)—and a potential correlation between

BRMS1 and RhoGDI2 (another metastasis suppressor) downregulation with higher metastatic potential was observed. Another bladder cancer model using MB49 cell line demonstrated suppression of experimental metastases by ectopic expression of BRMS1 (Ponnusamy et al., 2012); however, BRMS1 function in primary bladder tumors has not yet been investigated.

Several additional studies have correlated higher BRMS1 expression with lower metastatic propensity or diminished ability to accomplish several metastasis-associated phenotypes, like motility, invasion, resistance to anoikis, ultrastructure, cytoskeletal organization, etc. This pattern of expression is consistent with BRMS1 function as a metastasis suppressor, but also suggests that BRMS1 may have some clinical prognostic utility. Realization of any clinical utility requires some detailed analysis of the available data, quality reagents, and a more detailed understanding of the mechanism of action.

The quality of reagents, and in particular the utility of antibodies, has slowed some of the work to understand BRMS1 expression and function. In fact, some commercial antibodies that were described for BRMS1 were actually raised against the related protein BRMS1L. We generated and validated two monoclonal antibodies "in house" (IgG isotype (3a1.21) recognizes the N-terminal region of BRMS1 and an IgM isotype (1a5.7) recognizes the C-terminal region) that are effective for detecting and immunoprecipitating BRMS1 (Hurst et al., 2006, 2008; Hurst, Xie, Edmonds, & Welch, 2009). Correlative expression analyses with patient samples using these antibodies are ongoing.

Although BRMS1 clearly suppresses metastasis, determining the mechanism(s) by which it does so has proven extremely challenging (aside from confusion related to some reagents). Indeed, it is not even possible to conclusively state that BRMS1 blocks a particular step in the metastatic cascade. Rather, reexpression in cell lines reduces efficiency of cancer cells in completion of multiple different steps to varying degrees (Phadke, Vaidya, Nash, Hurst, & Welch, 2008), ultimately resulting in an 80–90% inhibition of metastasis in several different mouse models (Cook et al., 2012; Cui et al., 2012; Guo et al., 2015; Hedley, Vaidya, et al., 2008; Jiang, Xia, Feng, & Kong, 2007; Phadke et al., 2008; Ponnusamy et al., 2012; Samant et al., 2006, 2001; Seraj et al., 2000; Sheng, Zhou, Song, Zhou, & Liu, 2012; Shevde et al., 2002; Yang et al., 2008; Zhang, Lin, & Di, 2006). These data are summarized in Table 2. Depending upon the assay and cell lines used, BRMS1 can inhibit invasion, migration, adhesion, and promote apoptosis.

Table 2 Step(s) of Metastasis Inhibited by BRMS1

Step in Metastasis	References	Cancer Type	Key Findings
Invasion	Shevde et al. (2002)	Melanoma	BRMS1 decreased invasion into collagen
	Zhang, Huang, et al. (2006)	Ovarian carcinoma	Decreased Matrigel invasion of BRMS1-expressing cells
	Smith et al. (2009)	NSCLC	BRMS1 decreased transwell invasion and decreased invasiveness of spheroids implanted into a 3D collagen gel
	Mei et al. (2014)	Glioma	Decreased Matrigel invasion of cells expressing BRMS1
Migration	Zhang, Huang, et al. (2006)	Ovarian carcinoma	BRMS1 expression decreased motility in wound-healing assays
	Smith et al. (2009)	NSCLC	Decreased transwell migration of BRMS1-expressing cells
	Mei et al. (2014)	Glioma	Decreased transwell migration of cells expressing BRMS1
Adhesion	Zhang, Huang, et al. (2006)	Ovarian carcinoma	Decreased attachment to plates coated with matrix components
	Hedley, Vaidya, et al. (2008)	Breast cancer	Decreased adhesion to bone in vitro for BRMS1-expressing cells
	Khotskaya et al. (2014)	Breast cancer	Plating of BRMS1-expressing cells onto collagen or fibronectin failed to form adhesion projections
	Mei et al. (2014)	Glioma	Decreased adhesion of BRMS1-expressing cells to plates coated in fibronectin
Colonization	Phadke et al. (2008)	Breast cancer	BRMS1-expressing cells failed to proliferate after reaching secondary site
	Hedley, Vaidya, et al. (2008)	Breast cancer	Decreased ability of BRMS1 cells to survive after disseminating to lungs of mice
Promotion of apoptosis	Phadke et al. (2008)	Breast cancer	Increased cell death in the vasculature traveling from the primary tumor to secondary site for BRMS1 expressing cells; increased anoikis demonstrated by failure to survive on poly-hydroxyethyl methacrylate-coated dishes
	Hedley, Vaidya, et al. (2008)	Breast cancer	Increased sensitivity to hypoxic stress and anchorage-independent growth-induced anoikis
	You, He, Ding, and Zhang (2015)	NSCLC	BRMS1 expression sensitized cells to serum starvation-induced death

Pushkar Phadke et al. showed that *BRMS1*-expressing breast cancer cells reaching the secondary site fail to proliferate (Phadke et al., 2008). Although tumor dormancy was not definitively demonstrated in that study, the results suggest that late steps in metastasis, including colonization, are inhibited by BRMS1.

That growth is relatively unimpeded in one location but inhibited in another tissue suggests that BRMS1 is somehow involved in cellular response to extrinsic signals. Gap junctional intercellular communication (GJIC) is one such mechanism by which a cell responds to and communicates with the surrounding microenvironment. The first reported phenotypic change associated with BRMS1 reexpression was restoration of GJIC between breast cancer cells (Saunders et al., 2001). Subsequently, reduced GJIC between breast cancer cells and osteoblasts was identified when BRMS1 is expressed (Kapoor et al., 2004).

Sitaram Harihar, Daryll DeWald, and colleagues identified alteration of phosphoinositide signaling pathways by BRMS1. By selectively downregulating phosphoinositol 4-phosphate 5-kinase expression, BRMS1 selectively reduces phosphatidylinositol-4,5-bisphosphate (PtdIns(4,5)P_2) without changing levels of PtdIns(4)P and PtdIns(3)P (DeWald et al., 2005; Vaidya et al., 2008). Such changes reduce the substrate for phosphoinositide-3-kinase (PI3K), an important signaling molecule in many cancers. Additionally, PtdIns(4,5)P_2 is a key structural component of lipid rafts and signaling structures, such as primary cilia, suggesting that membrane structure is likely implicated in BRMS1-mediated metastasis suppression.

Kedar Vaidya and colleagues extended the lipid signaling data to demonstrate differential responses of breast cancer cells to growth factor signals. Briefly, they showed that each cell population expressing BRMS1 responded differently to individual growth factors (epidermal growth factor (EGF), hepatocyte growth factor (HGF) or platelet-derived growth factor (PDGF)). Some of the responses were due to differential receptor expression; others were due to changes in phosphoinositide signaling; while still others were due to alterations in other signaling cascades (Vaidya et al., 2008). Interestingly, GJIC was more closely associated with protein kinase A (PKA) activity than inhibition of PI3K.

BRMS1 was the inaugural member of a family of loosely related proteins, all associated with transcriptional regulation and epigenetics, that also includes BRMS1-like and SUDS3 (suppressor of defective silencing 3). The family is defined by the presence of a conserved region known as the

Sds3-like domain, but the functionality of the domain is uncertain except that it is thought to be important for protein–protein interactions. Interaction with HDAC1 has been demonstrated by coimmunoprecipitation (coIP), suggesting an association with similar chromatin remodeling complexes. BRMS1 is 23% identical and 49% similar in amino acid sequence to SUDS3 and is 57% identical and 79% similar to BRMS1L. BRMS1L (p40) was identified as a component of SIN3–HDAC (switch-independent 3–histone deacetylase) complexes and has been associated with other epigenetic interactomes. However, the roles of BRMS1L have not been clearly determined. SUDS3 (also known as SAP45 or mSds3) was identified first in yeast when mutation restored silencing at the *HMR* locus. coIP showed that SUDS3 is integral to formation of orthologous SIN3–HDAC chromatin remodeling complexes in multiple species (yeast, murine, and human). The yeast ortholog of SUDS3 promotes Sin3 complex integrity and is essential for HDAC activity. In mammalian cells, SUDS3 is essential for embryonic development and has been implicated in some cancers. Alexandra Silveira and colleagues showed that expression of SUDS3 in metastatic breast cancer cell lines did not suppress metastasis, suggesting that, although related, this family of proteins has distinct functions and that they do not functionally compensate for one another (Silveira, Hurst, Vaidya, Ayer, & Welch, 2009).

At both DNA and amino acid levels, BRMS1 shares a high degree of homology among multiple species. The *Drosophila* BRMS1 gene (*dBrms1*) regulates Notch signaling, indicating a functional conservation with *BRMS1L* (Zhang, Zhang, et al., 2014). Of particular note, the murine ortholog (*Brms1*) is 85% homologous at the DNA level and the amino acid sequence is 95% identical (Negrini et al., 1995). Interestingly, a genetic mapping study using inbred mouse strains showed that metastatic potential is linked to allelic variations of *Brms1* (Devilee et al., 1991; Hampton et al., 1994), but *Brms1* did not fulfill all of the genetic criteria as a metastasis modifier locus in those studies. Nonetheless, as predicted based upon our functional studies, differentially metastatic subpopulations derived from a single mammary carcinoma showed that Brms1 expression is inversely correlated with increased aggressiveness. Reexpression of Brms1 in murine mammary carcinomas confirmed the metastasis suppressor phenotype (Gudmundsson et al., 1995; Negrini et al., 1995). Leah Cook and colleagues showed that transgenic overexpression of *Brms1* ubiquitously in F_1 generation MMTV–PyMT crosses yielded significantly fewer lung metastases while not affecting primary tumor development (Cook et al., 2012). Surprisingly, selective expression of *Brms1* in the mammary gland (ie, driven by the MMTV promoter) did not significantly block metastasis to lung. Unfortunately, the reciprocal experiment (ie, genetic knockdown) was not successful since no viable pups were obtained (unpublished observations). Reagents to generate conditional knockout mice are being developed. Altogether, these results suggest that the mouse ortholog has similar functionalities as human BRMS1.

The *BRMS1* gene maps to chromosome 11q13.1–13.2 (Seraj et al., 2000). It lies within a region that is often lost in late stage breast cancers and is near sites that are among the most commonly associated with breast cancer progression (Welch & Wei, 1998). Within the 5′ upstream region, there are several putative regulatory elements including GATA-1, CREB, GATA-2, and CdxA. There is no TATA box suggesting that transcription of BRMS1 proceeds in a TATA-independent mechanism (Seraj et al., 2000). Two putative CpG islands were identified in the promoter region of *BRMS1* (nucleotides −3477 to −2214 and −531 to +608). The more distal CpG island is functional as determined by methylation-specific PCR showing methylation-dependent suppression of *BRMS1* expression (Metge et al., 2008). One mechanism of BRMS1 promoter methylation involves the recruitment of DNA methyltransferase 1 to the BRMS1 promoter by STAT3 and myocardin-related transcription factor A, resulting in promoter hypermethylation and gene silencing (Xing et al., 2015). Recently, in non-small cell lung carcinoma, David Jones' group showed that BRMS1 expression is inversely correlated with CpG methylation in clinical samples (Nagji, Liu, Stelow, Stukenborg, & Jones, 2010).

In addition to a wild-type transcript, several splice variants have been identified. The predominant form of *BRMS1* has 10 exons spanning 741 nucleotides. *BRMS1.v2* contains an alternative splice site in exon 10; *BRMS1.v3* is alternatively spliced in exon 5, lacks exons 6–9, and alternatively splices exon 10; and *BRMS1.v4* lacks exon 9 (Hurst, Xie, et al., 2009). These splice variants are differentially expressed in metastatic compared to nonmetastatic breast cancer cell lines, but the relevance of each variant is presently unknown since translation to protein has not yet been verified, nor has function been assessed for them.

BRMS1 protein is a 246 amino acid polypeptide and is predicted to be 28.5 kDa (Seraj et al., 2000). Electrophoretically, BRMS1 migrates closer to 35 kDa on SDS–PAGE (Hurst et al., 2008). This may be due to a glutamate rich N-terminal region. BRMS1 also includes two coiled–coil regions (aa51–81 and aa147–180) and two predicted nuclear localization sequences (aa198–205 and aa239–245). The proximal NLS, but not the distal NLS, is required for localization to the nucleus (Hurst et al., 2013).

Several phosphorylation sites are predicted using publically accessible algorithms. Until recently, there had been no reports of phosphorylation. However, Siti Roesley, Boris Sarcevic, and colleagues (Roesley et al., 2016) recently demonstrated that BRMS1 is phosphorylated by cyclin-dependent kinase 2 (cdk2) at serine-237, immediately upstream of the second NLS. Experiments are underway to assess the functional significance of the phosphorylation to metastasis suppression. BRMS1 has no predicted glycosylation sites.

Abundant proteomic and gene array evidence demonstrate that successful metastasis requires coordinated expression of multiple genes (Champine, Michaelson, Weimer, Welch, & DeWald, 2007; Cicek, Samant, Kinter, Welch, & Casey, 2004; Kang et al., 2003; Minn et al., 2005; Yang, Liu, & Yang, 2016), both coding and noncoding (see later). Proteomic and transcriptomic studies have identified possible BRMS1 targets. More than 500 nonrandom expression changes have been reported (Champine et al., 2007). Muzafer Cicek et al. first demonstrated selective expression changes associated with BRMS1 expression using 2D gel electrophoresis comparing breast cancer cells (Cicek et al., 2004). A similar study was done using melanoma cells (Rivera, Megias, & Bravo, 2007). Both identified annexins and glutathione-*S*-transferase, but few other overlaps were identified, suggesting that regulation may be cell type dependent.

Collectively, pathways involved in lipid metabolism and transport, secretion, and cellular architecture have been implicated (Kodura & Souchelnytskyi, 2015). Other BRMS1-selective targets include: osteopontin (OPN) (Hedley, Welch, et al., 2008; Samant et al., 2007), urokinase type plasminogen activator (uPA) (Cicek et al., 2009; Cicek, Fukuyama, Welch, Sizemore, & Casey, 2005), EGFR (Hurst et al., 2008; Vaidya et al., 2008), CXCR4 (Yang et al., 2008), and a number of noncoding, regulatory RNAs (Edmonds, Hurst, & Welch, 2009; Hurst, Edmonds, & Welch, 2009). Most have been validated as bona fide downstream mediators of BRMS1 that appear to be responsible for driving myriad phenotypic changes observed including: invasion (Mei et al., 2014; Pan, Yu, Gong, & Song, 2009; Samant et al., 2001), migration (Al-Alwan et al., 2011; Mei et al., 2014; Samant et al., 2001), adhesion (Guo et al., 2015; Jiang et al., 2007; Khotskaya et al., 2014; Kodura & Souchelnytskyi, 2015; Phadke et al., 2008; Samant et al., 2001; Zhang, Lin, & Di, 2006), epithelial–mesenchymal transition (EMT) (Hall et al., 2014; Liu et al.,

2015), restoration of GJIC (Bodenstine et al., 2010; Saunders et al., 2001), and anoikis (Hedley, Vaidya, et al., 2008; Khotskaya et al., 2014; Phadke et al., 2008; Wu et al., 2012). Not known are: How is BRMS1 accomplishing these changes? Which are the most relevant complexes altered by BRMS1? And what determines differences in BRMS1 regulation due to cellular origin?

To date, no evidence has been obtained showing that BRMS1 functions directly as a transcription factor. Two alternative mechanisms have been identified that could lead to gene expression changes that may be linked including interaction with chromatin remodeling complexes and inhibition of NFkB activity. These are discussed in greater detail later.

The first clues to BRMS1 action were inferred from studies identifying BRMS1-interacting proteins. Using protein–protein interaction assays such as yeast two-hybrid (Y2H), coIP, fluorescence resonance energy transfer (FRET) as well as affinity purification and mass spectroscopy and structural analyses, it was determined that BRMS1 has two coiled–coil regions (Hurst & Welch, 2011b). Among the many interesting protein partners are: chaperone proteins HSP90, HSP70, and MRJ (Hurst et al., 2006), subcellular localization/protein transport by sorting nexin 6 (SNX6) (Rivera, Megias, & Bravo, 2010) and karyopherin α5 (KPNA5) (Hurst et al., 2006), transcription factors N-myc interactor (NMI) and TAFII250 (Hurst et al., 2006), stress fiber smoothelin (SMTN) (Hurst et al., 2006), chromatin complex SWI/SNF member BAF57 (Hurst et al., 2006), and class II HDACs, HDAC4, 5, and 6 (Hurst et al., 2006). Most consistently, numerous data show that BRMS1 associates with chromatin remodeling complexes that repress transcription. As a result, BRMS1 is thought to regulate transcription via interaction with SIN3:HDAC chromatin remodeling complexes. The initial report of BRMS1 interacting with SIN3 was in 2004 by Bill Meehan et al. who described interactions with AT-Rich Interacting Domain 4A (ARID4A, previously known as retinoblastoma-binding protein 1, RBP1 or RBBP1) and SUDS3 using Y2H genetic screens (Meehan et al., 2004). Coprecipitation pulled down multiple other members of SIN3 complexes including SIN3A, SIN3B, SAP30, HDAC1, HDAC2, RBBP4, and RBBP7. In addition to detecting HDAC1/2 from coIP of BRMS1, deacetylase activity was measured (Meehan et al., 2004). Further confirming relevance of these interactions, transcription was found to be repressed by BRMS1 using a luciferase reporter assay. Importantly, the protein interactions identified have generally been paralleled in cell lines overexpressing BRMS1 as well as by many other groups who have identified *endogenous* BRMS1 complexes by mass spectroscopy. These studies include immunoprecipitation of ING1 (Doyon et al., 2006; Nikolaev, Papanikolaou, Li, Qin, & Gu, 2004), ING2 (Doyon et al., 2006; Smith, Martin-Brown, Florens, Washburn, & Workman, 2010), SAP25 (Shiio et al., 2006), and tandem affinity purification of SIN3B (Le Guezennec, Vermeulen, & Stunnenberg, 2006). Taken together, these studies provide a rational basis to explain widespread effects of BRMS1 on gene transcription.

In subsequent studies, mutational analysis of BRMS1 discovered that direct interactions of ARID4A or SUDS3 with BRMS1 are not essential for metastasis suppression (Hurst et al., 2008). Different domains of the BRMS1 protein bind each molecule; however, BRMS1 remained associated with SIN3 and HDAC1/2. Interestingly, disruption of each direct BRMS1 interaction using select mutants alters the gene expression profiles. These findings suggest that overall protein composition and interactions between proteins comprising SIN3–HDAC complexes determines activity. Our working hypothesis is that BRMS1 modifies the overall composition of SIN3 complexes leading to changes in specificity and resulting activity rather than simply promoting or inhibiting HDAC activity. It is interesting to note that inhibition of SIN3 function by disrupting transcription factor interaction within the PAH-2 region also results in decreased invasion and metastasis (Bansal et al., 2015; Farias et al., 2010; Kwon et al., 2015).

Although there are many shared proteins among chromatin remodeling complexes in mammals, they are also very diverse in composition and function as alluded to above. And, while there are similarities in lower organisms and yeast, complexity in mammals is logarithmically more. For example, in *Saccharomyces cerevisiae*, two distinct SIN3 complexes have been isolated (Rpd3L and Rpd3S) (Suryadinata, Sadowski, Steel, & Sarcevic, 2011), but no yeast ortholog of *BRMS1* has been identified. In mammalian cells, more than two SIN3 complexes exist and they do so in many sizes (several hundred kilodaltons to >2 MDa). Understanding the composition of these differently sized complexes will be important to determine how they function. The complexity contributes to the challenges in understanding how BRMS1, as a member of SIN3 complexes, truly functions (Hurst, 2012).

SIN3 scaffolds recruit HDAC-1 and -2 and other protein modifying enzymes to alter chromatin structure leading to transcriptional regulation of gene expression (Grzenda et al., 2009; Kadamb et al., 2013; Silverstein & Ekwall, 2005). Most chromatin remodeling studies focus on HAT or HDAC complexes and activity, frequently ignoring the fact that other proteins and DNA modifying enzymes, including glycosyltransferases and methyltransferases, are recruited. The addition and removal of acetyl groups, carbohydrate side chains and methyl groups significantly alter enzymatic and structural protein functions. Therefore, determining which genes are regulated and how they are regulated will be key to understanding how complicated phenotypes, like metastasis, are altered.

Presumably, recruitment of HDAC1–SIN3 complexes to specific genes by BRMS1 leads to inhibition of NFkB activity, as demonstrated using chromatin immunoprecipitation assays (Liu et al., 2011, 2015; Liu, Smith, & Jones, 2006). It has been suggested that HDAC1 deacetylates H3, reducing binding of the p65 subunit at NFkB promoter binding sites. Importantly, the effects are promoter specific. But since NFkB activity is often associated with increased tumor aggressiveness (Chen, Sosnoski, & Mastro, 2010; Cicek & Oursler, 2006; Fazilaty & Mehdipour, 2014; Karin, Cao, Greten, & Li, 2002), the links with BRMS1 as a (in)direct regulator and metastasis suppression are logical connections. A recent report demonstrated that BRMS1 inactivation of NFkB resulted in decreased hypoxia-inducible factor 1α transcription, as well as expression of TWIST1 and Snail, markedly decreasing breast cancer cell invasiveness and EMT (Cho, Yu, Cho, Park, & Lee, 2015). Sphingosine kinase 1 (SK1)/S1P signaling is also associated with NFkB activation and was recently shown to be inversely associated with BRMS1 expression (Ponnusamy et al., 2012). Also of interest, microRNA-146a (miR-146a), originally described as targeting NFkB by David Baltimore's group (Lu et al., 2010), is upregulated by NFkB which, in turn, targets translation of upstream signaling molecules IRAK1 and TRAF6 (Hurst, Edmonds, Scott, et al., 2009). BRMS1 also upregulates miR-146a. Overexpression of miR-146a or miR-146b in MDA-MB-231 cells suppresses metastasis (Hurst, Edmonds, Scott, et al., 2009). The paradox is that BRMS1 upregulates miR-146a but negatively regulates NFkB.

Other metastasis-associated microRNAs (metastamir, Edmonds et al., 2009; Hurst, Edmonds, & Welch, 2009) are also coordinately regulated by BRMS1 (Seraj et al., 2000). What is becoming increasingly clear is that BRMS1 is involved in parallel molecular pathways, including those involving metastamir. This is an important function, as recent studies have solidified the key roles of microRNAs in processes such as EMT and invasion (Zhang et al., 2015). The presumption is that the changes in microRNA expression involve SIN3 complexes, but those connections are not definitively established (Bodenstine et al., 2010; Champine et al., 2007; Cicek et al., 2004; Rivera et al., 2007). Clearly, however, understanding the biochemical mechanisms of BRMS1:SIN3:NFkB, how they are regulated by external signals in the microenvironment and defining differences between microenvironments (ie, orthotopic sites vs ectopic sites) will provide important clues related to molecular function as well as to the biological behaviors of tumor cells at different stages of the metastatic cascade.

All of these gene expression changes are moot if BRMS1 shows no usefulness clinically, either as a potential therapeutic target or, at least, a biomarker. While the correlations are imperfect, fortunately the predominance of data suggests that BRMS1 has potential as a clinically relevant biomarker for good prognosis (Table 3).

Recently, circulating tumor cells were isolated from operable breast cancer patients and the *BRMS1* promoter was found to be hypermethylated compared to the healthy control population (Samant et al., 2007). Other studies found that loss of BRMS1 protein has been correlated with reduced disease-free survival when patient samples were stratified by loss of estrogen or progesterone receptor (ER, PR) or expression of HER2 (Cicek et al., 2005). Laser capture microdissection is necessary to ensure purity of the material from a heterogeneous mass and was used to show that BRMS1 localization shifting from nuclear to cytoplasmic is associated with highly proliferative ER-negative breast cancers (Cicek et al., 2009). Decreased BRMS1 expression was identified in metastatic melanomas and is thought to contribute to angiogenesis and melanoma progression [56]. More recently, nonsmall cell lung cancers were shown to have increased *BRMS1* promoter methylation, leading to decreased BRMS1 expression (Liu et al., 2014; Nagji et al., 2010). While the predominance of data in clinical samples corroborates its functional role as a metastasis suppressor, there are clinical studies in which BRMS1 does not change (Kelly et al., 2005) or expression directly correlates with aggressiveness (Cui et al., 2012; Hicks et al., 2006; Lombardi et al., 2006). The reasons for these differences in clinical correlation are not fully defined. However, RNA and protein levels for BRMS1 are not directly correlated (Hurst, Xie, et al., 2009). Moreover, interpretation of many (if not most) expression studies are compromised by failure to determine the contribution of "contaminating" stromal cells' expression. Similarly, immunohistochemical studies have been scuppered because commercial antibodies have not been validated. Indeed, at least two antibodies recognize BRMS1 family members, not BRMS1 (yet the antibodies are sold as recognizing BRMS1).

Table 3 Studies Examining the Role(s) of BRMS1 in Human Cancers

Cancer Type	Measurement	Sample Information	Key Findings	References
Breast	mRNA and protein	36 cancer cases and normal surrounding tissue; tissue microarray of 209 cancer cases	BRMS1 mRNA and protein were increased in most cancer tissues (94% and 81%, respectively) compared to normal epithelium. Though BRMS1 was predominantly nuclear, some cancer specimens had greater cytoplasmic staining; this was associated with low ER, low PR, and high Ki-67 staining tumors	Frolova et al. (2009)
	Protein and copy number	238 newly diagnosed carcinomas for protein; 47 cancer cases for copy number	25% of cases were negative for BRMS1 protein. Negativity for BRMS1 was most common in tumors that were PR positive, HER2 positive, and from younger patients (<50 years old). Loss of BRMS1 protein expression correlated with reduced disease-free survival when data were stratified by loss of ER or PR, or overexpression of HER2	Hicks et al. (2006)
	mRNA	161 invasive carcinoma cases	High BRMS1 expression associated with better prognosis (disease-free survival). No relationship between BRMS1 expression and histological grade or axillary lymph node metastasis	Zhang, Yamashita, et al. (2006)
	mRNA	47 tumor cases, 14 peritumoral cases, and 15 metastatic cases	BRMS1 expression increased in tumor compared to normal tissue, and lower in lymph node samples compared to primary tumor. High BRMS1 expression correlated with shorter disease-free and overall survival	Lombardi et al. (2006)
	mRNA	11 normal breast, 16 fibroadenomas, 82 primary breast cancers, and 15 lymph node metastases	BRMS1 levels not significantly changed between normal breast, primary breast cancer, and lymph node metastases. BRMS1 expression was independent of tumor grade or node metastasis	Kelly et al. (2005)
	BRMS1 promoter methylation	CTC fractions from peripheral blood of 56 patients with operable breast cancer, 27 patients with metastases, and 23 healthy donors	BRMS1 promoter methylation observed in 32.1% of patients with operable cancer, compared to 44.4% of patients with metastases and 8.7% of healthy individuals	Chimonidou et al. (2011)

	BRMS1 promoter methylation	Tissues included 84 early breast cancers, 5 paired breast tumors and surrounding normal tissue, 14 noncancerous breast tissues, and 10 benign fibroadenomas. CTCs were isolated from peripheral blood of 39 of the early breast cancer patients	Methylation of the BRMS1 promoter in >1/3 of primary breast cancers, but not noncancerous or benign controls. BRMS1 promoter methylation was associated with reduced disease-free survival. BRMS1 promoter methylation was observed in a fraction of CTCs	Chimonidou, Kallergi, Georgoulias, Welch, and Lianidou (2013)
NSCLC	mRNA and protein	Cancer and adjacent normal tissues from 80 patients (protein) to 12 patients (mRNA)	BRMS1 mRNA and protein were both decreased in cancer compared to adjacent normal lung. Higher BRMS1 mRNA observed in tumors without lymphatic spread. Decreased BRMS1 expression was associated with a history of smoking. Expression of BRMS1 was associated with improved 5-year survival	Smith et al. (2009)
	BRMS1 promoter methylation	Tissue from 57 tumors and adjacent noncancerous tissue and cfDNA from 48 corresponding plasma samples; cfDNA from 74 advanced NSCLC; and cfDNA from 24 healthy donors	BRMS1 promoter methylation observed in NSCLC and corresponding cfDNA, as well as cfDNA from advanced NSCLC patients, but not cfDNA from healthy donors. Unmethylated BRMS1 promoter was associated with increased disease-free interval and overall survival	Balgkouranidou et al. (2014)
Melanoma	Protein	155 primary cancers, 69 metastases, and 15 nevi	Nuclear BRMS1 expression was observed in a greater proportion of nevi (87%) compared to melanomas (20%) and metastases (48%). Metastases had decreased cytoplasmic BRMS1 compared to nevi and primary tumors. High cytoplasmic BRMS1 level was associated with increased disease-free survival	Slipicevic et al. (2012)
	Protein	31 uveal melanoma cases	BRMS1 was expressed in 24/31 uveal melanoma specimens. There was no significant correlation between expression and survival	Ventura et al. (2014)

Continued

Table 3 Studies Examining the Role(s) of BRMS1 in Human Cancers—cont'd

Cancer Type	Measurement	Sample Information	Key Findings	References
Supraglottic laryngeal carcinoma	mRNA	66 cancer cases and adjacent normal mucosa	Lower expression of BRMS1 in cancer compared to adjacent tissue	Li et al. (2008)
Nasopharyngeal carcinoma	Protein	274 cancer cases	Low expression of BRMS1 associated with poor distant metastasis-free survival and poor overall survival	Cui et al. (2012)
Rectal	mRNA	80 cancer cases and 40 adjacent normal rectal mucosa specimens	Lower expression of BRMS1 in rectal cancer vs adjacent normal tissue. BRMS1 expression was related to lymph node metastasis, differentiation, and clinical disease stage	Zhang, Guan, et al. (2014)
Pheochromocytoma	mRNA	15 benign and 10 malignant pheochromocytomas	Significant downregulation of BRMS1 in malignant compared to benign tissue	Ohta et al. (2005)
Gallbladder adenocarcinoma	Protein	108 gallbladder adenocarcinomas, including 46 with matching peritumoral tissue; 15 gallbladder polyps; and 35 chronic cholecystitis specimens	Compared to peritumoral issue and polyps, fewer gallbladder adenocarcinoma specimens were positive for BRMS1 expression. BRMS1 positivity was associated with differentiation, lymph node metastasis, and invasion. Decreased BRMS1 was significantly associated with reduced overall survival	Yang et al. (2016)

CONCLUDING THOUGHTS

After the discovery of BRMS1 as a potent metastasis suppressor, the road to functional identification has been rocky at best without a clear GPS guided route to follow. Over the years, many processes important for metastasis have been demonstrated to be disrupted by BRMS1, yet the roles in normal biology for this enigmatic protein have not been fully elucidated. Uncovering the normal physiological roles for BRMS1 may be an important step to enable an escape from the "off-road" travels onto the superhighway toward understanding the clinical utility of metastasis-associated functionality. As such, it has become increasingly clear that epigenetic regulatory molecules important for normal development are also important for the process of metastasis (Micalizzi, Farabaugh, & Ford, 2010; Tam & Weinberg, 2013). It may very well be that BRMS1 involvement in metastasis is indicative of some critical roles necessary for normal organismal development. A more complete understanding of these roles will surely aid in our knowledge of metastasis and help in the design of new therapeutic, diagnostic, or prognostic strategies.

ACKNOWLEDGMENTS

We apologize to those whose work could not be cited due to space limitations. The Hurst and Welch labs have been generously funded by the National Cancer Institute (CA062168, CA087727, CA134981, CA089019, and CA168524), US Army Medical Research and Materiel Command, the National Foundation for Cancer Research, Susan G. Komen for the Cure, American Cancer Society (RSG-11-259-01-CSM), METAvivor Research and Support, Inc., and generous donations from a number of patients and their families.

REFERENCES

Al-Alwan, M., Olabi, S., Ghebeh, H., Barhoush, E., Tulbah, A., Al-Tweigeri, T., et al. (2011). Fascin is a key regulator of breast cancer invasion that acts via the modification of metastasis-associated molecules. *PLoS One*, *6*, e27339.

Balgkouranidou, I., Chimonidou, M., Milaki, G., Tsarouxa, E. G., Kakolyris, S., Welch, D. R., et al. (2014). Breast cancer metastasis suppressor-1 promoter methylation in cell-free DNA provides prognostic information in non-small cell lung cancer. *British Journal of Cancer*, *110*, 2054–2062.

Bansal, N., Petrie, K., Christova, R., Chung, C. Y., Leibovitch, B. A., Howell, L., et al. (2015). Targeting the SIN3A–PF1 interaction inhibits epithelial to mesenchymal transition and maintenance of a stem cell phenotype in triple negative breast cancer. *Oncotarget*, *6*, 34087–34105.

Bodenstine, T. M., Vaidya, K. S., Ismail, A., Beck, B. H., Cook, L. M., Diers, A. R., et al. (2010). Homotypic gap junctional communication associated with metastasis suppression increases with PKA activity and is unaffected by PI3K inhibition. *Cancer Research*, *70*, 10002–10011.

Bohl, C. R., Harihar, S., Denning, W. L., Sharma, R., & Welch, D. R. (2013). Metastasis suppressors in breast cancers: Mechanistic insights and clinical potential. *Journal of Molecular Medicine (Berlin, Germany)*, *92*, 13–30.

Bragado, P., Estrada, Y., Parikh, F., Krause, S., Capobianco, C., Farina, H. G., et al. (2013). TGF-beta2 dictates disseminated tumour cell fate in target organs through TGF-beta-RIII and p38alpha/beta signalling. *Nature Cell Biology*, *15*, 1351–1361.

Champine, P. J., Michaelson, J., Weimer, B., Welch, D. R., & DeWald, D. B. (2007). Microarray analysis reveals potential mechanisms of *BRMS1*-mediated metastasis suppression. *Clinical and Experimental Metastasis*, *24*, 551–565.

Chen, Y. C., Sosnoski, D. M., & Mastro, A. M. (2010). Breast cancer metastasis to the bone: Mechanisms of bone loss. *Breast Cancer Research*, *12*, 215.

Chimonidou, M., Kallergi, G., Georgoulias, V., Welch, D. R., & Lianidou, E. S. (2013). BRMS1 promoter methylation provides prognostic information in primary breast tumors. *Molecular Cancer Research*, *11*, 1248–1257.

Chimonidou, M., Strati, A., Tzitzira, A., Sotiropoulou, G., Malamos, N., Georgoulias, V., et al. (2011). DNA methylation of tumor suppressor and metastasis suppressor genes in circulating tumor cells. *Clinical Chemistry*, *57*, 1169–1177.

Cho, K. H., Yu, S. L., Cho, D. Y., Park, C. G., & Lee, H. Y. (2015). Breast cancer metastasis suppressor 1 (BRMS1) attenuates TGF-beta1-induced breast cancer cell aggressiveness through downregulating HIF-1alpha expression. *BMC Cancer*, *15*, 829.

Cicek, M., Fukuyama, R., Cicek, M. S., Sizemore, S., Welch, D. R., Sizemore, N., et al. (2009). BRMS1 contributes to the negative regulation of uPA gene expression through recruitment of HDAC1 to the NF-kappaB binding site of the uPA promoter. *Clinical and Experimental Metastasis*, *26*, 229–237.

Cicek, M., Fukuyama, R., Welch, D. R., Sizemore, N., & Casey, G. (2005). Breast cancer metastasis suppressor 1 inhibits gene expression by targeting nuclear factor-kB activity. *Cancer Research*, *65*, 3586–3595.

Cicek, M., & Oursler, M. J. (2006). Breast cancer bone metastasis and current small therapeutics. *Cancer and Metastasis Reviews*, *25*, 635–644.

Cicek, M., Samant, R. S., Kinter, M., Welch, D. R., & Casey, G. (2004). Identification of metastasis-associated proteins through protein analysis of metastatic MDA-MB-435 and metastasis-suppressed BRMS1 transfected-MDA-MB-435 cells. *Clinical and Experimental Metastasis*, *21*, 149–157.

Cook, L. M., Cao, X., Dowell, A. E., Debies, M. T., Edmonds, M. D., Beck, B. H., et al. (2012). Ubiquitous Brms1 expression is critical for mammary carcinoma metastasis suppression via promotion of apoptosis. *Clinical and Experimental Metastasis*, *29*, 315–325.

Crawford, N. P., Qian, X., Ziogas, A., Papageorge, A. G., Boersma, B. J., Walker, R. C., et al. (2007). Rrp1b, a new candidate susceptibility gene for breast cancer progression and metastasis. *PLoS Genetics*, *3*, e214.

Crawford, N. P. S., Yang, H. L., Mattaini, K. R., & Hunter, K. W. (2009). The metastasis efficiency modifier ribosomal RNA processing 1 homolog B (RRP1B) is a chromatin-associated factor. *Journal of Biological Chemistry*, *284*, 28660–28673.

Crea, F., Clermont, P. L., Parolia, A., Wang, Y., & Helgason, C. D. (2013). The non-coding transcriptome as a dynamic regulator of cancer metastasis. *Cancer Metastasis Reviews*, *33*(1), 1–16.

Cui, R. X., Liu, N., He, Q. M., Li, W. F., Huang, B. J., Sun, Y., et al. (2012). Low BRMS1 expression promotes nasopharyngeal carcinoma metastasis in vitro and in vivo and is associated with poor patient survival. *BMC Cancer*, *12*, 376.

Devilee, P., van den Broek, M., Mannens, M., Slater, R., Cornelisse, C. J., Westerveld, A., et al. (1991). Differences in patterns of allelic loss between two common types of adult cancer, breast and colon carcinoma, and Wilms' tumor of childhood. *International Journal of Cancer*, *47*, 817–821.

DeWald, D. B., Torabinejad, J., Samant, R. S., Johnston, D., Erin, N., Shope, J. C., et al. (2005). Metastasis suppression by breast cancer metastasis suppressor 1 involves reduction of phosphoinositide signaling in MDA-MB-435 breast carcinoma cells. *Cancer Research*, *65*, 713–717.

Doyon, Y., Cayrou, C., Ullah, M., Landry, A. J., Cote, V., Selleck, W., et al. (2006). ING tumor suppressor proteins are critical regulators of chromatin acetylation required for genome expression and perpetuation. *Molecular Cell*, *21*, 51–64.

Eccles, S. A., & Welch, D. R. (2007). Metastasis: Recent discoveries and novel treatment strategies. *Lancet*, *369*, 1742–1757.

Edmonds, M. D., Hurst, D. R., & Welch, D. R. (2009). Linking metastasis suppression with metastamiR regulation. *Cell Cycle*, *8*, 2673–2675.

Farias, E. F., Petrie, K., Leibovitch, B., Murtagh, J., Chornet, M. B., Schenk, T., et al. (2010). Interference with Sin3 function induces epigenetic reprogramming and differentiation in breast cancer cells. *Proceedings of the National Academy of Sciences of the United States of America*, *107*, 11811–11816.

Fazilaty, H., & Mehdipour, P. (2014). Genetics of breast cancer bone metastasis: A sequential multistep pattern. *Clinical & Experimental Metastasis*, *31*, 595–612.

Fidler, I. J., Kim, S. I., & Langley, R. R. (2007). The role of the organ microenvironment in the biology and therapy of cancer metastasis. *Journal of Cellular Biochemistry*, *101*, 927–936.

Frolova, N., Edmonds, M. D., Bodenstine, T. M., Seitz, R., Johnson, M. R., Feng, R., et al. (2009). A shift from nuclear to cytoplasmic Breast Cancer Metastasis Suppressor 1 expression is associated with highly proliferative estrogen receptor-negative breast cancers. *Tumor Biology*, *30*, 148–159.

Grzenda, A., Lomberk, G., Zhang, J. S., & Urrutia, R. (2009). Sin3: Master scaffold and transcriptional corepressor. *Biochimica et Biophysica Acta—Gene Regulatory Mechanisms*, *1789*, 443–450.

Gudmundsson, J., Barkardottir, R. B., Eiriksdottir, G., Baldursson, T., Arason, A., Egilsson, V., et al. (1995). Loss of heterozygosity at chromosome 11 in breast cancer: Association of prognostic factors with genetic alterations. *British Journal of Cancer*, *72*, 696–701.

Guo, X. L., Wang, Y. J., Cui, P. L., Wang, Y. B., Liang, P. X., Zhang, Y. N., et al. (2015). Effect of BRMS1 expression on proliferation, migration and adhesion of mouse forestomach carcinoma. *Asian Pacific Journal of Tropical Medicine*, *8*, 724–730.

Gupta, R. A., Shah, N., Wang, K. C., Kim, J., Horlings, H. M., Wong, D. J., et al. (2010). Long non-coding RNA HOTAIR reprograms chromatin state to promote cancer metastasis. *Nature*, *464*, 1071–1076.

Gutschner, T., Hammerle, M., Eissmann, M., Hsu, J., Kim, Y., Hung, G., et al. (2013). The noncoding RNA MALAT1 is a critical regulator of the metastasis phenotype of lung cancer cells. *Cancer Research*, *73*, 1180–1189.

Hall, E. H., Liu, Y., Xiao, A., Shock, L., Brautigan, D. L., Mayo, M. W., et al. (2014). Inhibition of breast cancer metastasis suppressor 1 promotes a mesenchymal phenotype in lung epithelial cells that express oncogenic K-RasV12 and loss of p53. *PLoS One*, *9*, e95869.

Hampton, G. M., Mannermaa, A., Winquist, R., Alavaikko, M., Blanco, G., Taskinen, P. J., et al. (1994). Loss of heterozygosity in sporadic human breast carcinoma: A common region between 11q22 and 11q23.3. *Cancer Research*, *54*, 4586–4589.

Hedley, B. D., Vaidya, K. S., Phadke, P., MacKenzie, L., Dales, D. W., Postenka, C. O., et al. (2008). BRMS1 suppresses breast cancer metastasis in multiple experimental models of metastasis by reducing solitary cell survival and inhibiting growth initiation. *Clinical and Experimental Metastasis*, *25*, 727–740.

Hedley, B. D., Welch, D. R., Allan, A. L., Al-Katib, W., Dales, D. W., Postenka, C. O., et al. (2008). Downregulation of osteopontin contributes to metastasis suppression by breast cancer metastasis suppressor 1. *International Journal of Cancer*, *123*, 526–534.
Hicks, D. G., Yoder, B. J., Short, S., Tarr, S., Prescott, N., Crowe, J. P., et al. (2006). Loss of BRMS1 protein expression predicts reduced disease-free survival in hormone receptor negative and HER2 positive subsets of breast cancer. *Clinical Cancer Research*, *12*, 6702–6708.
Hurst, D. R. (2012). Metastasis suppression by BRMS1 associated with SIN3 chromatin remodeling complexes. *Cancer Metastasis Reviews*, *31*, 641–651.
Hurst, D. R., Edmonds, M. D., Scott, G. K., Benz, C. C., Vaidya, K. S., & Welch, D. R. (2009a). Breast cancer metastasis suppressor 1 BRMS1 up-regulates miR-146 that suppresses breast cancer metastasis. *Cancer Research*, *69*, 1279–1283.
Hurst, D. R., Edmonds, M. D., & Welch, D. R. (2009b). Metastamir: The field of metastasis-regulatory microRNA is spreading. *Cancer Research*, *69*, 7495–7498.
Hurst, D. R., Mehta, A., Moore, B. P., Phadke, P. A., Meehan, W. J., Accavitti, M. A., et al. (2006). Breast cancer metastasis suppressor 1 (BRMS1) is stabilized by the Hsp90 chaperone. *Biochemical and Biophysical Research Communications*, *348*, 1429–1435.
Hurst, D. R., & Welch, D. R. (2011a). Metastasis suppressor genes: At the interface between the environment and tumor cell growth. *International Review of Cell and Molecular Biology*, *286*, 107–180.
Hurst, D. R., & Welch, D. R. (2011b). Unraveling the enigmatic complexities of BRMS1-mediated metastasis suppression. *FEBS Letters*, *585*, 3185–3190.
Hurst, D. R., Xie, Y., Edmonds, M. D., & Welch, D. R. (2009). Multiple forms of BRMS1 are differentially expressed in the MCF10 isogenic breast cancer progression model. *Clinical and Experimental Metastasis*, *26*, 89–96.
Hurst, D. R., Xie, Y., Thomas, J. W., Liu, J., Edmonds, M. D., Stewart, M. D., et al. (2013). The C-terminal putative nuclear localization sequence of BReast cancer Metastasis Suppressor 1, BRMS1, is necessary for metastasis suppression. *PLoS One*, *8*, e55966.
Hurst, D. R., Xie, Y., Vaidya, K. S., Mehta, A., Moore, B. P., Accavitti-Loper, M. A., et al. (2008). Alterations of BRMS1–ARID4A interaction modify gene expression but still suppress metastasis in human breast cancer cells. *Journal of Biological Chemistry*, *283*, 7438–7444.
Jiang, J., Xia, M., Feng, J. B., & Kong, B. H. (2007). Inhibitory effect of breast cancer metastasis suppressor l gene on metastasis of human ovarian cancer cell in vitro and in vivo. *Zhonghua Fu Chan Ke Za Zhi*, *42*, 398–402.
Jones, P. A., & Baylin, S. B. (2002). The fundamental role of epigenetic events in cancer. *Nature Reviews. Genetics*, *3*, 415–428.
Kadamb, R., Mittal, S., Bansal, N., Batra, H., & Saluja, D. (2013). Sin3: Insight into its transcription regulatory functions. *European Journal of Cell Biology*, *92*, 237–246.
Kang, Y. B., Siegel, P. M., Shu, W. P., Drobnjak, M., Kakonen, S. M., Cordón-Cardo, C., et al. (2003). A multigenic program mediating breast cancer metastasis to bone. *Cancer Cell*, *3*, 537–549.
Kapoor, P., Saunders, M. M., Li, Z., Zhou, Z., Schaeffer, N., Kunze, E. L., et al. (2004). Breast cancer metastatic potential: Correlation with increased heterotypic gap junctional intercellular communication between breast cancer cells and osteoblastic cells. *International Journal of Cancer*, *111*, 693–697.
Karin, M., Cao, Y., Greten, F. R., & Li, Z. W. (2002). NF-kappaB in cancer: From innocent bystander to major culprit. *Nature Reviews. Cancer*, *2*, 301–310.
Kelly, L. M., Buggy, Y., Hill, A., O'Donovan, N., Duggan, C., McDermott, E. W., et al. (2005). Expression of the breast cancer metastasis suppressor gene, BRMS1, in human breast carcinoma: Lack of correlation with metastasis to axillary lymph nodes. *Tumor Biology*, *26*, 213–216.

Khanna, C., Fan, T. M., Gorlick, R., Helman, L. J., Kleinerman, E. S., Adamson, P. C., et al. (2014). Toward a drug development path that targets metastatic progression in osteosarcoma. *Clinical Cancer Research*, *20*(16), 4200–4209.

Khotskaya, Y. B., Beck, B. H., Hurst, D. R., Han, Z., Xia, W., Hung, M. C., et al. (2014). Expression of metastasis suppressor BRMS1 in breast cancer cells results in a marked delay in cellular adhesion to matrix. *Molecular Carcinogenesis*, *53*(12), 1011–1026.

Kodura, M. A., & Souchelnytskyi, S. (2015). Breast carcinoma metastasis suppressor gene 1 (BRMS1): Update on its role as the suppressor of cancer metastases. *Cancer Metastasis Reviews*, *34*, 611–618.

Kogo, R., Shimamura, T., Mimori, K., Kawahara, K., Imoto, S., Sudo, T., et al. (2011). Long noncoding RNA HOTAIR regulates polycomb-dependent chromatin modification and is associated with poor prognosis in colorectal cancers. *Cancer Research*, *71*, 6320–6326.

Kwon, Y. J., Petrie, K., Leibovitch, B. A., Zeng, L., Mezei, M., Howell, L., et al. (2015). Selective inhibition of SIN3 corepressor with avermectins as a novel therapeutic strategy in triple-negative breast cancer. *Molecular Cancer Therapeutics*, *14*, 1824–1836.

Langley, R. R., & Fidler, I. J. (2011). The seed and soil hypothesis revisited—The role of tumor–stroma interactions in metastasis to different organs. *International Journal of Cancer*, *128*, 2527–2535.

Le Guezennec, X., Vermeulen, M., & Stunnenberg, H. G. (2006). Molecular characterization of Sin3 PAH-domain interactor specificity and identification of PAH partners. *Nucleic Acids Research*, *34*, 3929–3937.

Li, X., Guo, X., Li, F., Li, X., Guan, C., Yang, H., et al. (2008). Expression and clinical significance of breast cancer metastasis suppressor 1 mRNA in supraglottic laryngeal carcinoma. *Lin Chung Er Bi Yan Hou Tou Jing Wai Ke Za Zhi*, *22*, 241–244.

Li, W., & Kang, Y. (2014). A new Lnc in metastasis: Long noncoding RNA mediates the prometastatic functions of TGF-beta. *Cancer Cell*, *25*, 557–559.

Li, D. Q., Pakala, S. B., Nair, S. S., Eswaran, J., & Kumar, R. (2012). Metastasis-associated protein 1/nucleosome remodeling and histone deacetylase complex in cancer. *Cancer Research*, *72*, 387–394.

Liu, Y., Mayo, M. W., Nagji, A. S., Smith, P. W., Ramsey, C. S., Li, D., et al. (2011). Phosphorylation of RelA/p65 promotes DNMT-1 recruitment to chromatin and represses transcription of the tumor metastasis suppressor gene BRMS1. *Oncogene*, *31*, 1143–1154.

Liu, Y., Mayo, M. W., Xiao, A., Hall, E. H., Amin, E. B., Kadota, K., et al. (2015). Loss of BRMS1 promotes a mesenchymal phenotype through NF-kappaB-dependent regulation of Twist1. *Molecular and Cellular Biology*, *35*, 303–317.

Liu, Y., Smith, P. W., & Jones, D. R. (2006). Breast cancer metastasis suppressor 1 functions as a corepressor by enhancing histone deacetylase 1-mediated deacetylation of RelA/p65 and promoting apoptosis. *Molecular and Cellular Biology*, *26*, 8683–8696.

Liu, W., Vivian, C. J., Brinker, A. E., Hampton, K. R., Lianidou, E., & Welch, D. R. (2014). Microenvironmental influences on metastasis suppressor expression and function during a metastatic cell's journey. *Cancer Microenvironment*, *7*, 117–131.

Lombardi, G., Di Cristofano, C., Capodanno, A., Iorio, M. C., Aretini, P., Isola, P., et al. (2006). High level of messenger RNA for BRMS1 in primary breast carcinomas is associated with poor prognosis. *International Journal of Cancer*, *120*, 1169–1178.

Lu, L. F., Boldin, M. P., Chaudhry, A., Lin, L. L., Taganov, K. D., Hanada, T., et al. (2010). Function of miR-146a in controlling Treg cell-mediated regulation of Th1 responses. *Cell*, *142*, 914–929.

McDonel, P., Costello, I., & Hendrich, B. (2008). Keeping things quiet: Roles of NuRD and Sin3 co-repressor complexes during mammalian development. *The International Journal of Biochemistry & Cell Biology*, *48*, 108–116.

Meehan, W. J., Samant, R. S., Hopper, J. E., Carrozza, M. J., Shevde, L. A., Workman, J. L., et al. (2004). Breast cancer metastasis suppressor 1 (BRMS1) forms complexes with retinoblastoma-binding protein 1 (RBP1) and the mSin3 histone deacetylase complex and represses transcription. *Journal of Biological Chemistry, 279,* 1562–1569.

Mei, P., Bai, J., Shi, M., Liu, Q., Li, Z., Fan, Y., et al. (2014). BRMS1 suppresses glioma progression by regulating invasion, migration and adhesion of glioma cells. *PLoS One, 9,* e98544.

Metge, B. J., Frost, A. R., King, J. A., Dyess, D. L., Welch, D. R., Samant, R. S., et al. (2008). Epigenetic silencing contributes to the loss of BRMS1 expression in breast cancer. *Clinical and Experimental Metastasis, 25,* 753–763.

Micalizzi, D. S., Farabaugh, S. M., & Ford, H. L. (2010). Epithelial–mesenchymal transition in cancer: Parallels between normal development and tumor progression. *Journal of Mammary Gland Biology and Neoplasia, 15,* 117–134.

Minn, A. J., Kang, Y. B., Serganova, I., Gupta, G. P., Giri, D. D., Doubrovin, M., et al. (2005). Distinct organ-specific metastatic potential of individual breast cancer cells and primary tumors. *Journal of Clinical Investigation, 115,* 44–55.

Nagji, A. S., Liu, Y., Stelow, E. B., Stukenborg, G. J., & Jones, D. R. (2010). BRMS1 transcriptional repression correlates with CpG island methylation and advanced pathological stage in non-small cell lung cancer. *The Journal of Pathology, 221,* 229–237.

Negrini, M., Rasio, D., Hampton, G. M., Sabbioni, S., Rattan, S., Carter, S. L., et al. (1995). Definition and refinement of chromosome 11 regions of loss of heterozygosity in breast cancer: Identification of a new region at 11q23.3. *Cancer Research, 55,* 3003–3007.

Nguyen, D. X., Bos, P. D., & Massague, J. (2009). Metastasis: From dissemination to organ-specific colonization. *Nature Reviews. Cancer, 9,* 274–284.

Nikolaev, A. Y., Papanikolaou, N. A., Li, M., Qin, J., & Gu, W. (2004). Identification of a novel BRMS1-homologue protein p40 as a component of the mSin3A/p33(ING1b)/HDAC1 deacetylase complex. *Biochemical and Biophysical Research Communications, 323,* 1216–1222.

Obenauf, A. C., & Massague, J. (2015). Surviving at a distance: Organ specific metastasis. *Trends in Cancer, 1,* 76–91.

Ohta, S., Lai, E. W., Pang, A. L. Y., Brouwers, F. M., Chan, W. Y., Eisenhofer, G., et al. (2005). Downregulation of metastasis suppressor genes in malignant pheochromocytoma. *International Journal of Cancer, 114,* 139–143.

Pan, S., Yu, F., Gong, C., & Song, E. (2009). Tumor invasion and metastasis initiated by mir-106b in breast cancer by targeting BRMS1 and RB. *Cancer Research, 69,* 875S.

Phadke, P. A., Vaidya, K. S., Nash, K. T., Hurst, D. R., & Welch, D. R. (2008). BRMS1 suppresses breast cancer experimental metastasis to multiple organs by inhibiting several steps of the metastatic process. *American Journal of Pathology, 172,* 809–817.

Phillips, K. K., Welch, D. R., Miele, M. E., Lee, J.-H., Wei, L. L., & Weissman, B. E. (1996). Suppression of MDA-MB-435 breast carcinoma cell metastasis following the introduction of human chromosome 11. *Cancer Research, 56,* 1222–1226.

Ponnusamy, S., Selvam, S. P., Mehrotra, S., Kawamori, T., Snider, A. J., Obeid, L. M., et al. (2012). Communication between host organism and cancer cells is transduced by systemic sphingosine kinase 1/sphingosine 1-phosphate signalling to regulate tumour metastasis. *EMBO Molecular Medicine, 4,* 761–775.

Rivera, J., Megias, D., & Bravo, J. (2007). Proteomics-based strategy to delineate the molecular mechanisms of the metastasis suppressor gene BRMS1. *Journal of Proteome Research, 6,* 4006–4018.

Rivera, J., Megias, D., & Bravo, J. (2010). Sorting nexin 6 interacts with breast cancer metastasis suppressor-1 and promotes transcriptional repression. *Journal of Cellular Biochemistry, 111,* 1464–1472.

Rodriguez-Paredes, M., & Esteller, M. (2011). Cancer epigenetics reaches mainstream oncology. *Nature Medicine*, *17*, 330–339.

Roesley, S. N., Suryadinata, R., Morrish, E., Tan, A. R., Issa, S. M., Oakhill, J. S., et al. (2016). Cyclin-dependent kinase-mediated phosphorylation of breast cancer metastasis suppressor 1 (BRMS1) affects cell migration. *Cell Cycle*, *15*, 137–151.

Samant, R. S., Clark, D. W., Fillmore, R. A., Cicek, M., Metge, B. J., Chandramouli, K. H., et al. (2007). Breast cancer metastasis suppressor 1 (BRMS1) inhibits osteopontin transcription by abrogating NF-kappaB activation. *Molecular Cancer*, *6*, 6.

Samant, R. S., Debies, M. T., Hurst, D. R., Moore, B. P., Shevde, L. A., & Welch, D. R. (2006). Suppression of murine mammary carcinoma metastasis by the murine ortholog of breast cancer metastasis suppressor 1 (Brms1). *Cancer Letters*, *235*, 260–265.

Samant, R. S., Seraj, M. J., Saunders, M. M., Sakamaki, T., Shevde, L. A., Harms, J. F., et al. (2001). Analysis of mechanisms underlying *BRMS1* suppression of metastasis. *Clinical and Experimental Metastasis*, *18*, 683–693.

Saunders, M. M., Seraj, M. J., Li, Z. Y., Zhou, Z. Y., Winter, C. R., Welch, D. R., et al. (2001). Breast cancer metastatic potential correlates with a breakdown in homospecific and heterospecific gap junctional intercellular communication. *Cancer Research*, *61*, 1765–1767.

Seraj, M. J., Harding, M. A., Gildea, J. J., Welch, D. R., & Theodorescu, D. (2001). The relationship of *BRMS1* and *RhoGDI2* gene expression to metastatic potential in lineage related human bladder cancer cell lines. *Clinical and Experimental Metastasis*, *18*, 519–525.

Seraj, M. J., Samant, R. S., Verderame, M. F., & Welch, D. R. (2000). Functional evidence for a novel human breast carcinoma metastasis suppressor, *BRMS1*, encoded at chromosome 11q13. *Cancer Research*, *60*, 2764–2769.

Sethi, N., & Kang, Y. (2011). Unravelling the complexity of metastasis—Molecular understanding and targeted therapies. *Nature Reviews. Cancer*, *11*, 735–748.

Sheng, X. J., Zhou, Y. Q., Song, Q. Y., Zhou, D. M., & Liu, Q. C. (2012). Loss of breast cancer metastasis suppressor 1 promotes ovarian cancer cell metastasis by increasing chemokine receptor 4 expression. *Oncology Reports*, *27*, 1011–1018.

Shevde, L. A., Samant, R. S., Goldberg, S. F., Sikaneta, T., Alessandrini, A., Donahue, H. J., et al. (2002). Suppression of human melanoma metastasis by the metastasis suppressor gene, BRMS1. *Experimental Cell Research*, *273*, 229–239.

Shiio, Y., Rose, D. W., Aur, R., Donohoe, S., Aebersold, R., & Eisenman, R. N. (2006). Identification and characterization of SAP25, a novel component of the mSin3 corepressor complex. *Molecular and Cellular Biology*, *26*, 1386–1397.

Silveira, A. C., Hurst, D. R., Vaidya, K. S., Ayer, D. E., & Welch, D. R. (2009). Overexpression of the BRMS1 family member SUDS3 does not suppress metastasis of human cancer cells. *Cancer Letters*, *276*, 32–37.

Silverstein, R. A., & Ekwall, K. (2005). Sin3: A flexible regulator of global gene expression and genome stability. *Current Genetics*, *47*, 1–17.

Sleeman, J. P., Nazarenko, I., & Thiele, W. (2011). Do all roads lead to Rome? Routes to metastasis development. *International Journal of Cancer*, *128*, 2511–2526.

Slipicevic, A., Holm, R., Emilsen, E., Ree Rosnes, A. K., Welch, D. R., Maelandsmo, G. M., et al. (2012). Cytoplasmic BRMS1 expression in malignant melanoma is associated with increased disease-free survival. *BMC Cancer*, *12*, 73.

Smith, P. W., Liu, Y., Siefert, S. A., Moskaluk, C. A., Petroni, G. R., & Jones, D. R. (2009). Breast cancer metastasis suppressor 1 (BRMS1) suppresses metastasis and correlates with improved patient survival in non-small cell lung cancer. *Cancer Letters*, *276*, 196–203.

Smith, K. T., Martin-Brown, S. A., Florens, L., Washburn, M. P., & Workman, J. L. (2010). Deacetylase inhibitors dissociate the histone-targeting ING2 subunit from the Sin3 complex. *Chemistry and Biology*, *17*, 65–74.

Steeg, P. S., Anderson, R. L., Bar-Eli, M., Chambers, A. F., Eccles, S. A., Hunter, K., et al. (2009). Preclinical drug development must consider the impact on metastasis. *Clinical Cancer Research, 15*, 4529–4530.

Suryadinata, R., Sadowski, M., Steel, R., & Sarcevic, B. (2011). Cyclin-dependent kinase-mediated phosphorylation of RBP1 and pRb promotes their dissociation to mediate release of the SAP30-mSin3-HDAC transcriptional repressor complex. *Journal of Biological Chemistry, 286*, 5108–5118.

Tam, W. L., & Weinberg, R. A. (2013). The epigenetics of epithelial–mesenchymal plasticity in cancer. *Nature Medicine, 19*, 1438–1449.

Vaidya, K. S., Harihar, S., Stafford, L. J., Hurst, D. R., Hicks, D. G., Casey, G., et al. (2008). Breast cancer metastasis suppressor-1 differentially modulates growth factor signaling. *Journal of Biological Chemistry, 283*, 28354–28360.

Ventura, B. V., Quezada, C., Maloney, S. C., Fernandes, B. F., Antecka, E., Martins, C., et al. (2014). Expression of the metastasis suppressor BRMS1 in uveal melanoma. *ECancerMedicalScience, 8*, 410.

Wang, Y., Zhang, H., Chen, Y., Sun, Y., Yang, F., Yu, W., et al. (2009). LSD1 is a subunit of the NuRD complex and targets the metastasis programs in breast cancer. *Cell, 138*, 660–672.

Welch, D. R., & Wei, L. L. (1998). Genetic and epigenetic regulation of human breast cancer progression and metastasis. *Endocrine Related Cancer, 5*, 155–197.

Wu, Y., Jiang, W., Wang, Y., Wu, J., Saiyin, H., Qiao, X., et al. (2012). Breast cancer metastasis suppressor 1 regulates hepatocellular carcinoma cell apoptosis via suppressing osteopontin expression. *PLoS One*, 7, e42976.

Xing, W. J., Liao, X. H., Wang, N., Zhao, D. W., Zheng, L., Zheng, D. L., et al. (2015). MRTF-A and STAT3 promote MDA-MB-231 cell migration via hypermethylating BRSM1. *IUBMB Life, 67*, 202–217.

Xue, Y., Wong, J., Moreno, G. T., Young, M. K., Cote, J., & Wang, W. (1998). NURD, a novel complex with both ATP-dependent chromatin-remodeling and histone deacetylase activities. *Molecular Cell, 2*, 851–861.

Yang, Z., Liu, F., & Yang, Z. L. (2016). BRMS1 and HPA as progression, clinical biological behaviors, and poor prognosis-related biomarkers for gallbladder adenocarcinoma. *Applied Immunohistochemistry & Molecular Morphology, 24*(4), 275–282.

Yang, J., Zhang, B., Lin, Y., Yang, Y., Liu, X., & Lu, F. (2008). Breast cancer metastasis suppressor 1 inhibits SDF-1alpha-induced migration of non-small cell lung cancer by decreasing CXCR4 expression. *Cancer Letters, 269*, 46–56.

You, J., He, X., Ding, H., & Zhang, T. (2015). BRMS1 regulates apoptosis in non-small cell lung cancer cells. *Cell Biochemistry and Biophysics, 71*, 465–472.

Zhang, Y., Guan, J., Sun, Y., Chai, J., Zou, T., Gong, W., et al. (2014). Effect of BRMS1 on tumorigenicity and metastasis of human rectal cancer. *Cell Biochemistry and Biophysics, 70*(1), 505–509.

Zhang, L., Huang, J., Yang, N., Liang, S., Barchetti, A., Giannakakis, A., et al. (2006). Integrative genomic analysis of protein kinase C (PKC) family identifies PKCi as a biomarker and potential oncogene in ovarian carcinoma. *Cancer Research, 66*, 4627–4635.

Zhang, S., Lin, Q. D., & Di, W. (2006). Suppression of human ovarian carcinoma metastasis by the metastasis-suppressor gene, BRMS1. *International Journal of Gynecology and Cancer, 16*, 522–531.

Zhang, W., Qian, P., Zhang, X., Zhang, M., Wang, H., Wu, M., et al. (2015). Autocrine/paracrine human growth hormone-stimulated microRNA 96-182-183 cluster promotes epithelial–mesenchymal transition and invasion in breast cancer. *Journal of Biological Chemistry, 290*, 13812–13829.

Zhang, Z., Yamashita, H., Toyama, T., Yamamoto, Y., Kawasoe, T., & Iwase, H. (2006). Reduced expression of the breast cancer metastasis suppressor 1 mRNA is correlated with poor progress in breast cancer. *Clinical Cancer Research*, *12*, 6410–6414.

Zhang, Q., Zhang, Y., Wu, L., Yang, Y., Li, X., Gao, L., et al. (2014). dBrms1 acts as a positive regulator of notch signaling in drosophila wing. *Journal of Genetics and Genomics*, *41*, 317–325.

Zheng, P., Xiong, Q., Wu, Y., Chen, Y., Chen, Z., Fleming, J., et al. (2015). Quantitative proteomics analysis reveals novel insights into mechanisms of action of long noncoding RNA HOTAIR in HeLa cells. *Molecular & Cellular Proteomics: MCP*, *14*(6), 1447–1463.

CHAPTER FIVE

Immune Regulation of the Metastatic Process: Implications for Therapy

A. de Mingo Pulido, B. Ruffell[1]
H. Lee Moffitt Cancer Center and Research Institute, Tampa, FL, United States
[1]Corresponding author: e-mail address: brian.ruffell@moffitt.org

Contents

1. Introduction 140
2. Key Players 141
   2.1 Monocytes 141
   2.2 Macrophages 141
   2.3 Neutrophils 144
   2.4 T Lymphocytes 144
   2.5 Natural Killer Cells 145
3. Invasion and Intravasation 145
   3.1 Macrophages 145
   3.2 Neutrophils 147
   3.3 T Lymphocytes 148
4. Survival and Extravasation 148
   4.1 Monocytes 148
   4.2 Neutrophils 149
   4.3 T Lymphocytes 150
   4.4 NK Cells 151
5. Ectopic Growth 151
   5.1 Macrophages 151
   5.2 Neutrophils 152
   5.3 T Lymphocytes 153
   5.4 NK Cells 154
6. Therapeutic Implications 154
Acknowledgments 156
References 156

Abstract

Metastatic disease is the major cause of fatalities in cancer patients, but few therapies are designed to target the metastatic process. Cancer cells must perform a number of steps to successfully establish metastatic foci, including local invasion, intravasation, survival, extravasation, and growth in ectopic tissue. Due to the nonrandom distribution of

Advances in Cancer Research, Volume 132
ISSN 0065-230X
http://dx.doi.org/10.1016/bs.acr.2016.05.004

metastasis, it has long been recognized that the tissue microenvironment must be an important determinant of colonization. More recently it has been established in animal models that immune cells regulate the metastatic process, including a dominant role for monocytes and macrophages, and emerging roles for neutrophils and various lymphocyte populations. While most research has focused on the early dissemination process, patients usually present clinically with disseminated, if not macroscopic, disease. Identifying pathways by which immune cells regulate growth and therapeutic resistance within metastatic sites is therefore key to the development of pharmacological agents that will significantly extend patient survival.

1. INTRODUCTION

Although cancer is a disease defined by genetic mutations, genomic instability, and/or chromosomal alterations within malignant cells, solid tumors are comprised of multiple "host" stromal populations, including lymphatic and vascular endothelial cells, fibroblasts and other mesenchymal cells, and a diverse repertoire of immune cells. Cancer cells do not manifest into disease in isolation, but rather draft and corrupt resident and recruited normal cell types that can promote or restrain tumor progression in a highly context-dependent manner. The understanding of how of these stromal populations influence each step of tumor development from initiation to growth and progression has become more pronounced as the complexity of tumor modeling has increased (Hanahan & Coussens, 2012).

Cancer-related fatalities are usually associated with disseminated metastatic disease, not growth of the primary tumor (Valastyan & Weinberg, 2011), and thus understanding the process of metastasis is important for the development of efficacious therapeutic interventions. Metastasis occurs through a sequential cascade of steps: local invasion, intravasation, and survival in circulation, followed by extravasation, survival, and growth at the ectopic site. As with primary tumor development, each of these steps involves interplay between neoplastic cells and their microenvironment. Immune cells, in particular, play an important role navigating neoplastic cells through the endothelial layer, and interrupting this process diminishes the formation of metastatic lesions in animal models (Joyce & Pollard, 2009; Kitamura, Qian, & Pollard, 2015).

Unfortunately, the majority of human cancers have already disseminated when the primary tumor is detected, having entered the circulatory system prior to diagnosis (Friberg & Nystrom, 2015; Massagé & Obenauf, 2016).

It is therefore the later steps of the metastatic process that are most clinically relevant, but paradoxically are the least understood. A few preclinical studies have shown that whether quiescent cancer cells will establish overt metastatic lesions or remain in a dormant state during a latency period may depend upon immune surveillance (Eyles et al., 2010; Romero et al., 2014). Growth at the metastatic site can also be regulated by myeloid cells, as is well described for bone metastasis (Cook, Shay, Araujo, & Lynch, 2014). Finally, the primary tumor microenvironment dictates response to all forms of therapeutic intervention, and thus understanding the role of these cells in metastatic lesions could assist in the development of new treatment strategies (Coffelt & de Visser, 2015; Coussens, Zitvogel, & Palucka, 2013; Klemm & Joyce, 2015; Ruffell & Coussens, 2015). Here we will discuss the role of immune cells in regulating the metastatic process of solid tumors (Fig. 1), with a detailed examination of the establishment of overt metastasis and its implications for therapy. Due to the unique interplay between the immune system and hematological malignancies these will not be discussed (Meads, Gatenby, & Dalton, 2009; Nelson & Paulos, 2015).

2. KEY PLAYERS

2.1 Monocytes

Monocytes are usually associated with cancer only inasmuch as they are a precursor population to macrophages and dendritic cells (DCs) under inflammatory conditions. However, monocytes are found within the spleen and trafficking through tissue without differentiating into either terminal population (Jakubzick et al., 2013; Swirski et al., 2009). There is evidence that monocytes are functionally important during the early stage of tumor development (Juric et al., 2016), and in solid tumors a monocyte gene expression signature can be a negative prognostic indicator (Gentles et al., 2015). Much of the literature on myeloid-derived suppressor cells may also be attributable to $Ly6C^+$ monocytes within the spleens of tumor-bearing mice. Finally, recent studies have described opposing roles for $Ly6C^+$ classical/inflammatory and $Ly6C^-$ patrolling monocytes in mediating the early stages of lung colonization (Hanna et al., 2015; Qian et al., 2011).

2.2 Macrophages

Macrophages are present in all mammalian tissues, represented by unique phenotypic and functional resident populations that are critical for tissue

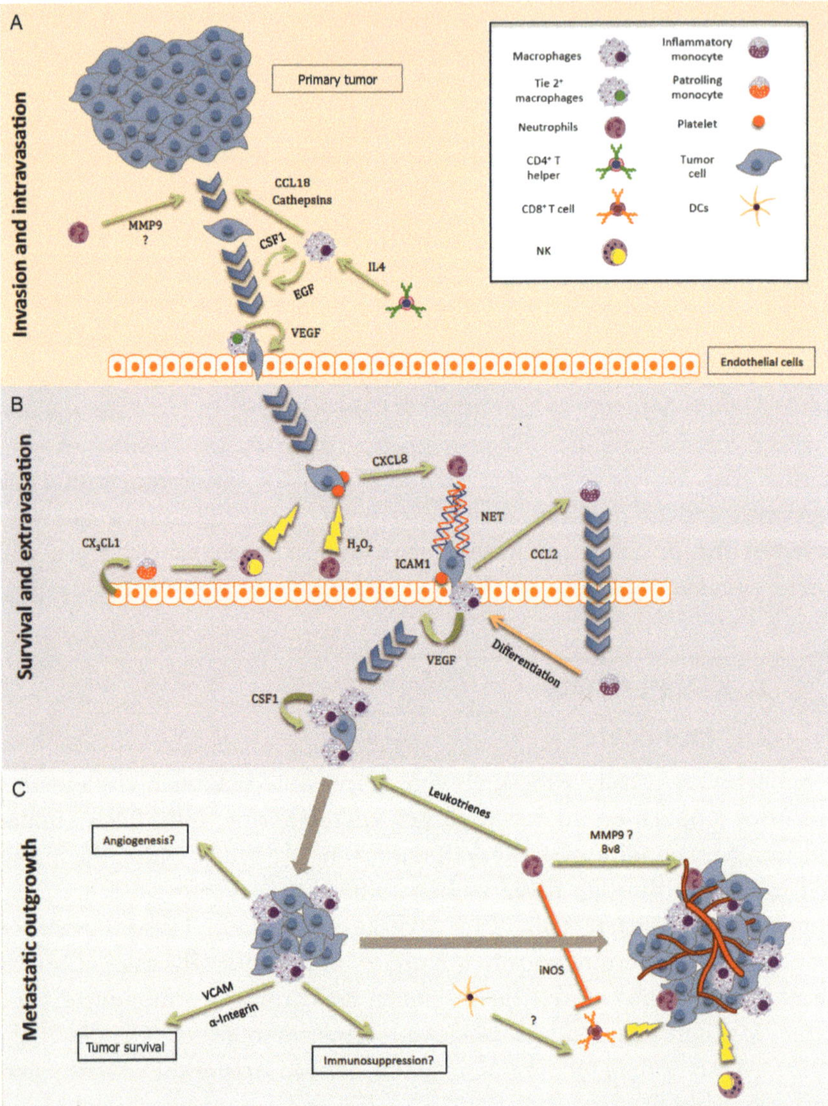

Fig. 1 Immune regulation of the metastatic cascade. (A) Invasion of tumor cells toward the vasculature is driven by macrophage expression of the chemoattractants EGF and CCL18 (in humans). Macrophage recruitment to the primary tumor and its ability to secrete EGF is dependent upon CSF-1 expression by the tumor cells, resulting in a paracrine loop. This loop is further augmented by the cytokine IL-4, expressed by T_H2-polarized $CD4^+$ T cells. IL-4 also drives cathepsin expression by macrophages, which promotes local invasion through an unidentified mechanism. Stromal-derived MMPs are also thought to enhance the invasion process by regulating the structure of the ECM and increasing the bioavailability of growth factors and chemokines. A subset of $Tie2^+$ macrophages associated with the endothelium is responsible for increasing

development and homeostasis. Under nonpathological conditions, resident macrophages in many tissues originate from embryonic progenitors and are conserved through local proliferation (Wynn, Chawla, & Pollard, 2013). Within tumors, macrophages originate from circulating monocytes and are one of the most prominent populations of immune cells. Although macrophages were originally thought to be part of an antitumor response, animal studies and clinical correlations indicate that macrophages promote tumorigenesis under most conditions by promoting invasion, angiogenesis, and survival, while also directly and indirectly suppressing a cytotoxic T cell response (Noy & Pollard, 2014; Ruffell, Affara, & Coussens, 2012). Exceptions to this are found in colorectal cancer and under select therapeutic conditions where the macrophage phenotype is more closely associated with classical activation (Klug et al., 2013; Zhang et al., 2012). Otherwise, targeting macrophages for depletion or repolarization can enhance response to multiple therapeutic modalities (Ruffell & Coussens, 2015). In the context of metastasis, macrophages can promote each of the steps of the metastatic cascade (Kitamura, Qian, Pollard, 2015), and a specific association between macrophages and invasion/metastasis has been noted in melanoma, breast, ovarian, colorectal, pancreatic neuroendocrine, and bladder cancer (Ruffell et al., 2012).

vascular permeability and allowing tumors cells to intravasate into the circulation. (B) Tumor cells need to survive in circulation and arrest within the lumen of the vasculature in order to invade ectopic organs. Attachment is enhanced by CD11b^{+} neutrophils recruited by IL-8 (CXCL8) or clotting factors/platelets. Neutrophils can then recruit additional cells by binding to ICAM1 on tumor cells, or releasing NETs under inflammatory conditions. Adherent tumor cells also recruit Ly6C^{+} monocytes via CCL2, leading to macrophage differentiation and secretion of VEGF. This increases vascular permeability and promotes tumor cell extravasation. Tumor cells can be directly killed by cytotoxic NK cells recruited to the endothelium by Ly6C^{-} patrolling monocytes responding to CX_3CL1, or by neutrophils stimulated with CCL2 to induce production of H_2O_2. (C) Macrophages are important in promoting the outgrowth of metastatic foci, and this is mediated at least in part by survival signals provided by VCAM-1 on tumor cells interacting with α4 integrins on macrophages. Additional roles for macrophages in angiogenesis and immunosuppression are likely based upon the phenotype in primary tumors, but have not been shown. Neutrophils support the early proliferative potential of metastatic cells through production of leukotrienes, as well as angiogenesis at later stages of growth through expression of Bv8 and possibly MMP9. Neutrophils in metastatic organs have the potential to suppress CD8^{+} T cell responses through iNOS, otherwise CD8^{+} T cells and NK cells can suppress metastasis through direct cytotoxicity. DCs within metastatic organs can suppress metastasis, likely by trafficking to draining lymph nodes and promoting a T cell response.

2.3 Neutrophils

Neutrophils rapidly infiltrate tissues when they are inflamed, and thus may control some of the earliest interactions of the host immune system with the emerging tumor (Blaisdell et al., 2015). Similar to macrophages, neutrophils were originally associated with an antitumor immune response consistent with their role during infection and potential for tissue destruction, but more recent studies demonstrating a role in the healing process highlight their ability to promote tumor growth (Kolaczkowska & Kubes, 2013). This has been shown in experimental tumor models where neutrophils can promote growth through angiogenesis and suppression of T cell immunity (Granot & Fridlender, 2015; Liang & Ferrara, 2016). In support of these preclinical findings, the presence of neutrophils is negatively associated with survival in breast cancer, lung adenocarcinoma, and melanoma (Gentles et al., 2015). This protumor role for neutrophils appears strongly influenced by transforming growth factor-β, which regulates the maturation state of the cell (Fridlender et al., 2009). Thus tumor-associated immature neutrophils can be immunosuppressive and proangiogenic, and at least in murine models represent an overlapping population with granulocytic myeloid-derived suppressor cells. More important than the primary tumor may be the role of neutrophils in the metastatic process, as there is evidence that neutrophils can promote local invasion, retention, and extravasation at ectopic sites, as well as survival and growth within the metastatic niche (Liang & Ferrara, 2016).

2.4 T Lymphocytes

T cells have diverse functions within tumors, reflective of the array of subtypes present (Ruffell, DeNardo, Affara, & Coussens, 2010). In general $CD8^+$, T_H1-polarized $CD4^+$, and γδ T cells suppress tumor growth, promote response to therapy, and are associated with improved outcome through production of interferon-γ (IFN-γ) and direct cytotoxic activity. However, there are exceptions to these statements in murine models and in clinical correlates (Ciampricotti, Hau, Doornebal, Jonkers, & de Visser, 2012; Gentles et al., 2015). The roles of T_H2, T_H17, and regulatory (T_{reg}) $CD4^+$ T cells are even more complex, and the presence of these cells is associated with both a positive and negative prognosis (Fridman, Pages, Sautes-Fridman, & Galon, 2012). This reflects both tissue and situational specificity. For example, T_{regs} promote the growth of mammary tumors, but suppress the development of colorectal cancer (Bos, Plitas, Rudra,

Lee, & Rudensky, 2013; Chaudhry et al., 2009). At the same time, T_{regs} can drive colon carcinogenesis in a model of inflammatory bowel disease (Geis et al., 2015). The function of T lymphocytes in the metastatic process is only beginning to be elucidated. As in the primary tumor, most studies find that $CD8^+$ T cells are important mediators of immune surveillance (Massagé & Obenauf, 2016). Studies have also found that interleukin (IL)-4 producing $CD4^+$ T cells and IL-17-producing γδ T cells promote metastasis by regulating the function of macrophages and neutrophils, respectively (Coffelt et al., 2015; DeNardo et al., 2009).

2.5 Natural Killer Cells

Natural killer (NK) cells were identified based on their ability to spontaneously lyse tumor cells in the absence of B and T lymphocytes, and share similarities with $CD8^+$ T cells in terms of their cytotoxic potential (Sabry & Lowdell, 2013). However, in primary tumors NK cells appear to play an important, but secondary, role to the adaptive immune response. This may reflect limitations in cell numbers, a susceptibility to immune suppression, or a lack of activating signals (Sabry & Lowdell, 2013; Waldhauer & Steinle, 2008). That said, NK cells could be critical mediators of antimetastatic immune surveillance (Massagé & Obenauf, 2016), and have recently been shown to restrain metastatic tumor cells in a dormant state (Romero et al., 2014).

3. INVASION AND INTRAVASATION

3.1 Macrophages

Macrophages were originally implicated in the metastatic process using mice with null recessive mutation in the colony-stimulating factor (*Csf1*) gene, resulting in the absence of mature macrophages in most tissues, including mammary tumors (Lin, Nguyen, Russell, & Pollard, 2001). Subsequent studies showed that CSF-1 expression by tumor cells—mediated in part by steroid receptor coactivator-1 (Wang et al., 2009)—was functionally involved in recruiting macrophages and inducing production and shedding of epidermal growth factor (EGF) (Ishihara et al., 2013), resulting in a paracrine loop that induced tumor cell migration toward the vasculature (Wyckoff et al., 2004, 2007). This paracrine loop can be further augmented by exogenous CXCL12 or heregulin B1 (Hernandez et al., 2009), although the mechanism and importance of these at endogenous levels is unclear.

These murine studies may be biased by high CSF-1 expression in the models examined, and a second paracrine loop has been described between granulocyte-macrophage CSF (GM-CSF) and CCL18 using human breast cancer cell lines and humanized mice (Su et al., 2014).

Macrophages are also involved in regulating the deposition and structure of the extracellular matrix (ECM) tracks used by tumor cells to migrate toward blood vessels (Condeelis & Segall, 2003). This has been demonstrated for SPARC (secreted protein, acidic rich in cysteine), also known as osteonectin: macrophage-derived SPARC doubles the migratory potential of 4T1 cells on fibronectin in vitro, and SPARC-deficient mice implanted with 4T1 mammary tumors display greatly reduced collagen and fibronectin deposition, along with reduced pulmonary metastatic burden (Sangaletti et al., 2008). Macrophage proteases have long been theoretically connected to an invasive tumor phenotype based upon their potential to degrade the ECM (Kessenbrock, Plaks, & Werb, 2010). While macrophage-derived matrix metalloproteinases (MMPs) have not been demonstrated to enhance cancer cell invasion in vivo, the absence of stromal urokinase/plasminogen activator (uPA) or cathepsin B, S, or Z all restrict local invasion and/or metastasis (Akkari et al., 2014; Almholt et al., 2005; Gocheva, Wang, et al., 2010; Vasiljeva et al., 2006). At least for the cathepsins this involves a reduction in the ability of macrophages to promote invasion in vitro (Akkari et al., 2014; Gocheva, Wang, et al., 2010). Cathepsins have also been shown to promote angiogenesis, but it is not clear whether this is linked to their role in metastasis (Gocheva, Chen, Peters, Reinheckel, & Joyce, 2010; Gocheva et al., 2006; Ruffell et al., 2013).

Finally, macrophages are important mediators of angiogenesis, as the absence or depletion of macrophages reduces tumor vascularization (Qian & Pollard, 2010; Ruffell & Coussens, 2015). More important than vascular density for the metastatic process may be the influence of macrophages on vascular structure. Specific deletion of vascular endothelial growth factor-A (*Vegfa*) in macrophages recapitulates the reduction in vessel density observed in MMTV-PyMT bearing the *Csf1* null mutation (Lin et al., 2006; Stockmann et al., 2008); however, deletion of *Vegfa* in macrophages actually reduces tissue hypoxia and increases tumor growth, a phenotype linked to vessel normalization and increased tissue perfusion. Poor pericyte coverage and hypoxia are associated with metastasis in this same model system (Cooke et al., 2012), and it is foreseeable that macrophages may enhance metastasis via this mechanism, rather than through an increase in the number of vessels. However, this has yet to be examined experimentally.

VEGF-A expression by a subset of vascular-associated macrophages expressing Tie2 has also been shown to promote cancer cell intravasation by increasing the permeability of the endothelium (Harney et al., 2015). This appears to be true in patients as well as the colocalization of macrophages, tumor cells, and endothelial cells is associated with breast cancer patients that develop systemic, but not lymph node metastasis (Robinson et al., 2009). This $Tie2^+$ macrophage population may be responsible for many of the angiogenic phenotypes associated with the general macrophage population. Critically, neutralizing angiopoietin 2—the ligand for Tie2—reduces the association of macrophages with the vasculature and diminishes pulmonary metastasis (Mazzieri et al., 2011). This includes inhibiting the later steps of the metastatic process, although this has not been directly linked to macrophage function.

3.2 Neutrophils

As with macrophages, tumor-associated neutrophils have the potential to promote invasion, angiogenesis, and ECM degradation based upon their expression of the requisite molecules (Liang & Ferrara, 2016). That said, little is know about the role of neutrophils in the early steps of the metastatic process, and there is a relative paucity of in vivo data demonstrating functionality. G-CSF expression results in a systemic expansion of the neutrophil population (Casbon et al., 2015) and multiple CXCR1/2 ligands have been implicated in neutrophil recruitment to tumors and increased metastasis, including CXCL1, CXCL2, CXCL5, and CXCL8/IL-8 (Acharyya et al., 2012; Bekes et al., 2011; Toh et al., 2011; Zhou et al., 2014). In some cases this is associated with an increase in MMP expression, and MMP9-deficient mice display reduced metastasis (Yan et al., 2010; Yang et al., 2008); however, to date the importance of MMPs in local invasion has only been demonstrated in xenotransplantation assays (Bekes et al., 2011). While a few studies have shown neutrophil depletion reduces metastasis, this is usually associated with an immunosuppressive phenotype, and it is not clear which stage of the metastatic process is affected (Coffelt et al., 2015; Simpson, Templeton, & Cross, 2012). For example, neutrophil depletion was reported to have no influence the number of circulating 4T1 cells (Granot et al., 2011), and in another orthotopic model of breast cancer neutrophil depletion reduces both lung and lymph node metastasis without affecting primary tumor growth (Coffelt et al., 2015). Neutrophils have been shown to promote migration of melanoma cells toward the endothelium during ultraviolet-induced inflammation (Bald et al., 2014), but the preponderance of evidence currently favors neutrophils being more involved in the later steps of the metastatic process (Acharyya et al., 2012; Coffelt et al., 2015; Cools-Lartigue et al., 2013; Granot et al., 2011).

3.3 T Lymphocytes

$CD4^+$ T cells were first described to promote the early steps of metastasis in the MMTV-PyMT model of breast cancer, with an approximate fivefold reduction in circulating tumor cells and pulmonary metastasis observed in CD4-deficient animals (DeNardo et al., 2009). This was mediated by T_H2-polarized $CD4^+$ T cells expressing IL-4 enhancing the protumor phenotype of macrophages, including augmentation of the CSF-1/EGF paracrine loop. IL-4 also promotes cathepsin protease activity in macrophages within pancreatic neuroendocrine tumors, leading to enhanced local invasion (Gocheva, Wang, et al., 2010). T_H2-polarization is specific to tumor $CD4^+$ T cells in the MMTV-PyMT model (DeNardo et al., 2009), and this appears to be driven by high levels of CCL5 chemokine expression within late stage tumors (Zhang et al., 2015). It is likely that IL-13 will have similar effects on macrophage polarization within tumors given that neutralization of either IL-4 or IL-13 produces an equivalent increase in the efficacy of paclitaxel chemotherapy (Shiao et al., 2015). $CD4^+$ T_H2 cells can thus promote invasion indirectly by modulating the phenotype of macrophages.

4. SURVIVAL AND EXTRAVASATION

4.1 Monocytes

Classical $Ly6C^+$ inflammatory and nonclassical $Ly6C^-$ patrolling monocytes appear to have opposing functions in murine models of lung metastasis. True to their name, $Ly6C^-CCR2^{lo}CX_3CR1^{hi}$ monocytes continually patrol the lumen of the endothelium under steady-state conditions, pausing in response to inflammatory signals that upregulate expression of CX_3CL1 on endothelial cells (Auffray et al., 2007; Carlin et al., 2013). During experimental metastasis $Ly6C^-$ monocyte patrolling is reduced within hours of intravenous injection of tumor cells, and the absence of $Ly6C^-$ monocytes increases the number of both experimental and spontaneous pulmonary foci (Hanna et al., 2015). This function may be induced by tumor cell death as $Ly6C^-$ monocytes were found to take up tumor cell debris, and another study has shown this occurs within the first 15 min following injection

(Headley et al., 2016). $Ly6C^-$ monocytes also promoted NK cell recruitment to the lungs, which was necessary for their ability to suppress metastasis of B16F10 melanoma cells (Hanna et al., 2015). Interestingly, $Ly6C^-$ monocytes have been shown to recruit neutrophils to the vessel lumen during sterile inflammation induced with a toll-like receptor 7 (TLR7) agonist (Carlin et al., 2013), hinting that these cells may incorporate local signals to differentially regulate inflammatory responses in the vasculature.

In contrast to the role of $Ly6C^-$ monocytes, studies have demonstrated a prometastatic role for $Ly6C^+CCR2^{hi}CX_3CR1^{lo}$ monocytes and their derivative macrophage population in the lung and liver (Kitamura, Qian, Pollard, 2015). Targeting CCR2/CCL2 prevents spontaneous liver metastasis in a pancreatic ductal adenocarcinoma model (Sanford et al., 2013) and diminishes the ability of mammary carcinoma cells to extravasate into the lung parenchyma and form overt metastatic lesions (Qian et al., 2011). In the lung this requires VEGFA expression by $Ly6C^+$ monocytes, which increases endothelial permeability and enhances tumor cell transmigration (Qian et al., 2011), similar to the role of VEGFA-expressing $Tie2^+$ macrophages in primary tumors (Harney et al., 2015). While it has been suggested that monocytes can promote tumor cell survival in circulation, this is most likely due to enhanced extravasation based upon the absence of a phenotype at 30 min postinjection (Gil-Bernabe et al., 2012; Hoos, Protsyuk, & Borsig, 2014). It remains unclear the degree to which $Ly6C^+$ monocytes, as opposed to macrophages, are responsible for the increase in endothelial permeability. Kinetic data suggest $Ly6C^+$ monocytes extravasate and differentiate into macrophages after their initial recruitment, especially monocytes that have phagocytosed tumor debris (Headley et al., 2016). CCL2 neutralization also reduces metastatic burden when initiated 2 days after tumor cell injection, a time point by which extravasation has already occurred. Similarly, there is a diminution in foci size when the presence or function of macrophages is interfered with (Kitamura, Qian, Soong, et al., 2015; Qian et al., 2009, 2011, 2015), indicative of an important role for macrophages in the growth phase of metastasis. Detailed imaging studies will be required to dissect the role of monocytes vs macrophages in extravasation and survival in ectopic organs.

4.2 Neutrophils

Several studies support a role for neutrophils in promoting cancer cell capture and retention on the endothelium. In a model of liver metastasis neutrophils were found to promote tumor cell adhesion to the endothelium via

CD11b, a process that is enhanced by the release of neutrophil extracellular traps (NETs) following LPS exposure (Cools-Lartigue et al., 2013; Spicer et al., 2012). While the use of the anti-Gr-1 antibody for depletion in these studies would also target $Ly6C^+$ monocytes, the reintroduction of purified neutrophils was able to reverse the effect of depletion, demonstrating the specificity of these findings (Spicer et al., 2012). Simultaneous intravenous injection of human neutrophils also supports lung metastasis by melanoma cells, with IL-8 (CXCL8) expression by cancer cells directing the interaction between the two cell types (Huh, Liang, Sharma, Dong, & Robertson, 2010). Finally, G-CSF-driven expansion of neutrophils in the lung promotes tumor cell extravasation via Bv8/prokineticin-2, a homologue of VEGF-A (Kowanetz et al., 2010).

In contrast to the aforementioned studies, neutrophils have also been shown to suppress pulmonary metastasis in the 4T1 and MMTV-PyMT/cMyc models of mammary carcinoma (Granot et al., 2011). This was mediated by a classical property of neutrophils, the ability to directly kill cells via release of H_2O_2. Adoptive transfer of neutrophils also suppressed experimental metastasis, and as neutrophils do not appear to be recruited into the lung during this process (Headley et al., 2016), it suggests targeted killing of arrested tumor cells in the vasculature. Indeed, neutrophil depletion reduces the frequency but not the size of metastatic foci. It is notable that the cytotoxic properties of human and mouse neutrophils were not observed under steady-state conditions, and were induced by CCL2, but not G-CSF (Granot et al., 2011). Chemokine expression patterns by tumor cells and other experimental variations may thus explain differences between studies showing a pro- or antimetastatic role for neutrophils.

4.3 T Lymphocytes

In addition to the MMTV-PyMT transgenic model, $CD4^+$ T cells are important in promoting lung metastasis in the MMTV-Erbb2 model of $Her2^+$ breast cancer (Tan et al., 2011). However, rather than T_H2-polarized cells, only reconstitution with $CD4^+CD25^+$ cells (presumably $FoxP3^+$ T_{regs}) could restore the metastatic phenotype in immunodeficient mice. This was due to preferential expression of receptor activator of nuclear factor-κB ligand (RANKL) by the T_{reg} population. The authors found their model was insensitive to RANKL-induced migration, and that instead a prosurvival phenotype in nonadherent cells in vitro and circulating tumor cells in vivo was sufficient to explain increased metastasis (Tan et al., 2011). It should be noted that intratumoral RANKL injection also promoted metastasis in this model, suggesting that the T_{regs} mediate their effect within the tumor, and not in circulation. However, these findings may be model and/or organ dependent, and RANKL can directly promote mouse and human epithelial cell migration and bone metastasis (Jones et al., 2006).

4.4 NK Cells

As mentioned, $Ly6C^-$ monocytes suppress metastasis by recruiting NK cells to the lung, although whether tumor cell killing occurs in the vasculature or parenchyma is unclear (Hanna et al., 2015). More direct evidence of a role for NK cells in the vasculature can be found in the enhancement of metastasis by platelets and clotting factors in the lung. NK cell depletion ablates the influence of clotting on metastasis, and enhances tumor cell survival at a 24-h time point even under normal conditions (Palumbo et al., 2005). Although direct killing of tumor cells has not been shown by intravital microscopy, these studies suggest that NK cells provide a level of immune surveillance against tumor cells adhering to the lumen of the endothelium.

5. ECTOPIC GROWTH

5.1 Macrophages

The formation of the premetastatic niche is dependent upon the recruitment and activation of $CD11b^+$ myeloid cells by endogenous TLR ligands or collagen crosslinking (Erler et al., 2009; Hiratsuka et al., 2008; Kim et al., 2009). $CD11b^+VEGFR1^+$ cells were originally described as the important population of this myeloid niche (Kaplan et al., 2005), and a subsequent study showed that VEGFR1 expression specifically by macrophages is critical for metastasis outgrowth through its ability to drive CSF-1 expression (Qian et al., 2015). The dominant myeloid population supporting survival and growth of metastatic cells thus appears to be recruited macrophage populations that have differentiated from $Ly6C^+$ monocytes. Significant differences between tumor cell survival in the presence or absence of recruited macrophages is observed between 12 and 24 h, indicating a prosurvival function for macrophages in multiple studies (Gil-Bernabe et al., 2012; Hoos et al., 2014; Qian & Pollard, 2010). Mechanistically this has been shown to result from prosurvival signals induced by vascular cell adhesion molecule-1 (VCAM-1) binding to $\alpha 4$ integrins on macrophages

(Chen, Zhang, & Massague, 2011). It seems likely that additional pathways that enhance tumor cell survival at the single cell level have yet to be identified.

There is also evidence that macrophages can promote the growth of metastatic cells, based upon the reduced rate of growth seen 48 h after experimental metastasis, and the ability of macrophage depletion to reduce foci size even when initiated 4 days after lung seeding (Qian et al., 2009). However, the mechanism underlying this observation remains unidentified, and there may be substantial differences between experimental and spontaneous metastasis. Also unclear is whether metastasis-associated macrophages posses the direct or indirect immunosuppressive phenotype of macrophages seen in primary tumors (Curiel et al., 2004; DeNardo et al., 2011; Kryczek et al., 2006; Ruffell et al., 2014). Finally, resident macrophage populations have not been extensively evaluated in the metastatic process, with the exception of the unique role of osteoclasts in bone reabsorption (Cook et al., 2014). All of these factors will be important to assess experimentally as macrophage-targeted agents are currently being tested clinically in patients with metastatic disease.

5.2 Neutrophils

As with macrophages, neutrophils are present in higher numbers in metastatic and premetastatic lungs and livers, driven by a G-CSF-dependent systemic expansion of the population (Casbon et al., 2015; Coffelt et al., 2015; Granot et al., 2011; Kowanetz et al., 2010; Wculek & Malanchi, 2015; Yan et al., 2010). Neutrophils have now been implicated in promoting the survival and outgrowth of extravasated tumor cells in the lung through two distinct mechanisms: immune suppression and proliferative potential. It is also possible that neutrophils may regulate angiogenesis and vascular structure through Bv8 and/or MMP9 (Kowanetz et al., 2010; Yan et al., 2010), but clear supportive data for this is lacking.

Neutrophils from metastatic lungs suppress proliferation and IFN-γ production by T cells, and at least in vitro this is dependent upon inducible nitric oxide synthase (iNOS) (Coffelt et al., 2015; Yan et al., 2010). Importantly, while neutrophil depletion via an anti-Ly6G antibody reduces metastasis, this was completely reversed by $CD8^+$ T cell depletion, demonstrating the importance of the immunosuppressive phenotype (Coffelt et al., 2015). However, it should be noted that neutrophil accumulation is observed in most organs, including the lymph node and spleen, and immune suppression within these secondary lymphoid organs, rather than within the metastatic environment may be important. Indeed, peripheral blood and splenic neutrophils from tumor-bearing mice are immunosuppressive (Casbon et al., 2015; Coffelt et al., 2015). This is consistent with the literature on Gr-1^{+} myeloid-derived suppressor cells and their ability to suppress systemic antitumor immune responses, leading to enhanced tumor growth and metastasis in many tumor models (Gabrilovich & Nagaraj, 2009).

In contrast to the above studies, one study has shown that neutrophil depletion can suppress spontaneous metastasis in immunodeficient mice (Wculek & Malanchi, 2015). A caveat to this is immunodeficient mice will lack many of the polarization signals that drive the phenotype of neutrophils in cancer. For example, IL-17 expression by either CD4^{+} T_H17 cells or $\gamma\delta$ T cells enhances G-CSF expression and phenotypic changes in neutrophils (Chung et al., 2013; Coffelt et al., 2015). Regardless, neutrophils and neutrophil culture medium were able to enhance the number of CD24^{+}CD90^{+} metastatic initiating cells present in the lung using multiple models of experimental metastasis in a leukotriene-dependent manner (Wculek & Malanchi, 2015). G-CSF-driven expansion of neutrophils can therefore promote multiple aspects of cancer cell growth within ectopic tissues.

5.3 T Lymphocytes

Immune surveillance and editing are well-established concepts in carcinogen-induced tumors, and mice are more susceptible to cancer in the absence of innate and/or adaptive components of the immune system that mediate cytotoxicity (Vesely, Kershaw, Schreiber, & Smyth, 2011). Increased tumor growth in immune deficient mice can lead to increased incidence of metastasis; however, less is known about the importance of immune surveillance in specifically suppressing the outgrowth of disseminated cancer cells. This may have been a limitation of the animal models usually employed, and there have now been studies demonstrating the importance of CD8^{+} T cells in maintaining metastatic dormancy using more unique systems. In a spontaneous model of melanoma wherein the cells disseminate but fail to develop into overt metastasis in most animals, CD8^{+} T cell depletion increased tumor cell proliferation and resulted in observable metastasis in the majority of mice (Eyles et al., 2010). Similar results were seen in a resectable fibrosarcoma model, with CD8^{+} T cell depletion increasing the incidence of metastasis from 0% to 100% (Romero et al., 2014). Interestingly, CD4^{+} T cell depletion in this model resulted in a

23% rate of incidence, and $CD4^+$ T cells have been shown to provide protection against development of hepatocellular carcinoma (Kang et al., 2011). Similar to primary tumors, T cells may provide an important but context specific role in immune surveillance in metastatic disease. Induction of G-CSF expression by IL-17-producing γδ T cells has already been discussed (Coffelt et al., 2015), but other cytokines and leukocytes are also likely to be involved in directing the immune response. For example, transportation of tumor antigen by resident $CD103^+$ DCs in the lung is protective against experimental metastasis (Headley et al., 2016). Given the potential importance of T cells in maintaining and/or eliminating dormant metastatic cells, understanding the kinetics and mechanism of this process is critical.

5.4 NK Cells

Numerous studies have demonstrated that NK cells protect against experimental and spontaneous metastasis in immunodeficient and immunocompetent murine models. This includes spontaneous metastasis to the lung, liver, and lymph node in models with no apparent role for NK cells in primary tumor growth (Milsom, Lee, Hackl, Man, & Kerbel, 2013; Olkhanud et al., 2009; Paolino et al., 2014; Smyth et al., 1999, 2001; Takeda et al., 2001). This protective effect is mediated by the direct cytotoxic activity of NK cells through expression of tumor necrosis factor-related apoptosis-inducing ligand (TRAIL) and perforin (Grosse-Wilde et al., 2008; Smyth et al., 1999; Takeda et al., 2001). As a comparison to the importance of cytotoxic T cells, NK cell depletion in the aforementioned resectable fibrosarcoma model resulted a metastatic incidence of 87% (Romero et al., 2014). Despite the number of studies demonstrating the importance of NK cells in suppressing metastasis, NK cell-based immunotherapies have not progressed to the same degree as those targeting T lymphocytes. This may be due to an overreliance on xenografts and other immunodeficient preclinical models, as well as the use of cell lines that display susceptibility to NK-mediated lysis due to low expression of major histocompatibility complex (MHC) I or sensitivity to TRAIL. Investigations employing transgenic models with spontaneous metastasis may help clarify the importance of NK cells and determine whether these cells represent a viable therapeutic target.

6. THERAPEUTIC IMPLICATIONS

Targeting key immune pathways to block the dissemination of tumor cells to ectopic organs has shown success in preclinical modeling. However, these pathways may be very context specific, as for example, liver metastasis by MC38 colon carcinoma cells is dependent upon CCL2, while metastasis of B16F10 melanoma cells is not (Zhao et al., 2013). Furthermore, while targeting CSF-1 or CCL2 can reduce metastasis in certain models, compensatory mechanisms mitigate the potential of these therapeutic approaches in others (Bonapace et al., 2014; Swierczak et al., 2014). More importantly, therapeutics targeting CSF-1/CSF-1R and CCL2/CCR2 are being used in patients with established metastatic disease, and it will be necessary to determine whether these and other pathways promote the growth of metastatic lesions or mediate therapeutic resistance, as opposed to cancer cell dissemination. As such, it is promising that CSF-1 is an important driver of metastatic outgrowth (Qian et al., 2015), and the size of metastatic foci is reduced by targeting the CSF-1/CSF-1R pathway in combination with paclitaxel chemotherapy (DeNardo et al., 2011; Ruffell et al., 2014). As a general concept it will be important to test immune-targeted agents in combination with other modalities; although early intervention approaches are often effective as single agents, late stage approaches rarely are.

Given the important role of $CD8^+$ T cells and NK cells in protecting against metastasis, augmenting the cytotoxic potential of these cells would appear to be the most promising route for therapeutic intervention. Immune checkpoint blockade therapy is already approved for the treatment of patients with advanced metastatic disease, and the goal is now is to search for approaches that can improve upon the relatively low response rates being observed (Topalian, Drake, & Pardoll, 2015). This includes targeting additional checkpoint molecules beyond programmed death-1 (PD-1) and cytotoxic T lymphocyte antigen-4 (CTLA-4), but also relieving immune suppression within metastatic tumors. Notably, the reduced lung metastatic burden seen following neutrophil depletion or CSF-1R inhibition is $CD8^+$ T cell dependent (Coffelt et al., 2015; DeNardo et al., 2011), and CSF-1R inhibition enhances the response of primary tumors to checkpoint blockade and adoptive cell transfer (Mok et al., 2014; Zhu et al., 2014). Whether these or other combinations will prove efficacious in a metastatic setting remains to be determined.

As agents that target the microenvironment move into the clinic, it is important to keep in mind that treatment approaches will need to be tailored to the specific organ/tissue. It is unclear the degree to which cancer cells reestablish their microenvironment in metastatic organs, but there is probably a complex interplay between the innate suppressive capacity of the cancer cells and the metastatic site. Even mathematical modeling that accounts simply for organ blood flow and T cell trafficking results in a complex interaction between the primary tumor and metastatic lesions during therapy (Poleszczuk et al., 2016). Lung, liver, brain, and bone metastasis are likely to require therapeutic modalities that reflect both the requirements for survival in these organs, as well as the unique roles that resident immune cells have in these tissues. Advanced preclinical modeling of metastatic disease that permit therapies to be evaluated in an adjuvant setting will greatly assist in therapeutic development. As the number of approved immunotherapies expands, this will also be a critical step in selecting rationale combinations that may increase the success rate of subsequent clinical trials.

ACKNOWLEDGMENTS

This work has been supported by a National Cancer Institute of the NIH Grant (R00CA185325-02), the Shula Breast Cancer Fund, and the Moffitt Lung Cancer Center of Excellence, all to B.R.

REFERENCES

Acharyya, S., Oskarsson, T., Vanharanta, S., Malladi, S., Kim, J., Morris, P. G., et al. (2012). A CXCL1 paracrine network links cancer chemoresistance and metastasis. *Cell*, *150*, 165–178.

Akkari, L., Gocheva, V., Kester, J. C., Hunter, K. E., Quick, M. L., Sevenich, L., et al. (2014). Distinct functions of macrophage-derived and cancer cell-derived cathepsin Z combine to promote tumor malignancy via interactions with the extracellular matrix. *Genes & Development*, *28*, 2134–2150.

Almholt, K., Lund, L. R., Rygaard, J., Nielsen, B. S., Dano, K., Romer, J., et al. (2005). Reduced metastasis of transgenic mammary cancer in urokinase-deficient mice. *International Journal of Cancer*, *113*, 525–532.

Auffray, C., Fogg, D., Garfa, M., Elain, G., Join-Lambert, O., Kayal, S., et al. (2007). Monitoring of blood vessels and tissues by a population of monocytes with patrolling behavior. *Science*, *317*, 666–670.

Bald, T., Quast, T., Landsberg, J., Rogava, M., Glodde, N., Lopez-Ramos, D., et al. (2014). Ultraviolet-radiation-induced inflammation promotes angiotropism and metastasis in melanoma. *Nature*, *507*, 109–113.

Bekes, E. M., Schweighofer, B., Kupriyanova, T. A., Zajac, E., Ardi, V. C., Quigley, J. P., et al. (2011). Tumor-recruited neutrophils and neutrophil TIMP-free MMP-9 regulate coordinately the levels of tumor angiogenesis and efficiency of malignant cell intravasation. *The American Journal of Pathology*, *179*, 1455–1470.

Blaisdell, A., Crequer, A., Columbus, D., Daikoku, T., Mittal, K., Dey, S. K., et al. (2015). Neutrophils oppose uterine epithelial carcinogenesis via debridement of hypoxic tumor cells. *Cancer Cell*, *28*, 785–799.

Bonapace, L., Coissieux, M. M., Wyckoff, J., Mertz, K. D., Varga, Z., Junt, T., et al. (2014). Cessation of CCL2 inhibition accelerates breast cancer metastasis by promoting angiogenesis. *Nature*, *515*, 130–133.

Bos, P. D., Plitas, G., Rudra, D., Lee, S. Y., & Rudensky, A. Y. (2013). Transient regulatory T cell ablation deters oncogene-driven breast cancer and enhances radiotherapy. *The Journal of Experimental Medicine*, *210*, 2435–2466.

Carlin, L. M., Stamatiades, E. G., Auffray, C., Hanna, R. N., Glover, L., Vizcay-Barrena, G., et al. (2013). Nr4a1-dependent Ly6C(low) monocytes monitor endothelial cells and orchestrate their disposal. *Cell, 153*, 362–375.

Casbon, A. J., Reynaud, D., Park, C., Khuc, E., Gan, D. D., Schepers, K., et al. (2015). Invasive breast cancer reprograms early myeloid differentiation in the bone marrow to generate immunosuppressive neutrophils. *Proceedings of the National Academy of Sciences of the United States of America, 112*, E566–E575.

Chaudhry, A., Rudra, D., Treuting, P., Samstein, R. M., Liang, Y., Kas, A., et al. (2009). CD4+ regulatory T cells control TH17 responses in a Stat3-dependent manner. *Science, 326*, 986–991.

Chen, Q., Zhang, X. H., & Massague, J. (2011). Macrophage binding to receptor VCAM-1 transmits survival signals in breast cancer cells that invade the lungs. *Cancer Cell, 20*, 538–549.

Chung, A. S., Wu, X., Zhuang, G., Ngu, H., Kasman, I., Zhang, J., et al. (2013). An interleukin-17-mediated paracrine network promotes tumor resistance to anti-angiogenic therapy. *Nature Medicine, 19*, 1114–1123.

Ciampricotti, M., Hau, C. S., Doornebal, C. W., Jonkers, J., & de Visser, K. E. (2012). Chemotherapy response of spontaneous mammary tumors is independent of the adaptive immune system. *Nature Medicine, 18*, 344–346. author reply 346.

Coffelt, S. B., & de Visser, K. E. (2015). Immune-mediated mechanisms influencing the efficacy of anticancer therapies. *Trends in Immunology, 36*, 198–216.

Coffelt, S. B., Kersten, K., Doornebal, C. W., Weiden, J., Vrijland, K., Hau, C. S., et al. (2015). IL-17-producing gammadelta T cells and neutrophils conspire to promote breast cancer metastasis. *Nature, 522*, 345–348.

Condeelis, J., & Segall, J. E. (2003). Intravital imaging of cell movement in tumours. *Nature Reviews. Cancer, 3*, 921–930.

Cook, L. M., Shay, G., Araujo, A., & Lynch, C. C. (2014). Integrating new discoveries into the "vicious cycle" paradigm of prostate to bone metastases. *Cancer Metastasis Reviews, 33*, 511–525.

Cooke, V. G., LeBleu, V. S., Keskin, D., Khan, Z., O'Connell, J. T., Teng, Y., et al. (2012). Pericyte depletion results in hypoxia-associated epithelial-to-mesenchymal transition and metastasis mediated by met signaling pathway. *Cancer Cell, 21*, 66–81.

Cools-Lartigue, J., Spicer, J., McDonald, B., Gowing, S., Chow, S., Giannias, B., et al. (2013). Neutrophil extracellular traps sequester circulating tumor cells and promote metastasis. *The Journal of Clinical Investigation, 123*, 3446–3458.

Coussens, L. M., Zitvogel, L., & Palucka, A. K. (2013). Neutralizing tumor-promoting chronic inflammation: A magic bullet? *Science, 339*, 286–291.

Curiel, T. J., Coukos, G., Zou, L., Alvarez, X., Cheng, P., Mottram, P., et al. (2004). Specific recruitment of regulatory T cells in ovarian carcinoma fosters immune privilege and predicts reduced survival. *Nature Medicine, 10*, 942–949.

DeNardo, D. G., Barreto, J. B., Andreu, P., Vasquez, L., Tawfik, D., Kolhatkar, N., et al. (2009). CD4(+) T cells regulate pulmonary metastasis of mammary carcinomas by enhancing protumor properties of macrophages. *Cancer Cell, 16*, 91–102.

DeNardo, D. G., Brennan, D. J., Rexhepaj, E., Ruffell, B., Shiao, S. L., Madden, S. F., et al. (2011). Leukocyte complexity predicts breast cancer survival and functionally regulates response to chemotherapy. *Cancer Discovery, 1*, 54–67.

Erler, J. T., Bennewith, K. L., Cox, T. R., Lang, G., Bird, D., Koong, A., et al. (2009). Hypoxia-induced lysyl oxidase is a critical mediator of bone marrow cell recruitment to form the premetastatic niche. *Cancer Cell, 15*, 35–44.

Eyles, J., Puaux, A. L., Wang, X., Toh, B., Prakash, C., Hong, M., et al. (2010). Tumor cells disseminate early, but immunosurveillance limits metastatic outgrowth, in a mouse model of melanoma. *The Journal of Clinical Investigation, 120*, 2030–2039.

Friberg, S., & Nystrom, A. (2015). Cancer metastases: Early dissemination and late recurrences. *Cancer Growth and Metastasis, 8*, 43–49.

Fridlender, Z. G., Sun, J., Kim, S., Kapoor, V., Cheng, G., Ling, L., et al. (2009). Polarization of tumor-associated neutrophil phenotype by TGF-beta: "N1" versus "N2" TAN. *Cancer Cell, 16*, 183–194.

Fridman, W. H., Pages, F., Sautes-Fridman, C., & Galon, J. (2012). The immune contexture in human tumours: Impact on clinical outcome. *Nature Reviews. Cancer, 12*, 298–306.

Gabrilovich, D. I., & Nagaraj, S. (2009). Myeloid-derived suppressor cells as regulators of the immune system. *Nature Reviews. Immunology, 9*, 162–174.

Geis, A. L., Fan, H., Wu, X., Wu, S., Huso, D. L., Wolfe, J. L., et al. (2015). Regulatory T-cell response to enterotoxigenic bacteroides fragilis colonization triggers IL17-dependent colon carcinogenesis. *Cancer Discovery, 5*, 1098–1109.

Gentles, A. J., Newman, A. M., Liu, C. L., Bratman, S. V., Feng, W., Kim, D., et al. (2015). The prognostic landscape of genes and infiltrating immune cells across human cancers. *Nature Medicine, 21*, 938–945.

Gil-Bernabe, A. M., Ferjancic, S., Tlalka, M., Zhao, L., Allen, P. D., Im, J. H., et al. (2012). Recruitment of monocytes/macrophages by tissue factor-mediated coagulation is essential for metastatic cell survival and premetastatic niche establishment in mice. *Blood, 119*, 3164–3175.

Gocheva, V., Chen, X., Peters, C., Reinheckel, T., & Joyce, J. A. (2010). Deletion of cathepsin H perturbs angiogenic switching, vascularization and growth of tumors in a mouse model of pancreatic islet cell cancer. *Biological Chemistry, 391*, 937–945.

Gocheva, V., Wang, H. W., Gadea, B. B., Shree, T., Hunter, K. E., Garfall, A. L., et al. (2010). IL-4 induces cathepsin protease activity in tumor-associated macrophages to promote cancer growth and invasion. *Genes & Development, 24*, 241–255.

Gocheva, V., Zeng, W., Ke, D., Klimstra, D., Reinheckel, T., Peters, C., et al. (2006). Distinct roles for cysteine cathepsin genes in multistage tumorigenesis. *Genes & Development, 20*, 543–556.

Granot, Z., & Fridlender, Z. G. (2015). Plasticity beyond cancer cells and the "Immunosuppressive Switch" *Cancer Research, 75*, 4441–4445.

Granot, Z., Henke, E., Comen, E. A., King, T. A., Norton, L., & Benezra, R. (2011). Tumor entrained neutrophils inhibit seeding in the premetastatic lung. *Cancer Cell, 20*, 300–314.

Grosse-Wilde, A., Voloshanenko, O., Bailey, S. L., Longton, G. M., Schaefer, U., Csernok, A. I., et al. (2008). TRAIL-R deficiency in mice enhances lymph node metastasis without affecting primary tumor development. *The Journal of Clinical Investigation, 118*, 100–110.

Hanahan, D., & Coussens, L. M. (2012). Accessories to the crime: Functions of cells recruited to the tumor microenvironment. *Cancer Cell, 21*, 309–322.

Hanna, R. N., Cekic, C., Sag, D., Tacke, R., Thomas, G. D., Nowyhed, H., et al. (2015). Patrolling monocytes control tumor metastasis to the lung. *Science, 350*, 985–990.

Harney, A. S., Arwert, E. N., Entenberg, D., Wang, Y., Guo, P., Qian, B. Z., et al. (2015). Real-time imaging reveals local, transient vascular permeability, and tumor cell intravasation stimulated by TIE2hi macrophage-derived VEGFA. *Cancer Discovery, 5*, 932–943.

Headley, M. B., Bins, A., Nip, A., Roberts, E. W., Looney, M. R., Gerard, A., et al. (2016). Visualization of immediate immune responses to pioneer metastatic cells in the lung. *Nature, 531*, 513–517.

Hernandez, L., Smirnova, T., Kedrin, D., Wyckoff, J., Zhu, L., Stanley, E. R., et al. (2009). The EGF/CSF-1 paracrine invasion loop can be triggered by heregulin beta1 and CXCL12. *Cancer Research, 69*, 3221–3227.

Hiratsuka, S., Watanabe, A., Sakurai, Y., Akashi-Takamura, S., Ishibashi, S., Miyake, K., et al. (2008). The S100A8-serum amyloid A3-TLR4 paracrine cascade establishes a pre-metastatic phase. *Nature Cell Biology*, *10*, 1349–1355.

Hoos, A., Protsyuk, D., & Borsig, L. (2014). Metastatic growth progression caused by PSGL-1-mediated recruitment of monocytes to metastatic sites. *Cancer Research*, *74*, 695–704.

Huh, S. J., Liang, S., Sharma, A., Dong, C., & Robertson, G. P. (2010). Transiently entrapped circulating tumor cells interact with neutrophils to facilitate lung metastasis development. *Cancer Research*, *70*, 6071–6082.

Ishihara, D., Dovas, A., Hernandez, L., Pozzuto, M., Wyckoff, J., Segall, J. E., et al. (2013). Wiskott-Aldrich syndrome protein regulates leukocyte-dependent breast cancer metastasis. *Cell Reports*, *4*, 429–436.

Jakubzick, C., Gautier, E. L., Gibbings, S. L., Sojka, D. K., Schlitzer, A., Johnson, T. E., et al. (2013). Minimal differentiation of classical monocytes as they survey steady-state tissues and transport antigen to lymph nodes. *Immunity*, *39*, 599–610.

Jones, D. H., Nakashima, T., Sanchez, O. H., Kozieradzki, I., Komarova, S. V., Sarosi, I., et al. (2006). Regulation of cancer cell migration and bone metastasis by RANKL. *Nature*, *440*, 692–696.

Joyce, J. A., & Pollard, J. W. (2009). Microenvironmental regulation of metastasis. *Nature Reviews. Cancer*, *9*, 239–252.

Juric, V., Ruffell, B., Evason, K. J., Hu, J., Che, L., Wang, L., et al. (2016). Monocyte-dependent liver injury promotes carcinogenesis in an oncogene-specific manner. *Journal of Hepatology*, *64*, 881–890.

Kang, T. W., Yevsa, T., Woller, N., Hoenicke, L., Wuestefeld, T., Dauch, D., et al. (2011). Senescence surveillance of pre-malignant hepatocytes limits liver cancer development. *Nature*, *479*, 547–551.

Kaplan, R. N., Riba, R. D., Zacharoulis, S., Bramley, A. H., Vincent, L., Costa, C., et al. (2005). VEGFR1-positive haematopoietic bone marrow progenitors initiate the pre-metastatic niche. *Nature*, *438*, 820–827.

Kessenbrock, K., Plaks, V., & Werb, Z. (2010). Matrix metalloproteinases: Regulators of the tumor microenvironment. *Cell*, *141*, 52–67.

Kim, S., Takahashi, H., Lin, W. W., Descargues, P., Grivennikov, S., Kim, Y., et al. (2009). Carcinoma-produced factors activate myeloid cells through TLR2 to stimulate metastasis. *Nature*, *457*, 102–106.

Kitamura, T., Qian, B. Z., & Pollard, J. W. (2015a). Immune cell promotion of metastasis. *Nature Reviews. Immunology*, *15*, 73–86.

Kitamura, T., Qian, B. Z., Soong, D., Cassetta, L., Noy, R., Sugano, G., et al. (2015b). CCL2-induced chemokine cascade promotes breast cancer metastasis by enhancing retention of metastasis-associated macrophages. *The Journal of Experimental Medicine*, *212*, 1043–1059.

Klemm, F., & Joyce, J. A. (2015). Microenvironmental regulation of therapeutic response in cancer. *Trends in Cell Biology*, *25*, 198–213.

Klug, F., Prakash, H., Huber, P. E., Seibel, T., Bender, N., Halama, N., et al. (2013). Low-dose irradiation programs macrophage differentiation to an iNOS(+)/M1 phenotype that orchestrates effective T cell immunotherapy. *Cancer Cell*, *24*, 589–602.

Kolaczkowska, E., & Kubes, P. (2013). Neutrophil recruitment and function in health and inflammation. *Nature Reviews. Immunology*, *13*, 159–175.

Kowanetz, M., Wu, X., Lee, J., Tan, M., Hagenbeek, T., Qu, X., et al. (2010). Granulocyte-colony stimulating factor promotes lung metastasis through mobilization of Ly6G+ Ly6C+ granulocytes. *Proceedings of the National Academy of Sciences of the United States of America*, *107*, 21248–21255.

Kryczek, I., Zou, L., Rodriguez, P., Zhu, G., Wei, S., Mottram, P., et al. (2006). B7-H4 expression identifies a novel suppressive macrophage population in human ovarian carcinoma. *The Journal of Experimental Medicine*, *203*, 871–881.

Liang, W., & Ferrara, N. (2016). The complex role of neutrophils in tumor angiogenesis and metastasis. *Cancer Immunology Research*, *4*, 83–91.

Lin, E. Y., Li, J. F., Gnatovskiy, L., Deng, Y., Zhu, L., Grzesik, D. A., et al. (2006). Macrophages regulate the angiogenic switch in a mouse model of breast cancer. *Cancer Research*, *66*, 11238–11246.

Lin, E. Y., Nguyen, A. V., Russell, R. G., & Pollard, J. W. (2001). Colony-stimulating factor 1 promotes progression of mammary tumors to malignancy. *The Journal of Experimental Medicine*, *193*, 727–740.

Massagé, J., & Obenauf, A. C. (2016). Metastatic colonization by circulating tumour cells. *Nature*, *529*, 298–306.

Mazzieri, R., Pucci, F., Moi, D., Zonari, E., Ranghetti, A., Berti, A., et al. (2011). Targeting the ANG2/TIE2 axis inhibits tumor growth and metastasis by impairing angiogenesis and disabling rebounds of proangiogenic myeloid cells. *Cancer Cell*, *19*, 512–526.

Meads, M. B., Gatenby, R. A., & Dalton, W. S. (2009). Environment-mediated drug resistance: A major contributor to minimal residual disease. *Nature Reviews. Cancer*, *9*, 665–674.

Milsom, C. C., Lee, C. R., Hackl, C., Man, S., & Kerbel, R. S. (2013). Differential post-surgical metastasis and survival in SCID, NOD-SCID and NOD-SCID-IL-2Rgamma (null) mice with parental and subline variants of human breast cancer: Implications for host defense mechanisms regulating metastasis. *PLoS One*, *8*, e71270.

Mok, S., Koya, R. C., Tsui, C., Xu, J., Robert, L., Wu, L., et al. (2014). Inhibition of CSF-1 receptor improves the antitumor efficacy of adoptive cell transfer immunotherapy. *Cancer Research*, *74*, 153–161.

Nelson, M. H., & Paulos, C. M. (2015). Novel immunotherapies for hematologic malignancies. *Immunological Reviews*, *263*, 90–105.

Noy, R., & Pollard, J. W. (2014). Tumor-associated macrophages: From mechanisms to therapy. *Immunity*, *41*, 49–61.

Olkhanud, P. B., Baatar, D., Bodogai, M., Hakim, F., Gress, R., Anderson, R. L., et al. (2009). Breast cancer lung metastasis requires expression of chemokine receptor CCR4 and regulatory T cells. *Cancer Research*, *69*, 5996–6004.

Palumbo, J. S., Talmage, K. E., Massari, J. V., La Jeunesse, C. M., Flick, M. J., Kombrinck, K. W., et al. (2005). Platelets and fibrin(ogen) increase metastatic potential by impeding natural killer cell-mediated elimination of tumor cells. *Blood*, *105*, 178–185.

Paolino, M., Choidas, A., Wallner, S., Pranjic, B., Uribesalgo, I., Loeser, S., et al. (2014). The E3 ligase Cbl-b and TAM receptors regulate cancer metastasis via natural killer cells. *Nature*, *507*, 508–512.

Poleszczuk, J. T., Luddy, K. A., Prokopiou, S., Robertson-Tessi, M., Moros, E. G., Fishman, M., et al. (2016). Abscopal benefits of localized radiotherapy depend on activated T-cell trafficking and distribution between metastatic lesions. *Cancer Research*, *76*, 1009–1018.

Qian, B., Deng, Y., Im, J. H., Muschel, R. J., Zou, Y., Li, J., et al. (2009). A distinct macrophage population mediates metastatic breast cancer cell extravasation, establishment and growth. *PLoS One*, *4*, e6562.

Qian, B. Z., Li, J., Zhang, H., Kitamura, T., Zhang, J., Campion, L. R., et al. (2011). CCL2 recruits inflammatory monocytes to facilitate breast-tumour metastasis. *Nature*, *475*, 222–225.

Qian, B. Z., & Pollard, J. W. (2010). Macrophage diversity enhances tumor progression and metastasis. *Cell*, *141*, 39–51.

Qian, B. Z., Zhang, H., Li, J., He, T., Yeo, E. J., Soong, D. Y., et al. (2015). FLT1 signaling in metastasis-associated macrophages activates an inflammatory signature that promotes breast cancer metastasis. *The Journal of Experimental Medicine, 212*, 1433–1448.

Robinson, B. D., Sica, G. L., Liu, Y. F., Rohan, T. E., Gertler, F. B., Condeelis, J. S., et al. (2009). Tumor microenvironment of metastasis in human breast carcinoma: A potential prognostic marker linked to hematogenous dissemination. *Clinical Cancer Research: An Official Journal of the American Association for Cancer Research, 15*, 2433–2441.

Romero, I., Garrido, C., Algarra, I., Collado, A., Garrido, F., & Garcia-Lora, A. M. (2014). T lymphocytes restrain spontaneous metastases in permanent dormancy. *Cancer Research, 74*, 1958–1968.

Ruffell, B., Affara, N. I., Cottone, L., Junankar, S., Johansson, M., DeNardo, D. G., et al. (2013). Cathepsin C is a tissue-specific regulator of squamous carcinogenesis. *Genes & Development, 27*, 2086–2098.

Ruffell, B., Affara, N. I., & Coussens, L. M. (2012). Differential macrophage programming in the tumor microenvironment. *Trends in Immunology, 33*, 119–126.

Ruffell, B., Chang-Strachan, D., Chan, V., Rosenbusch, A., Ho, C. M., Pryer, N., et al. (2014). Macrophage IL-10 blocks CD8+ T cell-dependent responses to chemotherapy by suppressing IL-12 expression in intratumoral dendritic cells. *Cancer Cell, 26*, 623–637.

Ruffell, B., & Coussens, L. M. (2015). Macrophages and therapeutic resistance in cancer. *Cancer Cell, 27*, 462–472.

Ruffell, B., DeNardo, D. G., Affara, N. I., & Coussens, L. M. (2010). Lymphocytes in cancer development: Polarization towards pro-tumor immunity. *Cytokine & Growth Factor Reviews, 21*, 3–10.

Sabry, M., & Lowdell, M. W. (2013). Tumor-Primed NK Cells: Waiting for the Green Light. *Frontiers in Immunology, 4*, 408.

Sanford, D. E., Belt, B. A., Panni, R. Z., Mayer, A., Deshpande, A. D., Carpenter, D., et al. (2013). Inflammatory monocyte mobilization decreases patient survival in pancreatic cancer: A role for targeting the CCL2/CCR2 axis. *Clinical Cancer Research: An Official Journal of the American Association for Cancer Research, 19*, 3404–3415.

Sangaletti, S., Di Carlo, E., Gariboldi, S., Miotti, S., Cappetti, B., Parenza, M., et al. (2008). Macrophage-derived SPARC bridges tumor cell-extracellular matrix interactions toward metastasis. *Cancer Research, 68*, 9050–9059.

Shiao, S. L., Ruffell, B., DeNardo, D. G., Faddegon, B. A., Park, C. C., & Coussens, L. M. (2015). TH2-polarized CD4(+) T cells and macrophages limit efficacy of radiotherapy. *Cancer Immunology Research, 3*, 518–525.

Simpson, K. D., Templeton, D. J., & Cross, J. V. (2012). Macrophage migration inhibitory factor promotes tumor growth and metastasis by inducing myeloid-derived suppressor cells in the tumor microenvironment. *Journal of Immunology, 189*, 5533–5540.

Smyth, M. J., Cretney, E., Takeda, K., Wiltrout, R. H., Sedger, L. M., Kayagaki, N., et al. (2001). Tumor necrosis factor-related apoptosis-inducing ligand (TRAIL) contributes to interferon gamma-dependent natural killer cell protection from tumor metastasis. *The Journal of Experimental Medicine, 193*, 661–670.

Smyth, M. J., Thia, K. Y., Cretney, E., Kelly, J. M., Snook, M. B., Forbes, C. A., et al. (1999). Perforin is a major contributor to NK cell control of tumor metastasis. *Journal of Immunology, 162*, 6658–6662.

Spicer, J. D., McDonald, B., Cools-Lartigue, J. J., Chow, S. C., Giannias, B., Kubes, P., et al. (2012). Neutrophils promote liver metastasis via Mac-1-mediated interactions with circulating tumor cells. *Cancer Research, 72*, 3919–3927.

Stockmann, C., Doedens, A., Weidemann, A., Zhang, N., Takeda, N., Greenberg, J. I., et al. (2008). Deletion of vascular endothelial growth factor in myeloid cells accelerates tumorigenesis. *Nature, 456*, 814–818.

Su, S., Liu, Q., Chen, J., Chen, J., Chen, F., He, C., et al. (2014). A positive feedback loop between mesenchymal-like cancer cells and macrophages is essential to breast cancer metastasis. *Cancer Cell*, *25*, 605–620.

Swierczak, A., Cook, A. D., Lenzo, J. C., Restall, C. M., Doherty, J. P., Anderson, R. L., et al. (2014). The promotion of breast cancer metastasis caused by inhibition of CSF-1R/CSF-1 signaling is blocked by targeting the G-CSF receptor. *Cancer Immunology Research*, *2*, 765–776.

Swirski, F. K., Nahrendorf, M., Etzrodt, M., Wildgruber, M., Cortez-Retamozo, V., Panizzi, P., et al. (2009). Identification of splenic reservoir monocytes and their deployment to inflammatory sites. *Science*, *325*, 612–616.

Takeda, K., Hayakawa, Y., Smyth, M. J., Kayagaki, N., Yamaguchi, N., Kakuta, S., et al. (2001). Involvement of tumor necrosis factor-related apoptosis-inducing ligand in surveillance of tumor metastasis by liver natural killer cells. *Nature Medicine*, *7*, 94–100.

Tan, W., Zhang, W., Strasner, A., Grivennikov, S., Cheng, J. Q., Hoffman, R. M., et al. (2011). Tumour-infiltrating regulatory T cells stimulate mammary cancer metastasis through RANKL-RANK signalling. *Nature*, *470*, 548–553.

Toh, B., Wang, X., Keeble, J., Sim, W. J., Khoo, K., Wong, W. C., et al. (2011). Mesenchymal transition and dissemination of cancer cells is driven by myeloid-derived suppressor cells infiltrating the primary tumor. *PLoS Biology*, *9*, e1001162.

Topalian, S. L., Drake, C. G., & Pardoll, D. M. (2015). Immune checkpoint blockade: A common denominator approach to cancer therapy. *Cancer Cell*, *27*, 450–461.

Valastyan, S., & Weinberg, R. A. (2011). Tumor metastasis: Molecular insights and evolving paradigms. *Cell*, *147*, 275–292.

Vasiljeva, O., Papazoglou, A., Kruger, A., Brodoefel, H., Korovin, M., Deussing, J., et al. (2006). Tumor cell-derived and macrophage-derived cathepsin B promotes progression and lung metastasis of mammary cancer. *Cancer Research*, *66*, 5242–5250.

Vesely, M. D., Kershaw, M. H., Schreiber, R. D., & Smyth, M. J. (2011). Natural innate and adaptive immunity to cancer. *Annual Review of Immunology*, *29*, 235–271.

Waldhauer, I., & Steinle, A. (2008). NK cells and cancer immunosurveillance. *Oncogene*, *27*, 5932–5943.

Wang, S., Yuan, Y., Liao, L., Kuang, S. Q., Tien, J. C., O'Malley, B. W., et al. (2009). Disruption of the SRC-1 gene in mice suppresses breast cancer metastasis without affecting primary tumor formation. *Proceedings of the National Academy of Sciences of the United States of America*, *106*, 151–156.

Wculek, S. K., & Malanchi, I. (2015). Neutrophils support lung colonization of metastasis-initiating breast cancer cells. *Nature*, *528*, 413–417.

Wyckoff, J. B., Wang, Y., Lin, E. Y., Li, J. F., Goswami, S., Stanley, E. R., et al. (2007). Direct visualization of macrophage-assisted tumor cell intravasation in mammary tumors. *Cancer Research*, *67*, 2649–2656.

Wyckoff, J., Wang, W., Lin, E. Y., Wang, Y., Pixley, F., Stanley, E. R., et al. (2004). A paracrine loop between tumor cells and macrophages is required for tumor cell migration in mammary tumors. *Cancer Research*, *64*, 7022–7029.

Wynn, T. A., Chawla, A., & Pollard, J. W. (2013). Macrophage biology in development, homeostasis and disease. *Nature*, *496*, 445–455.

Yan, H. H., Pickup, M., Pang, Y., Gorska, A. E., Li, Z., Chytil, A., et al. (2010). Gr-1+CD11b+myeloid cells tip the balance of immune protection to tumor promotion in the premetastatic lung. *Cancer Research*, *70*, 6139–6149.

Yang, L., Huang, J., Ren, X., Gorska, A. E., Chytil, A., Aakre, M., et al. (2008). Abrogation of TGF beta signaling in mammary carcinomas recruits Gr-1+CD11b+myeloid cells that promote metastasis. *Cancer Cell*, *13*, 23–35.

Zhang, Q. W., Liu, L., Gong, C. Y., Shi, H. S., Zeng, Y. H., Wang, X. Z., et al. (2012). Prognostic significance of tumor-associated macrophages in solid tumor: A meta-analysis of the literature. *PLoS One*, *7*, e50946.

Zhang, Q., Qin, J., Zhong, L., Gong, L., Zhang, B., Zhang, Y., et al. (2015). CCL5-mediated Th2 immune polarization promotes metastasis in luminal breast cancer. *Cancer Research*, *75*, 4312–4321.

Zhao, L., Lim, S. Y., Gordon-Weeks, A. N., Tapmeier, T. T., Im, J. H., Cao, Y., et al. (2013). Recruitment of a myeloid cell subset (CD11b/Gr1 mid) via CCL2/CCR2 promotes the development of colorectal cancer liver metastasis. *Hepatology*, *57*, 829–839.

Zhou, S. L., Dai, Z., Zhou, Z. J., Chen, Q., Wang, Z., Xiao, Y. S., et al. (2014). CXCL5 contributes to tumor metastasis and recurrence of intrahepatic cholangiocarcinoma by recruiting infiltrative intratumoral neutrophils. *Carcinogenesis*, *35*, 597–605.

Zhu, Y., Knolhoff, B. L., Meyer, M. A., Nywening, T. M., West, B. L., Luo, J., et al. (2014). CSF1/CSF1R blockade reprograms tumor-infiltrating macrophages and improves response to T-cell checkpoint immunotherapy in pancreatic cancer models. *Cancer Research*, *74*, 5057–5069.

CHAPTER SIX

MicroRNA and Metastasis

L. Ma[1]
The University of Texas MD Anderson Cancer Center, Houston, TX, United States
[1]Corresponding author: e-mail address: lma4@mdanderson.org

Contents

1. MicroRNAs 166
   1.1 Noncoding RNAs and the Discovery of MicroRNAs 166
   1.2 miRNA Biogenesis and Mechanisms of Action 169
   1.3 miRNA Functions 173
2. miRNAs in Cancer 176
   2.1 Deregulated Expression of miRNAs in Cancer 176
   2.2 miRNAs as Oncogenes or Tumor Suppressors 181
   2.3 The Complexity of miRNA Regulatory Networks in Cancer 184
3. miRNA Regulation of Metastasis 186
   3.1 Metastasis-Promoting miRNAs 187
   3.2 Metastasis-Suppressing miRNAs 195
4. miRNAs as Therapeutic Targets 197
   4.1 miRNA Mimics 197
   4.2 miRNA Inhibitors 198
   4.3 Combination Treatment 199

References 200

Abstract

Noncoding RNAs are important regulatory molecules of cellular processes. MicroRNAs (miRNAs) are small noncoding RNAs that bind to complementary sequences in the 3′ untranslated region of target mRNAs, leading to degradation of the target mRNAs and/or inhibition of their translation. Some miRNAs are essential for normal animal development; however, many other miRNAs are dispensable for development but play a critical role in pathological conditions, including tumorigenesis and metastasis. miRNA genes often reside at fragile chromosome sites and are deregulated in cancer. Some miRNAs function as oncogenes or tumor suppressors, collectively termed "oncomirs." Specific metastasis-regulating miRNAs, collectively termed "metastamirs," govern molecular processes and pathways in malignant progression in either a tumor cell-autonomous or a cell-nonautonomous manner. Recently, exosome-transferred miRNAs have emerged as mediators of the tumor-stroma cross talk. In this chapter, we focus on the functions, mechanisms of action, and therapeutic potential of miRNAs, particularly oncomirs and metastamirs.

Advances in Cancer Research, Volume 132
ISSN 0065-230X
http://dx.doi.org/10.1016/bs.acr.2016.07.004

1. MicroRNAs

1.1 Noncoding RNAs and the Discovery of MicroRNAs

Humans and other vertebrates have approximately the same number of protein-coding genes (~20,000) as *Caenorhabditis elegans* (*C. elegans*), less than those of plants (Arabidopsis, ~28,000; rice, ~40,000) and protozoa (~30,000). Many of the proteins are orthologous and have similar functions from worms to humans, and many proteins, such as cell cycle regulators, have their counterparts in the yeast. Then, where is the information that programs the developmental complexity of humans? The answer lies in the size of the genome. The proportion of nonprotein-coding (hereafter referred to as "noncoding") DNA broadly increases with the degree of developmental complexity, and humans have a genome size that is three orders of magnitude larger than that of prokaryotes (Mattick, 2004). Whereas more than 75% of prokaryotes' genome is protein-coding sequence, only 1.5% of the human genome encodes proteins and the remaining 98.5% is composed of regulatory sequences, introns, and noncoding RNA genes (Lander et al., 2001; Stein, 2004).

RNAs consist of messenger RNAs (mRNAs) and noncoding RNAs. mRNAs are protein-coding RNAs, and noncoding RNAs are transcripts with a structural, functional, or catalytic role. There are several different types of noncoding RNAs. Ribosomal RNAs (rRNAs) participate in protein synthesis. Transfer RNAs (tRNAs) serve as the interface between mRNAs and amino acids. Small nuclear RNAs (snRNAs) form part of the spliceosome. Small nucleolar RNAs (snoRNAs) are involved in the modification of rRNAs. Small interfering RNAs (siRNAs) are active RNA molecules in RNA interference (RNAi) that direct the cleavage of mRNAs through perfect base pairing, whereas microRNAs (miRNAs) regulate the expression of protein-coding genes mainly through imperfect base pairing. Long noncoding RNAs (lncRNAs) are transcripts that are more than 200 nucleotides in length and lack significant open reading frames (Huarte & Rinn, 2010; Wapinski & Chang, 2011).

Table 1 shows the timeline of key events in miRNA research. In 1993, Victor Ambros and Gary Ruvkun made the remarkable discovery that the *lin-4* gene, which is essential for controlling developmental timing in *C. elegans*, does not encode protein, but instead produces a 21-nucleotide RNA molecule (Lee, Feinbaum, & Ambros, 1993; Wightman, Ha, & Ruvkun, 1993). Now known as the founding member of the miRNA class,

Table 1 Timeline of Key Events in MicroRNA Research

Year	Contribution	Corresponding Authors
1993	The first miRNA: lin-4	Victor Ambros, Gary Ruvkun
2000	The second miRNA (also the first known human miRNA): let-7	Gary Ruvkun
2001	Identification of miRNAs as an extensive class of small noncoding RNAs	Thomas Tuschl, David Bartel, Victor Ambros
2002	The first evidence of miRNA deregulation in cancer: deletion of *mir-15a–16-1* in CLL	Carlo Croce
2005	The first functional evidence for cancer-promoting miRNAs: miR-17–92 (oncomir-1)	Gregory Hannon/Scott Hammond
2005	Identification of the mechanism of action of the first tumor-suppressing miRNA: let-7 targets *RAS*	Frank Slack
2005	The first in vivo delivery of miRNA antisense inhibitors: antagomirs	Markus Stoffel
2006	The first report that overexpression of a single miRNA causes cancer in transgenic mice: miR-155	Carlo Croce
2007	The first miRNA knockout mouse models	Eric Olson (miR-208), Allan Bradley/Martin Turner (miR-155), Klaus Rajewsky (miR-155), Deepak Srivastava (miR-1)
2007	The first metastasis-regulating miRNA: miR-10b	Robert Weinberg
2008	The first EMT-regulating miRNAs: miR-200, miR-205	Gregory Goodall, Marcus Peter
2009	The first report using adenovirus-associated vectors to deliver miRNA mimics to preclinical models of cancer: miR-26a mimics	Joshua Mendell
2010	The first miRNA knockout mouse model with a cancer phenotype: miR-15a–16 knockout	Riccardo Dalla-Favera

Continued

Table 1 Timeline of Key Events in MicroRNA Research—cont'd

Year	Contribution	Corresponding Authors
2010	The first systemic delivery of anti-miRs to preclinical models of metastatic cancer: miR-10b antagomirs	Robert Weinberg
2013	This first anti-miR-based drug advanced to clinical trials: miravirsen (miR-122 inhibitors)	Santaris Pharma
2013	The first miRNA-based drug advanced to cancer trials: MRX34 (miR-34 mimics)	Mirna Therapeutics, Inc.

CLL, chronic lymphocytic leukemia; *EMT*, epithelial–mesenchymal transition.

C. elegans lin-4 miRNA is processed from a less abundant precursor RNA with a stem-loop, hairpin-like structure (Fig. 1A). One of lin-4's target genes, *lin-14*, encodes a putative transcription factor. The lin-4 miRNA regulates lin-14 protein expression by binding to specific sequences in the 3′ untranslated region (UTR) of the *lin-14* mRNA (Fig. 1B) (Lee et al., 1993; Wightman et al., 1993). Upon lin-4 miRNA expression, lin-14 protein levels are reduced, whereas the transcription from the *lin-14* gene is not affected, suggesting that this is posttranscriptional regulation.

The second gene of this type, *let-7*, was found 7 years later by Gary Ruvkun's laboratory. Similar to *lin-4*, *let-7* also plays an essential role in worm development and encodes a 21-nucleotide untranslated RNA (Reinhart et al., 2000; Slack et al., 2000). However, unlike lin-4, let-7 is conserved in a wide range of animals including worms, flies, and humans. In the next year, several groups, including groups led by Thomas Tuschl, David Bartel, and Victor Ambros, simultaneously reported that there are many endogenously expressed small noncoding RNAs like let-7 (Lagos-Quintana, Rauhut, Lendeckel, & Tuschl, 2001; Lau, Lim, Weinstein, & Bartel, 2001; Lee & Ambros, 2001). They are highly conserved and are collectively named miRNAs.

In summary, miRNAs are small, single-stranded forms of RNA, approximately 22 nucleotides in length. They are generated from endogenously expressed hairpin-shaped precursor RNAs encoded by the genome. miRNAs normally downregulate the expression of protein-coding genes by repressing mRNA translation or by targeting mRNAs for degradation. The list of identified miRNAs has been growing rapidly. As of February

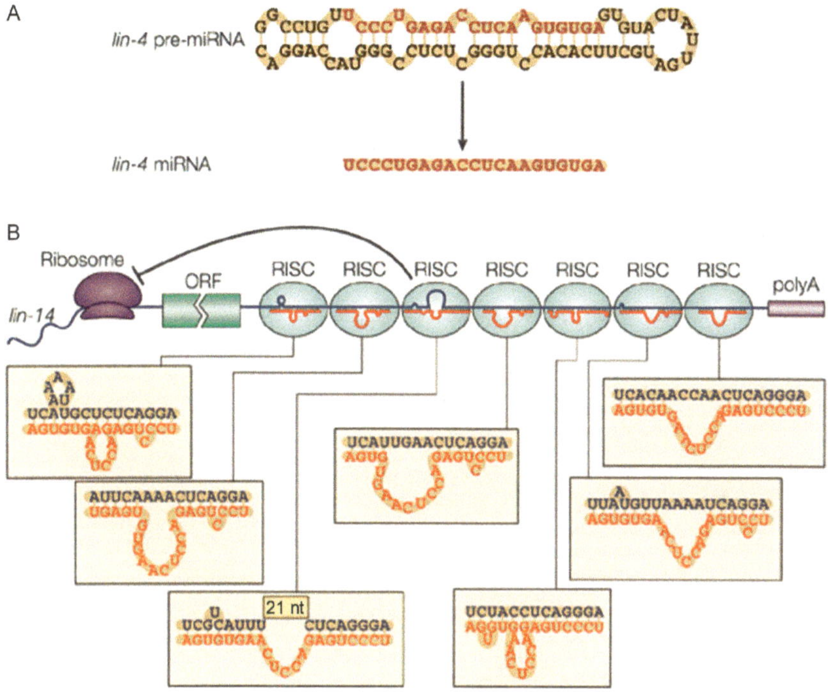

Fig. 1 The molecular hallmarks of lin-4, the founding member of the microRNA class. (A) The precursor structure and mature microRNA (miRNA) sequence of lin-4. (B) Sequence complementarity between lin-4 (*red*) and the 3′ untranslated region (UTR) of *lin-14* mRNA (*blue*). lin-4 is partially complementary to seven sites in the *lin-14* 3′ UTR; its binding to these sites of complementarity brings about repression of LIN-14 protein synthesis. *RISC*, RNA-induced silencing complex. *From He, L., & Hannon, G. J. (2004). MicroRNAs: Small RNAs with a big role in gene regulation.* Nature Reviews Genetics, 5, *522–553.*

2016, 1881 human miRNA and 1193 mouse miRNA entries have been included in the miRbase (http://www.mirbase.org). It has been estimated that more than 30% of human protein-coding genes are regulated by miRNAs (Bartel, 2004).

1.2 miRNA Biogenesis and Mechanisms of Action

Fig. 2 summarizes the steps of miRNA biogenesis (Winter, Jung, Keller, Gregory, & Diederichs, 2009). First, the primary transcripts, also called pri-miRNAs, are cleaved by Drosha to produce the hairpin-shaped miRNA precursors, also called pre-miRNAs. The hairpin precursors are then transported from the nucleus to the cytoplasm by Exportin-5, and subsequently

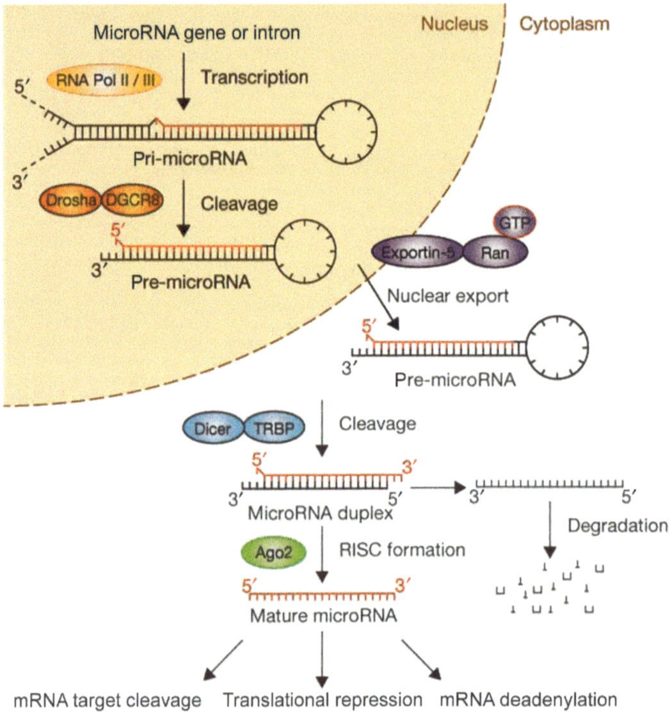

Fig. 2 The canonical pathway of microRNA processing. The canonical miRNA processing pathway includes the production of the primary miRNA transcript (pri-miRNA) by RNA polymerase II or III and cleavage of the pri-miRNA by the microprocessor complex Drosha-DGCR8 (Pasha) in the nucleus. The resulting precursor hairpin, the pre-miRNA, is exported from the nucleus by Exportin-5-Ran-GTP. In the cytoplasm, the RNase Dicer in complex with the double-stranded RNA-binding protein TRBP cleaves the pre-miRNA hairpin to its mature length. The functional strand of the mature miRNA is loaded together with Argonaute (Ago2) proteins into the RNA-induced silencing complex (RISC), where it guides RISC to silence target mRNAs through mRNA cleavage, translational repression, or deadenylation, whereas the passenger strand (black) is degraded. *From Winter, J., Jung, S., Keller, S., Gregory, R. I., & Diederichs, S. (2009). Many roads to maturity: MicroRNA biogenesis pathways and their regulation.* Nature Cell Biology, 11, *228–234.*

cleaved by Dicer to generate mature miRNAs. The mature miRNAs bind to either perfect, or in most cases, imperfect complementary sequences in the 3′ UTR of target mRNAs, leading to either degradation of target mRNAs or inhibition of their translation, or both.

The first step is the transcription of miRNA genes. The majority of miRNA genes are transcribed by RNA polymerase II, while some miRNA genes are transcribed by RNA polymerase III (Bartel, 2004). Intronic miRNA genes are located in the intron region of protein-coding or long noncoding transcripts. Exonic miRNA genes are located in the exon region of protein-coding or long noncoding transcripts. Intergenic miRNA genes are located in the intergenic region of the chromosome. For example, the *mir-15a–mir-16-1* cluster is located in the intron of a noncoding transcript named *DLEU2*; *mir-155* is embedded in the exon of a noncoding transcript named *BIC*; the gene encoding MCM7 protein encompasses a miRNA cluster, *mir-25–93–106b*, in one of its introns; and *mir-985* is embedded in the exon of a protein-coding gene named *CACNG8*. It should be noted that intronic and exonic miRNA genes can either share the same promoter and transcriptional start site as their host genes (the genes in which they are embedded) or can be transcribed from their own unique promoters (Corcoran et al., 2009).

The next step is the processing of miRNA primary transcripts, pri-miRNAs, by the Drosha RNase-III enzyme. In the nucleus, Drosha pairs with a binding partner, DGCR8, to form a protein complex called the microprocessor (Gregory et al., 2004). The microprocessor recognizes a pri-miRNA based on the hairpin terminal loop size, stem structure, and hairpin-flanking sequences. Once it binds to the substrate, Drosha cleaves the pri-miRNA and leaves a two-nucleotide overhang on the 3′ end of the product, ie, the hairpin precursor, or the pre-miRNA (Gregory et al., 2004; Han et al., 2004, 2006). Drosha is not yet found in plants and it is suspected that Dicer-like protein may do its job. Processing of pri-miRNAs can be regulated in different ways (Winter et al., 2009). For instance, some miRNAs require additional specificity factors (eg, p68 and p72) for efficient cleavage; interaction of *pri-mir-18a* with hnRNP A1 facilitates the cleavage of this pri-miRNA by Drosha; and TGF-β signaling induces binding of SMAD to *pri-mir-21*, which enhances its processing by Drosha. In some cases, splicing can replace Drosha processing, if the released intron has the length and hairpin structure of a pre-miRNA (Curtis, Sibley, & Wood, 2012).

The pre-miRNA released from the microprocessor is exported to the cytoplasm by Exportin-5 in a RanGTP-dependent manner (Bohnsack, Czaplinski, & Gorlich, 2004; Lund, Guttinger, Calado, Dahlberg, & Kutay, 2004; Yi, Qin, Macara, & Cullen, 2003). Ran is a small GTPase that exists in a GTP-bound state in the nucleus and a GDP-bound state in the cytoplasm. After nuclear export, the pre-miRNA is dissociated from Exportin-5 upon RanGTP hydrolysis. Subsequently, the pre-miRNA is further cleaved by Dicer, another type-III RNase. Dicer protein contains a putative helicase domain, a PAZ domain, tandem RNase-III domains, and a double-stranded RNA (dsRNA)-binding domain (Macrae et al., 2006). Dicer produces a ~22-nucleotide miRNA duplex with 3′ overhangs, 5′ phosphate groups, and in most cases, imperfect base pairing (Ha & Kim, 2014; Lin & Gregory, 2015). These mismatches cause one strand to be less stable and degraded, while the other strand is incorporated into an effector complex known as the RNA-induced silencing complex (RISC). The RISC binds to the target mRNA through base pairing between the miRNA and the 3′ UTR of the target mRNA.

If the base pairing between the miRNA and its target mRNA is perfect, the miRNA can cause degradation of the target mRNA just like siRNAs (Meister & Tuschl, 2004). How does gene silencing by dsRNA triggers occur? First, similar to Dicer processing of pre-miRNAs, long dsRNAs or short hairpin RNAs (shRNAs) are recognized by Dicer to generate 21- to 22-nucleotide siRNA duplexes with two-nucleotide overhangs at the 3′ end and phosphate groups at the 5′ end. In addition, synthetic 21-nucleotide siRNA duplexes can be introduced into cells and phosphorylated by a cellular kinase. Next, the siRNA duplex is incorporated into the RISC complex. The guide strand of the siRNA duplex is loaded onto Argonaute-2, which is the catalytic engine of the RISC, while the passenger strand is removed. Finally, this complex recognizes and cleaves the target mRNA. The RISC components can be recycled for multiple rounds of mRNA cleavage.

If the base pairing between the miRNA and its target mRNA is imperfect, the miRNA causes translational inhibition (Bartel, 2004), and recent studies demonstrated that miRNA-mediated translational repression is often followed by mRNA degradation (Eichhorn et al., 2014; Guo, Ingolia, Weissman, & Bartel, 2010). Different mechanistic models have been proposed for miRNA-mediated repression, from translational initiation, through elongation, ribosomal drop-off, to polypeptide degradation. However, a growing body of evidence points to the interference with the translational initiation process as the mechanism of action (Wilczynska & Bushell, 2015). In the current model, the recruitment of RISC-bound miRNA to an mRNA results in translational repression. This is mediated by the DEAD-box RNA helicase eIF4A2, which can clamp on to the 5′ UTR and inhibit ribosome subunit 40S scanning. Another RNA helicase, DDX6, is also likely to be involved in this process. Translational repression is followed by deadenylation of the mRNA by deadenylases and subsequent decapping and degradation.

The great majority of miRNAs bind to the 3′ UTR of target mRNAs. Through miRNA-mediated regulation as well as other mechanisms, such as alternative cleavage and polyadenylation, the 3′ UTR regulates mRNA stability and translational efficacy. In addition, alterative mechanisms of action of miRNAs have been reported. For instance, miR-134, miR-296, and miR-470 bind to the coding region of mRNAs encoding pluripotency factors Nanog, Oct4, and Sox2 and inhibit their translation (Tay, Zhang, Thomson, Lim, & Rigoutsos, 2008); miR-10a binds to the 5′ UTR of mRNAs encoding ribosomal proteins and enhances their translation (Orom, Nielsen, & Lund, 2008); miR-122 binds to the 5′ UTR of hepatitis C virus RNA and stimulates its replication (Jopling, Yi, Lancaster, Lemon, & Sarnow, 2005) and translation (Henke et al., 2008); miR-320 directs the recruitment of the RNAi protein Argonaute-1, the Polycomb group component EZH2, and trimethylated histone H3 lysine 27 (H3K27me3) to the *POLR3D* gene promoter to silence its transcription (Kim, Saetrom, Snove, & Rossi, 2008). However, it is not clear whether these alternative mechanisms hold true in general.

A number of computational programs, such as miRanda, TargetScan, PicTar, and RNAHybrid, have been developed to predict miRNA targets. Most of these algorithms scan the 3′ UTR of mRNAs for the presence of the putative binding site for a given miRNA. A popular miRNA target prediction program, TargetScan, defines a "seed" as positions 2–7 of a mature miRNA (Lewis, Burge, & Bartel, 2005). An miRNA family is composed of miRNAs with the same seed region. In addition to seed pairing, additional factors affect miRNA's targeting efficacy (Bartel, 2009; Grimson et al., 2007): (1) the number of miRNA-binding sites in the 3′ UTR correlates with targeting efficacy; (2) closely spaced sites often act synergistically; (3) additional Watson–Crick pairing at nucleotides 12–17 can enhance miRNA targeting; and (4) effective sites preferentially reside within an AU-rich context away from the center of the 3′ UTR, but not too close to the stop codon.

1.3 miRNA Functions

Because miRNAs can recognize and bind a wide spectrum of mRNA molecules, they are capable of regulating a diversity of cellular processes, such as cell cycle progression, apoptosis, differentiation, and stress response. By doing so, miRNAs play roles in organ development, metabolic homeostasis, tumor formation and metastasis, viral infection, immune response, and so on.

As mentioned in Section 1.1, lin-4 and let-7 are the founding members of the miRNA class, which were first discovered in *C. elegans*. When lin-4 or let-7 is mutated or deleted, worms exhibit strong developmental defects, demonstrating the essential roles of these two miRNAs in regulating the developmental timing of *C. elegans*. Early genetic studies also revealed additional functionally characterized miRNAs (Ambros, 2004). For example, miR-273 regulates neuronal cell fate in *C. elegans*; Bantam and miR-14 regulate cell death and proliferation in *Drosophila*; and miR-181 regulates hematopoietic cell fate in mice.

Although some miRNAs are essential regulators of development in worms, when Robert Horvitz's laboratory generated deletion mutants for 83% of known *C. elegans* miRNAs, only less than 10% of them were individually required for normal development or viability (Miska et al., 2007). This suggests significant functional redundancy among miRNAs or among genes and pathways regulated by miRNAs. What about the functions of miRNAs in the development of higher animals? Gregory Hannon and colleagues found that genetic deletion of Dicer, the enzyme that generates siRNAs and mature miRNAs, caused embryonic lethality in mice, suggesting that the RNAi machinery plays an essential role in early development (Bernstein et al., 2003). Similar to individual miRNA deletion in *C. elegans*, a number of knockout mouse studies demonstrated that many miRNAs are dispensable for development or viability (Vidigal & Ventura, 2015), whereas several particular miRNAs are essential for mouse development.

miRNA expression is often tissue type specific, which can be examined by in situ hybridization using miRNA-specific probes. For example, the expression of miR-124a is restricted to the brain and the spinal cord in the fish and mouse, and to the ventral nerve cord in the fruit fly; the expression of miR-1 is restricted to the heart in the mouse, and to the muscles in the fish and fly (Kloosterman & Plasterk, 2006). The evolutionarily conserved sequence and tissue expression pattern of miR-1 and miR-124a suggest that they may play roles in heart and brain development, respectively.

To investigate the role of miR-1 in cardiac development, Deepak Srivastava's group generated miR-1-2 knockout mice. Fifty percent of the mice lacking miR-1-2 died during embryonic development with defects in their ventricular septum, indicative of abnormal heart morphogenesis (Zhao et al., 2007). To determine the function of this miRNA in the adult heart, the group examined miR-1-2 knockout mice that survived to adulthood and observed physiological defects in the adult heart in the absence of miR-1-2, including an increase in cardiomyocyte proliferation and electrophysiological defects, eg, reduced heart rate and prolonged ventricular depolarization (Zhao et al., 2007). These phenotypes suggest that miR-1-2 not only controls cardiac development during embryogenesis but also regulates the machinery that establishes and maintains cardiac rhythm in adults. The group performed microarray analysis of miR-1-2-deficient heart and identified a list of genes regulated by miR-1-2, including genes that control cell cycle, cardiac growth and differentiation, cardiac conduction, ventricular repolarization, and ion channels. Among the genes upregulated by miR-1-2 deletion, they observed enrichment for genes with miR-1 binding site in the 3′ UTR, which are likely to be the targets of miR-1 (Zhao et al., 2007).

The miR-17–92 miRNA cluster has been reported by many groups to be overexpressed in lymphoma, lung cancer, and other tumor types (Olive, Li, & He, 2013). To study its physiological functions, Andrea Ventura, Tyler Jacks, and colleagues generated mice lacking the miR-17–92 miRNA cluster. These mutants displayed perinatal lethality and heart, lung, B-cell, and skeletal defects with complete penetrance, suggesting that this miRNA cluster is indispensable for early development (Ventura et al., 2008). miR-205 is a miRNA highly expressed in skin progenitors and stem cells. miR-205 knockout mice displayed neonatal lethality and severe skin defects, including compromised epidermal and hair follicle growth (Wang et al., 2013). On the other hand, however, many miRNA knockout mice do not show obvious phenotypic differences compared with wild-type mice under normal physiological conditions; when codeleted with their miRNA family members, some of the compound miRNA knockout mutants exhibit notable phenotypes, indicating functional redundancy (Vidigal & Ventura, 2015).

Moreover, some miRNAs are dispensable for development but are important in stress responses or pathological processes, such as cardiac stress, vascular and intestinal injuries, and oncogenic stress. Mice with deletion of these miRNAs show phenotypes in response to external or internal perturbations. Eric Olson and colleagues generated mice deficient in miR-208, a heart-specific miRNA (van Rooij et al., 2007). These mutants had no abnormalities. Only when the investigators subjected the knockout animals to cardiac stress, such as thoracic aortic banding, did they find that the heart of miR-208 null mice reacted differently to such perturbation: whereas wild-type mice displayed cardiac hypertrophy in response to the stress, miR-208 knockout mice showed no hypertrophy of cardiomyocytes.

Consistent with this phenotype, miR-208 regulates the expression of stress response genes (van Rooij et al., 2007). Another example is miR-21. As one of the best established oncogenic miRNAs, miR-21 is overexpressed in most types of cancer, including lung cancer, breast cancer, lymphoma, and other tumor types (Volinia et al., 2006). Is miR-21 essential for normal development? It appears that miR-21 knockout mice, also generated by Eric Olson's laboratory, developed normally without obvious phenotypes; however, genetic deletion of miR-21 partially protected mice against lung tumor formation driven by the *K-Ras* oncogene (Hatley et al., 2010).

In summary, miRNAs can buffer gene expression against internal and external perturbations; in many cases, deletion of individual miRNAs does not lead to phenotypic differences under normal physiological conditions, but under pathological conditions such as cardiac stress or oncogene activation, miRNAs may play a critical role. In Sections 2 and 3, we will focus on the role of miRNAs in cancer and metastasis.

2. miRNAs IN CANCER

2.1 Deregulated Expression of miRNAs in Cancer

The first evidence of aberrant miRNA expression in human cancers was described in 2002 by George Calin, Carlo Croce, and colleagues in B-cell chronic lymphocytic leukemia (CLL), in which chromosomal deletion at the 13q14 locus results in the loss of miR-15a and miR-16 expression (Calin et al., 2002). This region is deleted in more than half of B-cell CLL cases and does not contain protein-coding genes. This discovery of miRNAs linked to cancer prompted extensive investigation of miRNA expression in human tumors. Since then, there has been an explosion of studies that profiled miRNA expression in various cancer types and reported many miRNAs differentially expressed between normal issues and tumor tissues. Jun Lu, Todd Golub, and colleagues profiled the expression of 16,000 mRNAs and 217 miRNAs in 334 normal tissues and tumor samples; a number of miRNAs were found to be upregulated or downregulated in cancer, and the expression profile of these miRNAs classified cancer types better than that of mRNAs (Lu et al., 2005). miRNA expression profiling can be done either by miRNA microarray analysis or by miRNA qPCR array analysis, and more recently, high-throughput small RNA sequencing has enabled profiling of miRNA expression in a more dynamic range.

miRNA expression is altered by several mechanisms in human cancer, such as chromosomal abnormalities, ie, deletion, amplification, translocation, and mutation. Other mechanisms include transcriptional activation or repression, epigenetic changes, and defects in the miRNA biogenesis machinery. The Croce group analyzed miRNA's chromosomal location and reported that miRNA genes often reside in particular genomic regions that are prone to alterations in cancer (Calin et al., 2004). These include regions of deletion, which may harbor tumor suppressor genes; regions of amplification, which may contain oncogenes; and sites of chromosomal translocation or insertion of tumor-associated virus such as human papilloma virus.

Table 2 shows a list of miRNAs downregulated in cancer (Garzon, Marcucci, & Croce, 2010). As mentioned earlier, the *mir-15a–mir-16-1* cluster is embedded in 13q14, a region frequently deleted in CLL. Another example is the *mir-29* miRNA family located in 7q32 and 1q30, regions that are frequently deleted in acute myeloid leukemia. As listed in Table 2, in addition to genetic alterations, other mechanisms also contribute to the downregulation or loss of specific miRNAs. For instance, epigenetic silencing by CpG island hypermethylation has been observed for the *mir-34* miRNA family in multiple cancer types, which leads to miR-34 downregulation or loss of expression. Moreover, deregulation of miRNA expression can also be a result of increased or decreased transcription from the miRNA gene via aberrant transcription factor activity. For instance, the tumor-suppressing transcription factor p53 directly binds to the *mir-34* gene promoter and activates its transcription. Because p53 is frequently mutated or lost in human tumors, this could explain downregulation of miR-34 in these tumors. Similarly, the *mir-15a–mir-16-1* cluster and the *mir-29* family are also positively regulated by p53. On the other hand, some transcription factors are oncoproteins that are often amplified or overexpressed in human cancer. For example, MYC represses the transcription of *let-7a*, the *mir-29* family, and *mir-26a*, all of which are downregulated in cancer.

Some miRNAs are upregulated in cancer (Table 2). The miRNA cluster *mir-17–92* is located in chromosome 13q31, a region amplified in B-cell lymphoma and lung cancer, and its expression is elevated in these malignancies. Oncogenic transcription factors can directly activate the transcription of certain miRNA genes. For example, MYC and E2F transcriptionally activate the expression of the *mir-17–92* miRNA cluster, while NF-κB positively regulates the transcription of *mir-155*, which is upregulated in lymphoma, leukemia, breast, lung, and colon cancers.

Table 2 MicroRNAs Involved in Cancer

MicroRNA	Genomic Location	Expression in Patients	Deregulation Mechanism	Function	Targets	Experimental Data	Therapeutic Strategy
miR-15a–miR-16-1	13q14	Down in CLL, prostate cancer, and pituitary adenomas	Genomic loss, mutations, positive regulation by p53	Tumor suppressor	BCL-2, MCL1	In vitro overexpression induces apoptosis in CLL and prostate cancer cells; in vivo silencing causes CLL in mice	Mimics, vector based (viral), drugs
Let-7a-2	11q24	Down in lung, colon, breast, ovarian, and stomach cancer	Negative regulation by MYC	Tumor suppressor	KRAS, NRAS, CDK6, CDC25A, HMGA2, MYC	In vitro overexpression reduces cell growth in lung, breast, and colon cancer cells; In vivo overexpression reduces breast and lung tumor burden in mice	Mimics, vector based (viral), drugs
miR-29b-1–miR-29a miR-29b-2–miR-29c	7q32 1q32	Down in NPM1 wild-type AML, CLL, lung, and breast cancer, cholangiocarcinoma, lymphoma, hepatocarcinoma, and rhabdomyosarcoma	Genomic loss, negative regulation by MYC, positive regulation by p53	Tumor suppressor	MCL1, CDK6, TCL1, DNMT1, DNMT3α, DNMT3β	In vitro overexpression induces apoptosis, inhibits cell proliferation and induces DNA hypomethylation in several cancers; in vivo overexpression inhibits tumorigenicity in AML, liver, and lung cancer in mice	Mimics, vector based (viral), drugs

miR-34a, miR-34b, and miR-34c	1p36 11q23	Down in colon, lung, breast, kidney and bladder cancer, neuroblastoma, and melanoma cell lines	Methylation regulation, positive regulation by p53, deletion	Tumor suppressor	CDK4, CDK6, CCNE2, CCND1, MET, MYC, CREB, E2F3, BCL-2	In vitro overexpression induces cell cycle arrest, apoptosis, and inhibits cell proliferation	Mimics, vector based (viral), drugs
miR-26a	3p22	Down in liver cancer	Negative regulation by MYC	Tumor suppressor	CCND2, CCNE2	Restoration of miR-26 inhibits MYC-induced liver cancer	Vector based (viral)
miR-155	21q21	Up in high-risk CLL, AML, lung, colon, breast cancer, and in lymphomas	Positive regulation by NF-κB	Oncogene	SHIP1, CEBPB	Overexpression in HSC-induced myeloid proliferation and blocks erythropoiesis in mice; in vivo overexpression in lymphocytes induces pre-B lymphoma and leukemia in mice	Antisense oligonucleotides, miR-mask, miRNA sponges, drugs
miR-17–92	13q31	Up in lung, breast, colon and stomach cancer, myeloma, and t(11q23) AML	Amplification, positive regulation by E2F and MYC	Oncogene	BIM, PTEN, CDKN1A	Cooperates with MYC to induce lymphoma; in vivo overexpression in lymphocytes induces lymphoid proliferation and autoimmunity in mice	Antisense oligonucleotides, miR-mask, miRNA sponges, drugs

Continued

Table 2 MicroRNAs Involved in Cancer—cont'd

MicroRNA	Genomic Location	Expression in Patients	Deregulation Mechanism	Function	Targets	Experimental Data	Therapeutic Strategy
miR-21	17q23	Up in pancreas, breast, lung, prostate and stomach cancer, CLL, AML, myeloma, and glioblastoma	Positive regulation by IL-6 and GF1a	Oncogene	PDCD4, PTEN, TPM1	In vitro silencing enhances apoptosis in glioblastoma, lung, breast, and hepatocarcinoma cell lines	Antisense oligonucleotides, miR-mask, miRNA sponges, drugs
miR-372 miR-373	19q13	Up in testicular germ cell tumors and in breast cancer	Unknown	Oncogene	LATS2	Neutralizes the p53 pathway in vitro; in vivo overexpression stimulated cancer cell invasion	Antisense oligonucleotides, miR-mask, miRNA sponges, drugs

Three corrections (the genomic locations of *mir-15a–16-1*, *mir-29b-2–29c*, and *mir-17–92*) were made to the original table.

AML, acute myeloid leukemia; *BCL-2*, B-cell lymphoma protein-2; *CCN*, cyclin; *CDC*, cell division cycle; *CDKN1A*, cyclin-dependent kinase inhibitor 1A; *CEBPB*, CCAAT/enhancer-binding protein β; *CLL*, chronic lymphocytic leukemia; *CREB*, cAMP response element-binding protein; *DNMT*, DNA methyltransferase; *HMGA2*, high-mobility group AT-hook 2; *HSC*, hematopoietic stem cells; *IL-6*, interleukin-6; *KRAS*, v-Ki-ras2 Kirsten rat sarcoma viral oncogene homologue; *LATS2*, *LATS*, large tumor suppressor, homologue 2; *MCL1*, myeloid cell leukemia sequence 1 (BCL-2 related); *NF-κB*, nuclear factor-κB; *NPM1*, nucleophosmin (nucleolar phosphoprotein B23, numatrin); *NRAS*, neuroblastoma RAS viral (v-ras) oncogene homologue; *PDCD4*, programmed cell death 4; *PTEN*, phosphatase and tensin homologue; *SHIP1*, Src homology 2 domain-containing inositol 5-phosphatase 1; *TPM1*, tropomyosin 1.

Adapted from Garzon, R., Marcucci, G., & Croce, C. M. (2010). Targeting microRNAs in cancer: Rationale, strategies and challenges. *Nature Reviews Drug Discovery*, *9*, 775–789.

2.2 miRNAs as Oncogenes or Tumor Suppressors

Deregulated expression of miRNAs can be either a cause or a consequence of tumorigenesis. In this section, we discuss several miRNAs that do indeed play a causal role in tumor formation and growth, acting as either an oncogene or a tumor suppressor.

As mentioned in Section 2.1, the *mir-17–92* cluster is located in a chromosomal region amplified in human cancer. Consistent with this, real-time qPCR analysis revealed that the level of the *pri-mir-17–92* transcript is indeed increased in B-cell lymphomas and colon tumors from human patients, compared with the corresponding normal tissues (He et al., 2005). To determine the overexpression effect of this miRNA cluster on lymphomagenesis in vivo, Lin He, Gregory Hannon, Scott Hammond, and colleagues isolated fetal liver hematopoietic stem cells (HSCs) from Eμ-Myc transgenic mice, infected these HSCs with retrovirus expressing the miR-17–92 cluster and green fluorescent protein (GFP), and implanted the cells into recipient mice (Fig. 3A). Mice that received HSCs overexpressing the miRNA cluster had much worse tumor-free survival and overall survival (Fig. 3B); moreover, GFP imaging of tumor-bearing mice revealed that the miR-17–92 cluster-overexpressing tumors had a more disseminated phenotype (Fig. 3C) (He et al., 2005). These results suggest that the miR-17–92 miRNA cluster can promote Myc-induced lymphomagenesis. Three years later (in 2008), Klaus Rajewsky's group reported their studies on mice with transgenic overexpression of the miR-17–92 cluster in lymphocytes (Xiao et al., 2008). These mice developed lymphoproliferative disease and autoimmunity and died prematurely. Lymphocytes from miR-17–92 transgenic mice showed more proliferation and less cell death. They also found that the miR-17–92 miRNA suppresses the expression of the tumor suppressor PTEN and the proapoptotic protein Bim, which might explain the lymphoproliferative disease and autoimmunity in miR-17–92 transgenic mice and lymphoma development in patients with amplification of the *mir-17–92* gene. As mentioned in Section 1.3, mice lacking the miR-17–92 cluster died perinatally exhibiting multiple developmental defects including defective B-cell development. Taken together, this miRNA cluster plays an essential role in normal development, and deregulation of its expression contributes to tumorigenesis.

Overexpression of miR-155 is found in several types of lymphomas such as Hodgkin and Burkitt lymphomas. Carlo Croce's laboratory demonstrated the role of this miRNA in tumorigenesis by producing transgenic mice that specifically overexpress miR-155 in B cells (Costinean et al., 2006). These

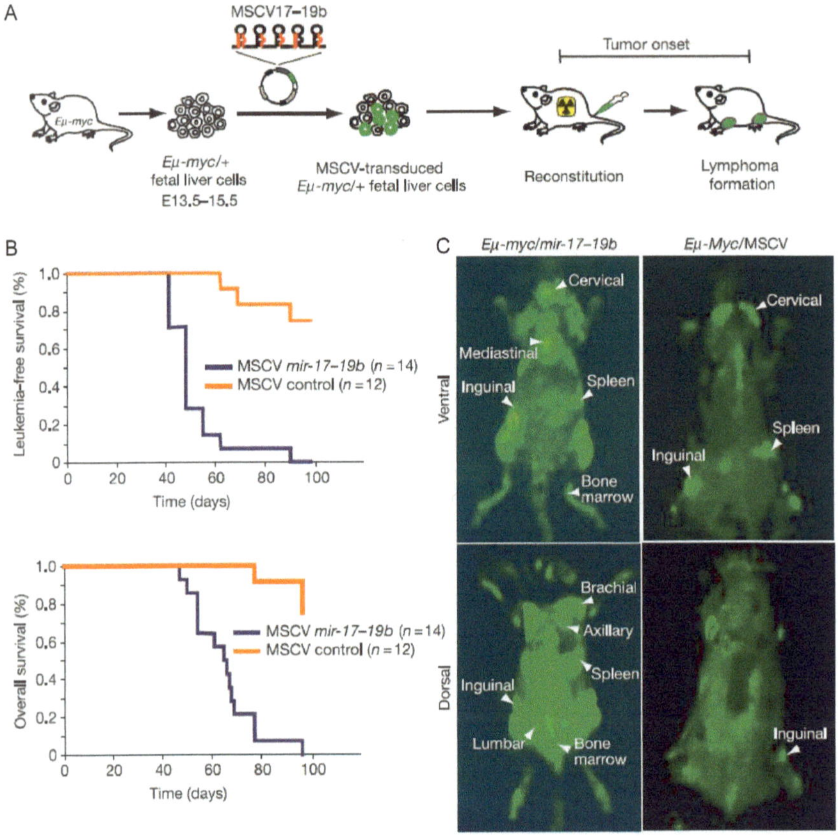

Fig. 3 Overexpression of the *mir-17–19b* cluster accelerates c-myc-induced lymphomagenesis in mice. (A) Schematic representation of the adoptive transfer protocol using *Eμ-myc* HSCs. (B) Mice reconstituted with HSCs expressing *mir-17–19b* in an MSCV retroviral vector (MSCV *mir-17–19b*) or infected with a control MSCV virus were monitored by blood smear analysis starting 5 weeks after transplantation. The Kaplan–Meier curves represent the percentage of leukemia-free survival or overall survival. (C) External GFP imaging of tumor-bearing mice with *Eμ-myc/mir-17–19b* or *Eμ-myc*/MSCV shows the overall distribution of tumor cells. *Eμ-myc/mir-17–19b* tumors show a more disseminated phenotype compared with control tumors. These animals are representative of their genotype. *From Lin, H., Thomson, J.M., Hemann, M.T., Hernando-Monge, E., Mu, D., Goodson, S., et al. (2005). A microRNA polycistron as a potential human oncogene.* Nature, 435, *828–833.*

transgenic mice developed preleukemic lymphoproliferative disease that progressed to B-cell leukemia and high-grade lymphoma. The spleen of miR-155 transgenic mice was enlarged due to increased lymphoid proliferation and expansion of leukemic and lymphoma cells. This was the first report showing that transgenic overexpression of a single miRNA can cause cancer.

As mentioned in Section 1.3, miR-21 is an oncomir that is overexpressed in most types of cancer. Using both miR-21 knockout and transgenic mice, Eric Olson and colleagues found that genetic deletion of miR-21 partially protected against K-Ras-induced lung tumor formation, and that overexpression of miR-21 by four- to sixfold over basal expression levels (driven by a ubiquitously expressed CAG promoter) enhanced tumorigenesis in the K-Ras model of nonsmall-cell lung cancer (Hatley et al., 2010). However, these miR-21 knockout and transgenic mutants showed no phenotypic differences compared with wild-type littermates. In contrast, Frank Slack's group developed a different miR-21 transgenic line in which Cre and Tet-off technologies were used to achieve tissue-specific and doxycycline-controlled expression of miR-21. This model exhibited 16-fold overexpression of miR-21 in hematopoietic tissues and development of pre-B-cell lymphoma, both of which could be completely reversed within 1 week of doxycycline treatment, demonstrating that miR-21 is responsible for both initiation and maintenance of lymphoma (Medina, Nolde, & Slack, 2010). Therefore, miR-21 is a bona fide oncogenic miRNA, and miR-21-driven tumors are addicted to this oncomir, which may be exploited therapeutically.

Some miRNAs function as tumor suppressors. As mentioned in Section 2.1, genomic deletion of the *mir-15a–mir-16-1* cluster in CLL was the first reported miRNA deregulation in cancer. In addition to CLL, this miRNA cluster is downregulated in prostate cancer and pituitary adenoma. Experiments in which miR-15a and miR-16 were ectopically expressed in leukemic cells revealed that this miRNA cluster promotes apoptosis (Cimmino et al., 2005). miRNA target prediction programs identified *BCL2*, a well-known antiapoptotic gene that is upregulated in a subset of patients with CLL, as the target of miR-15a and miR-16. Indeed, these two miRNAs directly bind to the *BCL2* 3′ UTR and inhibit its translation. A negative correlation was also identified between miR-15a–miR-16 miRNA expression and BCL2 protein expression in patients with CLL, suggesting that genomic deletion of the *mir-15a–mir-16-1* cluster in CLL results in derepression of BCL2 (Cimmino et al., 2005). The gold standard for a tumor suppressor is that the loss of function leads to tumor formation. Indeed, mice with genetic deletion of the *mir-15a–mir-16-1* cluster developed CLL-like disease and lymphoma (Klein et al., 2010).

As mentioned in Section 1.1, the first two reported miRNAs, lin-4 and let-7, were originally discovered in *C. elegans*, and both miRNAs control the timing of development. Let-7 is also the first known human miRNA. Frank Slack and colleagues identified several let-7 binding sites in the 3′ UTRs of the *K-Ras* and *N-Ras* oncogenes. They found that the 3′ UTR of *K-Ras* and *N-Ras* repressed the activity of a luciferase reporter in cells expressing let-7. Moreover, transfecting cells with the let-7 miRNA reduced the expression of Ras (Johnson et al., 2005). Several genes encoding the let-7 family of miRNAs reside in genomic regions that are deleted in cancer, and Slack's group reasoned that loss of let-7 could upregulate Ras oncoprotein and contribute to tumorigenesis. They analyzed let-7 expression in human tumors and found that let-7 was underexpressed in lung cancer compared with adjacent normal tissue; moreover, northern blot and western blot analyses revealed an inverse correlation between let-7 miRNA levels and Ras protein levels (Johnson et al., 2005). In addition to targeting *Ras*, let-7 also targets several other oncogenes such as *MYC*, *CDK6*, *HMGA2*, and *BACH1*. The Slack Lab further investigated whether let-7 inhibits lung tumor growth. They found that overexpression of let-7 inhibited the growth of tumor xenografts formed by lung cancer cell lines in immunodeficient mice (Esquela-Kerscher et al., 2008). These findings provide direct evidence that let-7 acts as a lung tumor suppressor.

The miR-34 family represents another example of tumor-suppressing miRNAs (He, He, & Hannon, 2007). miR-34 family members are highly conserved during evolution. *mir-34a* is located in the chromosome 1p36 region that is deleted in human tumors. Primary tumors and cancer cell lines often show downregulation of miR-34 expression. Several groups independently identified *mir-34* as a tumor suppressor and as a transcriptional target of p53, as evidenced by: (1) the expression of miR-34a, miR-34b, and miR-34c is robustly induced by DNA damage and oncogenic stress in a p53-dependent manner; (2) ectopic expression of miR-34 in cancer cells inhibits proliferation and activates cell death pathways; and (3) silencing miR-34 attenuates p53-mediated apoptosis. These studies implicated the miR-34 family of miRNAs in the p53 tumor suppressor pathway. Several regulators of cell cycle, apoptosis, and proliferation, including *CDK4*, *CDK6*, *CCNE2*, *CCND1*, *MET*, and *BCL2*, have been identified as the targets of miR-34.

2.3 The Complexity of miRNA Regulatory Networks in Cancer

The functions of a particular miRNA are often tissue specific, being dependent on the expression pattern of its target mRNAs in a given cell type.

Moreover, miRNA-targeted genes themselves may exert differential or even opposing effects in different cellular contexts. The importance of specific proteins and pathways in different cancer types is also tissue type dependent. Indeed, some cancer-implicated miRNAs can function as either an oncogene or a tumor suppressor in different contexts. For example, miR-21, miR-26a, miR-27a, and miR-29 are capable of both promoting and suppressing tumorigenesis in a tissue type-dependent manner. It has been shown that miR-26a inhibits tumorigenicity in multiple cancer settings, such as liver cancer, by targeting cyclins D2 and E2 (Kota et al., 2009); however, in other contexts, such as glioblastoma, miR-26a promotes tumorigenesis by targeting *PTEN* (Huse et al., 2009). To study the contexts in which the antitumor vs protumor function of miR-26 predominates in vivo, Joshua Mendell and colleagues generated miR-26a transgenic mice, which did not exhibit elevated tumorigenesis despite downregulation of Pten. Instead, transgenic overexpression of miR-26a inhibited intestinal adenoma formation in $Apc^{min/+}$ mice, which revealed a tumor-suppressing role of miR-26a in intestinal cancer that overrides its potential oncogenic activity (Zeitels et al., 2014).

RNAs that compete with each other through common miRNA recognition elements, termed competing endogenous RNAs (ceRNAs), have been proposed to regulate key oncogenes and tumor suppressor genes (Salmena, Poliseno, Tay, Kats, & Pandolfi, 2011). ceRNAs can be either mRNAs or RNAs transcribed from pseudogenes and other noncoding genes. For example, Pier Paolo Pandolfi's laboratory found that RNAs sharing miRNA-binding sites with *PTEN*, such as the *PTEN* pseudogene *PTENP1*, upregulate PTEN expression levels by acting as endogenous miRNA decoys or sponges (Karreth et al., 2011; Poliseno et al., 2010; Tay et al., 2011). Pandolfi's group also reported that mice with transgenic overexpression of the murine *Braf* pseudogene *Braf-rs1* developed lymphoma (Karreth et al., 2015); however, whether *Braf-rs1* triggers tumor formation through a ceRNA effect remains to be determined. While the evidence for the ceRNA hypothesis is emerging (Tay, Rinn, & Pandolfi, 2014), studies from other groups suggest that the targets of abundantly expressed miRNAs are likely not prone to derepression by ceRNA competition, due to the buffering capacity of the high miRNA and target pool concentrations, whereas active miRNAs with a low miRNA:target ratio might be amenable to titration by a high-affinity ceRNA (Bosson, Zamudio, & Sharp, 2014; Denzler, Agarwal, Stefano, Bartel, & Stoffel, 2014).

3. miRNA REGULATION OF METASTASIS

Although great advances have been made in combating cancer, particularly through the development of surgeries, chemotherapies, and radiation therapies that are highly effective against early-stage tumors, disease progression to metastasis remains a formidable and fatal challenge (Brabletz, Lyden, Steeg, & Werb, 2013; Eccles & Welch, 2007; Wan, Pantel, & Kang, 2013). The seeds of metastasis are often present early in the disease process and can be related to properties inherent in at least some cells in the primary tumor, such as chemoresistance, radioresistance, or a high potential for metastasis-promoting genetic and molecular alterations (Talmadge & Fidler, 2010; Wan et al., 2013). Accordingly, adjuvant therapies that are matched to these types of inherent properties are highly sought after and important to any further improvements in patient outcomes (Steeg & Theodorescu, 2008). In the current setting, running metastasis prevention trials on patients with early-stage cancer remains out of reach, because normal methods of conducting such trials are prohibitively lengthy and costly (Steeg, 2012). Furthermore, the time and cost of trials cannot be reduced without some knowledge of how to select patients with early-stage disease who are at elevated risk for metastatic disease. Finally, matching a patient's particular abnormalities with specific agents will remain impossible until we have developed better prognostic markers for metastasis, novel agents that target specific abnormalities, and predictive markers for treatment response (Sun & Ma, 2015).

Cancer cells can hijack developmental programs to acquire metastatic ability and become resistant to treatment. In particular, adherent epithelial tumor cells may convert to a motile mesenchymal state through epithelial–mesenchymal transition (EMT; Kalluri & Weinberg, 2009; Thiery, 2002). During EMT, cells lose adhesion and expression of epithelial markers such as E-cadherin and ZO-1, and simultaneously acquire motility and expression of mesenchymal markers such as N-cadherin, vimentin, and fibronectin (Kalluri & Weinberg, 2009). Both EMT and its reverse process, mesenchymal–epithelial transition (MET), play an essential role in developmental processes, such as neural crest development, mesoderm formation, heart valve development, and secondary palate formation (Yang & Weinberg, 2008). Recent studies suggest that primary tumor cells can resurrect the EMT program to invade and disseminate, and that the MET process enables already disseminated cancer cells to proliferate and colonize the distant organs (Tsai, Donaher, Murphy, Chau, & Yang, 2012; Tsai & Yang, 2013). In addition, induction of EMT in differentiated epithelial tumor cells can generate cells with properties of cancer stem cells (CSCs), defined operationally as tumor-initiating cells (Mani et al., 2008). Both EMT and CSC properties are regulated by pleiotropically acting molecules—transcription factors and miRNAs. Extracellular factors such as TGF-β have been shown to activate the expression of several transcription factors, including Snail, Slug, Twist, ZEB1, and ZEB2, which function as inducers of EMT and tumor metastasis (Yang & Weinberg, 2008). These transcription factors regulate the transcription of specific miRNA genes, and on the other hand, are regulated by specific miRNAs. Emerging evidence suggests that cancer cells exploit these transcription factors (Tsai & Yang, 2013) and miRNAs (Pencheva & Tavazoie, 2013; Piao & Ma, 2012; Zhang & Ma, 2012) to acquire plasticity and accomplish the invasion–metastasis cascade.

3.1 Metastasis-Promoting miRNAs

The key miRNAs involved in metastasis are listed in Table 3. The link between miRNAs and metastasis was first reported in 2007. Li Ma, Robert Weinberg, and colleagues compared miRNA expression between normal human mammary epithelial cells, nonmetastatic breast cancer cells, and metastatic breast cancer cells. This led to the identification of several metastasis-associated miRNAs. One of them, miR-10b, promoted lung metastasis from primary mammary tumors in xenograft models (Fig. 4A) (Ma, Teruya-Feldstein, & Weinberg, 2007). Conversely, silencing miR-10b inhibited lung metastasis in a mouse mammary tumor model (Fig. 4B) (Ma, Reinhardt, et al., 2010).

miR-10b promotes tumor progression and metastasis by targeting multiple genes in a variety of cancer types. Initially, miR-10b was found to be expressed at elevated levels in metastatic cell lines and primary tumors from patients with metastatic breast cancer (Edmonds et al., 2009; Ma et al., 2007). Subsequent analyses demonstrated that miR-10b expression is also associated with high-grade malignancy or metastasis in glioblastoma, pancreatic, bladder, liver, nasopharyngeal, and esophageal cancer, and that lymph node metastases express higher miR-10b levels than paired primary tumors in multiple types of human cancer (Baffa et al., 2009; Ma, 2010). Twist, an inducer of EMT and metastasis, binds to the *mir-10b* promoter and activates its transcription (Fig. 4C) (Ma et al., 2007). When overexpressed, miR-10b

Table 3 MicroRNAs Involved in Metastasis

MicroRNA	Function in Metastasis	Cancer Type	Targets
miR-10b	Promote	Breast, pancreatic, glioblastoma, bladder, hepatocellular, nasopharyngeal, esophageal	HOXD10, NF1, KLF4, PTEN
miR-9	Promote	Breast	CDH1, LIFR
miR-103/107	Promote	Breast	Dicer
miR-22	Promote	Breast	TET family
miR-182	Promote	Sarcoma, melanoma	RSU1, MTSS1, PAI1, TIMP1, FOXO3, MITF
miR-373	Promote	Breast, testicular germ cell	CD44, LATS2
miR-21	Promote	Breast, colorectal, gastric, lung, pancreatic, prostate, bladder, ovarian, hepatocellular	PDCD4, PTEN, TPM1, Maspin
miR-199a-3p, miR-199a-5p, miR-1908	Promote	Melanoma	DNAJA4, ApoE
miR-105	Promote	Breast	ZO-1
miR-122	Promote	Breast	PKM2
miR-19a	Promote	Breast	PTEN
let-7 family	Suppress	Breast, lung, hepatocellular, colorectal, gastric	RAS, MYC, CDK6, HMGA2, BACH1
miR-34 family	Suppress	Breast, prostate, lung, colorectal, pancreatic, hepatocellular, neuroblastoma	CD44, Tgif2, CDK4, CDK6, CCNE2, CCND1, MET, BCL2
miR-206	Suppress	Breast	NOTCH3
miR-335	Suppress	Breast	SOX4, TNC, PTPRN2
miR-126	Suppress	Breast	MERTK, PITPNC1, IGFBP2, SDF-1α
miR-200 family	Suppress or promote	Breast, lung	ZEB1, ZEB2, Sec23a
miR-29b	Suppress	Breast, prostate, hepatocellular	VEGFA, ANGPTL4, LOX, MMP2, MMP9, PDGF

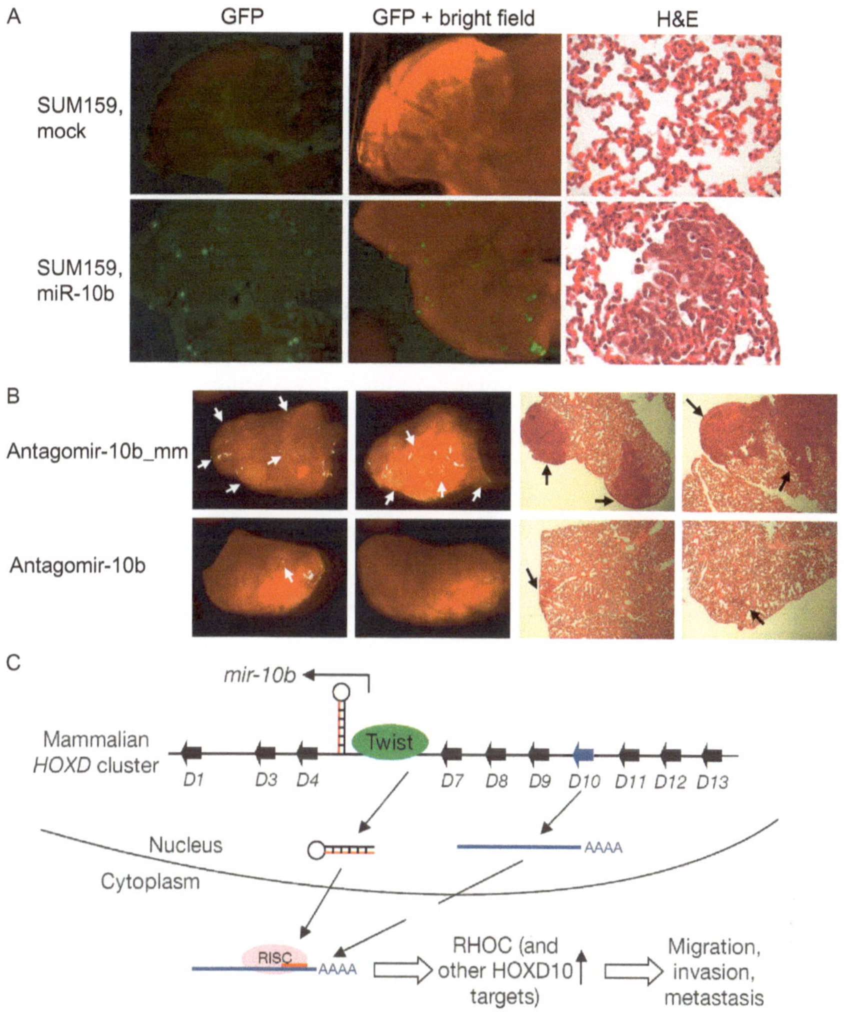

Fig. 4 miR-10b promotes breast cancer metastasis in mice. (A) Bright-field, GFP imaging, and H&E staining of the lungs from mice that received orthotopic injection of miR-10b-transduced or mock-infected SUM159 cells, at 11 weeks after transplantation. The MDH1-PGK-GFP 2.0 retroviral vector was used to express miR-10b. (B) Bright-field imaging and H&E staining of the lungs from 4T1 tumor-bearing mice treated with antagomir-10b or antagomir-10b_mm, at 4 weeks after orthotopic implantation. *Arrows* indicate lung metastases. Antagomir-10b_mm, a mutant miR-10b antagomir that harbors 12 mismatches within the complementary sequence to miR-10b and does not match any sequence in the mouse genome. (C) Model of the regulation and function of miR-10b in cancer metastasis. A Twist-induced miRNA, miR-10b, inhibits synthesis of the HOXD10 protein, permitting the expression of the prometastatic gene product, RHOC; this favors, in turn, cancer cell migration, invasion, and metastasis. In addition, miR-10b also targets other tumor suppressor genes or metastasis suppressor genes, such as *NF1*, *KLF4*, and *PTEN* (not shown). *RISC*, RNA-induced silencing complex. *Panels (A) and (C) are from Ma, L., Teruya-Feldstein, J., & Weinberg, R. A. (2007). Tumour invasion and metastasis initiated by microRNA-10b in breast cancer.* Nature, *449, 682–688. Panel (B) is from Ma, L., Reinhardt, F., Pan, E., Soutschek, J., Bhat, B., Marcusson, E. G., et al. (2010). Therapeutic silencing of miR-10b inhibits metastasis in a mouse mammary tumor model.* Nature Biotechnology, *28, 341–347.*

increases the motility and invasiveness of various cancer cell lines and induces metastasis in xenograft models by targeting multiple tumor suppressor genes or metastasis suppressor genes, such as *HOXD10*, *NF1*, *KLF4*, *PTEN*, and likely other genes; conversely, in multiple cancer types, silencing miR-10b expression inhibits cancer cell proliferation, migration, and invasion in vitro as well as tumor growth or metastasis in vivo, suggesting that miR-10b is a potential target for antitumor or antimetastatic therapy (Ma, 2010). Intriguingly, miR-10b is secreted by metastatic breast cancer cells via exosomes and, upon uptake, induces invasiveness of nonmalignant mammary epithelial cells (Singh, Pochampally, Watabe, Lu, & Mo, 2014).

A second metastasis-associated miRNA that Ma and Weinberg identified is miR-9, which also has multiple functional targets involved in metastasis (Fig. 5). Myc-induced expression of miR-9 targets *CDH1*, the mRNA encoding E-cadherin, and promotes EMT and metastasis in breast cancer cells expressing this adhesion molecule (Ma, Young, et al., 2010). Interestingly, miR-9 also induces metastasis in breast cancer cells that have lost E-cadherin. On a quest to find additional miR-9 targets, Li Ma's laboratory identified leukemia inhibitory factor receptor (LIFR) as a breast cancer metastasis suppressor targeted by this miRNA and as a prognostic marker for clinical outcomes (Chen et al., 2012). LIFR suppresses metastasis by activating a Hippo kinase cascade leading to phosphorylation, cytoplasmic retention, and functional inactivation of the Hippo effector YAP, which is a cancer-promoting transcriptional coactivator. LIFR-induced activation of Hippo signaling is mediated by the LIFR coreceptor gp130 and the adaptor protein Scribble (Chen et al., 2012). Oncomine, The Cancer Genome Atlas (TCGA), microarray, and tissue microarray analyses revealed that LIFR is downregulated in breast cancer, liver cancer, colon cancer, and many other types of cancer. Taken together, these findings suggest that miR-9 targets multiple metastasis suppressors. Furthermore, tumor cell-secreted miR-9 was found to promote endothelial cell migration and angiogenesis by activating the JAK-STAT pathway (Zhuang et al., 2012), demonstrating a cell-nonautonomous effect of this miRNA in the tumor microenvironment.

Whereas miR-9 directly targets E-cadherin, some miRNAs induce EMT indirectly. For instance, the miR-103/107 family attenuates miRNA biosynthesis by targeting Dicer, which in turn leads to downregulation of miR-200, a miRNA family that targets the mRNAs encoding the EMT-inducing transcription factors ZEB1 and ZEB2. By doing so, miR-103/107 induces EMT and promotes metastatic dissemination of otherwise

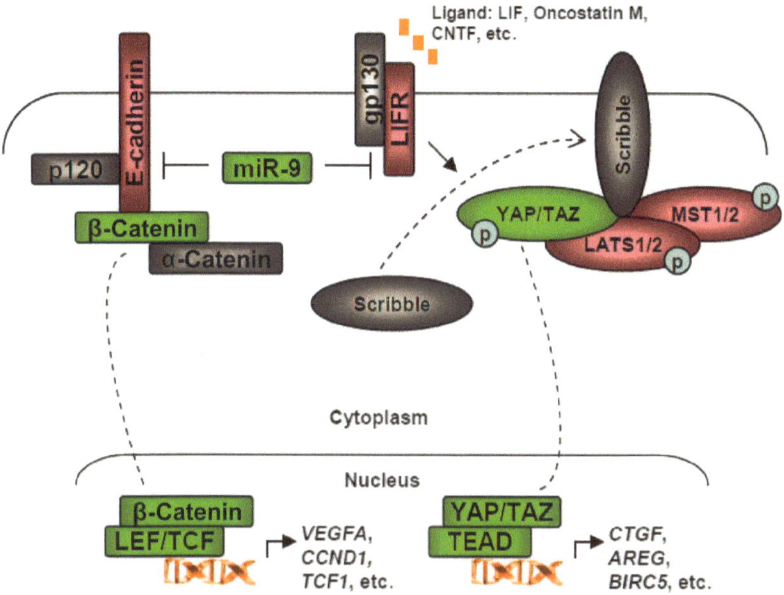

Fig. 5 Model of two metastasis suppressor pathways that are negatively regulated by miR-9 in breast cancer cells. miR-9 targets the mRNAs that encode two metastasis suppressors, E-cadherin and LIFR. E-cadherin maintains adherens junctions and sequesters β-catenin at the cytoplasmic membrane. LIFR promotes localization of Scribble to the cell membrane, which in turn activates Hippo signaling, leading to the phosphorylation and functional inactivation of the transcriptional coactivator YAP. *Green* indicates oncogenic and/or prometastatic factors; *pink* indicates tumor-suppressing and/or metastasis-suppressing factors. *MST1/2*, mammalian Hippo homologs 1 and 2; *LEF/TCF*, lymphoid enhancer-binding factor/T cell-specific factor; *LATS1/2*, large tumor suppressor homologs 1 and 2; *TEAD*, TEA domain. "p" in the *circles* indicates phosphorylation. *From Chen, D., Sun, Y., Wei, Y., Zhang, P., Rezaeian, A. H., Teruya-Feldstein, J., et al. (2012). LIFR is a breast cancer metastasis suppressor upstream of the Hippo-YAP pathway and a prognostic marker.* Nature Medicine, 18, *1511–1517.*

nonaggressive breast cancer cells (Martello et al., 2010). Pier Paolo Pandolfi and colleagues used genetically engineered mouse models to study an EMT- and metastasis-promoting miRNA (Song et al., 2013). Specifically, transgenic overexpression of miR-22 induced EMT and metastasis in the MMTV-neu mouse model of breast cancer. Mechanistically, miR-22 targets the TET family of methylcytosine dioxygenases, thereby inhibiting demethylation of the *mir-200* promoter. Also by using genetically engineered mice, David Kirsch's group identified miR-182 as a sarcoma metastasis-promoting miRNA (Sachdeva et al., 2014): genetic deletion of miR-182 suppressed, while transgenic overexpression of miR-182 increased cancer cell dissemination and lung metastasis in a sarcoma model after amputation of the tumor-bearing limb. These effects are mediated by concomitant suppression of four genes that inhibit migration or degradation of the extracellular matrix, *Rsu1*, *Mtss1*, *Pai1*, and *Timp1*. In addition, miR-182 was shown to promote melanoma metastasis by targeting *FOXO3* and *MITF* (Segura et al., 2009).

Certain oncoproteins (eg, HER2/ERBB2) not only initiate tumor formation but also promote migration, invasion, and metastasis. Similarly, some oncomirs also confer invasiveness and metastatic ability on cancer cells. miR-373 was initially identified in a forward genetic screen as an oncogenic miRNA in testicular germ cell tumors, which targets the tumor suppressor gene *LATS2* (Voorhoeve et al., 2006). A subsequent functional genomics screen identified this miRNA as a positive regulator of cell migration, which was further validated in xenograft models in which overexpression of miR-373 in nonmetastatic MCF-7 human breast cancer cells induced metastasis (Huang et al., 2008). As mentioned in Section 2.2, genetically engineered mice with targeted deletion or overexpression of miR-21 provided in vivo proof that this miRNA promotes tumorigenesis. miR-21 has also been shown to promote invasion, intravasation, and metastasis in breast cancer and colorectal cancer (Asangani et al., 2008; Zhu et al., 2008). Mechanistically, miR-21 targets a number of mRNAs encoding tumor suppressors and metastasis suppressors, including PDCD4, PTEN, TPM1, and Maspin.

As indicated earlier, a single miRNA can act as a driver of tumor metastasis through coordinated targeting of multiple genes. Conversely, multiple miRNAs can synergize to induce metastasis via convergent targeting of the same gene(s). For instance, gain- and loss-of-function studies demonstrated that miR-199a-3p, miR-199a-5p, and miR-1908 have a synergistic effect in promoting metastasis of melanoma cells (Pencheva et al., 2012). Mechanistically, these three miRNAs convergently target two genes that encode the heat shock factor DNAJA4 and the metabolic protein apolipoprotein-E (ApoE). DNAJA4 inhibits metastasis via upregulation of ApoE expression, establishing ApoE as the focal point of this convergent miRNA regulatory network. ApoE is secreted by melanoma cells and engages the LRP1 receptor on melanoma cells and the LRP8 receptor on endothelial cells, which leads to simultaneous inhibition of melanoma cell invasion and endothelial cell migration (Fig. 6A). Therefore, miRNA targets can regulate metastasis through both cell-intrinsic and cell-extrinsic mechanisms.

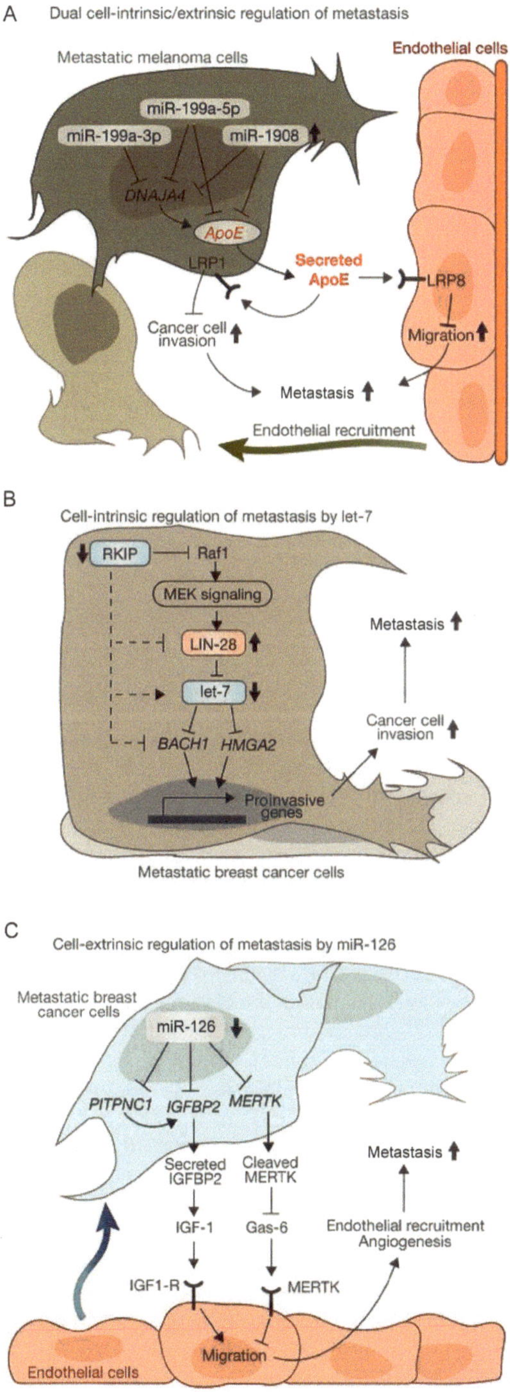

Fig. 6 See legend on next page.

As mentioned earlier for miR-10b and miR-9, miRNAs themselves can also act noncell autonomously. Tumor cell-secreted miRNAs have emerged as mediators of the cancer–host cross talk. miR-105 is highly expressed in metastatic breast cancer cells and secreted. Exosome-mediated transfer of tumor cell-secreted miR-105 targets the mRNA encoding the tight junction protein ZO-1 in endothelial cells and destroys the barrier to metastasis. Consequently, overexpression of miR-105 in otherwise non-metastatic breast cancer cells induced vascular permeability and metastasis in distant organs (Zhou et al., 2014). Another example is miR-122, which targets the mRNA encoding the glycolytic enzyme pyruvate kinase M2 (PKM2), leading to downregulation of GLUT1 expression and glucose uptake. Emily Wang's group demonstrated that miR-122 is secreted by cancer cells via exosomes (Fong et al., 2015). The export of miR-122 allows tumor cells to increase glucose uptake to support their growth, while simultaneously inhibiting glucose uptake in niche cells in distant organs, which facilitates metastatic colonization.

In addition to tumor cell-secreted miRNAs, some miRNAs are secreted by niche cells in the microenvironment and play an active role in metastatic progression. Dihua Yu's laboratory found that breast tumor cells lose PTEN

Fig. 6 Examples of cell-intrinsic, cell-extrinsic, and dual cell-intrinsic/extrinsic regulation of metastasis by microRNAs. (A) Three miRNAs—miR-199a-3p, miR-199a-5p, and miR-1908—promote melanoma metastasis by convergent targeting of *DNAJA4* and *ApoE*. DNAJA4 suppresses metastasis by inducing ApoE expression. Melanoma cell-secreted ApoE halts metastatic progression by both cell-autonomous and noncell-autonomous mechanisms. ApoE suppresses melanoma cell invasion by targeting melanoma LRP1 receptors, whereas its inhibition of endothelial cell migration results from its engagement of endothelial LRP8 receptors. (B) The let-7 regulatory network suppresses metastasis by inhibiting the cell-intrinsic phenotype of cancer cell invasion. Direct targeting of *HMGA2* and *BACH1* by let-7 leads to transcriptional suppression of a set of proinvasive genes. Let-7 is regulated by LIN-28, a miRNA-binding protein that destabilizes the let-7 pre-miRNA and ultimately promotes metastasis. RKIP, an indirect repressor of LIN-28, suppresses metastasis through upregulation of *let-7*. (C) miR-126 halts metastasis by impairing the cell-extrinsic ability of breast cancer cells to recruit endothelial cells into the metastatic niche. Cancer-expressed miR-126 coordinately targets *PITPNC1*, *IGFBP2*, and *MERTK*. Cancer cell-secreted IGFBP2 promotes endothelial cell migration by binding to endothelial IGF1-R, and PITPNC1 induces IGFBP2 expression. In parallel to this IGFBP2-driven pathway, cancer cell-cleaved MERTK ectodomain increases endothelial cell migration by binding and sequestering Gas-6, an extracellular factor that inhibits endothelial cell migration through its action on endothelial MERTK receptors. *From Pencheva, N., & Tavazoie, S. F. (2013). Control of metastatic progression by microRNA regulatory networks.* Nature Cell Biology, 15, *546–554.*

expression after dissemination to the brain, but not to other organs, while brain-metastatic tumor cells that have experienced PTEN loss restore PTEN levels once they leave the brain. This brain microenvironment-dependent, reversible PTEN downregulation is caused by *PTEN*-targeting miRNAs, such as miR-19a, from astrocyte-derived exosomes (Zhang, Zhang, et al., 2015). The adaptive PTEN loss in brain-metastatic tumor cells leads to NF-κB and AKT activation and increased secretion of the chemokine CCL2, which recruits IBA1-expressing myeloid cells to promote the outgrowth of brain-metastatic tumor cells via enhanced proliferation and reduced apoptosis. These findings suggest a pivotal role of the coevolution between metastatic tumor cells and the foreign microenvironment during adaptive metastatic outgrowth.

3.2 Metastasis-Suppressing miRNAs

CSCs are responsible for generating both primary and metastatic tumors (Al-Hajj, Wicha, Benito-Hernandez, Morrison, & Clarke, 2003; Liu et al., 2010; Malanchi et al., 2011). miRNAs that inhibit CSC properties have been found to suppress both tumorigenesis and metastasis. As mentioned in Section 2.2, the let-7 and miR-34 families of miRNAs are tumor suppressors. These miRNAs are downregulated in a variety of cancer cells and stem cells. For instance, the expression level of let-7 is markedly reduced in breast CSCs and increases with differentiation. Silencing let-7 expression in non-CSCs induced self-renewal ability; conversely, restoring let-7 expression in breast CSCs suppressed their tumorigenicity and metastatic ability. These effects are mediated by the oncogenic and proinvasive targets of let-7, including Ras, HMGA2, and BACH1 (Fig. 6B) (Dangi-Garimella et al., 2009; Yu et al., 2007; Yun et al., 2011). Similarly, miR-34a is underexpressed in the prostate CSC population, which can be enriched by the cell surface protein CD44, a marker of breast and prostate CSCs. Enforced expression of miR-34a in CD44-positive prostate cancer cells repressed tumor regeneration and metastasis, while silencing miR-34a expression in CD44-negative prostate cancer cells promoted tumor growth and metastasis. Interestingly, CD44 is a direct and functional target of miR-34a (Liu et al., 2011).

The action of miR-34a goes beyond cell-autonomous effects on cancer cells, as this miRNA also impedes breast cancer bone metastasis by inhibiting osteoclastogenesis (Krzeszinski et al., 2014). The expression of miR-34a is downregulated during osteoclast differentiation. Transgenic mice with overexpression of miR-34a in osteoclasts showed reduced bone resorption and increased bone mass, whereas miR-34a knockout mice had increased bone resorption and reduced bone mass. Mechanistically, transforming growth factor-β-induced factor 2 (Tgif2), which is a proosteoclastogenic protein, is a direct and functional target of miR-34a. Intriguingly, ovariectomy-induced osteoporosis and bone metastasis (from breast and skin tumors) were attenuated in transgenic mice overexpressing miR-34a in osteoclasts (Krzeszinski et al., 2014).

Therefore, similar to prometastatic miRNAs, metastasis-suppressing miRNAs can also act in a cell-autonomous or a cell-nonautonomous manner. This has been demonstrated by further evidence. Through miRNA expression profiling analysis of the highly metastatic subline of the MDA-MB-231 human breast cancer cells, Sohail Tavazoie, Joan Massagué, and colleagues identified miR-206, miR-335, and miR-126 as antimetastatic miRNAs (Tavazoie et al., 2008). Mechanistically, miR-206 induces apoptosis and inhibits migration by targeting *NOTCH3* in cancer cells (Song, Zhang, & Wang, 2009), and miR-335 targets *SOX4*, *TNC*, and *PTPRN2* to remodel the extracellular matrix to block tumor cell migration and invasion (Tavazoie et al., 2008). On the other hand, miR-126 inhibits metastasis by targeting *MERTK*, *PITPNC1*, and *IGFBP2* in cancer cells, which in turn halts the ability of tumor cells to recruit endothelial cells to the metastatic niche. The MERTK ectodomain cleaved from tumor cells increases endothelial cell migration by binding and sequestering Gas-6, an extracellular factor that inhibits endothelial cell migration by binding to MERTK receptors on endothelial cells. PITPNC1 upregulates IGFBP2 expression. When secreted, IGFBP2 binds to endothelial IGF1 receptor and promotes endothelial cell migration (Fig. 6C) (Png, Halberg, Yoshida, & Tavazoie, 2011). Moreover, miR-126 inhibits the recruitment of mesenchymal stem cells and inflammatory monocytes to the primary tumor environment by suppressing stromal cell-derived factor 1α (SDF-1α) expression, providing an additional mechanism for the metastasis-suppressing function of this miRNA (Zhang et al., 2013).

Some metastasis-suppressing miRNAs, such as miR-205 and the miR-200 family (miR-200a, miR-200b, miR-200c, miR-141, and miR-429, which share a consensus seed sequence), are markers of epithelial tissues and inhibitors of EMT (Gregory et al., 2008; Park, Gaur, Lengyel, & Peter, 2008) and stem cell properties (Shimono et al., 2009). miR-205 and the miR-200 family members promote MET and inhibit EMT by directly targeting the mRNAs that encode ZEB1 and ZEB2; conversely, ZEB1 directly represses the transcription of *mir-200* and *mir-205* gene promoters (Burk et al., 2008; Gregory et al., 2011; Zhang, Wang, et al., 2014). Consistent with the proinvasive role of EMT, ectopic expression of miR-200 inhibited metastasis from primary xenograft tumors (Gibbons et al., 2009). On the other hand, however, the fact that miR-200 promotes MET suggests a potential positive role in metastatic outgrowth. Indeed, enforced expression of miR-200 in the 4TO7 mouse mammary tumor cells promoted MET and formation of macroscopic metastases in the lung and liver after these cells were injected intravenously into syngeneic mice (Dykxhoorn et al., 2009; Korpal et al., 2011). Collectively, these findings suggest a model in which the miR-200 miRNA family inhibits EMT, invasion, and metastatic dissemination, but enhances metastatic colonization after cancer cells have disseminated to distant organs.

Zena Werb's group provided further evidence for EMT-repressing miRNAs as metastasis suppressors (Chou et al., 2013). GATA3-induced expression of miR-29b inhibits EMT and breast cancer metastasis by targeting a network of prometastatic genes and microenvironmental genes involved in angiogenesis, collagen remodeling, and matrix degradation, including *VEGFA*, *ANGPTL4*, *LOX*, *MMP2*, *MMP9*, and *PDGF*, suggesting that miR-29b modulates both cancer cell plasticity and the tumor microenvironment to repress metastasis. Consistently, depletion of miR-29 in breast cancer cells promoted EMT and metastasis. The metastasis-suppressing effect of miR-29b has also been observed in prostate cancer and liver cancer.

4. miRNAs AS THERAPEUTIC TARGETS

Because miRNAs target multiple genes in one or more pathways, targeting a single miRNA is expected to influence the expression of multiple genes and their associated signaling pathways. Although no miRNAs have been approved by the FDA as drugs, progress is being made in developing them as therapeutic strategies. Growing numbers of studies have demonstrated the efficacy of miRNA-based therapeutic agents in preclinical models. Moreover, specific miRNA mimics or inhibitors have entered clinical trials for treating cancer or infectious disease.

4.1 miRNA Mimics

Among tumor-suppressing miRNAs, let-7, miR-34, and miR-26a have been most extensively investigated for their therapeutic potential. Intravenous administration of adeno-associated virus carrying miR-26a dramatically inhibited tumorigenesis in a Myc-induced mouse model of hepatocellular carcinoma (Kota et al., 2009). However, it is unclear whether miRNAs expressed from a viral vector represent a viable strategy in the cancer clinic. Using an established genetic model of lung cancer, Frank Slack and colleagues showed that intranasal let-7 administration inhibited lung tumor formation in animals expressing a mutant K-Ras oncoprotein (Esquela-Kerscher et al., 2008), and that systemic delivery of neutral lipid emulsions of chemically synthesized let-7 and miR-34 mimics significantly reduced tumor burdens in the K-Ras-induced mouse model of lung cancer (Trang et al., 2011). Similarly, treatment with miR-34 mimics inhibited tumor growth in xenograft models of colon cancer (Tazawa, Tsuchiya, Izumiya, & Nakagama, 2007) and pancreatic cancer (Pramanik et al., 2011). In an orthotopic prostate cancer model, tail vein injection of miR-34a complexed with a lipid-based delivery agent suppressed metastasis and extended survival (Liu et al., 2011). Moreover, tumor-targeted delivery of miR-34a mimics coupled to nanoparticles coated with an antibody against the neuroblastoma-specific antigen disialoganglioside GD2 halted neuroblastoma growth in an orthotopic model (Tivnan et al., 2012). Consistent with the inhibitory role of miR-34a in osteoclastogenesis, intravenous administration of chitosan-encapsulated miR-34a mimics attenuated ovariectomy-induced osteoporosis and bone metastasis in mice bearing breast cancer or melanoma (Krzeszinski et al., 2014). Notably, in 2013, a liposome-formulated miR-34 mimic, named MRX34 (Mirna Therapeutics, Inc.), entered clinical trials for the treatment of patients with advanced liver cancer, representing the first miRNA-based therapeutic agent advanced to cancer trials (Ling, Fabbri, & Calin, 2013). Therefore, miRNAs are en route to the clinic as new anticancer drugs.

4.2 miRNA Inhibitors

Cancer-promoting miRNAs can be targeted by antisense RNA oligonucleotides (anti-miRs), such as antagomirs and locked nucleic acids (LNAs). Antagomirs are cholesterol-conjugated antisense miRNA inhibitors with a 2′-O-methyl linkage and phosphorothioate modification (Krutzfeldt et al., 2005). LNA-based anti-miRs are antisense RNAs with several nucleotides replaced by bicyclic RNA analogs in a "locked" conformation (Elmen et al., 2008). These modifications serve to enhance the stability and delivery of the anti-miR and increase its affinity to the targeted miRNA. In a 4T1 mouse mammary tumor model (4T1 cells are highly metastatic and express high levels of Twist and miR-10b), intravenous administration of miR-10b antagomirs had a sequence-specific metastasis-suppressing effect without affecting primary mammary tumor growth (Ma, Reinhardt, et al., 2010). This is the first report showing the proof of principle that antagomirs can be efficiently delivered to rapidly growing metastatic tumor cells in vivo and can prevent metastasis formation by otherwise highly malignant cells. Furthermore, combining nanoparticle-encapsulated miR-10b antisense inhibitors with low-dose doxorubicin treatment achieved complete durable regression of metastases in a xenograft model of breast cancer (Yoo et al., 2015). In mice that received intrasplenic injection of melanoma cells, intraperitoneal injection of anti-miR-182 oligonucleotides synthesized with 2′ sugar modifications and a phosphorothioate backbone reduced burdens of liver metastases (Huynh et al., 2011). Similarly, in an experimental metastasis model in which melanoma cells were injected into mice via the tail vein, systemic delivery of a cocktail of LNAs targeting miR-199a-3p, miR-199a-5p, and miR-1908 inhibited lung metastasis formation (Pencheva et al., 2012).

Cancer-secreted miRNAs also represent potential therapeutic targets. For instance, therapeutic silencing of miR-105 inhibited breast cancer metastasis in mice (Zhou et al., 2014), and treatment with miR-122 inhibitors restored glucose uptake by niche cells in distant organs and reduced the incidence of metastasis in preclinical models of breast cancer (Fong et al., 2015). As mentioned in Section 1.2, miR-122 stimulates hepatitis C virus RNA replication and translation. Interestingly, LNA-based miR-122 anti-miRs (miravirsen, Santaris Pharma) have shown therapeutic benefits in a Phase 2 clinical trial for treating patients with hepatitis C virus infection (Janssen et al., 2013), whereas miRNA inhibitors are yet to be advanced to cancer trials.

4.3 Combination Treatment

Tumor cells that are resistant to treatment give rise to local recurrence and metastatic relapse. miRNAs and anti-miRs have been shown to increase the sensitivity to chemotherapeutic agents, radiation treatment, or agents of targeted therapy in cultured cell lines and animal models. An adenoviral vector carrying the tumor-suppressing miRNA miR-145 sensitized breast cancer cells to 5-fluorouracil treatment both in vitro and in vivo (Kim et al., 2011). In a xenograft model of triple-negative breast cancer, miR-30c sensitized tumor cells to doxorubicin treatment (Bockhorn et al., 2013). Treatment with miR-21 inhibitors was shown to sensitize breast tumor cells to topotecan and taxol (Mei et al., 2010); moreover, in a U87 xenograft model of glioblastoma, combination treatment of LNA-anti-miR-21 with the cytotoxic agent TNF-related apoptosis-inducing ligand (TRAIL) resulted in tumor eradication (Corsten et al., 2007). Recently, Li Ma and colleagues found that radiation treatment of breast cancer cells leads to therapy-induced radioresistance through ATM-mediated phosphorylation and stabilization of the EMT-inducing transcription factor ZEB1 (Zhang, Wei, et al., 2014). Other studies also demonstrated that ZEB1 is upregulated in therapy-resistant tumor cells and promotes radioresistance, chemoresistance, and drug resistance in various cancer types (Zhang, Sun, & Ma, 2015). Interestingly, therapeutic delivery of nanoliposome-encapsulated miR-205 or miR-200c mimics, which directly targets *ZEB1*, sensitized breast cancer cells and lung cancer cells to radiation treatment in xenograft models (Cortez et al., 2014; Zhang, Wang, et al., 2014). Collectively, these findings suggest that combination of miRNA-based agents with other therapeutic approaches may improve cancer treatment.

REFERENCES

Al-Hajj, M., Wicha, M. S., Benito-Hernandez, A., Morrison, S. J., & Clarke, M. F. (2003). Prospective identification of tumorigenic breast cancer cells. *Proceedings of the National Academy of Sciences of the United States of America*, *100*, 3983–3988.

Ambros, V. (2004). The functions of animal microRNAs. *Nature*, *431*, 350–355.

Asangani, I. A., Rasheed, S. A., Nikolova, D. A., Leupold, J. H., Colburn, N. H., Post, S., et al. (2008). MicroRNA-21 (miR-21) post-transcriptionally downregulates tumor suppressor Pdcd4 and stimulates invasion, intravasation and metastasis in colorectal cancer. *Oncogene*, *27*, 2128–2136.

Baffa, R., Fassan, M., Volinia, S., O'Hara, B., Liu, C. G., Palazzo, J. P., et al. (2009). MicroRNA expression profiling of human metastatic cancers identifies cancer gene targets. *The Journal of Pathology*, *219*, 214–221.

Bartel, D. P. (2004). MicroRNAs: Genomics, biogenesis, mechanism, and function. *Cell*, *116*, 281–297.

Bartel, D. P. (2009). MicroRNAs: Target recognition and regulatory functions. *Cell*, *136*, 215–233.

Bernstein, E., Kim, S. Y., Carmell, M. A., Murchison, E. P., Alcorn, H., Li, M. Z., et al. (2003). Dicer is essential for mouse development. *Nature Genetics*, *35*, 215–217.

Bockhorn, J., Dalton, R., Nwachukwu, C., Huang, S., Prat, A., Yee, K., et al. (2013). MicroRNA-30c inhibits human breast tumour chemotherapy resistance by regulating TWF1 and IL-11. *Nature Communications*, *4*, 1393.

Bohnsack, M. T., Czaplinski, K., & Gorlich, D. (2004). Exportin 5 is a RanGTP-dependent dsRNA-binding protein that mediates nuclear export of pre-miRNAs. *RNA*, *10*, 185–191.

Bosson, A. D., Zamudio, J. R., & Sharp, P. A. (2014). Endogenous miRNA and target concentrations determine susceptibility to potential ceRNA competition. *Molecular Cell*, *56*, 347–359.

Brabletz, T., Lyden, D., Steeg, P. S., & Werb, Z. (2013). Roadblocks to translational advances on metastasis research. *Nature Medicine*, *19*, 1104–1109.

Burk, U., Schubert, J., Wellner, U., Schmalhofer, O., Vincan, E., Spaderna, S., et al. (2008). A reciprocal repression between ZEB1 and members of the miR-200 family promotes EMT and invasion in cancer cells. *EMBO Reports, 9*, 582–589.

Calin, G. A., Dumitru, C. D., Shimizu, M., Bichi, R., Zupo, S., Noch, E., et al. (2002). Frequent deletions and down-regulation of micro-RNA genes miR15 and miR16 at 13q14 in chronic lymphocytic leukemia. *Proceedings of the National Academy of Sciences of the United States of America, 99*, 15524–15529.

Calin, G. A., Sevignani, C., Dumitru, C. D., Hyslop, T., Noch, E., Yendamuri, S., et al. (2004). Human microRNA genes are frequently located at fragile sites and genomic regions involved in cancers. *Proceedings of the National Academy of Sciences of the United States of America, 101*, 2999–3004.

Chen, D., Sun, Y., Wei, Y., Zhang, P., Rezaeian, A. H., Teruya-Feldstein, J., et al. (2012). LIFR is a breast cancer metastasis suppressor upstream of the Hippo-YAP pathway and a prognostic marker. *Nature Medicine, 18*, 1511–1517.

Chou, J., Lin, J. H., Brenot, A., Kim, J. W., Provot, S., & Werb, Z. (2013). GATA3 suppresses metastasis and modulates the tumour microenvironment by regulating microRNA-29b expression. *Nature Cell Biology, 15*, 201–213.

Cimmino, A., Calin, G. A., Fabbri, M., Iorio, M. V., Ferracin, M., Shimizu, M., et al. (2005). miR-15 and miR-16 induce apoptosis by targeting BCL2. *Proceedings of the National Academy of Sciences of the United States of America, 102*, 13944–13949.

Corcoran, D. L., Pandit, K. V., Gordon, B., Bhattacharjee, A., Kaminski, N., & Benos, P. V. (2009). Features of mammalian microRNA promoters emerge from polymerase II chromatin immunoprecipitation data. *PloS One, 4*, e5279.

Corsten, M. F., Miranda, R., Kasmieh, R., Krichevsky, A. M., Weissleder, R., & Shah, K. (2007). MicroRNA-21 knockdown disrupts glioma growth in vivo and displays synergistic cytotoxicity with neural precursor cell delivered S-TRAIL in human gliomas. *Cancer Research, 67*, 8994–9000.

Cortez, M. A., Valdecanas, D., Zhang, X., Zhan, Y., Bhardwaj, V., Calin, G. A., et al. (2014). Therapeutic delivery of miR-200c enhances radiosensitivity in lung cancer. *Molecular Therapy, 22*, 1494–1503.

Costinean, S., Zanesi, N., Pekarsky, Y., Tili, E., Volinia, S., Heerema, N., et al. (2006). Pre-B cell proliferation and lymphoblastic leukemia/high-grade lymphoma in E(mu)-miR155 transgenic mice. *Proceedings of the National Academy of Sciences of the United States of America, 103*, 7024–7029.

Curtis, H. J., Sibley, C. R., & Wood, M. J. (2012). Mirtrons, an emerging class of atypical miRNA. *Wiley Interdisciplinary Reviews. RNA, 3*, 617–632.

Dangi-Garimella, S., Yun, J., Eves, E. M., Newman, M., Erkeland, S. J., Hammond, S. M., et al. (2009). Raf kinase inhibitory protein suppresses a metastasis signalling cascade involving LIN28 and let-7. *The EMBO Journal, 28*, 347–358.

Denzler, R., Agarwal, V., Stefano, J., Bartel, D. P., & Stoffel, M. (2014). Assessing the ceRNA hypothesis with quantitative measurements of miRNA and target abundance. *Molecular Cell, 54*, 766–776.

Dykxhoorn, D. M., Wu, Y., Xie, H., Yu, F., Lal, A., Petrocca, F., et al. (2009). miR-200 enhances mouse breast cancer cell colonization to form distant metastases. *PloS One, 4*, e7181.

Eccles, S. A., & Welch, D. R. (2007). Metastasis: Recent discoveries and novel treatment strategies. *Lancet, 369*, 1742–1757.

Edmonds, M. D., Hurst, D. R., Vaidya, K. S., Stafford, L. J., Chen, D., & Welch, D. R. (2009). Breast cancer metastasis suppressor 1 coordinately regulates metastasis-associated microRNA expression. *International Journal of Cancer, 125*, 1778–1785.

Eichhorn, S. W., Guo, H., McGeary, S. E., Rodriguez-Mias, R. A., Shin, C., Baek, D., et al. (2014). mRNA destabilization is the dominant effect of mammalian microRNAs by the time substantial repression ensues. *Molecular Cell, 56*, 104–115.

Elmen, J., Lindow, M., Schutz, S., Lawrence, M., Petri, A., Obad, S., et al. (2008). LNA-mediated microRNA silencing in non-human primates. *Nature*, *452*, 896–899.

Esquela-Kerscher, A., Trang, P., Wiggins, J. F., Patrawala, L., Cheng, A., Ford, L., et al. (2008). The let-7 microRNA reduces tumor growth in mouse models of lung cancer. *Cell Cycle*, *7*, 759–764.

Fong, M. Y., Zhou, W., Liu, L., Alontaga, A. Y., Chandra, M., Ashby, J., et al. (2015). Breast-cancer-secreted miR-122 reprograms glucose metabolism in premetastatic niche to promote metastasis. *Nature Cell Biology*, *17*, 183–194.

Garzon, R., Marcucci, G., & Croce, C. M. (2010). Targeting microRNAs in cancer: Rationale, strategies and challenges. *Nature Reviews. Drug Discovery*, *9*, 775–789.

Gibbons, D. L., Lin, W., Creighton, C. J., Rizvi, Z. H., Gregory, P. A., Goodall, G. J., et al. (2009). Contextual extracellular cues promote tumor cell EMT and metastasis by regulating miR-200 family expression. *Genes & Development*, *23*, 2140–2151.

Gregory, P. A., Bert, A. G., Paterson, E. L., Barry, S. C., Tsykin, A., Farshid, G., et al. (2008). The miR-200 family and miR-205 regulate epithelial to mesenchymal transition by targeting ZEB1 and SIP1. *Nature Cell Biology*, *10*, 593–601.

Gregory, P. A., Bracken, C. P., Smith, E., Bert, A. G., Wright, J. A., Roslan, S., et al. (2011). An autocrine TGF-beta/ZEB/miR-200 signaling network regulates establishment and maintenance of epithelial-mesenchymal transition. *Molecular Biology of the Cell*, *22*, 1686–1698.

Gregory, R. I., Yan, K. P., Amuthan, G., Chendrimada, T., Doratotaj, B., Cooch, N., et al. (2004). The microprocessor complex mediates the genesis of microRNAs. *Nature*, *432*, 235–240.

Grimson, A., Farh, K. K., Johnston, W. K., Garrett-Engele, P., Lim, L. P., & Bartel, D. P. (2007). MicroRNA targeting specificity in mammals: Determinants beyond seed pairing. *Molecular Cell*, *27*, 91–105.

Guo, H., Ingolia, N. T., Weissman, J. S., & Bartel, D. P. (2010). Mammalian microRNAs predominantly act to decrease target mRNA levels. *Nature*, *466*, 835–840.

Ha, M., & Kim, V. N. (2014). Regulation of microRNA biogenesis. *Nature Reviews. Molecular Cell Biology*, *15*, 509–524.

Han, J., Lee, Y., Yeom, K. H., Kim, Y. K., Jin, H., & Kim, V. N. (2004). The Drosha-DGCR8 complex in primary microRNA processing. *Genes & Development*, *18*, 3016–3027.

Han, J., Lee, Y., Yeom, K. H., Nam, J. W., Heo, I., Rhee, J. K., et al. (2006). Molecular basis for the recognition of primary microRNAs by the Drosha-DGCR8 complex. *Cell*, *125*, 887–901.

Hatley, M. E., Patrick, D. M., Garcia, M. R., Richardson, J. A., Bassel-Duby, R., van Rooij, E., et al. (2010). Modulation of K-Ras-dependent lung tumorigenesis by microRNA-21. *Cancer Cell*, *18*, 282–293.

He, X., He, L., & Hannon, G. J. (2007). The guardian's little helper: MicroRNAs in the p53 tumor suppressor network. *Cancer Research*, *67*, 11099–11101.

He, L., Thomson, J. M., Hemann, M. T., Hernando-Monge, E., Mu, D., Goodson, S., et al. (2005). A microRNA polycistron as a potential human oncogene. *Nature*, *435*, 828–833.

Henke, J. I., Goergen, D., Zheng, J., Song, Y., Schuttler, C. G., Fehr, C., et al. (2008). microRNA-122 stimulates translation of hepatitis C virus RNA. *The EMBO Journal*, *27*, 3300–3310.

Huang, Q., Gumireddy, K., Schrier, M., le Sage, C., Nagel, R., Nair, S., et al. (2008). The microRNAs miR-373 and miR-520c promote tumour invasion and metastasis. *Nature Cell Biology*, *10*, 202–210.

Huarte, M., & Rinn, J. L. (2010). Large non-coding RNAs: Missing links in cancer? *Human Molecular Genetics*, *19*, R152–R161.

Huse, J. T., Brennan, C., Hambardzumyan, D., Wee, B., Pena, J., Rouhanifard, S. H., et al. (2009). The PTEN-regulating microRNA miR-26a is amplified in high-grade glioma and facilitates gliomagenesis in vivo. *Genes & Development*, *23*, 1327–1337.

Huynh, C., Segura, M. F., Gaziel-Sovran, A., Menendez, S., Darvishian, F., Chiriboga, L., et al. (2011). Efficient in vivo microRNA targeting of liver metastasis. *Oncogene*, *30*, 1481–1488.

Janssen, H. L., Reesink, H. W., Lawitz, E. J., Zeuzem, S., Rodriguez-Torres, M., Patel, K., et al. (2013). Treatment of HCV infection by targeting microRNA. *The New England Journal of Medicine*, *368*, 1685–1694.

Johnson, S. M., Grosshans, H., Shingara, J., Byrom, M., Jarvis, R., Cheng, A., et al. (2005). RAS is regulated by the let-7 microRNA family. *Cell*, *120*, 635–647.

Jopling, C. L., Yi, M., Lancaster, A. M., Lemon, S. M., & Sarnow, P. (2005). Modulation of hepatitis C virus RNA abundance by a liver-specific MicroRNA. *Science*, *309*, 1577–1581.

Kalluri, R., & Weinberg, R. A. (2009). The basics of epithelial-mesenchymal transition. *The Journal of Clinical Investigation*, *119*, 1420–1428.

Karreth, F. A., Reschke, M., Ruocco, A., Ng, C., Chapuy, B., Leopold, V., et al. (2015). The BRAF pseudogene functions as a competitive endogenous RNA and induces lymphoma in vivo. *Cell*, *161*, 319–332.

Karreth, F. A., Tay, Y., Perna, D., Ala, U., Tan, S. M., Rust, A. G., et al. (2011). In vivo identification of tumor-suppressive PTEN ceRNAs in an oncogenic BRAF-induced mouse model of melanoma. *Cell*, *147*, 382–395.

Kim, D. H., Saetrom, P., Snove, O., Jr., & Rossi, J. J. (2008). MicroRNA-directed transcriptional gene silencing in mammalian cells. *Proceedings of the National Academy of Sciences of the United States of America*, *105*, 16230–16235.

Kim, S. J., Oh, J. S., Shin, J. Y., Lee, K. D., Sung, K. W., Nam, S. J., et al. (2011). Development of microRNA-145 for therapeutic application in breast cancer. *Journal of Controlled Release*, *155*, 427–434.

Klein, U., Lia, M., Crespo, M., Siegel, R., Shen, Q., Mo, T., et al. (2010). The DLEU2/miR-15a/16-1 cluster controls B cell proliferation and its deletion leads to chronic lymphocytic leukemia. *Cancer Cell*, *17*, 28–40.

Kloosterman, W. P., & Plasterk, R. H. (2006). The diverse functions of microRNAs in animal development and disease. *Developmental Cell*, *11*, 441–450.

Korpal, M., Ell, B. J., Buffa, F. M., Ibrahim, T., Blanco, M. A., Celia-Terrassa, T., et al. (2011). Direct targeting of Sec23a by miR-200s influences cancer cell secretome and promotes metastatic colonization. *Nature Medicine*, *17*, 1101–1108.

Kota, J., Chivukula, R. R., O'Donnell, K. A., Wentzel, E. A., Montgomery, C. L., Hwang, H. W., et al. (2009). Therapeutic microRNA delivery suppresses tumorigenesis in a murine liver cancer model. *Cell*, *137*, 1005–1017.

Krutzfeldt, J., Rajewsky, N., Braich, R., Rajeev, K. G., Tuschl, T., Manoharan, M., et al. (2005). Silencing of microRNAs in vivo with 'antagomirs'. *Nature*, *438*, 685–689.

Krzeszinski, J. Y., Wei, W., Huynh, H., Jin, Z., Wang, X., Chang, T. C., et al. (2014). miR-34a blocks osteoporosis and bone metastasis by inhibiting osteoclastogenesis and Tgif2. *Nature*, *512*, 431–435.

Lagos-Quintana, M., Rauhut, R., Lendeckel, W., & Tuschl, T. (2001). Identification of novel genes coding for small expressed RNAs. *Science*, *294*, 853–858.

Lander, E. S., Linton, L. M., Birren, B., Nusbaum, C., Zody, M. C., Baldwin, J., et al. (2001). Initial sequencing and analysis of the human genome. *Nature*, *409*, 860–921.

Lau, N. C., Lim, L. P., Weinstein, E. G., & Bartel, D. P. (2001). An abundant class of tiny RNAs with probable regulatory roles in Caenorhabditis elegans. *Science*, *294*, 858–862.

Lee, R. C., & Ambros, V. (2001). An extensive class of small RNAs in Caenorhabditis elegans. *Science*, *294*, 862–864.

Lee, R. C., Feinbaum, R. L., & Ambros, V. (1993). The C. elegans heterochronic gene lin-4 encodes small RNAs with antisense complementarity to lin-14. *Cell*, *75*, 843–854.

Lewis, B. P., Burge, C. B., & Bartel, D. P. (2005). Conserved seed pairing, often flanked by adenosines, indicates that thousands of human genes are microRNA targets. *Cell*, *120*, 15–20.

Lin, S., & Gregory, R. I. (2015). MicroRNA biogenesis pathways in cancer. *Nature Reviews. Cancer*, *15*, 321–333.

Ling, H., Fabbri, M., & Calin, G. A. (2013). MicroRNAs and other non-coding RNAs as targets for anticancer drug development. *Nature Reviews. Drug Discovery*, *12*, 847–865.

Liu, C., Kelnar, K., Liu, B., Chen, X., Calhoun-Davis, T., Li, H., et al. (2011). The microRNA miR-34a inhibits prostate cancer stem cells and metastasis by directly repressing CD44. *Nature Medicine*, *17*, 211–215.

Liu, H., Patel, M. R., Prescher, J. A., Patsialou, A., Qian, D., Lin, J., et al. (2010). Cancer stem cells from human breast tumors are involved in spontaneous metastases in orthotopic mouse models. *Proceedings of the National Academy of Sciences of the United States of America*, *107*, 18115–18120.

Lu, J., Getz, G., Miska, E. A., Alvarez-Saavedra, E., Lamb, J., Peck, D., et al. (2005). MicroRNA expression profiles classify human cancers. *Nature*, *435*, 834–838.

Lund, E., Guttinger, S., Calado, A., Dahlberg, J. E., & Kutay, U. (2004). Nuclear export of microRNA precursors. *Science*, *303*, 95–98.

Ma, L. (2010). Role of miR-10b in breast cancer metastasis. *Breast Cancer Research*, *12*, 210.

Ma, L., Reinhardt, F., Pan, E., Soutschek, J., Bhat, B., Marcusson, E. G., et al. (2010). Therapeutic silencing of miR-10b inhibits metastasis in a mouse mammary tumor model. *Nature Biotechnology*, *28*, 341–347.

Ma, L., Teruya-Feldstein, J., & Weinberg, R. A. (2007). Tumour invasion and metastasis initiated by microRNA-10b in breast cancer. *Nature*, *449*, 682–688.

Ma, L., Young, J., Prabhala, H., Pan, E., Mestdagh, P., Muth, D., et al. (2010). miR-9, a MYC/MYCN-activated microRNA, regulates E-cadherin and cancer metastasis. *Nature Cell Biology*, *12*, 247–256.

Macrae, I. J., Zhou, K., Li, F., Repic, A., Brooks, A. N., Cande, W. Z., et al. (2006). Structural basis for double-stranded RNA processing by Dicer. *Science*, *311*, 195–198.

Malanchi, I., Santamaria-Martinez, A., Susanto, E., Peng, H., Lehr, H. A., Delaloye, J. F., et al. (2011). Interactions between cancer stem cells and their niche govern metastatic colonization. *Nature*, *481*, 85–89.

Mani, S. A., Guo, W., Liao, M. J., Eaton, E. N., Ayyanan, A., Zhou, A. Y., et al. (2008). The epithelial-mesenchymal transition generates cells with properties of stem cells. *Cell*, *133*, 704–715.

Martello, G., Rosato, A., Ferrari, F., Manfrin, A., Cordenonsi, M., Dupont, S., et al. (2010). A microRNA targeting dicer for metastasis control. *Cell*, *141*, 1195–1207.

Mattick, J. S. (2004). RNA regulation: A new genetics? *Nature Reviews. Genetics*, *5*, 316–323.

Medina, P. P., Nolde, M., & Slack, F. J. (2010). OncomiR addiction in an in vivo model of microRNA-21-induced pre-B-cell lymphoma. *Nature*, *467*, 86–90.

Mei, M., Ren, Y., Zhou, X., Yuan, X. B., Han, L., Wang, G. X., et al. (2010). Downregulation of miR-21 enhances chemotherapeutic effect of taxol in breast carcinoma cells. *Technology in Cancer Research & Treatment*, *9*, 77–86.

Meister, G., & Tuschl, T. (2004). Mechanisms of gene silencing by double-stranded RNA. *Nature*, *431*, 343–349.

Miska, E. A., Alvarez-Saavedra, E., Abbott, A. L., Lau, N. C., Hellman, A. B., McGonagle, S. M., et al. (2007). Most Caenorhabditis elegans microRNAs are individually not essential for development or viability. *PLoS Genetics*, *3*, e215.

Olive, V., Li, Q., & He, L. (2013). mir-17-92: A polycistronic oncomir with pleiotropic functions. *Immunological Reviews*, *253*, 158–166.

Orom, U. A., Nielsen, F. C., & Lund, A. H. (2008). MicroRNA-10a binds the 5′UTR of ribosomal protein mRNAs and enhances their translation. *Molecular Cell, 30*, 460–471.
Park, S. M., Gaur, A. B., Lengyel, E., & Peter, M. E. (2008). The miR-200 family determines the epithelial phenotype of cancer cells by targeting the E-cadherin repressors ZEB1 and ZEB2. *Genes & Development, 22*, 894–907.
Pencheva, N., & Tavazoie, S. F. (2013). Control of metastatic progression by microRNA regulatory networks. *Nature Cell Biology, 15*, 546–554.
Pencheva, N., Tran, H., Buss, C., Huh, D., Drobnjak, M., Busam, K., et al. (2012). Convergent multi-miRNA targeting of ApoE drives LRP1/LRP8-dependent melanoma metastasis and angiogenesis. *Cell, 151*, 1068–1082.
Piao, H. L., & Ma, L. (2012). Non-coding RNAs as regulators of mammary development and breast cancer. *Journal of Mammary Gland Biology and Neoplasia, 17*, 33–42.
Png, K. J., Halberg, N., Yoshida, M., & Tavazoie, S. F. (2011). A microRNA regulon that mediates endothelial recruitment and metastasis by cancer cells. *Nature, 481*, 190–194.
Poliseno, L., Salmena, L., Zhang, J., Carver, B., Haveman, W. J., & Pandolfi, P. P. (2010). A coding-independent function of gene and pseudogene mRNAs regulates tumour biology. *Nature, 465*, 1033–1038.
Pramanik, D., Campbell, N. R., Karikari, C., Chivukula, R., Kent, O. A., Mendell, J. T., et al. (2011). Restitution of tumor suppressor microRNAs using a systemic nanovector inhibits pancreatic cancer growth in mice. *Molecular Cancer Therapeutics, 10*, 1470–1480.
Reinhart, B. J., Slack, F. J., Basson, M., Pasquinelli, A. E., Bettinger, J. C., Rougvie, A. E., et al. (2000). The 21-nucleotide let-7 RNA regulates developmental timing in Caenorhabditis elegans. *Nature, 403*, 901–906.
Sachdeva, M., Mito, J. K., Lee, C. L., Zhang, M., Li, Z., Dodd, R. D., et al. (2014). MicroRNA-182 drives metastasis of primary sarcomas by targeting multiple genes. *The Journal of Clinical Investigation, 124*, 4305–4319.
Salmena, L., Poliseno, L., Tay, Y., Kats, L., & Pandolfi, P. P. (2011). A ceRNA hypothesis: The Rosetta stone of a hidden RNA language? *Cell, 146*, 353–358.
Segura, M. F., Hanniford, D., Menendez, S., Reavie, L., Zou, X., Alvarez-Diaz, S., et al. (2009). Aberrant miR-182 expression promotes melanoma metastasis by repressing FOXO3 and microphthalmia-associated transcription factor. *Proceedings of the National Academy of Sciences of the United States of America, 106*, 1814–1819.
Shimono, Y., Zabala, M., Cho, R. W., Lobo, N., Dalerba, P., Qian, D., et al. (2009). Downregulation of miRNA-200c links breast cancer stem cells with normal stem cells. *Cell, 138*, 592–603.
Singh, R., Pochampally, R., Watabe, K., Lu, Z., & Mo, Y. Y. (2014). Exosome-mediated transfer of miR-10b promotes cell invasion in breast cancer. *Molecular Cancer, 13*, 256.
Slack, F. J., Basson, M., Liu, Z., Ambros, V., Horvitz, H. R., & Ruvkun, G. (2000). The lin-41 RBCC gene acts in the C. elegans heterochronic pathway between the let-7 regulatory RNA and the LIN-29 transcription factor. *Molecular Cell, 5*, 659–669.
Song, G., Zhang, Y., & Wang, L. (2009). MicroRNA-206 targets notch3, activates apoptosis, and inhibits tumor cell migration and focus formation. *The Journal of Biological Chemistry, 284*, 31921–31927.
Song, S. J., Poliseno, L., Song, M. S., Ala, U., Webster, K., Ng, C., et al. (2013). MicroRNA-antagonism regulates breast cancer stemness and metastasis via TET-family-dependent chromatin remodeling. *Cell, 154*, 311–324.
Steeg, P. S. (2012). Perspective: The right trials. *Nature, 485*, S58–S59.
Steeg, P. S., & Theodorescu, D. (2008). Metastasis: A therapeutic target for cancer. *Nature Clinical Practice. Oncology, 5*, 206–219.
Stein, L. D. (2004). Human genome: End of the beginning. *Nature, 431*, 915–916.
Sun, Y., & Ma, L. (2015). The emerging molecular machinery and therapeutic targets of metastasis. *Trends in Pharmacological Sciences, 36*, 349–359.

Talmadge, J. E., & Fidler, I. J. (2010). AACR centennial series: The biology of cancer metastasis: Historical perspective. *Cancer Research, 70*, 5649–5669.

Tavazoie, S. F., Alarcon, C., Oskarsson, T., Padua, D., Wang, Q., Bos, P. D., et al. (2008). Endogenous human microRNAs that suppress breast cancer metastasis. *Nature, 451*, 147–152.

Tay, Y., Kats, L., Salmena, L., Weiss, D., Tan, S. M., Ala, U., et al. (2011). Coding-independent regulation of the tumor suppressor PTEN by competing endogenous mRNAs. *Cell, 147*, 344–357.

Tay, Y., Rinn, J., & Pandolfi, P. P. (2014). The multilayered complexity of ceRNA crosstalk and competition. *Nature, 505*, 344–352.

Tay, Y., Zhang, J., Thomson, A. M., Lim, B., & Rigoutsos, I. (2008). MicroRNAs to Nanog, Oct4 and Sox2 coding regions modulate embryonic stem cell differentiation. *Nature, 455*, 1124–1128.

Tazawa, H., Tsuchiya, N., Izumiya, M., & Nakagama, H. (2007). Tumor-suppressive miR-34a induces senescence-like growth arrest through modulation of the E2F pathway in human colon cancer cells. *Proceedings of the National Academy of Sciences of the United States of America, 104*, 15472–15477.

Thiery, J. P. (2002). Epithelial-mesenchymal transitions in tumour progression. *Nature Reviews. Cancer, 2*, 442–454.

Tivnan, A., Orr, W. S., Gubala, V., Nooney, R., Williams, D. E., McDonagh, C., et al. (2012). Inhibition of neuroblastoma tumor growth by targeted delivery of microRNA-34a using anti-disialoganglioside GD2 coated nanoparticles. *PloS One, 7*, e38129.

Trang, P., Wiggins, J. F., Daige, C. L., Cho, C., Omotola, M., Brown, D., et al. (2011). Systemic delivery of tumor suppressor microRNA mimics using a neutral lipid emulsion inhibits lung tumors in mice. *Molecular Therapy, 19*, 1116–1122.

Tsai, J. H., Donaher, J. L., Murphy, D. A., Chau, S., & Yang, J. (2012). Spatiotemporal regulation of epithelial-mesenchymal transition is essential for squamous cell carcinoma metastasis. *Cancer Cell, 22*, 725–736.

Tsai, J. H., & Yang, J. (2013). Epithelial-mesenchymal plasticity in carcinoma metastasis. *Genes & Development, 27*, 2192–2206.

van Rooij, E., Sutherland, L. B., Qi, X., Richardson, J. A., Hill, J., & Olson, E. N. (2007). Control of stress-dependent cardiac growth and gene expression by a microRNA. *Science, 316*, 575–579.

Ventura, A., Young, A. G., Winslow, M. M., Lintault, L., Meissner, A., Erkeland, S. J., et al. (2008). Targeted deletion reveals essential and overlapping functions of the miR-17 through 92 family of miRNA clusters. *Cell, 132*, 875–886.

Vidigal, J. A., & Ventura, A. (2015). The biological functions of miRNAs: Lessons from in vivo studies. *Trends in Cell Biology, 25*, 137–147.

Volinia, S., Calin, G. A., Liu, C. G., Ambs, S., Cimmino, A., Petrocca, F., et al. (2006). A microRNA expression signature of human solid tumors defines cancer gene targets. *Proceedings of the National Academy of Sciences of the United States of America, 103*, 2257–2261.

Voorhoeve, P. M., le Sage, C., Schrier, M., Gillis, A. J., Stoop, H., Nagel, R., et al. (2006). A genetic screen implicates miRNA-372 and miRNA-373 as oncogenes in testicular germ cell tumors. *Cell, 124*, 1169–1181.

Wan, L., Pantel, K., & Kang, Y. (2013). Tumor metastasis: Moving new biological insights into the clinic. *Nature Medicine, 19*, 1450–1464.

Wang, D., Zhang, Z., O'Loughlin, E., Wang, L., Fan, X., Lai, E. C., et al. (2013). MicroRNA-205 controls neonatal expansion of skin stem cells by modulating the PI(3)K pathway. *Nature Cell Biology, 15*, 1153–1163.

Wapinski, O., & Chang, H. Y. (2011). Long noncoding RNAs and human disease. *Trends in Cell Biology, 21*, 354–361.

Wightman, B., Ha, I., & Ruvkun, G. (1993). Posttranscriptional regulation of the heterochronic gene lin-14 by lin-4 mediates temporal pattern formation in C. elegans. *Cell*, *75*, 855–862.

Wilczynska, A., & Bushell, M. (2015). The complexity of miRNA-mediated repression. *Cell Death and Differentiation*, *22*, 22–33.

Winter, J., Jung, S., Keller, S., Gregory, R. I., & Diederichs, S. (2009). Many roads to maturity: MicroRNA biogenesis pathways and their regulation. *Nature Cell Biology*, *11*, 228–234.

Xiao, C., Srinivasan, L., Calado, D. P., Patterson, H. C., Zhang, B., Wang, J., et al. (2008). Lymphoproliferative disease and autoimmunity in mice with increased miR-17-92 expression in lymphocytes. *Nature Immunology*, *9*, 405–414.

Yang, J., & Weinberg, R. A. (2008). Epithelial-mesenchymal transition: At the crossroads of development and tumor metastasis. *Developmental Cell*, *14*, 818–829.

Yi, R., Qin, Y., Macara, I. G., & Cullen, B. R. (2003). Exportin-5 mediates the nuclear export of pre-microRNAs and short hairpin RNAs. *Genes & Development*, *17*, 3011–3016.

Yoo, B., Kavishwar, A., Ross, A., Wang, P., Tabassum, D. P., Polyak, K., et al. (2015). Combining miR-10b-targeted nanotherapy with low-dose doxorubicin elicits durable regressions of metastatic breast cancer. *Cancer Research*, *75*, 4407–4415.

Yu, F., Yao, H., Zhu, P., Zhang, X., Pan, Q., Gong, C., et al. (2007). let-7 regulates self renewal and tumorigenicity of breast cancer cells. *Cell*, *131*, 1109–1123.

Yun, J., Frankenberger, C. A., Kuo, W. L., Boelens, M. C., Eves, E. M., Cheng, N., et al. (2011). Signalling pathway for RKIP and Let-7 regulates and predicts metastatic breast cancer. *The EMBO Journal*, *30*, 4500–4514.

Zeitels, L. R., Acharya, A., Shi, G., Chivukula, D., Chivukula, R. R., Anandam, J. L., et al. (2014). Tumor suppression by miR-26 overrides potential oncogenic activity in intestinal tumorigenesis. *Genes & Development*, *28*, 2585–2590.

Zhang, J., & Ma, L. (2012). MicroRNA control of epithelial-mesenchymal transition and metastasis. *Cancer Metastasis Reviews*, *31*, 653–662.

Zhang, L., Zhang, S., Yao, J., Lowery, F. J., Zhang, Q., Huang, W. C., et al. (2015). Microenvironment-induced PTEN loss by exosomal microRNA primes brain metastasis outgrowth. *Nature*, *527*, 100–104.

Zhang, P., Sun, Y., & Ma, L. (2015). ZEB1: At the crossroads of epithelial-mesenchymal transition, metastasis and therapy resistance. *Cell Cycle*, *14*, 481–487.

Zhang, P., Wang, L., Rodriguez-Aguayo, C., Yuan, Y., Debeb, B. G., Chen, D., et al. (2014). miR-205 acts as a tumour radiosensitizer by targeting ZEB1 and Ubc13. *Nature Communications*, *5*, 5671.

Zhang, P., Wei, Y., Wang, L., Debeb, B. G., Yuan, Y., Zhang, J., et al. (2014). ATM-mediated stabilization of ZEB1 promotes DNA damage response and radioresistance through CHK1. *Nature Cell Biology*, *16*, 864–875.

Zhang, Y., Yang, P., Sun, T., Li, D., Xu, X., Rui, Y., et al. (2013). miR-126 and miR-126* repress recruitment of mesenchymal stem cells and inflammatory monocytes to inhibit breast cancer metastasis. *Nature Cell Biology*, *15*, 284–294.

Zhao, Y., Ransom, J. F., Li, A., Vedantham, V., von Drehle, M., Muth, A. N., et al. (2007). Dysregulation of cardiogenesis, cardiac conduction, and cell cycle in mice lacking miRNA-1-2. *Cell*, *129*, 303–317.

Zhou, W., Fong, M. Y., Min, Y., Somlo, G., Liu, L., Palomares, M. R., et al. (2014). Cancer-secreted miR-105 destroys vascular endothelial barriers to promote metastasis. *Cancer Cell*, *25*, 501–515.

Zhu, S., Wu, H., Wu, F., Nie, D., Sheng, S., & Mo, Y. Y. (2008). MicroRNA-21 targets tumor suppressor genes in invasion and metastasis. *Cell Research*, *18*, 350–359.

Zhuang, G., Wu, X., Jiang, Z., Kasman, I., Yao, J., Guan, Y., et al. (2012). Tumour-secreted miR-9 promotes endothelial cell migration and angiogenesis by activating the JAK-STAT pathway. *The EMBO Journal*, *31*, 3513–3523.

CHAPTER SEVEN

Plasticity of Cancer Cell Invasion—Mechanisms and Implications for Therapy

V. te Boekhorst*, P. Friedl*,†,‡,1

*David H. Koch Center for Applied Research of Genitourinary Cancers, The University of Texas MD Anderson Cancer Center, Houston, TX, United States
†Radboud University Medical Centre, Nijmegen, The Netherlands
‡Cancer Genomics Center (CGC.nl), Utrecht, The Netherlands
[1]Corresponding author: e-mail addresses: pfriedl@mdanderson.org; peter.friedl@radboudumc.nl

Contents

1. Introduction 210
2. Cancer Types and Their Migration Modes 212
   2.1 Amoeboid Migration 213
   2.2 Mesenchymal Migration 215
   2.3 Collective Cell Migration 216
   2.4 Diversity of Metastatic Evasion—Individual and Clustered CTCs 218
3. Cancer Cell Migration as a Plastic and Adaptive Process 219
   3.1 Plasticity of Tumor Cell Migration Programs 219
   3.2 Microenvironmental Regulators of Cancer Cell Migration Programs 223
4. Therapeutic Targeting of Cancer Cell Migration—Implications for Anticancer Therapy 229
   4.1 Interference with Cytoskeletal Organization and Function 231
   4.2 Adhesion Systems 233
   4.3 Proteases 237
   4.4 Migration Inhibition by Molecular Targeted Therapies 238
   4.5 Challenges in Deriving and Interpreting "Antimigration" Therapies 239
5. Conclusions 240
6. Future Implications 241
Acknowledgments 241
References 241

Abstract

Cancer cell migration is a plastic and adaptive process integrating cytoskeletal dynamics, cell–extracellular matrix and cell–cell adhesion, as well as tissue remodeling. In response to molecular and physical microenvironmental cues during metastatic dissemination, cancer cells exploit a versatile repertoire of invasion and dissemination strategies, including collective and single-cell migration programs. This diversity

Advances in Cancer Research, Volume 132
ISSN 0065-230X
http://dx.doi.org/10.1016/bs.acr.2016.07.005

generates molecular and physical heterogeneity of migration mechanisms and metastatic routes, and provides a basis for adaptation in response to microenvironmental and therapeutic challenge. We here summarize how cytoskeletal dynamics, protease systems, cell–matrix and cell–cell adhesion pathways control cancer cell invasion programs, and how reciprocal interaction of tumor cells with the microenvironment contributes to plasticity of invasion and dissemination strategies. We discuss the potential and future implications of predicted "antimigration" therapies that target cytoskeletal dynamics, adhesion, and protease systems to interfere with metastatic dissemination, and the options for integrating antimigration therapy into the spectrum of targeted molecular therapies.

LIST OF ABBREVIATIONS

CCL/CXCL chemokine ligand
cMET/HGFR hepatocyte growth factor receptor
CSF-1 colony-stimulating factor-1
EGF/R epidermal growth factor/receptor
ERK extracellular-signal-regulated kinase
FGF/R fibroblast growth factor/receptor
HGF hepatocyte growth factor
IGFR insulin-like growth factor receptor
IL interleukin
JAK janus kinase
JNK c-Jun N-terminal kinase
MAPK mitogen-activated protein kinases
MLK mixed-lineage kinase
MMP matrix metalloproteinase
mTOR mechanistic target of rapamycin
NF-kB nuclear factor-kappa beta
PAK p21-associated kinase
PI3K phosphoinositide 3-kinase
ROCK Rho kinase
SDF-1 stromal cell-derived factor 1
SFKs Src family kinases
STAT signal transducer and activator of transcription
TGFβ transforming growth factor beta
TNFα tumor necrosis factor alpha
VEGF/R vascular endothelial growth factor/receptor

1. INTRODUCTION

Cell migration is an evolutionarily conserved, multifaceted process which allows individual cells and cell groups to move, change position and build, or maintain tissues and organs for embryonic development as well as homeostasis and regeneration (Sonnemann & Bement, 2011; Weijer, 2009). Virtually any nucleated cell, after promigratory stimulation, is able to transiently or continuously migrate, including stem cells, epithelial and endothelial cells, stromal and neuronal cells, and leukocytes (Friedl, 2004). In progressing cancer disease, tumor cells can acquire similar migration ability, which enables tumor cells to change position within a given tissue and spread from a tumor mass (Friedl & Alexander, 2011; Massagué & Obenauf, 2016; Sahai, 2007). Common to all cell types, cell migration results from an integrated multistep process mediated by molecular and physical programs, including cytoskeletal polarity and dynamics, cell–cell and cell–matrix adhesion, and pericellular proteolysis (Friedl & Wolf, 2003). Moving cells are responsive and adjust to external environmental cues, such as tissue dimension, structure and substrate type, and molecular triggers, including extracellular matrix (ECM), cytokines, and chemokines (Friedl & Wolf, 2010). In addition, by mechanically and chemically modifying structural proteins of tissues, moving cells can mediate structural alterations and shaping of tissue structures and thereby, depending on context, contribute to tissue formation or degeneration.

Both, the primary tumor and metastatic lesions can release disseminating cells which penetrate local tissues, enter blood vessels and circulate systemically (Fidler, 2003). The movement of cancer cells and metastatic dissemination occur from early to late stages of cancer disease (Hüsemann et al., 2008; Klein, 2009). Thus, integral to cancer progression, metastatic evasion, seeding and reseeding all contribute to the initiation and amplification of systemic metastasis (Wan, Pantel, & Kang, 2013). As an underlying mechanism, cancer metastasis depends upon the ability of cancer cells to switch from migratory, hence disseminating to sessile, hence locally growing state, likely in response to microenvironmental cues.

To initiate and complete the metastatic cascade, two principal types of migration programs are required, (i) the ability to cross tissue barriers, such as traversing across basement membranes (BM) and vascular walls, and (ii) interstitial invasion, by movement along macromolecular structures and through gaps and spaces present in connective tissue (Alexander, Weigelin, Winkler, & Friedl, 2013; Fidler, 2003; Valastyan & Weinberg, 2011). Traversing across tissue barriers is required for tumor cells to migrate between tissue compartments, such as penetration through the BM underlying the epithelium, from which epithelial tumors originate, or to intra- and extravasate blood vessels (Glentis, Gurchenkov, & Vignjevic, 2014). For interstitial invasion, after BM penetration or when tumors primarily originate in the interstitium (eg, sarcoma, hematopoietic, and glial tumors), cells directly engage with interstitial matrix and move along tissue interfaces and cell surfaces (Alexander et al., 2013). For systemic metastasis, moving cells undergo alternating phases of tissue barrier penetration, interstitial migration and migration arrest, which requires a high degree of versatility and flexibility of migration abilities as well as responsiveness to tissue cues. Thus, each step of the metastatic cascade evokes different mechanical and chemical activities of the tumor cell in a spatiotemporal-controlled manner, including adhesion and mechanotransduction, proteolysis, and cell deformation (Chambers, Groom, & MacDonald, 2002; Nguyen, Bos, & Massagué, 2009).

Depending on the tumor type, stimulation, and tissue context, tumor cells apply a range of molecular and mechanical mechanisms to move and integrate tissue-derived signals (Friedl & Wolf, 2003; Sahai, 2007). Tumor cells can either disseminate as individual cells, via mesenchymal or amoeboid single-cell migration, or by collective migration, whereby cell–cell adhesion and cooperation remain intact (Friedl & Wolf, 2003). Induced by environmental triggers, tumor cells may additionally adapt and change their migration mechanism and transit between migration programs, including collective-to-single-cell or mesenchymal-to-amoeboid transitions (Friedl & Alexander, 2011). Such plasticity of migration modes represents a putatively important coping strategy, which allows tumor cells to retain migration capability and escape from regions of environmental challenge. The adaptation of cancer cell migration strategies likely recapitulates mechanical and molecular programs engaged in stem cells and primordial tissue dynamics during embryonic development (Friedl & Gilmour, 2009; Paluch & Raz, 2013). We here review central mechanisms and plasticity programs of cancer cell migration. We summarize cellular, molecular, and environmental determinants that underlie plasticity of tumor cell migration, established in experimental in vitro and preclinical models, and discuss their clinical relevance. For key pathways underlying tumor cell migration, we discuss therapeutic options suited to limit metastatic evasion as well as challenges originating from signaling cross-talk between cell migration and survival/resistance signaling and their relevance for cancer progression.

2. CANCER TYPES AND THEIR MIGRATION MODES

The types and molecular programs underlying cancer cell migration, originally identified using experimental 2D and 3D live-cell migration assays in vitro, have been confirmed by in vivo imaging in animal models and validated by inference with the histological patterns of invasion zones and the organization of circulating tumor cells (CTCs) in clinical specimens. In aggregate, these analyses suggest that both, tumor type and its differentiation state determine adaptive migration programs by which cancer cells navigate through each step of the metastatic cascade.

2.1 Amoeboid Migration

Amoeboid single tumor cell migration is classified from roundish but highly deformable cell morphology typically detected in myeloic leukemia and lymphoma cells, small-cell lung carcinoma and small-cell prostate cancer, as well as induced experimental models of melanoma and epithelial tumors, such as breast cancer (Condeelis & Segall, 2003; Madsen & Sahai, 2010; Pinner & Sahai, 2008). Histologically amoeboid movement is characterized by roundish single cells, which individually and often diffusely infiltrate the tissue. In leukemia, amoeboid tumor cells recirculate in large numbers and represent the predominant cell type in the blood (Hutchinson et al., 2014).

Amoeboid movement is initiated and maintained by asymmetric cell interactions with tissue structures via membrane protrusions at the leading edge. Subsequently, amoeboid translocation is mediated by relatively low-adhesive and poorly proteolytic cell–ECM/substrate interactions, which are maintained by dynamic shape change and cell contractility. Thereby, amoeboid movement mediates particularly flexible cell–tissue interactions which depend on shape change and intercalation for propagation through tissue gaps and discontinuities (elbowing) (Fig. 1E and F) (Friedl, Entschladen, Conrad, Niggemann, & Zänker, 1998; Lorentzen, Bamber, Sadok, Elson-Schwab, & Marshall, 2011; Paluch & Raz, 2013; Wyckoff, Pinner, Gschmeissner, Condeelis, & Sahai, 2006). The roundish shape, types of cell protrusions, and cytoskeletal flexibility are jointly regulated by the small GTPases RhoA and Rac. Both cell polarization toward direction of migration and retraction of the cell rear depend upon cortical actomyosin contractility, controlled by RhoA via Rho kinase (ROCK) signaling and myosin II activity (O'Connor & Chen, 2013; Paluch & Raz, 2013). Cortical actin network contraction additionally increases hydrostatic pressure in the cell which triggers bleb-like outward protrusion of the plasma membrane (blebbing) through which cells protrude towards and intercalate surrounding 3D tissue structures (Blaser et al., 2006; Paluch, Piel, Prost, Bornens, & Sykes, 2005; Paluch & Raz, 2013). Other protrusions at the leading edge, including pseudopods, lamellipods, and filopods, depend upon actin polymerization mediated by local Rac activity and provide polarized

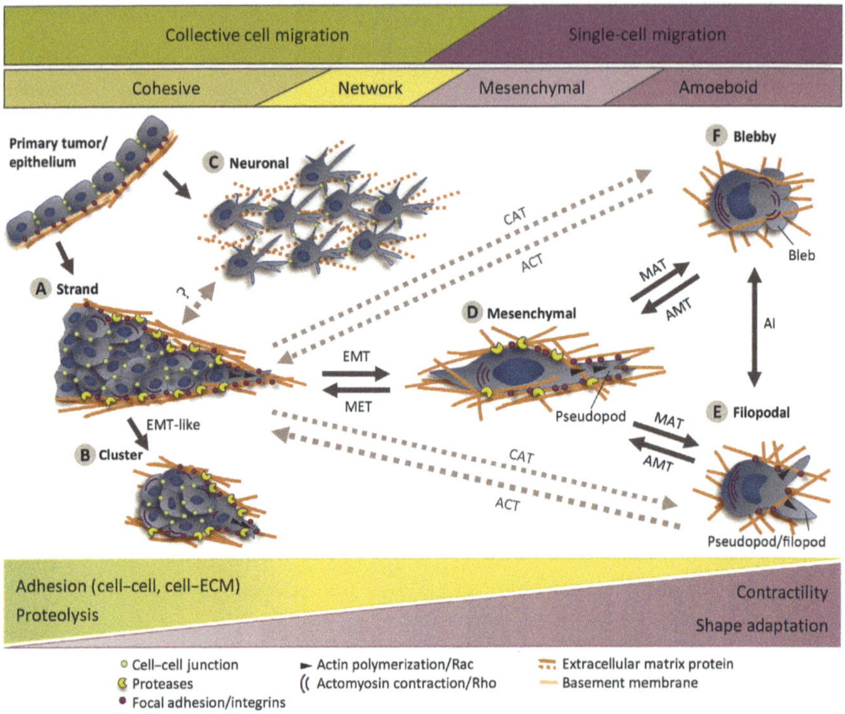

Fig. 1 Mechanisms, modes and plasticity of cancer cell migration strategies. Different modes and mechanisms of cancer cell migration include collective and single-cell migration strategies. Collective tumor migration modes include: (A) collectively invading strands with high intercellular adhesion strength, pericellular proteolysis and matrix remodeling, and integrin-mediated adhesion and traction toward ECM; (B) detached cohesive, proteolytic, and traction-force generating tumor cell clusters; and (C) neuronal-type networks of cells moving with transient and adaptive filament-based cell–cell junctions. Single tumor cell migration modes include: (D) mesenchymal migration depending on pericellular proteolysis and integrin-mediated adhesion/traction; (E) filopodal amoeboid migration mediated by Rac and intermediate/weak adhesions; and (F) blebby amoeboid migration mediated by contractile Rho-directed actomyosin contraction and poorly/nonadhesive bleb-like protrusions. Transitions between cancer cell migration strategies in dependence of cell–cell adhesion, ECM-binding activity, and actin dynamics. Plasticity programs occur between collective migration strategies (strand vs clusters), including strand-to-cluster transition (EMT-like); collective and single-cell migration, including epithelial-to-mesenchymal transition (EMT) and collective-to-amoeboid transition (CAT); or in reverse direction mesenchymal-to-epithelial transition (MET) and amoeboid-to-collective transition (ACT). Single-cell plasticity depends upon the level of protease-mediated ECM remodeling, Rho/Rac actomyosin dynamics and ECM adhesion strength, which define transitions from mesenchymal to amoeboid (MAT) and from amoeboid-to-mesenchymal (AMT) migration strategy. Rho/Rac levels determine amoeboid interconversion (AI) and the stability of filopodal or blebby amoeboid migration strategy. *Arrows*, uni- or bidirectional transitions from one migration mode to another (plasticity). *Continuous lines*, well-characterized transitions; *dotted lines*, less-well-characterized transitions.

adhesion to 2D and 3D substrates (Sanz-Moreno & Marshall, 2010). While the cell front is engaged with tissue structures, lateral and rear regions of the cell undergo contraction controlled by RhoA/ROCK and myosin II activity, which mediates shortening of the cell length axis and the rear (Bastounis et al., 2014), and carries the nucleus forward (Renkawitz et al., 2009; Wolf et al., 2013; Wyckoff et al., 2006).

Thus, shape adaptation mediated by Rho/ROCK-driven actin contraction and deformability are key mechanisms enabling amoeboid tumor cell migration modes (Sanz-Moreno & Marshall, 2010; Wyckoff et al., 2006). In addition, the adaptive shape change enables cells to move along and between tissue discontinuities without proteolytic degradation of extracellular matrix (Wolf et al., 2013). Amoeboid migration arguably represents the simplest, most flexible and therefore potentially the most efficient strategy to move through tissues and between tissue barriers.

2.2 Mesenchymal Migration

Mesenchymal movement is often found in sarcomas, gliomas, and epithelial cancer cells after undergoing epithelial-to-mesenchymal transition (EMT), and further is present in tumor types of generally low differentiation (spindle-cell tumors) (Cates et al., 2008; Chanrion et al., 2014). Histologically mesenchymal movement can be inferred from elongated, spindle-like cell shape with oval nuclear morphology (Moreno-Bueno et al., 2009). Often mesenchymal tumor cells express EMT markers, including nuclear Oct-4, Twist, Snail/Slug, and cytoplasmic vimentin, which allow their detection in tissue samples as grouped or individual cells infiltrating remodeled, often desmoplastic tissue (Zeisberg & Neilson, 2009).

During migration, mesenchymal tumor cells maintain an elongated, polarized cell shape, with a protruding pseudopod and/or multiple filopods, which adhere to and pull on ECM substrate and determine the direction of migration (Starke, Maaser, Wehrle-Haller, & Friedl, 2013). At the leading edge, Rac-induced actin assembly and integrin binding to the substrate induce cell polarization and protrusion (Geiger, Spatz, & Bershadsky, 2009). Concomitant with integrin engagement, FAK and Src kinases induce maturation of focal ECM adhesion and mechanotransduction, which, together with contractile cortical actin, mediates high contractile tension and pulling forces toward ECM structures (Guarino, 2010; Lawson et al., 2012; Parsons & Parsons, 2004). At the rear, RhoA-induced actomyosin contractility coincides with turnover of integrin-mediated focal adhesions to the ECM, which reduces anchorage of the cell rear and leads to forward sliding of the cell body (Huveneers & Danen, 2009; Sadok & Marshall, 2014). Thus, alternating Rac-induced cell elongation at the leading edge and Rho-mediated rear retraction enables cycles of cell-matrix adhesion, pulling and relaxation on tissue structures (Starke et al., 2013; van Helvert & Friedl, 2016). Adjacent to focal interactions to the ECM, surface matrix metalloproteases (MMPs) remove ECM structures and barriers, such as collagen fibers, which leads to tissue remodeling and the generation of spatially widened tissue tracks along the migration path (Fig. 1D) (Friedl et al., 1997; Wolf et al., 2007).

Studies using 2D and 3D migration assays have shown, integrin adhesion receptors equip cancer cells with a repertoire of substrate-recognition and attachment abilities, thus providing flexibility in substrate choice. For example, α2β1, α1β1, and α11β1 integrins mediate binding to collagen and αVβ3, αVβ5, and α5β1 binding to fibronectin during migration (Balcioglu, van Hoorn, Donato, Schmidt, & Danen, 2015; Hynes, 2002). This ability to recognize different substrates is likely critical for cancer cell dissemination and metastasis, as series of different ECM interaction systems engage with different tissue components. Examples include cell-binding to interstitial collagens when invading interstitial tissue, laminins when moving along BM, or fibronectin when entering inflamed microenvironments, with concomitant coengagement of cell–ECM adhesion systems in complex environments (Gritsenko, Ilina, & Friedl, 2012; Lu, Weaver, & Werb, 2012; Weigelin, Bakker, & Friedl, 2012).

Mesenchymal migration as a molecularly and mechanistically complex mechanotransduction program thus supports both cell migration and tissue remodeling in a combined process. As a consequence of coordinated adhesion and proteolysis systems, mesenchymal movement allows tumor cells to even penetrate tissues of very high density (Wolf et al., 2013) as well as the passage of vascular BM (Vitale, Avizienyte, Brunton, & Frame, 2008; Wang & McNiven, 2012).

2.3 Collective Cell Migration

Collective tumor cell migration is a important invasion strategy in most epithelial tumors such as breast cancer, squamous cell carcinoma, colon cancer and others, as well as in certain mesenchymal tumors (Cheung, Gabrielson, Werb, & Ewald, 2013; Friedl & Gilmour, 2009; Friedl et al., 1995). Histologically, collective invasion is characterized by direct, next-neighbor position of tumor cells which move through interstitial stroma outside the tumor core but retain cell–cell junctions, including adherens junctions and membrane-localized β-catenin (Cheung et al., 2013; Friedl, Locker, Sahai, & Segall, 2012). Collective migration is best identified by careful 3D reconstruction of histological samples (Bronsert et al., 2014), but also can be inferred from the presence of roundish cell nests and elongated strands in conventional histology (Friedl et al., 2012).

During collective migration in experimental 2D and 3D assays, cell–cell contacts coordinate and polarize migrating tumor cells into a multicellular functional unit (super cell) (Friedl et al., 2012; Khalil & Friedl, 2010). Distinct levels of actin dynamics, substrate adhesion and ECM remodeling functions are shared and coordinated between neighboring cells and define leader and follower cell behaviors. At the invasion front, particularly polarized and migratory leader cells engage with surrounding tissue structures via Rac-driven filopodal protrusions and integrin-mediated substrate adhesion, while remaining connected to follower inner and outer strand cells (Khalil & Friedl, 2010; Yamaguchi, Mizutani, Kawabata, & Haga, 2015). Cell–cell connections depend on sufficiently stable cell–cell adhesion to follower cells to withstand dragging forces generated by leader cells and Rho-mediated actin contraction across multiple cell bodies (Bazellières et al., 2015; Reffay et al., 2014). Guided by cell–cell contacts, inner and outer strand cells follow along the same path of the leading front (Cheung & Ewald, 2014). Both, leader and follower cells engage integrins and/or other adhesion systems to generate force, and exert proteolytic matrix remodeling, particularly through MMPs, to digest tissue barriers and generate a path of least resistance along which the cells move forward (Fig. 1A) (Hegerfeldt, Tusch, Bröcker, & Friedl, 2002; Mayor & Etienne-Manneville, 2016; Wolf et al., 2007). Cell–cell contacts are typically formed by cadherins (eg, E-, N-, P-cadherin) and cortical actin strings, which together form and mediate the stability of adherens junctions (Haeger, Wolf, Zegers, & Friedl, 2015; Meng & Takeichi, 2009). Other cell–cell adhesion systems include immunoglobulin superfamily members and ephrins/EpH receptor adhesions that mediate more labile or transient cell–cell interactions (Kania & Klein, 2016; Wai Wong, Dye, & Coombe, 2012), as well as connexins, which enable communication through gap junctions and signal transduction between connected tumor cells (Li, Zhou, & Donahue, 2008).

Collective invasion is adaptive, in a cell type and tissue-responsive manner. Tumor cells either locally invade as (1) collective sheets or strands while remaining attached to the tumor lesion, typically detected in epithelial tumors, such as breast or squamous cell carcinoma (Fig. 1A) (Bronsert et al., 2014); (2) isolated clusters detached from the primary/metastatic lesion such as epithelial tumors and melanoma (Fig. 1B) (Friedl et al., 1995); (3) neuronal-like networks of connected cells, detected in neuroectodermal tumors, such as glioblastoma (Osswald et al., 2015) (Fig. 1C); or (4) as "jammed" collective cohorts induced by spatially narrow tissue boundaries (confinement) of otherwise transiently/loosely connected (single) cells, such as grouped melanoma and sarcoma cells (Haeger, Krause, Wolf, & Friedl, 2014).

Compared to single-cell migration, collectivity of migrating cancer cells likely bears advantages for surviving the metastatic cascade, including: (i) a highly promigratory and prosurvival environment between connected cells by secretion of growth factors, chemokines and proteases; (ii) the passive displacement of otherwise poorly mobile or even immobile but potentially highly proliferative cells inside the strands by highly mobile neighbor or leader cells; and (iii) protection of cells located in inner regions of groups from environmental assault, such as reduced immune cell attack or lower shear stress and nuclear damage in the tissue or vasculature (Denais et al., 2016; Friedl & Gilmour, 2009).

2.4 Diversity of Metastatic Evasion—Individual and Clustered CTCs

Clinically, indications for both individual and collective metastasis can be detected at all stages of disease progression, evidenced by histologically diverse patterns of local tissue invasion and the circulation of both individual and clustered tumor cells (Aceto et al., 2014; Liotta, Saidel, & Kleinerman, 1976; Massagué & Obenauf, 2016). In primary lesions of epithelial tumors, collective invasion is very prominent (Friedl et al., 2012), but only a minority of clustered and a majority of individual CTCs are detected in peripheral blood of epithelial cancer patients (Kraan et al., 2011). Thus, the most prevalent migration mode in tissue may differ from the most likely strategy for entering and surviving the circulation.

Given their distinct mechanisms for propagating through tissue, individual or collective tumor cell migration likely generate distinct routes of entry into the circulation and strategies for metastatic organ colonization (Aceto et al., 2014; Cheung et al., 2016), and several facts argue in favor for clinical relevance of each dissemination program. Clustered CTCs are many-fold more likely to initiate organ colonization compared to individually circulating cells (Aceto et al., 2014; Cheung et al., 2016; Liotta et al., 1976). Alternatively, in epithelial tumors single CTCs may express EMT and stemness markers (Armstrong et al., 2011; Satelli et al., 2015), suggesting a mesenchymal route for hematogenous spread and survival. In hematological neoplasias, including lymphoma and leukemia, amoeboid single-cell migration coincides with anchorage-independent survival and proliferation and an ability for particularly efficient systemic dissemination and recirculation (Trendowski, 2014), not unlike recirculating leukocytes (Hutchinson et al., 2014). Thus, the likelihood for individual vs collective metastasis may originate from distinct mechanisms, types, and phases of disease. The migration mode(s) most suited to enter and support survival in the circulation as well as particular niches in the tumor stroma, vascular beds, and distant organs most favorable for each migration mode and resulting metastatic efficacy remain to be identified.

3. CANCER CELL MIGRATION AS A PLASTIC AND ADAPTIVE PROCESS

By which strategy and how efficiently cancer cells move, and whether or not tumor cells transition between migration modes depends on molecular and physical input moving cells receive from their environment.

3.1 Plasticity of Tumor Cell Migration Programs

The adaptation and interconversion of ongoing migration strategies is regulated by the induction of different signaling programs, here summarized as plasticity of cancer cell dissemination (Fig. 1).

3.1.1 Plasticity of Single Cancer Cell Migration

Known transitions between single-cell migration programs include the interconversion between mesenchymal and amoeboid migration and difficult-to-classify, intermediate states between amoeboid migration types. Molecular interference has revealed that the transition from mesenchymal-to-amoeboid migration (Fig. 1, MAT) is favored by a range of experimental conditions, including (1) reduced cell–ECM adhesion, (2) loss of ECM proteolysis, (3) enhanced RhoA- and myosin II-mediated actomyosin contractility, (4) microtubule (MT) destabilization, which all result in the movement of rounded cells and in increased ability of cell-shape adaptation (Belletti et al., 2008, 2010; Berton et al., 2009; Sahai & Marshall, 2003; Sanz-Moreno et al., 2008; Taddei et al., 2011; Wolf et al., 2003). Particularly reduced ECM adhesiveness and elevated cell contractility support the morphological adaptation from an elongated to a roundish-ellipsoid cell body and transition of migration mode. By gaining amoeboid abilities, the increase in deformability and shape change, despite reduced proteolytic ECM clearance, enables efficient passage and circumvention of tissue barriers (Wolf et al., 2003).

In a reverse process, amoeboid-migrating cells may convert to mesenchymal migration (Fig. 1, AMT) when (1) ECM adhesion, (2) proteolytic ECM remodeling, and (3) Rac-induced actin assembly and protrusion formation are increased (Panková, Rösel, Novotný, & Brábek, 2010; Sanz-Moreno et al., 2008). As consequence of these molecular changes, AMT can be identified in moving cells by a morphological transition from round, spherical to elongated, spindle-cell shape.

Amoeboid interconversion (Fig. 1, AI) includes a transition between amoeboid-blebby and amoeboid-filopodal states. Both states may be transient, based on oscillatory engagement of Rho and Rac activity at the leading edge, and result in difficult-to-categorize phenotypes. When Rac is active, filopodal membrane protrusions predominate, while dominant RhoA activity supports protrusive blebs at the leading edge (Fackler & Grosse, 2008; Petrie & Yamada, 2012). These different protrusions likely generate a range of distinct interaction types to tissue components, the differential mechanics and consequences of which remain to be clarified.

Interconversions of migration programs likely provide particular flexibility for moving cancer cells when confronted with different types of tissues, vascular walls and BM. MAT, AMT, and AI may further be associated with the regulation of stemness and altered ability of metastasis formation (Taddei et al., 2014), yet their relative contributions to any step of the metastatic cascade remain to be clarified.

3.1.2 Collective-to-Single-Cell Transition

The transition from collective-to-single-cell migration occurs frequently during the progression of epithelial tumors when (1) cell–cell adhesion systems are downregulated or (2) when tissue confinement is relieved during transit from dense to loose tissue topology.

The EMT (Fig. 1A, EMT) is the result of a signaling program which downregulates epithelial features, particularly cell–cell junctions and apicobasal anchorage to BM, and upregulates stromal-type adhesion systems and cytoskeletal dynamics (Polyak & Weinberg, 2009; Xu, Lamouille, & Derynck, 2009). EMT results in the progression of many, if not all, epithelial cancers, including colon, lung, prostate, and breast cancer (Savagner, 2010), and strongly enhances cancer invasion, metastatic progression, morphological heterogeneity, and correlates with poor prognosis (Polyak & Weinberg, 2009).

With EMT, cells weaken or fully resolve cell–cell junctions, including adherens junctions, desmosomes, and tight junctions, upregulate stromal protease as well as integrin adhesion systems (eg, from β1 to β3 switching, enhanced αV signaling), and shift Rho-mediated actomyosin contractility from cell–cell junctions toward cell–matrix interactions (Mamuya & Duncan, 2012; Parvani, Galliher-Beckley, Schiemann, & Schiemann, 2013). These molecular reprogramming events result in deregulated cell–cell contacts, loss of apicobasal but gain of front-rear polarity, and ultimately favor the transition from epithelial or collective to mesenchymal features (Bousquet et al., 2016; Grünert, Jechlinger, & Beug, 2003; Moustakas & Heldin, 2007). Jointly, altered integrin-mediated focal adhesion dynamics, cytoskeletal reorganization, and enhanced ECM remodeling facilitate cell dispersal and invasive migration into tissues (Xu, Boudreau, & Bissell, 2009; Xu, Lamouille, et al., 2009). In addition to cell individualization, recent observational and modeling work indicates a high likelihood for mixed behaviors after EMT, including intermediate (eg, metastable or hybrid) phenotypes, such as detached collective or loosely connected migrating groups (Jolly et al., 2015; Savagner, 2010). With such EMT-associated reprogramming, or partial EMT, emerging moving cell clusters (Fig. 1C, EMT-like) may still maintain cell–cell contacts but already alter their differentiation state and gain pluripotent potency (Jolly et al., 2015; Savagner, 2010). Thus, EMT and EMT-like processes reflect molecular plasticity with direct and indirect, reversible or irreversible impact on cancer cell migration modes. The relevance of EMT-like processes in evoking intermediate phenotypes in cancer progression, eg, circulating cluster formation and survival, remains to be determined (Jolly et al., 2015).

Besides EMT, the downregulation of cell–cell adhesions alone can cause collective-to-single-cell transition. As examples, experimental interference with β1 integrin, which lowers both cell–matrix and cell–cell adhesion stability, leads to disintegration of melanoma clusters and transition to amoeboid single-cell migration (Hegerfeldt et al., 2002). Similarly, enhancing EGF signaling promotes cell evasion from the solid tumor followed by amoeboid dissemination in vivo (Wyckoff et al., 2004). In addition, growth factor signaling can cause single-cell detachment independent of E-cadherin regulation. As example, hepatocyte growth factor (HGF)-induced signaling can weaken cell–cell junctions and promotes single-cell dissemination in the absence of EMT (de Rooij, Kerstens, Danuser, Schwartz, & Waterman-Storer, 2005).

Biophysical mechanisms controlling collective-to-single-cell transitions depend upon the space and geometries encountered by moving cells in different tissues. Collectively moving mesenchymal cells undergo conversion to single-cell dispersal when encountering loosely organized tissue, allowing migrating cells to abandon weak cell–cell interactions and disperse individually, a process termed "unjamming" (Haeger et al., 2014). In vivo, unjamming occurs when cancer cells migrate from confined into open space, eg, from dense collagen meshworks into loose interstitial tissue (Weigelin et al., 2012; Weigelin, Bakker, & Friedl, 2016).

3.1.3 Collective Plasticity

Plasticity of collective migration modes occurs frequently in epithelial cancers and describes different cohesive and structural organizations of invading cell groups (Friedl & Gilmour, 2009). Different collective modes include highly differentiated epithelial cohorts with conserved lumen formation; collective strands without lumen but with cohesive multicellular organization; small clusters after detachment or cell aggregation, and chain migration in strand-like fashion with preserved but minimal cell–cell contact (Friedl et al., 1995; 2012).

Collective plasticity is a consequence of varying degrees of apicobasal polarity within the group as well as varying strength of intercellular adhesion. High apicobasal polarity leads to inner lumen formation in collective epithelial strands, often accompanied by secretion and interaction with BM proteins at the basal side; this recapitulaes tubulogenesis such as during glandular duct formation of otherwise quiescent and immobilized epithelia (Cheung & Ewald, 2014). Intercellular adhesion strength can be modified by up- or downregulation of adherens junctions, eg, by partial EMT, which may coincide with different levels of cohesive behavior (Batlle & Wilkinson, 2012; Bazellières et al., 2015; Peglion, Llense, & Etienne-Manneville, 2014). Collective plasticity likely underlies the formation of CTC clusters (microemboli) after collective transmigration of the vessel wall and intraluminal disintegration of the cluster by molecular regulation or shear force exerted by the blood stream (Armstrong et al., 2011; Cheung & Ewald, 2016). A morphological framework for the plasticity of collective invasion and its implications for transitions between tissues remains to be established.

The ability of tumor cells to switch transiently or permanently between molecular pathways that determine migration ability and strategy, ie, pericellular proteolysis, adhesion to ECM, and cell–cell interaction, represents an array of coping strategies that enable tumor cells to engage with virtually any tissue type and respond to numerous microenvironmental challenges.

3.2 Microenvironmental Regulators of Cancer Cell Migration Programs

Migrating tumor cells integrate structural and chemical signals from the tumor microenvironment and adjust their migration strategy accordingly. Virtually any mechanical cue and/or signaling molecule encountered in tissues can modulate how cancer cells migrate.

3.2.1 ECM Organization

Interstitial tissues and organ sites represent complex microenvironments and interfaces, with a high variety of structural, mechanical and molecular properties that guide, rather than impede, cancer cell migration before or during tumor-induced microenvironmental perturbation (Gritsenko et al., 2012). Tissue-intrinsic structural and mechanical properties include geometry, spacing, degree of ECM fiber orientation, and ECM stiffness (te Boekhorst, Preziosi, & Friedl, 2016). Tissues comprise 2D surfaces (eg, BM), random or aligned 3D fibrillar collagen in respectively loose or dense interstitial tissue, and 3D space bordered by 2D surfaces of varying width, textures and composition, such as tissue tracks along muscle fibers, nerves, and blood vessels (Alexander et al., 2013; Joyce & Pollard, 2009; Weigelin et al., 2016). These molecular and mechanical cues generate different degrees of barriers and guidance and in cooperation impact on migration programs, by either reinforcing an ongoing migration strategy or by inducing modulation followed by a plasticity response (Table 1) (Haeger et al., 2014; Kumar et al., 2016; Liu et al., 2015; Wolf et al., 2013; Xu, Boudreau, et al., 2009). For example, space confinement together with low ECM adhesion supports amoeboid migration of cells that on unconfined 2D substrate migrate by adhesive, mesenchymal mechanisms (Liu et al., 2015). Alternatively, aligned matrix fibers, which provide a track of low resistance, support cancer cell migration independent of proteolytic ability, whereas migration in dense tissue with very small pores is protease dependent (Kumar et al., 2016; Wolf et al., 2013).

In progressing tumors, the organization of the peritumoral stroma underlies significant structural and molecular remodeling induced by both cancer cells and tissue-resident cells, which can influence invasion programs. Cancer-associated fibroblasts (CAFs) with generally high proteolytic ability can generate tissue tracks, which support collective migration of tumor cells (Gaggioli et al., 2007). Moreover, by deposition, stretching and cross-linking of structural matrix proteins (eg, FN, collagen type I and III), CAFs can promote a stiffness-related invasion response in many cancer types

Table 1 Modulation of Cancer Cell Migration Modes by ECM Organization

Modulator	Tumor Type	Impact/Plasticity Response	References
Geometry, confinement	Melanoma, sarcoma	Collective invasion along linear confined spaces along myofibers, blood vessels, nerves, and adipocyte surfaces, but not through complex-shaped low-density stroma; single-cell migration in fibrillar collagen-rich tissue of high porosity; tissue geometry dictates alternation between collective and single-cell migration	Haeger et al. (2014), Weigelin et al. (2012, 2016)
Confinement (space)	Melanoma, Fibrosarcoma	Enhanced collective migration of otherwise single-cell migration (cell jamming) by confined space, including tube-like collagen tissue tracks in vitro or tissue conduits between muscle and nerve fibers in vivo	Haeger et al. (2014) and Weigelin et al. (2012)
	Normal human dermal fibroblasts	Induction of MAT and conversion to low ECM adhesion with increasing confinement	Liu et al. (2015)
ECM alignment	Mouse breast cancer models; breast cancer	Enhanced invasiveness, in particular low-proteolytic migration by collagen alignment (linearization) and collagen cross-linking (thickening) at the tumor–stromal interface	Condeelis and Segall (2003), Kumar, Kapoor, Desai, Inamdar, and Sen (2016), and Provenzano et al. (2006)

ECM meshworks	Breast cancer	Disadvantage for single-cell migration and support for collective migration in interstitial tissue	Kumar et al. (2016)
Stiffness	Transformed breast epithelial; glioma, glioblastoma	Increased invasion and migration speed with increasing substrate stiffness, via elevation of integrin-ECM adhesion and PI3K signaling	Paszek et al. (2005), Levental et al. (2009), and Pathak and Kumar (2012)
Substrate type	Lung cancer, pancreatic cancer	Induction of EMT via TGFβ signaling; cooperative α2β1 integrin and discoidin receptor 1 (DDR) signaling upon binding to collagen type I	Shintani, Fukumoto, et al. (2008) and Shintani, Maeda, Chaika, Johnson, and Wheelock (2008)
	Gallbladder cancer, ovarian carcinoma	Enhanced migration and invasion mediated by adhesion to fibronectin and downstream induction of promigratory signaling (FAK/PI3K/Akt, mTOR)	Cao et al. (2015) and Yousif (2014)
Density	Breast cancer, melanoma, fibrosarcoma	Enhanced invasiveness of otherwise poorly invasive breast cancer cells; transition from collective to single-cell migration; dense collagen favoring collective migration; loose tissue promoting single-cell dissemination	Haeger et al. (2014) and Provenzano et al. (2008)
Tissue remodeling	Fibrosarcoma, breast cancer	Transition from single-cell to collective migration induced by MT1-MMP mediated collagenolysis and formation of a path of least resistance (track generation); facilitated local invasion by tumor cell-induced fiber alignment	Kumar et al. (2016) and Wolf et al. (2007)

(Lu, Weaver, et al., 2012; Paszek et al., 2005). Thus, by combined mechanical, spatial, and associated signaling parameters, the tumor stroma coevolves as an inductive scaffold and reciprocally controls the type, extent, and plasticity response of cancer invasion (Egeblad, Rasch, & Weaver, 2010; Provenzano et al., 2006).

3.2.2 Growth Factors and Chemokines

Besides ECM guidance cues, the activated tumor stroma expresses a variable repertoire of growth factors and chemokines which, by diffusion or as ECM-tethered immobilized ligands, impact cancer cell invasion and dissemination programs (Joyce & Pollard, 2009; Lu, Weaver, et al., 2012). Growth factors, including EGF, HGF, VEGF, IGF, TGFβ, and FGF as well as chemokines, including CXCL10, CXCL12, CCL21, or CCL25, among many other factors, induce promigratory chemotactic signaling, EMT and other plasticity programs in tumor cells (Table 2) (Friedl & Alexander, 2011; Goel & Mercurio, 2013; Hollier, Kricker, Van Lonkhuyzen, Leavesley, & Upton, 2008; Kermorgant, Aparicio, Dessirier, Lewin, & Lehy, 2001; Roussos, Condeelis, & Patsialou, 2011; Wells, Chao, Grahovac, Wu, & Lauffenburger, 2011). As example, CAFs release TGFβ and PDGF which induce EMT and promote invasion of cancer cells (Matise et al., 2012). Cancer-activated adipocytes secrete cytokines (eg, TNFα, leptin, and IL-6) which induce EMT (Wolfson, Eades, & Zhou, 2015), upregulate the release of proteases (eg, MMP-2, MMP-11) and thereby promote both migration plasticity and invasion in breast cancer cells (Dirat et al., 2011; Fujisaki et al., 2015; Lee, Jung, & Koo, 2015). Extracellular growth factors and chemokines induce signaling including mitogen-activated protein kinases (MAPKs), cMET, JAK/PI3K/JNK, mTOR, Src family kinase (SFKs), and RhoA/Rac signaling in both migrating cancer cells and stromal cells, which leads to the reprogramming of signaling networks and, accordingly, cell functions (Avizienyte & Frame, 2005; Donà et al., 2013; El Haibi et al., 2010; Friedl & Alexander, 2011; Heit, Tavener, Raharjo, & Kubes, 2002). For example, cooperation of IGFR and TGFβ signaling induces EMT in breast cancer cells (Walsh & Damjanovski, 2011), and IGFR signaling induces protease expression via PI3K signaling, which supports a proinvasive behavior in lung carcinoma cells (Zhang & Brodt, 2003). Macrophage-induced paracrine loops between CXCL12 and EGF signaling guides the invasion of breast cancer cells to nearby blood vessels (Goswami et al., 2005; Wyckoff et al., 2004) or promotes breast cancer invasion via alternate signaling routes, eg, CXCR2/FGF and Wnt signaling (Bohrer & Schwertfeger, 2012; Pukrop et al., 2006).

Table 2 Modulation of Cancer Cell Migration Modes by Growth Factors and Chemokines

Modulator	Tumor Type	Impact/Plasticity Response	References
HGF/cMET	Prostate cancer, breast cancer	Induction of invasiveness, EMT and anchorage-independent growth via activation of PI3K, JNK, and MAPK signaling, induction of Rac and Rho, and interaction with β4 integrin, CD44 receptors and diverse cytoskeletal proteins mediating ECM adhesion	Han, Luo, Zhao, Li, and Jiang (2014), Orian-Rousseau et al. (2007), Spina et al. (2015), and Trusolino, Bertotti, and Comoglio (2010)
FGF-2/FGFR4	Colorectal cancer	Enhanced invasiveness via Src signaling and integrin-mediated adhesion to collagen; MT1-MMP activation; induces EMT in cooperation with TGFβ	Knuchel, Anderle, Werfelli, Diamantis, and Rüegg (2015), Lamouille, Xu, and Derynck (2014), Shirakihara et al. (2011), and Sugiyama et al. (2010)
CXCL12/CXCR4	Colorectal cancer, breast cancer	Induction of migration, EMT, and intravasation via cross talk with integrin, TGFβ, and Wnt signaling	Hu et al. (2014), Jin, Brockmeier, Otterbach, and Metzen (2012), and Wang et al. (2014)
TGFβ	NSCLC, gastric cancer, metastatic breast cancer, prostate cancer, colorectal cancer, etc.	Induction of invasion, EMT, metastasis; counteracts collective behaviors; additional proinvasive effects through myofibroblast activation and matrix remodeling	Giampieri et al. (2009), Giampieri, Pinner, and Sahai (2010), Katsuno, Lamouille, and Derynck (2013), Kojima et al. (2010), Margadant and Sonnenberg (2010), Matise et al. (2012), Oyanagi et al. (2014), and Padua and Massagué (2009)

Continued

Table 2 Modulation of Cancer Cell Migration Modes by Growth Factors and Chemokines—cont'd

Modulator	Tumor Type	Impact/Plasticity Response	References
VEGF	Prostate cancer; glioblastoma multiforme	Induction of invasion, migration, and EMT; upregulation of N-cadherin, snail and vimentin; downregulation of E-cadherin (prostate cancer cells); inhibition of migration and EMT via MET/VEGFR2 signaling (glioma cells)	Gonzalez-Moreno et al. (2010) and Lu, Chang, et al. (2012)
IL-6	Head and neck cancer, breast cancer, ovarian cancer, bone metastasis	Enhanced migration, EMT, and metastasis via VEGF cross talk or induction of JAK-STAT3/Snail signaling	De Luca, Lamura, Gallo, Maffia, and Normanno (2012), Guo, Xu, Lu, Duan, and Zhang (2012), and Yadav, Kumar, Datta, Teknos, and Kumar (2011)
CXCL12/ CXCR4	Prostate cancer; mammary adenocarcinoma	Induction of migration and metastasis via ERK1/2, JNK, Akt signaling, and MMP-9 activation; increase in invasion and remodeling of tumor microenvironment in vivo, including enhanced macrophage and blood vessel abundance	Boimel et al. (2012) and Brand et al. (2005)
PKCα	Melanoma, breast cancer	Induction of MAT in vitro; PKCα inhibition mediates AMT	Vaškovičová et al. (2015)

Growth factor and chemokine repertoires as well as their receptors are not stable, but transient and adaptive parameters. ECM-bound chemokines and growth factors and their receptors can be proteolytically cleaved and either liberated and activated, or degraded, and further regulation occurs by internalization from the membrane and recycling (Kessenbrock, Plaks, & Werb, 2010; Murphy, 2008). In addition, the extent of tumor-associated inflammation modulates pro-inflammatory cytokine pools (Landskron, De la Fuente, Thuwajit, Thuwajit, & Hermoso, 2014; López-Novoa & Nieto, 2009).

3.2.3 Metabolic Control

Metabolic perturbation of the tumor and its microenvironment strongly impact the migratory behaviors and adaptation of cancer cells (Table 3) (Clark & Vignjevic, 2015; López-Novoa & Nieto, 2009; Lu & Kang, 2010). As example, low oxygen tension (hypoxia) causes metabolic reprogramming of tumor cells (Warburg effect) (Rankin & Giaccia, 2016), and the activation of migration programs, often via EMT signaling (Webb, Chimenti, Jacobson, & Barber, 2011). Tumor hypoxia, via stabilization of hypoxia-inducible transcription factors, induces EMT, including downregulation of adherens junctions, upregulation of migration-enhancing vimentin, and Rac activation (Gilkes et al., 2014; McInroy & Määttä, 2007; Vuoriluoto et al., 2011). Such reprogramming can lead to enhanced invasion, including collective-to-single-cell transition (Zhang, Huang, Li, et al., 2013).

The combined action of tissue organization, composition, and molecular cross talk between growth factor and cytokine signaling, protease activity and metabolic reprogramming cooperate and progress over weeks and months; jointly, they impose molecular adaptation in cancer cells and, directly and indirectly, diversify strategies of metastatic dissemination over time. The relevance of microenvironmental regulation of migration programs for different steps of the metastatic cascade remains to be determined.

4. THERAPEUTIC TARGETING OF CANCER CELL MIGRATION—IMPLICATIONS FOR ANTICANCER THERAPY

The prevalence of systemic dissemination followed by metastatic seeding and reseeding from early- to late-stage disease indicates that

Table 3 Metabolic Modulators of Cancer Cell Migration Modes

Modulator	Tumor Type	Impact/Plasticity Response	References
Hypoxia	Hepatocellular carcinoma; breast cancer, NSCLC	Induction of migration by EMT and partial EMT, depending on cell type and differentiation status	Gilkes, Semenza, and Wirtz (2014), Lundgren, Nordenskjöld, and Landberg (2009), Renaud, Guenot, Falcoz, Massard, and Beau-Faller (2014), and Zhang, Huang, Zhang, et al. (2013)
Extracellular acidosis	Melanoma, breast cancer, colorectal cancer, and others	Enhanced or impaired invasion and migration; increase in MMP/cathepsin-mediated matrix remodeling; altered cell-ECM adhesion and cytoskeletal (actomyosin) dynamics; dependence on substrate and cell type	Estrella et al. (2013), Kato et al. (2007), Martínez-Zaguilán et al. (1996), Stock and Schwab (2009), and Stock et al. (2005)
Peritumoral inflammation	Renal cancer, MMTV transgenic mouse model for mammary tumors	Accelerated migration/intravasation in vivo; association with paracrine CSF-1/EGF signaling loops with perivascular macrophages; EMT induction by TGFβ/TNFα and NF-kB signaling	Condeelis and Pollard (2006), Goswami et al. (2005), Lin, Nguyen, Russell, and Pollard (2001), López-Novoa and Nieto (2009), Wyckoff et al. (2007), and Wyckoff et al. (2004)
Fibrosis, desmoplasia	Pancreatic cancer, lung cancer, breast cancer	Combined effects with tissue remodeling (compare Table 2); enhanced migration, EMT and metastatic evasion by cytokine abundance (eg, TNFα, IL-6, TGFβ, others); cross talk with integrin signaling	López-Novoa and Nieto (2009), Lu, Chang, et al. (2012), and Margadant and Sonnenberg (2010)

interfering with cancer cell migration could serve as preventive or intervention therapy to limit metastatic progression (Cheung & Ewald, 2016; Massagué & Obenauf, 2016). Experimental strategies with the potential to interfere with tumor cell migration programs include inhibitors targeting (i) the actin and MT cytoskeleton, (ii) integrins and downstream adhesion signaling, and (iii) proteases to combat growth factor processing and tissue remodeling.

4.1 Interference with Cytoskeletal Organization and Function

4.1.1 Actin Dynamics

Members of the Rho family of small GTPases, including RhoA, Rac1, and Cdc42, and their downstream effector kinases, such as ROCK, p21-associated kinase (PAK), and MLK, have been implicated as regulators of cytoskeletal dynamics underlying cancer cell migration and metastasis (Guan et al., 2013; Olson & Sahai, 2009; Sadok & Marshall, 2014). Aberrant activity and expression levels of Rho GTPases and their effector kinases deregulate adhesion and cytoskeletal organization and promote invasion in several human cancers, including melanoma, T-cell lymphoma, colon, breast, lung, and gastric carcinoma (Kakiuchi et al., 2014; Kamai et al., 2003; Krauthammer et al., 2012; Y. Lin & Zheng, 2015; Lochhead, Wickman, Mezna, & Olson, 2010; Stransky et al., 2011; Yoo et al., 2014). Beyond their contribution to cell migration, Rho kinases further regulate cell division, apoptosis, and oncogenic transformation (Rath & Olson, 2012). Therefore, targeting Rho GTPase pathways may inhibit cancer cell dynamics, growth, and survival simultaneously.

Rho GTPases themselves are difficult to target directly, therefore inhibitors interfering with Rho and Rac downstream effector kinases, such as ROCK and PAK, were developed (Lin & Zheng, 2015; Prudnikova, Rawat, & Chernoff, 2015). In vitro and preclinical analyses suggest ROCK inhibition as relevant and efficient therapeutic strategy to inhibit cancer cell migration. Accordingly, ROCK inhibition by Y-27632, Fasudil or other small inhibitors, including WF-536, H1152, and RKI-1447, efficiently reduces cell migration, invasion, and growth in vitro and experimental metastasis in vivo in melanoma, lung, and mammary tumor models (Lin & Zheng, 2015; Matsuoka & Yashiro, 2014; Nakajima, Hayashi, Egi, et al., 2003; Nakajima, Hayashi, Katayama, et al., 2003; Patel et al., 2012; Prudnikova et al., 2015; Rath & Olson, 2012; Ying et al., 2006).

CCT129254 and AT13148, new-generation AKT kinase inhibitors which also inhibit ROCK (Lin & Zheng, 2015; Prudnikova et al., 2015), compromise actomyosin contractility in both amoeboid and mesenchymal migration of melanoma cells in vitro and in vivo (Sadok et al., 2015). CCT129254 further reduces spontaneous metastasis and impairs metastatic outgrowth in a mouse model for melanoma (Sadok et al., 2015). With the potential to target both mesenchymal and amoeboid migration, ROCK inhibition may be particularly suited to limit therapy-induced migratory plasticity, and awaits further exploration for efficacy in clinical trials (Sadok et al., 2015; Yap et al., 2012).

Alternative strategies targeting the actin cytoskeleton are currently in preclinical development (Bonello, Stehn, & Gunning, 2009; Nürnberg, Kollmannsperger, & Grosse, 2014). Tropomyosin inhibitors disrupt actin filament integrity and thereby deregulate cell migration and survival (Stehn et al., 2013). For example, in melanoma and neuroblastoma mouse models, treatment with tropomyosin inhibitors inhibits cell migration and growth (Bonello et al., 2016; Stehn et al., 2013). Furthermore, small molecule inhibitors of actin-related protein 2/3 (Arp2/3) (Baggett et al., 2012; Hetrick, Han, Helgeson, & Nolen, 2013), which besides other functions controls lamellipod protrusion during migration (Rotty, Wu, & Bear, 2013), interfere with actin assembly and tumor cell motility (Ilatovskaya et al., 2013; Nürnberg et al., 2014).

4.1.2 Microtubules

MT-targeting agents, including MT-destabilizing agents (eg, vincristine, vinblastine, and combretastatins) and MT-stabilizing agents (eg, paclitaxel, docetaxel, and epothilones) (Fanale et al., 2015; Jordan & Wilson, 2004), are widely used as chemotherapeutics of hematological malignancies and solid tumors (Fanale et al., 2015; Mukhtar, Adhami, & Mukhtar, 2014). MTs control chromosomal segregation and cytokinesis during mitosis in both cancer and stromal cells and contribute to overall tumor growth (Mukhtar et al., 2014). Consequently, MT inhibitors interfere with cell cycle progression and induce apoptosis in cancer cells in vitro (Mukhtar et al., 2014). However, when probed in vivo, mitotic arrest induced by MT inhibitors is moderate, with ongoing proliferation of tumor cell subsets in intact tumors (Komlodi-Pasztor, Sackett, Wilkerson, & Fojo, 2011; Mitchison, 2012; Zasadil et al., 2014).

Besides cell cycle regulation, MTs further control cell migration by several mechanisms. By dynamic instability and by enabling delivery of proteins to cell–substrate interaction sites, MTs establish front-rear polarity, initialize protrusion formation, and control cell–ECM adhesion dynamics and intracellular signaling involved in the migration cycle (Etienne-Manneville, 2013). In transformed fibroblasts and sarcoma cells, MT stabilization induced by the cell cycle regulator $p27^{kip1}$, which disables the MT-destabilizing protein stathmin, favors cell elongation and mesenchymal migration; consequently, decreased MT stability in cells that lack $p27^{kip1}$ favors RhoA signaling and amoeboid movement (Belletti et al., 2008, 2010; Berton et al., 2009). Consistently, treatment of gastric cancer cells with the microtubule-destabilizing agent vincristine induces RhoA/ROCK signaling and stimulates an amoeboid migration program (Eitaki, Yamamori, Meike, Yasui, & Inanami, 2012). Likewise, MT destabilization by combrestatin CA-4-P activates RhoA, induces cytoskeletal reorganization, and destabilizes cell–cell junctions in endothelial cells, with low-adhesive amoeboid cell behavior as outcome (Kanthou & Tozer, 2002). To which extent clinically used MT-targeting agents induce plasticity of invasion in vivo and may impair the clinical efficacy of MT inhibitors remains to be investigated; however, emerging evidence suggests that resistant tumor cell subsets emerge during MT inhibition therapy which develop altered invasive ability (Fanale et al., 2015; Holohan, Van Schaeybroeck, Longley, & Johnston, 2013; Kavallaris, 2010; Kavallaris et al., 2001). For example, resistance to paclitaxel in prostate cancer is associated with a more aggressive and invasive phenotype, parallel to EMT induction (Kim et al., 2013). Whether such plasticity responses can be minimized by a combination of MT-targeting agents with inhibitors of actin and cell adhesion dynamics is currently in preclinical investigation (Dorff & Quinn, 2013; Eitaki et al., 2012).

4.2 Adhesion Systems

4.2.1 Integrins

The multiple functions of integrins in mediating local invasion and metastasis formation, as well as their contribution to cell growth and survival have motivated developments of integrin antagonists to interfere with cancer progression (Blandin et al., 2015; Desgrosellier & Cheresh, 2010). In vitro and vivo inhibition of integrin subsets reveals their critical role in mediating cancer cell adhesion, migration, and metastasis formation in mouse models in several cancer types, including melanoma ($\alpha5\beta1$, $av\beta3$), colorectal cancer ($\alpha6\beta4$, $\alpha v\beta6$), gastric carcinoma ($\alpha6\beta4$), breast and prostate cancer ($\alpha6\beta1$, $\alpha6\beta4$, $\alpha v\beta3$), glioblastoma ($\alpha v\beta3$, $\alpha v\beta5$), and ovarian cancer cells ($\alpha v\beta3$, $\alpha4\beta1$) (Bates et al., 2005; Chao, Lotz, Clarke, & Mercurio, 1996;

Desgrosellier & Cheresh, 2010; Friedl & Wolf, 2003; McCabe, De, Vasanji, Brainard, & Byzova, 2007; Strobel & Cannistra, 1999; Trikha et al., 1997). Besides their functions in tumor cells, integrins expressed by stromal cells, including endothelial cells, fibroblasts, pericytes, and immune cells, contribute to cancer progression (Jin, Aiyer, et al., 2006; Jin, Su, Garmy-Susini, Kleeman, & Varner, 2006).

Multiple integrin inhibitors, including RGD-related peptides which interfere with α5β1, avβ3, and αvβ5, and humanized integrin-blocking antibodies targeting αvβ3 and αvβ5 integrin have been and are still considered in clinical trials (Carter, 2010; MacDonald et al., 2008; McNeel et al., 2005; Mullamitha et al., 2007; Reardon et al., 2008). Despite promising preclinical results, clinical trials on targeting integrins show only a moderate to insufficient therapeutic effect. For example, targeting αvβ3 integrin with vitaxin/etaracizumab in melanoma and other solid tumors (Hersey et al., 2010) or inhibition of αvβ3/αvβ5 integrins by cilengitide in addition to chemotherapy in glioblastoma (Stupp et al., 2014) or recurrent/metastatic squamous cell carcinoma (Vermorken et al., 2014) failed to improve progression-free or overall survival.

The importance of integrin signaling in both tumor and stromal cells infers a complex mode of action of integrin targeting therapy. In mouse models, targeting of αvβ3/αvβ5 integrin can promote angiogenesis and tumor growth, thus disabling potential inhibitory effects of anti-integrin therapy (Desgrosellier & Cheresh, 2010; Reynolds et al., 2002; Taverna et al., 2004). Moreover, the efficacy of integrin monotherapy may suffer from the variability of integrin functions in tumor cells. First, integrin expression varies between tumor types and tumor subclones even within the same tumor lesion (Hoffmann et al., 2005; Taherian, Li, Liu, & Haas, 2011), suggesting the presence of cell subsets with primary unresponsiveness to integrin targeting. Second, the diversity of integrin functions, ranging from migration promotion to stable anchorage, may account for unexpected outcome. In breast cancer cells, for example, α2β1 integrin inhibits migration- and anchorage-independent growth in vitro, and reduces metastasis formation in vivo (Ramirez et al., 2011; Zutter, Santoro, Staatz, & Tsung, 1995), and downregulation of α2 integrin promotes migration and anchorage-independent growth in vitro, and increases tumor growth and metastasis in vivo (Ramirez et al., 2011). Third, compensatory upregulation of alternative integrin subtypes that overlap in ligand spectrum and downstream signaling can rescue integrin function, a common phenomenon in experimental models (Rossi et al., 2010). Fourth, moving cancer cells may switch between integrin-dependent and -independent adhesion systems to maintain migration, thus bypassing the requirement of integrins for cell adhesion and migration. As examples, tumor cells exposed to 3D confinement in vitro are able to move by largely integrin-independent migration modes (Liu et al., 2015) and experimental inhibition of β1 integrins in tumor cells can induce a switch from one migration strategy to another, such as the collective-to-amoeboid transition in primary melanoma explant culture (Hegerfeldt et al., 2002).

While integrin monotherapy is likely obsolete, taking these considerations into account, combinatorial targeting of integrins together with other established oncogenic pathways, eg, ERBB-2 and EGF signaling, could improve therapeutic benefits (Desgrosellier & Cheresh, 2010; Guo et al., 2006; Ricono et al., 2009). The in vivo effects of combining integrin targeting with other strategies, such as cytoskeletal or protease inhibitors, remain to be determined.

4.2.2 Focal Adhesion Kinase

As central downstream effector of integrin signaling, focal adhesion kinase (FAK) also regulates cancer cell adhesion and migration (Chan, Cortesio, & Huttenlocher, 2009; Lawson et al., 2012), and further provides cross talk with oncogenic growth and survival signaling pathways (Sulzmaier, Jean, & Schlaepfer, 2014; Zhao & Guan, 2009). Mechanistically, FAK inhibitors, which impair either FAK kinase or scaffolding function, interfere with (i) integrin function/signaling, and (ii) FAK physical engagement with receptor tyrosine kinases (eg, EGFR) and transcriptional regulators (eg, p53) involved in migration, proliferation, and survival signaling (Sulzmaier et al., 2014; Tavora et al., 2010). In experimental migration assays, FAK inhibition reduces tumor cell migration and invasion as well as survival and anchorage-independent growth in different cancer models, including neuroblastoma, glioblastoma, and kidney cancer cells (Megison et al., 2014; Megison, Stewart, Nabers, Gillory, & Beierle, 2013; Shi et al., 2007; Tancioni et al., 2014). Accordingly, inhibition of FAK in mouse models strongly impairs tumor growth, metastasis and induces stromal normalization in several cancer types (Cabrita et al., 2011; Jean et al., 2014; Walsh et al., 2010; Wendt & Schiemann, 2009).

Given the efficacy of FAK inhibitors on both tumor cells and the tumor stroma, FAK inhibitors are currently being tested in preclinical and clinical trials for several tumor types, including pancreatic cancer, prostate cancer, ovarian cancer, and nonhematological malignancies (Infante et al., 2012;

Megison et al., 2014, 2013; Roberts et al., 2008; Shi et al., 2007; Sulzmaier et al., 2014). Preliminary results suggest that FAK inhibitors PF-00562271 and VS-6063 (PF-04554878) induce stable disease in some cases of advanced solid tumors, including colorectal cancer, prostate cancer, and invasive ductal breast carcinoma (Infante et al., 2012; Jones et al., 2015). The relative contribution of compromised migration and metastasis programs relative to growth and survival functions during FAK inhibition and its efficacy when applied to metastasis prevention schemes remains to be explored.

4.2.3 Src Family Kinases

SFKs, including c-Src, Lyn, Fyn, and Yes, execute pleiotropic functions in regulating adhesion-dependent cell migration, growth, and survival (Guarino, 2010; Mayer & Krop, 2010). Not surprisingly, deregulated SFK expression and activity have been implicated in the development and metastatic progression of multiple solid cancers (Armaiz-Pena et al., 2013; Zhang & Yu, 2012). As a downstream effector kinase of integrin signaling, Src, via RhoA, Rac, MAPK, and PI3K/AKT, controls actomyosin contractility, focal adhesion turnover, and proteolytic breakdown of ECM (Guarino, 2010; Vitale et al., 2008; Zhang, Huang, Zhang, et al., 2013). Further, serving as a downstream effector of cadherin/p120 catenin, Src also regulates cell–cell junction stability in epithelial cancers (Dohn, Brown, & Reynolds, 2009; Veracini et al., 2015). Whether SFKs stabilize or weaken cell–cell contacts is dependent on the tumor type. In a pancreatic cancer model, active Src tends to weaken cell–cell junctions and promotes single-cell migration (Nagathihalli & Merchant, 2012), whereas activation of Src and Yes stabilizes cell–cell junctions and supports collective migration in a squamous cell carcinoma model (Veracini et al., 2015). Thus, Src-related signaling can promote either collective or single-cell migration depending on tumor type and tissue context.

In experimental assays, Src inhibition impairs cell invasion and migration in different cancer types, including epithelial cancer, melanoma, and sarcoma cells (Buettner, Mesa, Vultur, Lee, & Jove, 2008; Je, Ym, Ji, Cho, & Lee, 2014; Sievers et al., 2015). In vivo, Src inhibition strongly reduces tumor growth, metastasis formation, and metastasis outgrowth in models for breast cancer, thyroid cancer, colon cancer, and pancreatic adenocarcinoma (Chan et al., 2012; Jallal et al., 2007; Kopetz et al., 2009; Trevino et al., 2006; Zhang, Huang, Li, et al., 2013). Beyond migration inhibition, interference with Src reduces cell proliferation, induces apoptosis, and also sensitizes cancer cells to chemotherapy and inhibitors of growth factor signaling (eg, EGFR, IGFR) (Chen et al., 2015; Kopetz et al., 2009; Min et al., 2015).

In clinical trials, monotherapy with SFK inhibitors (eg, imatinib, dasatinib saracatinib, bosutinib) induces stable disease in only few cases with otherwise marginal or absent treatment response in advanced solid cancers, including melanoma, breast, prostate, metastatic colorectal, and small-cell lung cancer (Lara et al., 2009; Mackay et al., 2012; Mayer & Krop, 2010; Sharma et al., 2012; Zhang & Yu, 2012). Likewise, combined SFK and MT inhibition failed to improve survival in patients with castration-resistant prostate cancer (Araujo et al., 2009, 2013). Given the clear conceptual and preclinical effects of Src targeting, improved biomarker identification and patient stratification for Src dependence and compensatory oncogenic signaling pathways are currently being explored (Arcaroli et al., 2010; Chen et al., 2009). By resolving cell–cell junctions, inhibition of Src-related signaling can release single cells of otherwise collectively migrating squamous cell carcinoma cells (Veracini et al., 2015), hinting toward potentially cofounding effects of Src targeting in otherwise cohesive tumor types. Whether resistance or therapeutic inefficacy of Src inhibition involves migratory escape is an emerging area of interest.

4.3 Proteases

Given the multiple tumor-progression promoting functions of proteases, including ECM remodeling, proteolytic migration, and the regulation of growth factor and receptor repertoires, MMPs and other proteases (eg, uPA, cathepsins), have been evaluated as promising anticancer targets (Cathcart, Pulkoski-Gross, & Cao, 2015; Deryugina & Quigley, 2006; Itoh et al., 2006). In vitro, interference with MMP activity strongly reduces proteolytic cancer cell invasion and migration when tissue density is high. As examples, inhibition of MT1-MMP, MMP-2, or MMP-9 activity impairs invasion and migration in breast cancer, fibrosarcoma, gastric adenocarcinoma, and glioma cells (Björklund, Heikkilä, & Koivunen, 2004; Itoh et al., 2006; Lakka et al., 2002; Mehner et al., 2014; Ueda, Kajita, Suenaga, Fujii, & Seiki, 2003). In preclinical mouse models, MMP inhibition or genetic interference attenuates tumor growth and metastasis in glioma, and breast cancer (Lakka et al., 2002; Mehner et al., 2014). However, in clinical trials MMP inhibitors failed to improve patient survival or to delay disease progression in several epithelial and mesenchymal tumor types (Cathcart et al., 2015; Fisher & Mobashery, 2006; Hirte et al., 2006; Sparano et al., 2004).

This poor efficacy of MMP inhibition therapy can be explained by different mechanisms of escape as well as conceptual considerations. In experimental models, cancer cells can retain their migratory ability despite MMP inhibition, eg, in moderate density or aligned interstitial tissue (Table 1) or by switching to a protease-independent amoeboid migration mode (Wolf et al., 2013). Hence the efficacy of protease inhibition in slowing metastatic progression may be compromised by the ability of cells to engage in nonproteolytic movements. When combined with Rho/ROCK inhibitors, MMP inhibitors may prevent migratory escape and consequently lower metastatic progression (Mardilovich, Olson, & Baugh, 2012; Somlyo et al., 2003). In addition, while many MMPs are prooncogenic, other MMPs exert tumor suppressive functions, thus rendering general broad-spectrum MMP inhibition obsolete (Garg et al., 2010; López-Otín & Matrisian, 2007). For example, MMP-3 promotes invasion and EMT in breast cancer, but can serve as tumor suppressor in squamous cell carcinoma (McCawley, Crawford, King, Mudgett, & Matrisian, 2004). In certain cancers, including lymphoma, ovarian, and oesophagal cancer, protease inhibitors not only fail to impair, but even enhance migration and accelerate metastatic liver colonization in mouse models (Della Porta et al., 1999; Krüger et al., 2001). Although metastatic progression during targeted-protease therapy in early-stage cancer patients was not tested clinically, the discrepancy between preclinical and clinical results indicates that the complexity and plasticity of protease networks in metastatic disease progression require further conceptual refinement (Deu, Verdoes, & Bogyo, 2012; Haris et al., 2014; Klingler & Hardt, 2012).

4.4 Migration Inhibition by Molecular Targeted Therapies

Molecular therapies targeting intracellular kinase systems often inhibit cancer cell migration, among several other cell functions. Although these "antimigration" bystander effects are not systematically being addressed, the emerging efficiency profile of new drugs may comprise significant antimetastasis components that warrant mechanistic analysis.

4.4.1 TGFβ

As an inducer of EMT, TGFβ promotes cancer cell migration and metastatic growth in several cancer models (Giampieri et al., 2009, 2010; Padua & Massagué, 2009). Additional pleiotropic roles of TGFβ in cancer progression and metastasis formation include mediating angiogenesis, fibrosis, and immune responses (Neuzillet et al., 2015). In vitro, the inhibition of TGFβ reduces migration of epithelial tumor cells when tested in monoculture, but not in coculture with fibroblasts which release HGF (Gaspar et al., 2007; Oyanagi et al., 2014). In preclinical studies, TGFβ inhibition reduces lung and liver metastasis and prolongs survival of mice in colorectal and pancreatic cancer models (Calon et al., 2012; Gaspar et al., 2007; Melisi et al., 2008; Schlingensiepen et al., 2011; Zhang, Halder, Zhang, & Datta, 2009). Consequently, therapeutic inhibition of TGFβ pathways is currently being evaluated in clinical trials (Neuzillet et al., 2015).

4.4.2 cMET

HGFR/cMET signaling has been implicated in cancer cell invasion, proliferation, angiogenesis, and metastasis formation (Merlin et al., 2009; Trusolino et al., 2010), and cMET inhibitors are being tested for targeting invasive and metastatic cancer disease (de Bono & Yap, 2011). cMET inhibition reduces cancer cell proliferation and invasion in vitro, and inhibits tumor growth and metastasis formation in vivo (Corso et al., 2008; Petrelli et al., 2006), with particular efficacy in cancer cells resistant to VEGF inhibition (Sennino et al., 2012). In clinical trials, c-MET inhibition (eg, by tivantinib) delays disease progression in certain cancer types, including metastatic colorectal cancer, gastric cancer, and hepatocellular carcinoma (Kang et al., 2014; Santoro et al., 2013), but not in other cancer types (eg, NSCLC) (Scagliotti et al., 2015; Sharma & Adjei, 2011; Spigel et al., 2013).

To what extent and under which conditions promigratory effects of TGFβ, HGF/cMet and other molecular targeting strategies contribute to antimetastatic outcome remains to be defined.

4.5 Challenges in Deriving and Interpreting "Antimigration" Therapies

Overall, although conceptually attractive, attempts to target cell migration and thereby limit metastasis has not yet reached a satisfying proof-of-concept stage. Several reasons may account for the inefficiency in moving interference strategies from bench to bedside.

4.5.1 Technical Monitoring of Metastasis

Mechanistic interpretation and validation of experimental results in vivo, using preclinical mouse models, is hampered by the multistep complexity of metastasis. Most interference regimens target critical cell functions simultaneously, including cell migration, proliferation, and survival signaling machineries, which cooperate in promoting metastasis. Such oncogenic cross talk complicates the mechanistic understanding of outcome after therapeutically targeting cancer cell migration. This renders most studies inconclusive for translating an observed effect (eg, reduced number of metastases in the lungs) into mechanistic insight, unless every step of the metastatic cascade is monitored, and rate-limiting steps and underlying mechanisms (eg, migration arrest vs growth arrest vs apoptosis induction) are being identified.

4.5.2 Variability and Plasticity of Tumor Models and Cell Subsets

The heterogeneity of molecular processes and dissemination strategies between tumor types and even within the same tumor lesions challenge "one-strategy-fits-all" type molecular targeting (eg, adhesion, proteases, or signaling pathways). Stratifying patient subsets prior to molecular-targeted therapies, as mono- or combination therapy candidates, may improve initial response rates. However, molecular and physical drug-induced reprogramming of cancer invasion and dissemination strategies, resulting from their natural plasticity repertoire and/or cross-talk between migration with proliferation and survival signaling, are currently impossible to predict. Thus, longitudinal biomarkers, which can be monitored to derive adaptations of migration and dissemination programs and the associated molecular signature changes need to be identified and validated in defined migration models and applied in both preclinical and clinical studies.

5. CONCLUSIONS

By combining in vitro, preclinical in vivo and clinical histological and CTC data in different cancer types, preliminary concepts are emerging about the potential contribution of different cancer cell migration modes to local tissue invasion and metastatic disease progression. While collective migration occurs predominantly in epithelial and neurological tumors, single-cell migration is a general hallmark of dedifferentiated cancers and hematologic neoplasia. The ability to transition between both mesenchymal and amoeboid as well as different types of collective invasion modes equips moving cancer cells with a versatile repertoire of dissemination and adaptation strategies.

To successfully interfere with cancer cell migration and its contribution to the metastatic cascade, and to also cope with plasticity responses, likely multiple cellular pathways need to be targeted in parallel, including cell–cell, cell–ECM, and cytoskeletal signaling pathways. In addition, modulatory mechanosignaling from the tumor stroma may offer intervention strategies, which additionally address growth factors and ECM substrate changes by combinatorial targeting with antimigration strategies. To this end, since mechanotransduction and effector pathways control cancer cell migration in cooperation with survival and proliferation signaling (Alexander & Friedl, 2012), mechanistically singular "antimigration" therapies may not exist but leverage from molecular cross talk with therapeutically targeted oncogenic signaling.

6. FUTURE IMPLICATIONS

To better understand whether current anticancer therapies are suited to target cancer migration and metastasis, several measures may be suited to improve mechanistic insight from clinical trials. First, coclinical studies will allow to refine mechanistic effects of molecular-targeted therapies and their impact on the metastatic cascade. This will require orthotopic implantation models in small animals using standard cell lines, patient-derived cells and, ideally, genetically engineered mouse models which reflect the diversity and plasticity of spontaneous cancers. This will allow read-out of each step of the metastatic cascade, including local invasion, intravasation/circulation, and metastatic seeding. Second, routine analyses of the frequency and aggregation type of CTCs at all stages of disease, including the disease-free interval and during treatment phases, are required to deduce and refine concepts on the mechanisms of intra- and extravasation and suited molecular interference. Third, morphometric histological analysis of tumor–stromal interfaces and invasive tumor margins of patient samples of both the primary tumor and follow-up biopsies of metastatic lesions, will allow the spectrum of local invasion and dissemination modes to be mapped, before, during, and after therapeutic intervention. Jointly, these measures will help to stratify the relative contribution of invasion and dissemination modes to metastatic progression and their plasticity response to therapeutic challenge.

ACKNOWLEDGMENTS

We thank Michael Weiger for critical reading of the manuscript. V.tB. is recipient of the Rosali B. Hite Fellowship/MD Anderson Cancer Center; P.F. is supported by the European Research Council (617430-DEEPINSIGHT), NWO-Vici (918.11.626), Horizon 2020 consortium MULTIMOT (634107-2), and the Cancer Genomics Center (CGC.nl).

REFERENCES

Aceto, N., Bardia, A., Miyamoto, D. T., Donaldson, M. C., Wittner, B. S., Spencer, J. A., … Maheswaran, S. (2014). Circulating tumor cell clusters are oligoclonal precursors of breast cancer metastasis. *Cell*, *158*(5), 1110–1122. http://doi.org/10.1016/j.cell.2014.07.013.

Alexander, S., & Friedl, P. (2012). Cancer invasion and resistance: Interconnected processes of disease progression and therapy failure. *Trends in Molecular Medicine, 18*(1), 13–26. http://dx.doi.org/10.1016/j.molmed.2011.11.003.

Alexander, S., Weigelin, B., Winkler, F., & Friedl, P. (2013). Preclinical intravital microscopy of the tumour-stroma interface: Invasion, metastasis, and therapy response. *Current Opinion in Cell Biology, 25*(5), 659–671. http://doi.org/10.1016/j.ceb.2013.07.001.

Araujo, J., Armstrong, A. J., Braud, E. L., Posadas, E., Lonberg, M., Gallick, G. E., ... Logothetis, C. J. (2009). Dasatinib and docetaxel combination treatment for patients with castration-resistant progressive prostate cancer: A phase I/II study (CA180086). *Journal of Oncology, 27*(15S), 5061. Retrieved from, http://hwmaint.meeting.ascopubs.org/cgi/content/abstract/27/15S/5061, ASCO Meeting Abstracts.

Araujo, J. C., Trudel, G. C., Saad, F., Armstrong, A. J., Yu, E. Y., Bellmunt, J., ... Logothetis, C. J. (2013). Docetaxel and dasatinib or placebo in men with metastatic castration-resistant prostate cancer (READY): A randomised, double-blind phase 3 trial. *The Lancet Oncology, 14*(13), 1307–1316. http://doi.org/10.1016/S1470-2045(13)70479-0.

Arcaroli, J. J., Touban, B. M., Tan, A. C., Varella-Garcia, M., Powell, R. W., Eckhardt, S. G., ... Messersmith, W. A. (2010). Gene array and fluorescence in situ hybridization biomarkers of activity of saracatinib (AZD0530), a Src inhibitor, in a preclinical model of colorectal cancer. *Clinical Cancer Research: An Official Journal of the American Association for Cancer Research, 16*(16), 4165–4177. http://doi.org/10.1158/1078-0432.CCR-10-0066.

Armaiz-Pena, G. N., Allen, J. K., Cruz, A., Stone, R. L., Nick, A. M., Lin, Y. G., ... Sood, A. K. (2013). Src activation by β-adrenoreceptors is a key switch for tumour metastasis. *Nature Communications, 4*, 1403. http://doi.org/10.1038/ncomms2413.

Armstrong, A. J., Marengo, M. S., Oltean, S., Kemeny, G., Bitting, R. L., Turnbull, J. D., ... Garcia-Blanco, M. A. (2011). Circulating tumor cells from patients with advanced prostate and breast cancer display both epithelial and mesenchymal markers. *Molecular Cancer Research: MCR, 9*(8), 997–1007. http://dx.doi.org/10.1158/1541-7786.MCR-10-0490.

Avizienyte, E., & Frame, M. C. (2005). Src and FAK signalling controls adhesion fate and the epithelial-to-mesenchymal transition. *Current Opinion in Cell Biology, 17*(5), 542–547. http://doi.org/10.1016/j.ceb.2005.08.007.

Baggett, A. W., Cournia, Z., Han, M. S., Patargias, G., Glass, A. C., Liu, S.-Y., & Nolen, B. J. (2012). Structural characterization and computer-aided optimization of a small-molecule inhibitor of the Arp2/3 complex, a key regulator of the actin cytoskeleton. *ChemMedChem*, 7(7), 1286–1294. http://doi.org/10.1002/cmdc.201200104.

Balcioglu, H. E., van Hoorn, H., Donato, D. M., Schmidt, T., & Danen, E. H. J. (2015). The integrin expression profile modulates orientation and dynamics of force transmission at cell-matrix adhesions. *Journal of Cell Science, 128*(7), 1316–1326. http://doi.org/10.1242/jcs.156950.

Bastounis, E., Meili, R., Álvarez-González, B., Francois, J., del Álamo, J. C., Firtel, R. A., & Lasheras, J. C. (2014). Both contractile axial and lateral traction force dynamics drive amoeboid cell motility. *The Journal of Cell Biology, 204*(6), 1045–1061. http://doi.org/10.1083/jcb.201307106.

Bates, R. C., Bellovin, D. I., Brown, C., Maynard, E., Wu, B., Kawakatsu, H., ... Mercurio, A. M. (2005). Transcriptional activation of integrin β6 during the epithelial-mesenchymal transition defines a novel prognostic indicator of aggressive colon carcinoma. *Journal of Clinical Investigation, 115*(2), 339–347. http://doi.org/10.1172/JCI23183.

Batlle, E., & Wilkinson, D. G. (2012). Molecular mechanisms of cell segregation and boundary formation in development and tumorigenesis. *Cold Spring Harbor Perspectives in Biology, 4*(1), a008227. http://doi.org/10.1101/cshperspect.a008227.

Bazellières, E., Conte, V., Elosegui-Artola, A., Serra-Picamal, X., Bintanel-Morcillo, M., Roca-Cusachs, P., … Trepat, X. (2015). Control of cell-cell forces and collective cell dynamics by the intercellular adhesome. *Nature Cell Biology, 17*(4), 409–420. http://doi.org/10.1038/ncb3135.

Belletti, B., Nicoloso, M. S., Schiappacassi, M., Berton, S., Lovat, F., Wolf, K., … Baldassarre, G. (2008). Stathmin activity influences sarcoma cell shape, motility, and metastatic potential. *Molecular Biology of the Cell, 19*(5), 2003–2013. http://doi.org/10.1091/mbc.E07-09-0894.

Belletti, B., Pellizzari, I., Berton, S., Fabris, L., Wolf, K., Lovat, F., … Baldassarre, G. (2010). p27kip1 controls cell morphology and motility by regulating microtubule-dependent lipid raft recycling. *Molecular and Cellular Biology, 30*(9), 2229–2240. http://doi.org/10.1128/MCB.00723-09.

Berton, S., Belletti, B., Wolf, K., Canzonieri, V., Lovat, F., Vecchione, A., … Baldassarre, G. (2009). The tumor suppressor functions of p27(kip1) include control of the mesenchymal/amoeboid transition. *Molecular and Cellular Biology, 29*(18), 5031–5045. http://doi.org/10.1128/MCB.00144-09.

Björklund, M., Heikkilä, P., & Koivunen, E. (2004). Peptide inhibition of catalytic and noncatalytic activities of matrix metalloproteinase-9 blocks tumor cell migration and invasion. *The Journal of Biological Chemistry, 279*(28), 29589–29597. http://doi.org/10.1074/jbc.M401601200.

Blandin, A.-F., Renner, G., Lehmann, M., Lelong-Rebel, I., Martin, S., & Dontenwill, M. (2015). β1 integrins as therapeutic targets to disrupt Hallmarks of cancer. *Frontiers in Pharmacology, 6*, 279. http://doi.org/10.3389/fphar.2015.00279.

Blaser, H., Reichman-Fried, M., Castanon, I., Dumstrei, K., Marlow, F. L., Kawakami, K., … Raz, E. (2006). Migration of zebrafish primordial germ cells: A role for myosin contraction and cytoplasmic flow. *Developmental Cell, 11*(5), 613–627. http://doi.org/10.1016/j.devcel.2006.09.023.

Bohrer, L. R., & Schwertfeger, K. L. (2012). Macrophages promote fibroblast growth factor receptor-driven tumor cell migration and invasion in a CXCR2-dependent manner. *Molecular Cancer Research: MCR, 10*(10), 1294–1305. http://doi.org/10.1158/1541-7786.MCR-12-0275.

Boimel, P. J., Smirnova, T., Zhou, Z. N., Wyckoff, J., Park, H., Coniglio, S. J., … Segall, J. E. (2012). Contribution of CXCL12 secretion to invasion of breast cancer cells. *Breast Cancer Research: BCR, 14*(1), R23. http://doi.org/10.1186/bcr3108.

Bonello, T. T., Janco, M., Hook, J., Byun, A., Appaduray, M., Dedova, I., … Gunning, P. W. (2016). A small molecule inhibitor of tropomyosin dissociates actin binding from tropomyosin-directed regulation of actin dynamics. *Scientific Reports, 6*, 19816. http://doi.org/10.1038/srep19816.

Bonello, T. T., Stehn, J. R., & Gunning, P. W. (2009). New approaches to targeting the actin cytoskeleton for chemotherapy. *Future Medicinal Chemistry, 1*(7), 1311–1331. http://doi.org/10.4155/fmc.09.99.

Bousquet, E., Calvayrac, O., Mazières, J., Lajoie-Mazenc, I., Boubekeur, N., Favre, G., & Pradines, A. (2016). RhoB loss induces Rac1-dependent mesenchymal cell invasion in lung cells through PP2A inhibition. *Oncogene, 35*(14), 1760–1769. http://doi.org/10.1038/onc.2015.240.

Brand, S., Dambacher, J., Beigel, F., Olszak, T., Diebold, J., Otte, J.-M., … Eichhorst, S. T. (2005). CXCR4 and CXCL12 are inversely expressed in colorectal cancer cells and modulate cancer cell migration, invasion and MMP-9 activation. *Experimental Cell Research, 310*(1), 117–130. http://doi.org/10.1016/j.yexcr.2005.07.006.

Bronsert, P., Enderle-Ammour, K., Bader, M., Timme, S., Kuehs, M., Csanadi, A., … Wellner, U. F. (2014). Cancer cell invasion and EMT marker expression: A three-dimensional study of the human cancer-host interface. *The Journal of Pathology, 234*(3), 410–422. http://doi.org/10.1002/path.4416.

Buettner, R., Mesa, T., Vultur, A., Lee, F., & Jove, R. (2008). Inhibition of Src family kinases with dasatinib blocks migration and invasion of human melanoma cells. *Molecular Cancer Research: MCR*, *6*(11), 1766–1774. http://doi.org/10.1158/1541-7786.MCR-08-0169.

Cabrita, M. A., Jones, L. M., Quizi, J. L., Sabourin, L. A., McKay, B. C., & Addison, C. L. (2011). Focal adhesion kinase inhibitors are potent anti-angiogenic agents. *Molecular Oncology*, *5*(6), 517–526. http://doi.org/10.1016/j.molonc.2011.10.004.

Calon, A., Espinet, E., Palomo-Ponce, S., Tauriello, D. V. F., Iglesias, M., Céspedes, M. V., … Batlle, E. (2012). Dependency of colorectal cancer on a TGF-β-driven program in stromal cells for metastasis initiation. *Cancer Cell*, *22*(5), 571–584. http://doi.org/10.1016/j.ccr.2012.08.013.

Cao, Y., Liu, X., Lu, W., Chen, Y., Wu, X., Li, M., … Liu, Y. (2015). Fibronectin promotes cell proliferation and invasion through mTOR signaling pathway activation in gallbladder cancer. *Cancer Letters*, *360*(2), 141–150. http://doi.org/10.1016/j.canlet.2015.01.041.

Carter, A. (2010). Integrins as target: First phase III trial launches, but questions remain. *Journal of the National Cancer Institute*, *102*(10), 675–677. http://doi.org/10.1093/jnci/djq186.

Cates, J. M. M., Dupont, W. D., Barnes, J. W., Edmunds, H. S., Fasig, J. H., Olson, S. J., & Black, C. C. (2008). Markers of epithelial-mesenchymal transition and epithelial differentiation in sarcomatoid carcinoma: Utility in the differential diagnosis with sarcoma. *Applied Immunohistochemistry & Molecular Morphology: AIMM/Official Publication of the Society for Applied Immunohistochemistry*, *16*(3), 251–262. http://doi.org/10.1097/PAI.0b013e318156e9b4.

Cathcart, J., Pulkoski-Gross, A., & Cao, J. (2015). Targeting matrix metalloproteinases in cancer: Bringing new life to old ideas. *Genes & Diseases*, *2*(1), 26–34. http://doi.org/10.1016/j.gendis.2014.12.002.

Chambers, A. F., Groom, A. C., & MacDonald, I. C. (2002). Dissemination and growth of cancer cells in metastatic sites. Nature Reviews. *Cancer*, *2*(8), 563–572. http://doi.org/10.1038/nrc865.

Chan, K. T., Cortesio, C. L., & Huttenlocher, A. (2009). FAK alters invadopodia and focal adhesion composition and dynamics to regulate breast cancer invasion. *The Journal of Cell Biology*, *185*(2), 357–370. http://doi.org/10.1083/jcb.200809110.

Chan, C. M., Jing, X., Pike, L. A., Zhou, Q., Lim, D.-J., Sams, S. B., … Schweppe, R. E. (2012). Targeted inhibition of Src kinase with dasatinib blocks thyroid cancer growth and metastasis. *Clinical Cancer Research: An Official Journal of the American Association for Cancer Research*, *18*(13), 3580–3591. http://doi.org/10.1158/1078-0432.CCR-11-3359.

Chanrion, M., Kuperstein, I., Barrière, C., El Marjou, F., Cohen, D., Vignjevic, D., … Robine, S. (2014). Concomitant Notch activation and p53 deletion trigger epithelial-to-mesenchymal transition and metastasis in mouse gut. *Nature Communications*, *5*, 5005. http://doi.org/10.1038/ncomms6005.

Chao, C., Lotz, M. M., Clarke, A. C., & Mercurio, A. M. (1996). A function for the integrin alpha6beta4 in the invasive properties of colorectal carcinoma cells. *Cancer Research*, *56*(20), 4811–4819. Retrieved from, http://www.ncbi.nlm.nih.gov/pubmed/8841003.

Chen, Y., Guggisberg, N., Jorda, M., Gonzalez-Angulo, A., Hennessy, B., Mills, G. B., … Slingerland, J. M. (2009). Combined Src and aromatase inhibition impairs human breast cancer growth in vivo and bypass pathways are activated in AZD0530-resistant tumors. *Clinical Cancer Research: An Official Journal of the American Association for Cancer Research*, *15*(10), 3396–3405. http://doi.org/10.1158/1078-0432.CCR-08-3127.

Chen, J., Lan, T., Zhang, W., Dong, L., Kang, N., Fu, M., … Zhan, Q. (2015). Dasatinib enhances cisplatin sensitivity in human esophageal squamous cell carcinoma (ESCC) cells via suppression of PI3K/AKT and Stat3 pathways. *Archives of Biochemistry and Biophysics*, *575*, 38–45. http://doi.org/10.1016/j.abb.2014.11.008.

Cheung, K. J., & Ewald, A. J. (2014). Illuminating breast cancer invasion: Diverse roles for cell-cell interactions. *Current Opinion in Cell Biology*, *30*, 99–111. http://doi.org/10.1016/j.ceb.2014.07.003.

Cheung, K. J., & Ewald, A. J. (2016). A collective route to metastasis: Seeding by tumor cell clusters. *Science*, *352*(6282), 167–169. http://doi.org/10.1126/science.aaf6546.

Cheung, K. J., Gabrielson, E., Werb, Z., & Ewald, A. J. (2013). Collective invasion in breast cancer requires a conserved basal epithelial program. *Cell*, *155*(7), 1639–1651. http://doi.org/10.1016/j.cell.2013.11.029.

Cheung, K. J., Padmanaban, V., Silvestri, V., Schipper, K., Cohen, J. D., Fairchild, A. N., … Ewald, A. J. (2016). Polyclonal breast cancer metastases arise from collective dissemination of keratin 14-expressing tumor cell clusters. *Proceedings of the National Academy of Sciences*, *113*(7), 201508541. http://doi.org/10.1073/pnas.1508541113.

Clark, A. G., & Vignjevic, D. M. (2015). Modes of cancer cell invasion and the role of the microenvironment. *Current Opinion in Cell Biology*, *36*, 13–22. http://doi.org/10.1016/j.ceb.2015.06.004.

Condeelis, J., & Pollard, J. W. (2006). Macrophages: Obligate partners for tumor cell migration, invasion, and metastasis. *Cell*, *124*(2), 263–266. http://doi.org/10.1016/j.cell.2006.01.007.

Condeelis, J., & Segall, J. E. (2003). Intravital imaging of cell movement in tumours. *Nature Reviews. Cancer*, *3*(12), 921–930. http://doi.org/10.1038/nrc1231.

Corso, S., Migliore, C., Ghiso, E., De Rosa, G., Comoglio, P. M., & Giordano, S. (2008). Silencing the MET oncogene leads to regression of experimental tumors and metastases. *Oncogene*, *27*(5), 684–693. http://doi.org/10.1038/sj.onc.1210697.

de Bono, J. S., & Yap, T. A. (2011). c-MET: An exciting new target for anticancer therapy. *Therapeutic Advances in Medical Oncology*, *3*(1 Suppl.), S3–S5. http://doi.org/10.1177/1758834011423402.

De Luca, A., Lamura, L., Gallo, M., Maffia, V., & Normanno, N. (2012). Mesenchymal stem cell-derived interleukin-6 and vascular endothelial growth factor promote breast cancer cell migration. *Journal of Cellular Biochemistry*, *113*(11), 3363–3370. http://doi.org/10.1002/jcb.24212.

de Rooij, J., Kerstens, A., Danuser, G., Schwartz, M. A., & Waterman-Storer, C. M. (2005). Integrin-dependent actomyosin contraction regulates epithelial cell scattering. *The Journal of Cell Biology*, *171*(1), 153–164. http://doi.org/10.1083/jcb.200506152.

Della Porta, P., Soeltl, R., Krell, H. W., Collins, K., O'Donoghue, M., Schmitt, M., & Krüger, A. (1999). Combined treatment with serine protease inhibitor aprotinin and matrix metalloproteinase inhibitor Batimastat (BB-94) does not prevent invasion of human esophageal and ovarian carcinoma cells in vivo. *Anticancer Research*, *19*(5B), 3809–3816. Retrieved from, http://www.ncbi.nlm.nih.gov/pubmed/10628317.

Denais, C. M., Gilbert, R. M., Isermann, P., McGregor, A. L., te Lindert, M., Weigelin, B., … Lammerding, J. (2016). Nuclear envelope rupture and repair during cancer cell migration. *Science*, *352*(6283), 353–358. http://doi.org/10.1126/science.aad7297.

Deryugina, E. I., & Quigley, J. P. (2006). Matrix metalloproteinases and tumor metastasis. *Cancer Metastasis Reviews*, *25*(1), 9–34. http://doi.org/10.1007/s10555-006-7886-9.

Desgrosellier, J. S., & Cheresh, D. A. (2010). Integrins in cancer: Biological implications and therapeutic opportunities. *Nature Reviews. Cancer*, *10*(1), 9–22. http://doi.org/10.1038/nrc2748.

Deu, E., Verdoes, M., & Bogyo, M. (2012). New approaches for dissecting protease functions to improve probe development and drug discovery. *Nature Structural & Molecular Biology*, *19*(1), 9–16. http://doi.org/10.1038/nsmb.2203.

Dirat, B., Bochet, L., Dabek, M., Daviaud, D., Dauvillier, S., Majed, B., ... Muller, C. (2011). Cancer-associated adipocytes exhibit an activated phenotype and contribute to breast cancer invasion. *Cancer Research, 71*(7), 2455–2465. http://doi.org/10.1158/0008-5472.CAN-10-3323.

Dohn, M. R., Brown, M. V., & Reynolds, A. B. (2009). An essential role for p120-catenin in Src- and Rac1-mediated anchorage-independent cell growth. *The Journal of Cell Biology, 184*(3), 437–450. http://doi.org/10.1083/jcb.200807096.

Donà, E., Barry, J. D., Valentin, G., Quirin, C., Khmelinskii, A., Kunze, A., ... Gilmour, D. (2013). Directional tissue migration through a self-generated chemokine gradient. *Nature, 503*(7475), 285–289. http://doi.org/10.1038/nature12635.

Dorff, T. B., & Quinn, D. I. (2013). Speeding dating for docetaxel and recent debutantes in castration-resistant prostate cancer: "Plus or minus" may be a minus. *Annals of Oncology: Official Journal of the European Society for Medical Oncology/ESMO, 24*(2), 270–272. http://doi.org/10.1093/annonc/mds641.

Egeblad, M., Rasch, M. G., & Weaver, V. M. (2010). Dynamic interplay between the collagen scaffold and tumor evolution. *Current Opinion in Cell Biology, 22*(5), 697–706. http://doi.org/10.1016/j.ceb.2010.08.015.

Eitaki, M., Yamamori, T., Meike, S., Yasui, H., & Inanami, O. (2012). Vincristine enhances amoeboid-like motility via GEF-H1/RhoA/ROCK/Myosin light chain signaling in MKN45 cells. *BMC Cancer, 12*, 469. http://doi.org/10.1186/1471-2407-12-469.

El Haibi, C. P., Sharma, P. K., Singh, R., Johnson, P. R., Suttles, J., Singh, S., & Lillard, J. W. (2010). PI3Kp110-, Src-, FAK-dependent and DOCK2-independent migration and invasion of CXCL13-stimulated prostate cancer cells. *Molecular Cancer, 9*, 85. http://doi.org/10.1186/1476-4598-9-85.

Estrella, V., Chen, T., Lloyd, M., Wojtkowiak, J., Cornnell, H. H., Ibrahim-Hashim, A., ... Gillies, R. J. (2013). Acidity generated by the tumor microenvironment drives local invasion. *Cancer Research, 73*(5), 1524–1535. http://doi.org/10.1158/0008-5472.CAN-12-2796.

Etienne-Manneville, S. (2013). Microtubules in cell migration. *Annual Review of Cell and Developmental Biology, 29*, 471–499. http://doi.org/10.1146/annurev-cellbio-101011-155711.

Fackler, O. T., & Grosse, R. (2008). Cell motility through plasma membrane blebbing. *The Journal of Cell Biology, 181*(6), 879–884. http://doi.org/10.1083/jcb.200802081.

Fanale, D., Bronte, G., Passiglia, F., Calò, V., Castiglia, M., Di Piazza, F., ... Bazan, V. (2015). Stabilizing versus destabilizing the microtubules: A double-edge sword for an effective cancer treatment option? *Analytical Cellular Pathology (Amsterdam), 2015*, 690916. http://doi.org/10.1155/2015/690916.

Fidler, I. J. (2003). The pathogenesis of cancer metastasis: The "seed and soil" hypothesis revisited. Nature Reviews. *Cancer, 3*(6), 453–458. http://doi.org/10.1038/nrc1098.

Fisher, J. F., & Mobashery, S. (2006). Recent advances in MMP inhibitor design. *Cancer Metastasis Reviews, 25*(1), 115–136. http://doi.org/10.1007/s10555-006-7894-9.

Friedl, P. (2004). Prespecification and plasticity: Shifting mechanisms of cell migration. *Current Opinion in Cell Biology, 16*(1), 14–23. http://doi.org/10.1016/j.ceb.2003.11.001.

Friedl, P., & Alexander, S. (2011). Cancer invasion and the microenvironment: Plasticity and reciprocity. *Cell, 147*(5), 992–1009. http://doi.org/10.1016/j.cell.2011.11.016.

Friedl, P., Entschladen, F., Conrad, C., Niggemann, B., & Zänker, K. S. (1998). CD4+ T lymphocytes migrating in three-dimensional collagen lattices lack focal adhesions and utilize beta1 integrin-independent strategies for polarization, interaction with collagen fibers and locomotion. *European Journal of Immunology, 28*(8), 2331–2343. http://www.ncbi.nlm.nih.gov/pubmed/9710211.

Friedl, P., & Gilmour, D. (2009). Collective cell migration in morphogenesis, regeneration and cancer. *Nature Reviews. Molecular Cell Biology, 10*(7), 445–457. http://doi.org/10.1038/nrm2720.

Friedl, P., Locker, J., Sahai, E., & Segall, J. E. (2012). Classifying collective cancer cell invasion. *Nature Cell Biology*, *14*(8), 777–783. http://doi.org/10.1038/ncb2548.

Friedl, P., Maaser, K., Klein, C. E., Niggemann, B., Krohne, G., & Zänker, K. S. (1997). Migration of highly aggressive MV3 melanoma cells in 3-dimensional collagen lattices results in local matrix reorganization and shedding of alpha2 and beta1 integrins and CD44. *Cancer Research*, *57*(10), 2061–2070. Retrieved from, http://www.ncbi.nlm.nih.gov/pubmed/9158006.

Friedl, P., Noble, P. B., Walton, P. A., Laird, D. W., Chauvin, P. J., Tabah, R. J., … Zanker, K. S. (1995). Migration of coordinated cell clusters in mesenchymal and epithelial cancer explants in vitro. *Cancer Research*, *55*(20), 4557–4560. Retrieved from, http://cancerres.aacrjournals.org/content/55/20/4557?ijkey=0e566f8ae834977630d86d11aa2d2198e41273f3&keytype2=tf_ipsecsha.

Friedl, P., & Wolf, K. (2003). Tumour-cell invasion and migration: Diversity and escape mechanisms. *Nature Reviews. Cancer*, *3*(5), 362–374. http://doi.org/10.1038/nrc1075.

Friedl, P., & Wolf, K. (2010). Plasticity of cell migration: A multiscale tuning model. *The Journal of Cell Biology*, *188*(1), 11–19. http://doi.org/10.1083/jcb.200909003.

Fujisaki, K., Fujimoto, H., Sangai, T., Nagashima, T., Sakakibara, M., Shiina, N., … Miyazaki, M. (2015). Cancer-mediated adipose reversion promotes cancer cell migration via IL-6 and MCP-1. *Breast Cancer Research and Treatment*, *150*(2), 255–263. http://doi.org/10.1007/s10549-015-3318-2.

Gaggioli, C., Hooper, S., Hidalgo-Carcedo, C., Grosse, R., Marshall, J. F., Harrington, K., & Sahai, E. (2007). Fibroblast-led collective invasion of carcinoma cells with differing roles for RhoGTPases in leading and following cells. *Nature Cell Biology*, *9*(12), 1392–1400. http://doi.org/10.1038/ncb1658.

Garg, P., Sarma, D., Jeppsson, S., Patel, N. R., Gewirtz, A. T., Merlin, D., & Sitaraman, S. V. (2010). Matrix metalloproteinase-9 functions as a tumor suppressor in colitis-associated cancer. *Cancer Research*, *70*(2), 792–801. http://doi.org/10.1158/0008-5472.CAN-09-3166.

Gaspar, N. J., Li, L., Kapoun, A. M., Medicherla, S., Reddy, M., Li, G., … Wong, D. H. (2007). Inhibition of transforming growth factor beta signaling reduces pancreatic adenocarcinoma growth and invasiveness. *Molecular Pharmacology*, *72*(1), 152–161. http://doi.org/10.1124/mol.106.029025.

Geiger, B., Spatz, J. P., & Bershadsky, A. D. (2009). Environmental sensing through focal adhesions. *Nature Reviews. Molecular Cell Biology*, *10*(1), 21–33. http://doi.org/10.1038/nrm2593.

Giampieri, S., Manning, C., Hooper, S., Jones, L., Hill, C. S., & Sahai, E. (2009). Localized and reversible TGFbeta signalling switches breast cancer cells from cohesive to single cell motility. *Nature Cell Biology*, *11*(11), 1287–1296. http://doi.org/10.1038/ncb1973.

Giampieri, S., Pinner, S., & Sahai, E. (2010). Intravital imaging illuminates transforming growth factor beta signaling switches during metastasis. *Cancer Research*, *70*(9), 3435–3439. http://doi.org/10.1158/0008-5472.CAN-10-0466.

Gilkes, D. M., Semenza, G. L., & Wirtz, D. (2014). Hypoxia and the extracellular matrix: Drivers of tumour metastasis. *Nature Reviews. Cancer*, *14*(6), 430–439. http://doi.org/10.1038/nrc3726.

Glentis, A., Gurchenkov, V., & Vignjevic, D. M. (2014). Assembly, heterogeneity, and breaching of the basement membranes. *Cell Adhesion & Migration*, *8*(3), 236–245. http://doi.org/10.4161/cam.28733.

Goel, H. L., & Mercurio, A. M. (2013). VEGF targets the tumour cell. *Nature Reviews. Cancer*, *13*(12), 871–882. http://doi.org/10.1038/nrc3627.

Gonzalez-Moreno, O., Lecanda, J., Green, J. E., Segura, V., Catena, R., Serrano, D., & Calvo, A. (2010). VEGF elicits epithelial-mesenchymal transition (EMT) in prostate intraepithelial neoplasia (PIN)-like cells via an autocrine loop. *Experimental Cell Research*, *316*(4), 554–567. http://doi.org/10.1016/j.yexcr.2009.11.020.

Goswami, S., Sahai, E., Wyckoff, J. B., Cammer, M., Cox, D., Pixley, F. J., … Condeelis, J. S. (2005). Macrophages promote the invasion of breast carcinoma cells via a colony-stimulating factor-1/epidermal growth factor paracrine loop. *Cancer Research*, *65*(12), 5278–5283. http://doi.org/10.1158/0008-5472.CAN-04-1853.
Gritsenko, P. G., Ilina, O., & Friedl, P. (2012). Interstitial guidance of cancer invasion. *The Journal of Pathology*, *226*(2), 185–199. http://doi.org/10.1002/path.3031.
Grünert, S., Jechlinger, M., & Beug, H. (2003). Diverse cellular and molecular mechanisms contribute to epithelial plasticity and metastasis. *Nature Reviews. Molecular Cell Biology*, *4*(8), 657–665. http://doi.org/10.1038/nrm1175.
Guan, R., Xu, X., Chen, M., Hu, H., Ge, H., Wen, S., … Pi, R. (2013). Advances in the studies of roles of Rho/Rho-kinase in diseases and the development of its inhibitors. *European Journal of Medicinal Chemistry*, *70*, 613–622. http://doi.org/10.1016/j.ejmech.2013.10.048.
Guarino, M. (2010). Src signaling in cancer invasion. *Journal of Cellular Physiology*, *223*(1), 14–26. http://doi.org/10.1002/jcp.22011.
Guo, W., Pylayeva, Y., Pepe, A., Yoshioka, T., Muller, W. J., Inghirami, G., & Giancotti, F. G. (2006). Beta 4 integrin amplifies ErbB2 signaling to promote mammary tumorigenesis. *Cell*, *126*(3), 489–502. http://doi.org/10.1016/j.cell.2006.05.047.
Guo, Y., Xu, F., Lu, T., Duan, Z., & Zhang, Z. (2012). Interleukin-6 signaling pathway in targeted therapy for cancer. *Cancer Treatment Reviews*, *38*(7), 904–910. http://doi.org/10.1016/j.ctrv.2012.04.007.
Haeger, A., Krause, M., Wolf, K., & Friedl, P. (2014). Cell jamming: Collective invasion of mesenchymal tumor cells imposed by tissue confinement. *Biochimica et Biophysica Acta*, *1840*(8), 2386–2395. http://doi.org/10.1016/j.bbagen.2014.03.020.
Haeger, A., Wolf, K., Zegers, M. M., & Friedl, P. (2015). Collective cell migration: Guidance principles and hierarchies. *Trends in Cell Biology*, *25*(9), 556–566. http://doi.org/10.1016/j.tcb.2015.06.003.
Han, Y., Luo, Y., Zhao, J., Li, M., & Jiang, Y. (2014). Overexpression of c-Met increases the tumor invasion of human prostate LNCaP cancer cells in vitro and in vivo. *Oncology Letters*, *8*(4), 1618–1624. http://doi.org/10.3892/ol.2014.2390.
Haris, M., Singh, A., Mohammed, I., Ittyerah, R., Nath, K., Nanga, R. P. R., … Reddy, R. (2014). In vivo magnetic resonance imaging of tumor protease activity. *Scientific Reports*, *4*, 6081. http://doi.org/10.1038/srep06081.
Hegerfeldt, Y., Tusch, M., Bröcker, E.-B., & Friedl, P. (2002). Collective cell movement in primary melanoma explants: Plasticity of cell-cell interaction, beta1-integrin function, and migration strategies. *Cancer Research*, *62*(7), 2125–2130. Retrieved from, http://www.ncbi.nlm.nih.gov/pubmed/11929834.
Heit, B., Tavener, S., Raharjo, E., & Kubes, P. (2002). An intracellular signaling hierarchy determines direction of migration in opposing chemotactic gradients. *The Journal of Cell Biology*, *159*(1), 91–102. http://doi.org/10.1083/jcb.200202114.
Hersey, P., Sosman, J., O'Day, S., Richards, J., Bedikian, A., Gonzalez, R., … Kirkwood, J. M. (2010). A randomized phase 2 study of etaracizumab, a monoclonal antibody against integrin alpha(v)beta(3), + or - dacarbazine in patients with stage IV metastatic melanoma. *Cancer*, *116*(6), 1526–1534. http://doi.org/10.1002/cncr.24821.
Hetrick, B., Han, M. S., Helgeson, L. A., & Nolen, B. J. (2013). Small molecules CK-666 and CK-869 inhibit actin-related protein 2/3 complex by blocking an activating conformational change. *Chemistry & Biology*, *20*(5), 701–712. http://doi.org/10.1016/j.chembiol.2013.03.019.
Hirte, H., Vergote, I. B., Jeffrey, J. R., Grimshaw, R. N., Coppieters, S., Schwartz, B., … Seymour, L. (2006). A phase III randomized trial of BAY 12-9566 (tanomastat) as maintenance therapy in patients with advanced ovarian cancer responsive to primary surgery and paclitaxel/platinum containing chemotherapy: A National Cancer Institute of

Canada Clinical Trials G. *Gynecologic Oncology, 102*(2), 300–308. http://doi.org/10.1016/j.ygyno.2005.12.020.
Hoffmann, S., Maschuw, K., Hassan, I., Reckzeh, B., Wunderlich, A., Lingelbach, S., & Zielke, A. (2005). Differential pattern of integrin receptor expression in differentiated and anaplastic thyroid cancer cell lines. *Thyroid: Official Journal of the American Thyroid Association, 15*(9), 1011–1020. http://doi.org/10.1089/thy.2005.15.1011.
Hollier, B. G., Kricker, J. A., Van Lonkhuyzen, D. R., Leavesley, D. I., & Upton, Z. (2008). Substrate-bound insulin-like growth factor (IGF)-I-IGF binding protein-vitronectin-stimulated breast cell migration is enhanced by coactivation of the phosphatidylinositide 3-kinase/AKT pathway by alphav-integrins and the IGF-I receptor. *Endocrinology, 149*(3), 1075–1090. http://doi.org/10.1210/en.2007-0740.
Holohan, C., Van Schaeybroeck, S., Longley, D. B., & Johnston, P. G. (2013). Cancer drug resistance: An evolving paradigm. *Nature Reviews. Cancer, 13*(10), 714–726. http://doi.org/10.1038/nrc3599.
Hu, T.-H., Yao, Y., Yu, S., Han, L.-L., Wang, W.-J., Guo, H., … Nan, K.-J. (2014). SDF-1/CXCR4 promotes epithelial-mesenchymal transition and progression of colorectal cancer by activation of the Wnt/β-catenin signaling pathway. *Cancer Letters, 354*(2), 417–426. http://doi.org/10.1016/j.canlet.2014.08.012.
Hüsemann, Y., Geigl, J. B., Schubert, F., Musiani, P., Meyer, M., Burghart, E., … Klein, C. A. (2008). Systemic spread is an early step in breast cancer. *Cancer Cell, 13*(1), 58–68. http://doi.org/10.1016/j.ccr.2007.12.003.
Hutchinson, C. V., Natarajan, S., Johnson, S. M., Adams, J. A., Rees-Unwin, K. S., & Burthem, J. (2014). Lymphocytes from chronic lymphocytic leukaemia undergo ABL1-linked amoeboid motility and homotypic interaction as an early adaptive change to ex vivo culture. *Experimental Hematology & Oncology, 3*(1), 7. http://doi.org/10.1186/2162-3619-3-7.
Huveneers, S., & Danen, E. H. J. (2009). Adhesion signaling—Crosstalk between integrins, Src and Rho. *Journal of Cell Science, 122*(Pt. 8), 1059–1069. http://doi.org/10.1242/jcs.039446.
Hynes, R. O. (2002). Integrins: Bidirectional, allosteric signaling machines. *Cell, 110*(6), 673–687. Retrieved from, http://www.ncbi.nlm.nih.gov/pubmed/12297042.
Ilatovskaya, D. V., Chubinskiy-Nadezhdin, V., Pavlov, T. S., Shuyskiy, L. S., Tomilin, V., Palygin, O., … Negulyaev, Y. A. (2013). Arp2/3 complex inhibitors adversely affect actin cytoskeleton remodeling in the cultured murine kidney collecting duct M-1 cells. *Cell and Tissue Research, 354*(3), 783–792. http://doi.org/10.1007/s00441-013-1710-y.
Infante, J. R., Camidge, D. R., Mileshkin, L. R., Chen, E. X., Hicks, R. J., Rischin, D., … Siu, L. L. (2012). Safety, pharmacokinetic, and pharmacodynamic phase I dose-escalation trial of PF-00562271, an inhibitor of focal adhesion kinase, in advanced solid tumors. *Journal of Clinical Oncology: Official Journal of the American Society of Clinical Oncology, 30*(13), 1527–1533. http://doi.org/10.1200/JCO.2011.38.9346.
Itoh, Y., Ito, N., Nagase, H., Evans, R. D., Bird, S. A., & Seiki, M. (2006). Cell surface collagenolysis requires homodimerization of the membrane-bound collagenase MT1-MMP. *Molecular Biology of the Cell, 17*(12), 5390–5399. http://doi.org/10.1091/mbc.E06-08-0740.
Jallal, H., Valentino, M.-L., Chen, G., Boschelli, F., Ali, S., & Rabbani, S. A. (2007). A Src/Abl kinase inhibitor, SKI-606, blocks breast cancer invasion, growth, and metastasis in vitro and in vivo. *Cancer Research, 67*(4), 1580–1588. http://doi.org/10.1158/0008-5472.CAN-06-2027.
Je, D. W., Ym, O., Ji, Y. G., Cho, Y., & Lee, D. H. (2014). The inhibition of SRC family kinase suppresses pancreatic cancer cell proliferation, migration, and invasion. *Pancreas, 43*(5), 768–776. http://doi.org/10.1097/MPA.0000000000000103.
Jean, C., Chen, X. L., Nam, J.-O., Tancioni, I., Uryu, S., Lawson, C., … Schlaepfer, D. D. (2014). Inhibition of endothelial FAK activity prevents tumor metastasis by enhancing

barrier function. *The Journal of Cell Biology*, *204*(2), 247–263. http://doi.org/10.1083/jcb.201307067.

Jin, H., Aiyer, A., Su, J., Borgstrom, P., Stupack, D., Friedlander, M., & Varner, J. (2006). A homing mechanism for bone marrow-derived progenitor cell recruitment to the neovasculature. *The Journal of Clinical Investigation*, *116*(3), 652–662. http://doi.org/10.1172/JCI24751.

Jin, F., Brockmeier, U., Otterbach, F., & Metzen, E. (2012). New insight into the SDF-1/CXCR4 axis in a breast carcinoma model: Hypoxia-induced endothelial SDF-1 and tumor cell CXCR4 are required for tumor cell intravasation. *Molecular Cancer Research: MCR*, *10*(8), 1021–1031. http://doi.org/10.1158/1541-7786.MCR-11-0498.

Jin, H., Su, J., Garmy-Susini, B., Kleeman, J., & Varner, J. (2006). Integrin alpha4beta1 promotes monocyte trafficking and angiogenesis in tumors. *Cancer Research*, *66*(4), 2146–2152. http://doi.org/10.1158/0008-5472.CAN-05-2704.

Jolly, M. K., Boareto, M., Huang, B., Jia, D., Lu, M., Ben-Jacob, E., … Levine, H. (2015). Implications of the hybrid epithelial/mesenchymal phenotype in metastasis. *Frontiers in Oncology*, *5*, 155. http://doi.org/10.3389/fonc.2015.00155.

Jones, S. F., Siu, L. L., Bendell, J. C., Cleary, J. M., Razak, A. R. A., Infante, J. R., … Shapiro, G. I. (2015). A phase I study of VS-6063, a second-generation focal adhesion kinase inhibitor, in patients with advanced solid tumors. *Investigational New Drugs*, *33*(5), 1100–1107. http://doi.org/10.1007/s10637-015-0282-y.

Jordan, M. A., & Wilson, L. (2004). Microtubules as a target for anticancer drugs. *Nature Reviews. Cancer*, *4*(4), 253–265. http://doi.org/10.1038/nrc1317.

Joyce, J. A., & Pollard, J. W. (2009). Microenvironmental regulation of metastasis. *Nature Reviews. Cancer*, *9*(4), 239–252. http://doi.org/10.1038/nrc2618.

Kakiuchi, M., Nishizawa, T., Ueda, H., Gotoh, K., Tanaka, A., Hayashi, A., … Ishikawa, S. (2014). Recurrent gain-of-function mutations of RHOA in diffuse-type gastric carcinoma. *Nature Genetics*, *46*(6), 583–587. http://doi.org/10.1038/ng.2984.

Kamai, T., Tsujii, T., Arai, K., Takagi, K., Asami, H., Ito, Y., & Oshima, H. (2003). Significant association of Rho/ROCK pathway with invasion and metastasis of bladder cancer. *Clinical Cancer Research: An Official Journal of the American Association for Cancer Research*, *9*(7), 2632–2641. Retrieved from, http://www.ncbi.nlm.nih.gov/pubmed/12855641.

Kang, Y.-K., Muro, K., Ryu, M.-H., Yasui, H., Nishina, T., Ryoo, B.-Y., … Boku, N. (2014). A phase II trial of a selective c-Met inhibitor tivantinib (ARQ 197) monotherapy as a second- or third-line therapy in the patients with metastatic gastric cancer. *Investigational New Drugs*, *32*(2), 355–361. http://doi.org/10.1007/s10637-013-0057-2.

Kania, A., & Klein, R. (2016). Mechanisms of ephrin-Eph signalling in development, physiology and disease. *Nature Reviews. Molecular Cell Biology*, *17*(4), 240–256. http://doi.org/10.1038/nrm.2015.16.

Kanthou, C., & Tozer, G. M. (2002). The tumor vascular targeting agent combretastatin A-4-phosphate induces reorganization of the actin cytoskeleton and early membrane blebbing in human endothelial cells. *Blood*, *99*(6), 2060–2069. Retrieved from, http://www.ncbi.nlm.nih.gov/pubmed/11877280.

Kato, Y., Ozawa, S., Tsukuda, M., Kubota, E., Miyazaki, K., St-Pierre, Y., & Hata, R.-I. (2007). Acidic extracellular pH increases calcium influx-triggered phospholipase D activity along with acidic sphingomyelinase activation to induce matrix metalloproteinase-9 expression in mouse metastatic melanoma. *The FEBS Journal*, *274*(12), 3171–3183. http://doi.org/10.1111/j.1742-4658.2007.05848.x.

Katsuno, Y., Lamouille, S., & Derynck, R. (2013). TGF-β signaling and epithelial-mesenchymal transition in cancer progression. *Current Opinion in Oncology*, *25*(1), 76–84. http://doi.org/10.1097/CCO.0b013e32835b6371.

Kavallaris, M. (2010). Microtubules and resistance to tubulin-binding agents. *Nature Reviews. Cancer*, *10*(3), 194–204. http://doi.org/10.1038/nrc2803.
Kavallaris, M., Tait, A. S., Walsh, B. J., He, L., Horwitz, S. B., Norris, M. D., & Haber, M. (2001). Multiple microtubule alterations are associated with Vinca alkaloid resistance in human leukemia cells. *Cancer Research*, *61*(15), 5803–5809. Retrieved from, http://cancerres.aacrjournals.org/content/61/15/5803.short.
Kermorgant, S., Aparicio, T., Dessirier, V., Lewin, M. J., & Lehy, T. (2001). Hepatocyte growth factor induces colonic cancer cell invasiveness via enhanced motility and protease overproduction. Evidence for PI3 kinase and PKC involvement. *Carcinogenesis*, *22*(7), 1035–1042. Retrieved from, http://www.ncbi.nlm.nih.gov/pubmed/11408346.
Kessenbrock, K., Plaks, V., & Werb, Z. (2010). Matrix metalloproteinases: Regulators of the tumor microenvironment. *Cell*, *141*(1), 52–67. http://doi.org/10.1016/j.cell.2010.03.015.
Khalil, A. A., & Friedl, P. (2010). Determinants of leader cells in collective cell migration. *Integrative Biology: Quantitative Biosciences from Nano to Macro*, *2*(11-12), 568–574. http://doi.org/10.1039/c0ib00052c.
Kim, J. J., Yin, B., Christudass, C. S., Terada, N., Rajagopalan, K., Fabry, B., … Mooney, S. M. (2013). Acquisition of paclitaxel resistance is associated with a more aggressive and invasive phenotype in prostate cancer. *Journal of Cellular Biochemistry*, *114*(6), 1286–1293. http://doi.org/10.1002/jcb.24464.
Klein, C. A. (2009). Parallel progression of primary tumours and metastases. *Nature Reviews. Cancer*, *9*(4), 302–312. http://doi.org/10.1038/nrc2627.
Klingler, D., & Hardt, M. (2012). Profiling protease activities by dynamic proteomics workflows. *Proteomics*, *12*(4-5), 587–596. http://doi.org/10.1002/pmic.201100399.
Knuchel, S., Anderle, P., Werfelli, P., Diamantis, E., & Rüegg, C. (2015). Fibroblast surface-associated FGF-2 promotes contact-dependent colorectal cancer cell migration and invasion through FGFR-SRC signaling and integrin αvβ5-mediated adhesion. *Oncotarget*, *6*(16), 14300–14317. http://doi.org/10.18632/oncotarget.3883.
Kojima, Y., Acar, A., Eaton, E. N., Mellody, K. T., Scheel, C., Ben-Porath, I., … Orimo, A. (2010). Autocrine TGF- and stromal cell-derived factor-1 (SDF-1) signaling drives the evolution of tumor-promoting mammary stromal myofibroblasts. *Proceedings of the National Academy of Sciences*, *107*(46), 20009–20014. http://doi.org/10.1073/pnas.1013805107.
Komlodi-Pasztor, E., Sackett, D., Wilkerson, J., & Fojo, T. (2011). Mitosis is not a key target of microtubule agents in patient tumors. *Nature Reviews. Clinical Oncology*, *8*(4), 244–250. http://doi.org/10.1038/nrclinonc.2010.228.
Kopetz, S., Lesslie, D. P., Dallas, N. A., Park, S. I., Johnson, M., Parikh, N. U., … Gallick, G. E. (2009). Synergistic activity of the SRC family kinase inhibitor dasatinib and oxaliplatin in colon carcinoma cells is mediated by oxidative stress. *Cancer Research*, *69*(9), 3842–3849. http://doi.org/10.1158/0008-5472.CAN-08-2246.
Kraan, J., Sleijfer, S., Strijbos, M. H., Ignatiadis, M., Peeters, D., Pierga, J.-Y., … Gratama, J. W. (2011). External quality assurance of circulating tumor cell enumeration using the Cell Search(®) system: A feasibility study. *Cytometry. Part B, Clinical Cytometry*, *80*(2), 112–118. http://doi.org/10.1002/cyto.b.20573.
Krauthammer, M., Kong, Y., Ha, B. H., Evans, P., Bacchiocchi, A., McCusker, J. P., … Halaban, R. (2012). Exome sequencing identifies recurrent somatic RAC1 mutations in melanoma. *Nature Genetics*, *44*(9), 1006–1014. http://doi.org/10.1038/ng.2359.
Krüger, A., Soeltl, R., Sopov, I., Kopitz, C., Arlt, M., Magdolen, V., … Schmitt, M. (2001). Hydroxamate-type matrix metalloproteinase inhibitor batimastat promotes liver metastasis. *Cancer Research*, *61*(4), 1272–1275. Retrieved from, http://www.ncbi.nlm.nih.gov/pubmed/11245418.

Kumar, S., Kapoor, A., Desai, S., Inamdar, M. M., & Sen, S. (2016). Proteolytic and non-proteolytic regulation of collective cell invasion: Tuning by ECM density and organization. *Scientific Reports*, *6*, 19905. http://doi.org/10.1038/srep19905.

Lakka, S. S., Rajan, M., Gondi, C., Yanamandra, N., Chandrasekar, N., Jasti, S. L., … Rao, J. S. (2002). Adenovirus-mediated expression of antisense MMP-9 in glioma cells inhibits tumor growth and invasion. *Oncogene*, *21*(52), 8011–8019. http://doi.org/10.1038/sj.onc.1205894.

Lamouille, S., Xu, J., & Derynck, R. (2014). Molecular mechanisms of epithelial-mesenchymal transition. *Nature Reviews. Molecular Cell Biology*, *15*(3), 178–196. http://doi.org/10.1038/nrm3758.

Landskron, G., De la Fuente, M., Thuwajit, P., Thuwajit, C., & Hermoso, M. A. (2014). Chronic inflammation and cytokines in the tumor microenvironment. *International Journal of Immunology Research*, *2014*(2), 149185. http://doi.org/10.1155/2014/149185.

Lara, P. N., Longmate, J., Evans, C. P., Quinn, D. I., Twardowski, P., Chatta, G., … Gandara, D. R. (2009). A phase II trial of the Src-kinase inhibitor AZD0530 in patients with advanced castration-resistant prostate cancer: A California Cancer Consortium study. *Anti-Cancer Drugs*, *20*(3), 179–184. http://doi.org/10.1097/CAD.0b013e328325a867.

Lawson, C., Lim, S.-T., Uryu, S., Chen, X. L., Calderwood, D. A., & Schlaepfer, D. D. (2012). FAK promotes recruitment of talin to nascent adhesions to control cell motility. *The Journal of Cell Biology*, *196*(2), 223–232. http://doi.org/10.1083/jcb.201108078.

Lee, Y., Jung, W. H., & Koo, J. S. (2015). Adipocytes can induce epithelial-mesenchymal transition in breast cancer cells. *Breast Cancer Research and Treatment*, *153*(2), 323–335. http://doi.org/10.1007/s10549-015-3550-9.

Levental, K. R., Yu, H., Kass, L., Lakins, J. N., Egeblad, M., Erler, J. T., … Weaver, V. M. (2009). Matrix crosslinking forces tumor progression by enhancing integrin signaling. *Cell*, *139*(5), 891–906. http://doi.org/10.1016/j.cell.2009.10.027.

Li, Z., Zhou, Z., & Donahue, H. J. (2008). Alterations in Cx43 and OB-cadherin affect breast cancer cell metastatic potential. *Clinical & Experimental Metastasis*, *25*(3), 265–272. http://doi.org/10.1007/s10585-007-9140-4.

Lin, E. Y., Nguyen, A. V., Russell, R. G., & Pollard, J. W. (2001). Colony-stimulating factor 1 promotes progression of mammary tumors to malignancy. *The Journal of Experimental Medicine*, *193*(6), 727–740. http://doi.org/10.1084/jem.193.6.727.

Lin, Y., & Zheng, Y. (2015). Approaches of targeting Rho GTPases in cancer drug discovery. *Expert Opinion on Drug Discovery*, *10*, 991–1010. Informa Healthcare.

Liotta, L. A., Saidel, M. G., & Kleinerman, J. (1976). The significance of hematogenous tumor cell clumps in the metastatic process. *Cancer Research*, *36*(3), 889–894. Retrieved from, http://www.ncbi.nlm.nih.gov/pubmed/1253177.

Liu, Y.-J., Le Berre, M., Lautenschlaeger, F., Maiuri, P., Callan-Jones, A., Heuzé, M., … Piel, M. (2015). Confinement and low adhesion induce fast amoeboid migration of slow mesenchymal cells. *Cell*, *160*(4), 659–672. http://doi.org/10.1016/j.cell.2015.01.007.

Lochhead, P. A., Wickman, G., Mezna, M., & Olson, M. F. (2010). Activating ROCK1 somatic mutations in human cancer. *Oncogene*, *29*(17), 2591–2598. http://doi.org/10.1038/onc.2010.3.

López-Novoa, J. M., & Nieto, M. A. (2009). Inflammation and EMT: An alliance towards organ fibrosis and cancer progression. *EMBO Molecular Medicine*, *1*(6–7), 303–314. http://doi.org/10.1002/emmm.200900043.

López-Otín, C., & Matrisian, L. M. (2007). Emerging roles of proteases in tumour suppression. *Nature Reviews. Cancer*, 7(10), 800–808. http://doi.org/10.1038/nrc2228.

Lorentzen, A., Bamber, J., Sadok, A., Elson-Schwab, I., & Marshall, C. J. (2011). An ezrin-rich, rigid uropod-like structure directs movement of amoeboid blebbing cells. *Journal of Cell Science*, *124*(Pt. 8), 1256–1267. http://doi.org/10.1242/jcs.074849.

Lu, K. V., Chang, J. P., Parachoniak, C. A., Pandika, M. M., Aghi, M. K., Meyronet, D., ... Bergers, G. (2012). VEGF inhibits tumor cell invasion and mesenchymal transition through a MET/VEGFR2 complex. *Cancer Cell, 22*(1), 21–35. http://doi.org/10.1016/j.ccr.2012.05.037.

Lu, X., & Kang, Y. (2010). Hypoxia and hypoxia-inducible factors: Master regulators of metastasis. *Clinical Cancer Research: An Official Journal of the American Association for Cancer Research, 16*(24), 5928–5935. http://doi.org/10.1158/1078-0432.CCR-10-1360.

Lu, P., Weaver, V. M., & Werb, Z. (2012). The extracellular matrix: A dynamic niche in cancer progression. *The Journal of Cell Biology, 196*(4), 395–406. http://doi.org/10.1083/jcb.201102147.

Lundgren, K., Nordenskjöld, B., & Landberg, G. (2009). Hypoxia, Snail and incomplete epithelial-mesenchymal transition in breast cancer. *British Journal of Cancer, 101*(10), 1769–1781. http://doi.org/10.1038/sj.bjc.6605369.

MacDonald, T. J., Stewart, C. F., Kocak, M., Goldman, S., Ellenbogen, R. G., Phillips, P., ... Kun, L. E. (2008). Phase I clinical trial of cilengitide in children with refractory brain tumors: Pediatric Brain Tumor Consortium Study PBTC-012. *Journal of Clinical Oncology: Official Journal of the American Society of Clinical Oncology, 26*(6), 919–924. http://doi.org/10.1200/JCO.2007.14.1812.

Mackay, H. J., Au, H. J., McWhirter, E., Alcindor, T., Jarvi, A., MacAlpine, K., ... Oza, A. M. (2012). A phase II trial of the Src kinase inhibitor saracatinib (AZD0530) in patients with metastatic or locally advanced gastric or gastro esophageal junction (GEJ) adenocarcinoma: A trial of the PMH phase II consortium. *Investigational New Drugs, 30*(3), 1158–1163. http://doi.org/10.1007/s10637-011-9650-4.

Madsen, C. D., & Sahai, E. (2010). Cancer dissemination—Lessons from leukocytes. *Developmental Cell, 19*(1), 13–26. http://doi.org/10.1016/j.devcel.2010.06.013.

Mamuya, F. A., & Duncan, M. K. (2012). aV integrins and TGF-β-induced EMT: A circle of regulation. *Journal of Cellular and Molecular Medicine, 16*(3), 445–455. http://doi.org/10.1111/j.1582-4934.2011.01419.x.

Mardilovich, K., Olson, M. F., & Baugh, M. (2012). Targeting Rho GTPase signaling for cancer therapy. *Future Oncology (London, England), 8*(2), 165–177. http://doi.org/10.2217/fon.11.143.

Margadant, C., & Sonnenberg, A. (2010). Integrin-TGF-beta crosstalk in fibrosis, cancer and wound healing. *EMBO Reports, 11*(2), 97–105. http://doi.org/10.1038/embor.2009.276.

Martínez-Zaguilán, R., Seftor, E. A., Seftor, R. E., Chu, Y. W., Gillies, R. J., & Hendrix, M. J. (1996). Acidic pH enhances the invasive behavior of human melanoma cells. *Clinical & Experimental Metastasis, 14*(2), 176–186. Retrieved from, http://www.ncbi.nlm.nih.gov/pubmed/8605731.

Massagué, J., & Obenauf, A. C. (2016). Metastatic colonization by circulating tumour cells. *Nature, 529*(7586), 298–306. http://doi.org/10.1038/nature17038.

Matise, L. A., Palmer, T. D., Ashby, W. J., Nashabi, A., Chytil, A., Aakre, M., ... Moses, H. L. (2012). Lack of transforming growth factor-β signaling promotes collective cancer cell invasion through tumor-stromal crosstalk. *Breast Cancer Research: BCR, 14*(4), R98. http://doi.org/10.1186/bcr3217.

Matsuoka, T., & Yashiro, M. (2014). Rho/ROCK signaling in motility and metastasis of gastric cancer. *World Journal of Gastroenterology, 20*(38), 13756–13766. http://doi.org/10.3748/wjg.v20.i38.13756.

Mayer, E. L., & Krop, I. E. (2010). Advances in targeting SRC in the treatment of breast cancer and other solid malignancies. *Clinical Cancer Research: An Official Journal of the American Association for Cancer Research, 16*(14), 3526–3532. http://doi.org/10.1158/1078-0432.CCR-09-1834.

Mayor, R., & Etienne-Manneville, S. (2016). The front and rear of collective cell migration. *Nature Reviews. Molecular Cell Biology*, *17*(2), 97–109. http://doi.org/10.1038/nrm.2015.14.

McCabe, N. P., De, S., Vasanji, A., Brainard, J., & Byzova, T. V. (2007). Prostate cancer specific integrin alphavbeta3 modulates bone metastatic growth and tissue remodeling. *Oncogene*, *26*(42), 6238–6243. http://doi.org/10.1038/sj.onc.1210429.

McCawley, L. J., Crawford, H. C., King, L. E., Mudgett, J., & Matrisian, L. M. (2004). A protective role for matrix metalloproteinase-3 in squamous cell carcinoma. *Cancer Research*, *64*(19), 6965–6972. http://doi.org/10.1158/0008-5472.CAN-04-0910.

McInroy, L., & Määttä, A. (2007). Down-regulation of vimentin expression inhibits carcinoma cell migration and adhesion. *Biochemical and Biophysical Research Communications*, *360*(1), 109–114. http://doi.org/10.1016/j.bbrc.2007.06.036.

McNeel, D. G., Eickhoff, J., Lee, F. T., King, D. M., Alberti, D., Thomas, J. P., … Wilding, G. (2005). Phase I trial of a monoclonal antibody specific for alphavbeta3 integrin (MEDI-522) in patients with advanced malignancies, including an assessment of effect on tumor perfusion. *Clinical Cancer Research: An Official Journal of the American Association for Cancer Research*, *11*(21), 7851–7860. http://doi.org/10.1158/1078-0432.CCR-05-0262.

Megison, M. L., Gillory, L. A., Stewart, J. E., Nabers, H. C., Mrozcek-Musulman, E., & Beierle, E. A. (2014). FAK inhibition abrogates the malignant phenotype in aggressive pediatric renal tumors. *Molecular Cancer Research: MCR*, *12*(4), 514–526. http://doi.org/10.1158/1541-7786.MCR-13-0505.

Megison, M. L., Stewart, J. E., Nabers, H. C., Gillory, L. A., & Beierle, E. A. (2013). FAK inhibition decreases cell invasion, migration and metastasis in MYCN amplified neuroblastoma. *Clinical & Experimental Metastasis*, *30*(5), 555–568. http://doi.org/10.1007/s10585-012-9560-7.

Mehner, C., Hockla, A., Miller, E., Ran, S., Radisky, D. C., & Radisky, E. S. (2014). Tumor cell-produced matrix metalloproteinase 9 (MMP-9) drives malignant progression and metastasis of basal-like triple negative breast cancer. *Oncotarget*, *5*(9), 2736–2749. http://doi.org/10.18632/oncotarget.1932.

Melisi, D., Ishiyama, S., Sclabas, G. M., Fleming, J. B., Xia, Q., Tortora, G., … Chiao, P. J. (2008). LY2109761, a novel transforming growth factor beta receptor type I and type II dual inhibitor, as a therapeutic approach to suppressing pancreatic cancer metastasis. *Molecular Cancer Therapeutics*, 7(4), 829–840. http://doi.org/10.1158/1535-7163.MCT-07-0337.

Meng, W., & Takeichi, M. (2009). Adherens junction: Molecular architecture and regulation. *Cold Spring Harbor Perspectives in Biology*, *1*(6), a002899. http://doi.org/10.1101/cshperspect.a002899.

Merlin, S., Pietronave, S., Locarno, D., Valente, G., Follenzi, A., & Prat, M. (2009). Deletion of the ectodomain unleashes the transforming, invasive, and tumorigenic potential of the MET oncogene. *Cancer Science*, *100*(4), 633–638. http://doi.org/10.1111/j.1349-7006.2008.01079.x.

Min, H.-Y., Yun, H. J., Lee, J.-S., Lee, H.-J., Cho, J., Jang, H.-J., … Lee, H.-Y. (2015). Targeting the insulin-like growth factor receptor and Src signaling network for the treatment of non-small cell lung cancer. *Molecular Cancer*, *14*(1), 113. http://doi.org/10.1186/s12943-015-0392-3.

Mitchison, T. J. (2012). The proliferation rate paradox in antimitotic chemotherapy. *Molecular Biology of the Cell*, *23*(1), 1–6. http://doi.org/10.1091/mbc.E10-04-0335.

Moreno-Bueno, G., Peinado, H., Molina, P., Olmeda, D., Cubillo, E., Santos, V., … Cano, A. (2009). The morphological and molecular features of the epithelial-to-mesenchymal transition. *Nature Protocols*, *4*(11), 1591–1613. http://doi.org/10.1038/nprot.2009.152.

Moustakas, A., & Heldin, C.-H. (2007). Signaling networks guiding epithelial-mesenchymal transitions during embryogenesis and cancer progression. *Cancer Science*, *98*(10), 1512–1520. http://doi.org/10.1111/j.1349-7006.2007.00550.x.

Mukhtar, E., Adhami, V. M., & Mukhtar, H. (2014). Targeting microtubules by natural agents for cancer therapy. *Molecular Cancer Therapeutics*, *13*(2), 275–284. http://doi.org/10.1158/1535-7163.MCT-13-0791.

Mullamitha, S. A., Ton, N. C., Parker, G. J. M., Jackson, A., Julyan, P. J., Roberts, C., … Jayson, G. C. (2007). Phase I evaluation of a fully human anti-alphav integrin monoclonal antibody (CNTO 95) in patients with advanced solid tumors. *Clinical Cancer Research: An Official Journal of the American Association for Cancer Research*, *13*(7), 2128–2135. http://doi.org/10.1158/1078-0432.CCR-06-2779.

Murphy, G. (2008). The ADAMs: Signalling scissors in the tumour microenvironment. *Nature Reviews. Cancer*, *8*(12), 929–941. http://doi.org/10.1038/nrc2459.

Nagathihalli, N. S., & Merchant, N. B. (2012). Src-mediated regulation of E-cadherin and EMT in pancreatic cancer. *Frontiers in Bioscience (Landmark Edition)*, *17*, 2059–2069. Retrieved from, http://www.ncbi.nlm.nih.gov/pubmed/22652764.

Nakajima, M., Hayashi, K., Egi, Y., Katayama, K., Amano, Y., Uehata, M., … Kondo, T. (2003). Effect of Wf-536, a novel ROCK inhibitor, against metastasis of B16 melanoma. *Cancer Chemotherapy and Pharmacology*, *52*(4), 319–324. http://doi.org/10.1007/s00280-003-0641-9.

Nakajima, M., Hayashi, K., Katayama, K., Amano, Y., Egi, Y., Uehata, M., … Kondo, T. (2003). Wf-536 prevents tumor metastasis by inhibiting both tumor motility and angiogenic actions. *European Journal of Pharmacology*, *459*(2–3), 113–120. Retrieved from, http://www.ncbi.nlm.nih.gov/pubmed/12524136.

Neuzillet, C., Tijeras-Raballand, A., Cohen, R., Cros, J., Faivre, S., Raymond, E., & de Gramont, A. (2015). Targeting the TGFβ pathway for cancer therapy. *Pharmacology & Therapeutics*, *147*, 22–31. http://doi.org/10.1016/j.pharmthera.2014.11.001.

Nguyen, D. X., Bos, P. D., & Massagué, J. (2009). Metastasis: From dissemination to organ-specific colonization. *Nature Reviews. Cancer*, *9*(4), 274–284. http://doi.org/10.1038/nrc2622.

Nürnberg, A., Kollmannsperger, A., & Grosse, R. (2014). Pharmacological inhibition of actin assembly to target tumor cell motility. *Reviews of Physiology, Biochemistry and Pharmacology*, *166*, 23–42. http://doi.org/10.1007/112_2013_16.

O'Connor, K., & Chen, M. (2013). Dynamic functions of RhoA in tumor cell migration and invasion. *Small GTPases*, *4*(3), 141–147. http://doi.org/10.4161/sgtp.25131.

Olson, M. F., & Sahai, E. (2009). The actin cytoskeleton in cancer cell motility. *Clinical & Experimental Metastasis*, *26*(4), 273–287. http://doi.org/10.1007/s10585-008-9174-2.

Orian-Rousseau, V., Morrison, H., Matzke, A., Kastilan, T., Pace, G., Herrlich, P., & Ponta, H. (2007). Hepatocyte growth factor-induced Ras activation requires ERM proteins linked to both CD44v6 and F-actin. *Molecular Biology of the Cell*, *18*(1), 76–83. http://doi.org/10.1091/mbc.E06-08-0674.

Osswald, M., Jung, E., Sahm, F., Solecki, G., Venkataramani, V., Blaes, J., … Winkler, F. (2015). Brain tumour cells interconnect to a functional and resistant network. *Nature*, *528*(7580), 93–98. http://doi.org/10.1038/nature16071.

Oyanagi, J., Kojima, N., Sato, H., Higashi, S., Kikuchi, K., Sakai, K., … Miyazaki, K. (2014). Inhibition of transforming growth factor-β signaling potentiates tumor cell invasion into collagen matrix induced by fibroblast-derived hepatocyte growth factor. *Experimental Cell Research*, *326*(2), 267–279. http://doi.org/10.1016/j.yexcr.2014.04.009.

Padua, D., & Massagué, J. (2009). Roles of TGFbeta in metastasis. *Cell Research*, *19*(1), 89–102. http://doi.org/10.1038/cr.2008.316.

Paluch, E., Piel, M., Prost, J., Bornens, M., & Sykes, C. (2005). Cortical actomyosin breakage triggers shape oscillations in cells and cell fragments. *Biophysical Journal, 89*(1), 724–733. http://doi.org/10.1529/biophysj.105.060590.

Paluch, E. K., & Raz, E. (2013). The role and regulation of blebs in cell migration. *Current Opinion in Cell Biology, 25*(5), 582–590. http://doi.org/10.1016/j.ceb.2013.05.005.

Panková, K., Rösel, D., Novotný, M., & Brábek, J. (2010). The molecular mechanisms of transition between mesenchymal and amoeboid invasiveness in tumor cells. *Cellular and Molecular Life Sciences: CMLS, 67*(1), 63–71. http://doi.org/10.1007/s00018-009-0132-1.

Parsons, S. J., & Parsons, J. T. (2004). Src family kinases, key regulators of signal transduction. *Oncogene, 23*(48), 7906–7909. http://doi.org/10.1038/sj.onc.1208160.

Parvani, J. G., Galliher-Beckley, A. J., Schiemann, B. J., & Schiemann, W. P. (2013). Targeted inactivation of β1 integrin induces β3 integrin switching, which drives breast cancer metastasis by TGF-β. *Molecular Biology of the Cell, 24*(21), 3449–3459. http://doi.org/10.1091/mbc.E12-10-0776.

Paszek, M. J., Zahir, N., Johnson, K. R., Lakins, J. N., Rozenberg, G. I., Gefen, A., … Weaver, V. M. (2005). Tensional homeostasis and the malignant phenotype. *Cancer Cell, 8*(3), 241–254. http://doi.org/10.1016/j.ccr.2005.08.010.

Patel, R. A., Forinash, K. D., Pireddu, R., Sun, Y., Sun, N., Martin, M. P., … Sebti, S. M. (2012). RKI-1447 is a potent inhibitor of the Rho-associated ROCK kinases with anti-invasive and antitumor activities in breast cancer. *Cancer Research, 72*(19), 5025–5034. http://doi.org/10.1158/0008-5472.CAN-12-0954.

Pathak, A., & Kumar, S. (2012). Independent regulation of tumor cell migration by matrix stiffness and confinement. *Proceedings of the National Academy of Sciences of the United States of America, 109*(26), 10334–10339. http://doi.org/10.1073/pnas.1118073109.

Peglion, F., Llense, F., & Etienne-Manneville, S. (2014). Adherens junction treadmilling during collective migration. *Nature Cell Biology, 16*(7), 639–651. http://doi.org/10.1038/ncb2985.

Petrelli, A., Circosta, P., Granziero, L., Mazzone, M., Pisacane, A., Fenoglio, S., … Giordano, S. (2006). Ab-induced ectodomain shedding mediates hepatocyte growth factor receptor down-regulation and hampers biological activity. *Proceedings of the National Academy of Sciences of the United States of America, 103*(13), 5090–5095. http://doi.org/10.1073/pnas.0508156103.

Petrie, R. J., & Yamada, K. M. (2012). At the leading edge of three-dimensional cell migration. *Journal of Cell Science, 125*(Pt. 24), 5917–5926. http://doi.org/10.1242/jcs.093732.

Pinner, S., & Sahai, E. (2008). Imaging amoeboid cancer cell motility in vivo. *Journal of Microscopy, 231*(3), 441–445. http://doi.org/10.1111/j.1365-2818.2008.02056.x.

Polyak, K., & Weinberg, R. A. (2009). Transitions between epithelial and mesenchymal states: Acquisition of malignant and stem cell traits. *Nature Reviews. Cancer, 9*(4), 265–273. http://doi.org/10.1038/nrc2620.

Provenzano, P. P., Eliceiri, K. W., Campbell, J. M., Inman, D. R., White, J. G., & Keely, P. J. (2006). Collagen reorganization at the tumor-stromal interface facilitates local invasion. *BMC Medicine, 4*(1), 38. http://doi.org/10.1186/1741-7015-4-38.

Provenzano, P. P., Inman, D. R., Eliceiri, K. W., Knittel, J. G., Yan, L., Rueden, C. T., … Keely, P. J. (2008). Collagen density promotes mammary tumor initiation and progression. *BMC Medicine, 6*, 11. http://doi.org/10.1186/1741-7015-6-11.

Prudnikova, T. Y., Rawat, S. J., & Chernoff, J. (2015). Molecular pathways: Targeting the kinase effectors of RHO-family GTPases. *Clinical Cancer Research: An Official Journal of the American Association for Cancer Research, 21*(1), 24–29. http://doi.org/10.1158/1078-0432.CCR-14-0827.

Pukrop, T., Klemm, F., Hagemann, T., Gradl, D., Schulz, M., Siemes, S., … Binder, C. (2006). Wnt 5a signaling is critical for macrophage-induced invasion of breast cancer cell lines. *Proceedings of the National Academy of Sciences of the United States of America, 103*(14), 5454–5459. http://doi.org/10.1073/pnas.0509703103.

Ramirez, N. E., Zhang, Z., Madamanchi, A., Boyd, K. L., O'Rear, L. D., Nashabi, A., ... Zutter, M. M. (2011). The $\alpha_2\beta_1$ integrin is a metastasis suppressor in mouse models and human cancer. *The Journal of Clinical Investigation*, *121*(1), 226–237. http://doi.org/10.1172/JCI42328.

Rankin, E. B., & Giaccia, A. J. (2016). Hypoxic control of metastasis. *Science*, *352*(6282), 175–180. http://doi.org/10.1126/science.aaf4405.

Rath, N., & Olson, M. F. (2012). Rho-associated kinases in tumorigenesis: Re-considering ROCK inhibition for cancer therapy. *EMBO Reports*, *13*(10), 900–908. http://doi.org/10.1038/embor.2012.127.

Reardon, D. A., Fink, K. L., Mikkelsen, T., Cloughesy, T. F., O'Neill, A., Plotkin, S., ... Nabors, L. B. (2008). Randomized phase II study of cilengitide, an integrin-targeting arginine-glycine-aspartic acid peptide, in recurrent glioblastoma multiforme. *Journal of Clinical Oncology: Official Journal of the American Society of Clinical Oncology*, *26*(34), 5610–5617. http://doi.org/10.1200/JCO.2008.16.7510.

Reffay, M., Parrini, M. C., Cochet-Escartin, O., Ladoux, B., Buguin, A., Coscoy, S., ... Silberzan, P. (2014). Interplay of RhoA and mechanical forces in collective cell migration driven by leader cells. *Nature Cell Biology*, *16*(3), 217–223. http://doi.org/10.1038/ncb2917.

Renaud, S., Guenot, D., Falcoz, P.-E., Massard, G., & Beau-Faller, M. (2014). Role of hypoxia in epithelial-to-mesenchymal transition (EMT) in non-small cell lung cancer (NSCLC). *The European Respiratory Journal*, *44*(Suppl. 58), P814. Retrieved from, http://erj.ersjournals.com/content/44/Suppl_58/P814.

Renkawitz, J., Schumann, K., Weber, M., Lämmermann, T., Pflicke, H., Piel, M., ... Sixt, M. (2009). Adaptive force transmission in amoeboid cell migration. *Nature Cell Biology*, *11*(12), 1438–1443. http://doi.org/10.1038/ncb1992.

Reynolds, L. E., Wyder, L., Lively, J. C., Taverna, D., Robinson, S. D., Huang, X., ... Hodivala-Dilke, K. M. (2002). Enhanced pathological angiogenesis in mice lacking beta3 integrin or beta3 and beta5 integrins. *Nature Medicine*, *8*(1), 27–34. http://doi.org/10.1038/nm0102-27.

Ricono, J. M., Huang, M., Barnes, L. A., Lau, S. K., Weis, S. M., Schlaepfer, D. D., ... Cheresh, D. A. (2009). Specific cross-talk between epidermal growth factor receptor and integrin alphavbeta5 promotes carcinoma cell invasion and metastasis. *Cancer Research*, *69*(4), 1383–1391. http://doi.org/10.1158/0008-5472.CAN-08-3612.

Roberts, W. G., Ung, E., Whalen, P., Cooper, B., Hulford, C., Autry, C., ... Vajdos, F. (2008). Antitumor activity and pharmacology of a selective focal adhesion kinase inhibitor, PF-562,271. *Cancer Research*, *68*(6), 1935–1944. http://doi.org/10.1158/0008-5472.CAN-07-5155.

Rossi, F., Yozgat, Y., de Stanchina, E., Veach, D., Clarkson, B., Manova, K., ... Besmer, P. (2010). Imatinib upregulates compensatory integrin signaling in a mouse model of gastrointestinal stromal tumor and is more effective when combined with dasatinib. *Molecular Cancer Research: MCR*, *8*(9), 1271–1283. http://doi.org/10.1158/1541-7786.MCR-10-0065.

Rotty, J. D., Wu, C., & Bear, J. E. (2013). New insights into the regulation and cellular functions of the ARP2/3 complex. *Nature Reviews. Molecular Cell Biology*, *14*(1), 7–12. http://doi.org/10.1038/nrm3492.

Roussos, E. T., Condeelis, J. S., & Patsialou, A. (2011). Chemotaxis in cancer. *Nature Reviews. Cancer*, *11*(8), 573–587. http://doi.org/10.1038/nrc3078.

Sadok, A., & Marshall, C. J. (2014). Rho GTPases: Masters of cell migration. *Small GTPases*, *5*, e29710. http://doi.org/10.4161/sgtp.29710.

Sadok, A., McCarthy, A., Caldwell, J., Collins, I., Garrett, M. D., Yeo, M., ... Marshall, C. J. (2015). Rho kinase inhibitors block melanoma cell migration and inhibit metastasis. *Cancer Research*, *75*(11), 2272–2284. http://doi.org/10.1158/0008-5472.CAN-14-2156.

Sahai, E. (2007). Illuminating the metastatic process. *Nature Reviews. Cancer*, 7(10), 737–749. http://doi.org/10.1038/nrc2229.

Sahai, E., & Marshall, C. J. (2003). Differing modes of tumour cell invasion have distinct requirements for Rho/ROCK signalling and extracellular proteolysis. *Nature Cell Biology*, *5*(8), 711–719. http://doi.org/10.1038/ncb1019.

Santoro, A., Rimassa, L., Borbath, I., Daniele, B., Salvagni, S., Van Laethem, J. L., … Porta, C. (2013). Tivantinib for second-line treatment of advanced hepatocellular carcinoma: A randomised, placebo-controlled phase 2 study. *The Lancet Oncology*, *14*(1), 55–63. http://dx.doi.org/10.1016/S1470-2045(12)70490-4.

Sanz-Moreno, V., Gadea, G., Ahn, J., Paterson, H., Marra, P., Pinner, S., … Marshall, C. J. (2008). Rac activation and inactivation control plasticity of tumor cell movement. *Cell*, *135*(3), 510–523. http://doi.org/10.1016/j.cell.2008.09.043.

Sanz-Moreno, V., & Marshall, C. J. (2010). The plasticity of cytoskeletal dynamics underlying neoplastic cell migration. *Current Opinion in Cell Biology*, *22*(5), 690–696. http://doi.org/10.1016/j.ceb.2010.08.020.

Satelli, A., Mitra, A., Brownlee, Z., Xia, X., Bellister, S., Overman, M. J., … Li, S. (2015). Epithelial-mesenchymal transitioned circulating tumor cells capture for detecting tumor progression. *Clinical Cancer Research: An Official Journal of the American Association for Cancer Research*, *21*(4), 899–906. http://doi.org/10.1158/1078-0432.CCR-14-0894.

Savagner, P. (2010). The epithelial-mesenchymal transition (EMT) phenomenon. *Annals of Oncology: Official Journal of the European Society for Medical Oncology/ESMO*, *21*(Suppl. 7), vii89–vii92. http://doi.org/10.1093/annonc/mdq292.

Scagliotti, G., von Pawel, J., Novello, S., Ramlau, R., Favaretto, A., Barlesi, F., … Schwartz, B. (2015). Phase III multinational, randomized, double-blind, placebo-controlled study of Tivantinib (ARQ 197) Plus Erlotinib versus Erlotinib alone in previously treated patients with locally advanced or metastatic nonsquamous non-small-cell lung cancer. *Journal of Clinical Oncology: Official Journal of the American Society of Clinical Oncology*, *33*(24), 2667–2674. http://doi.org/10.1200/JCO.2014.60.7317.

Schlingensiepen, K.-H., Jaschinski, F., Lang, S. A., Moser, C., Geissler, E. K., Schlitt, H. J., … Schneider, A. (2011). Transforming growth factor-beta 2 gene silencing with trabedersen (AP 12009) in pancreatic cancer. *Cancer Science*, *102*(6), 1193–1200. http://doi.org/10.1111/j.1349-7006.2011.01917.x.

Sennino, B., Ishiguro-Oonuma, T., Wei, Y., Naylor, R. M., Williamson, C. W., Bhagwandin, V., … McDonald, D. M. (2012). Suppression of tumor invasion and metastasis by concurrent inhibition of c-Met and VEGF signaling in pancreatic neuroendocrine tumors. *Cancer Discovery*, *2*(3), 270–287. http://doi.org/10.1158/2159-8290.CD-11-0240.

Sharma, N., & Adjei, A. A. (2011). In the clinic: Ongoing clinical trials evaluating c-MET-inhibiting drugs. *Therapeutic Advances in Medical Oncology*, *3*(1 Suppl.), S37–S50. http://doi.org/10.1177/1758834011423403.

Sharma, M. R., Wroblewski, K., Polite, B. N., Knost, J. A., Wallace, J. A., Modi, S., … Kindler, H. L. (2012). Dasatinib in previously treated metastatic colorectal cancer: A phase II trial of the University of Chicago Phase II Consortium. *Investigational New Drugs*, *30*(3), 1211–1215. http://doi.org/10.1007/s10637-011-9681-x.

Shi, Q., Hjelmeland, A. B., Keir, S. T., Song, L., Wickman, S., Jackson, D., … Rich, J. N. (2007). A novel low-molecular weight inhibitor of focal adhesion kinase, TAE226, inhibits glioma growth. *Molecular Carcinogenesis*, *46*(6), 488–496. http://doi.org/10.1002/mc.20297.

Shintani, Y., Fukumoto, Y., Chaika, N., Svoboda, R., Wheelock, M. J., & Johnson, K. R. (2008). Collagen I–mediated up-regulation of N-cadherin requires cooperative signals from integrins and discoidin domain receptor 1. *The Journal of Cell Biology*, *180*(6), 1277–1289. http://doi.org/10.1083/jcb.200708137.

Shintani, Y., Maeda, M., Chaika, N., Johnson, K. R., & Wheelock, M. J. (2008). Collagen I promotes epithelial-to-mesenchymal transition in lung cancer cells via transforming growth factor-beta signaling. *American Journal of Respiratory Cell and Molecular Biology*, *38*(1), 95–104. http://doi.org/10.1165/rcmb.2007-0071OC.

Shirakihara, T., Horiguchi, K., Miyazawa, K., Ehata, S., Shibata, T., Morita, I., … Saitoh, M. (2011). TGF-β regulates isoform switching of FGF receptors and epithelial-mesenchymal transition. *The EMBO Journal*, *30*(4), 783–795. http://doi.org/10.1038/emboj.2010.351.

Sievers, E., Trautmann, M., Kindler, D., Huss, S., Gruenewald, I., Dirksen, U., … Hartmann, W. (2015). SRC inhibition represents a potential therapeutic strategy in liposarcoma. *International Journal of Cancer*, *137*(11), 2578–2588. http://doi.org/10.1002/ijc.29645.

Somlyo, A. V., Phelps, C., Dipierro, C., Eto, M., Read, P., Barrett, M., … Somlyo, A. P. (2003). Rho kinase and matrix metalloproteinase inhibitors cooperate to inhibit angiogenesis and growth of human prostate cancer xenotransplants. *FASEB Journal: Official Publication of the Federation of American Societies for Experimental Biology*, *17*(2), 223–234. http://doi.org/10.1096/fj.02-0655com.

Sonnemann, K. J., & Bement, W. M. (2011). Wound repair: Toward understanding and integration of single-cell and multicellular wound responses. *Annual Review of Cell and Developmental Biology*, *27*, 237–263. http://doi.org/10.1146/annurev-cellbio-092910-154251.

Sparano, J. A., Bernardo, P., Stephenson, P., Gradishar, W. J., Ingle, J. N., Zucker, S., & Davidson, N. E. (2004). Randomized phase III trial of marimastat versus placebo in patients with metastatic breast cancer who have responding or stable disease after first-line chemotherapy: Eastern Cooperative Oncology Group trial E2196. *Journal of Clinical Oncology: Official Journal of the American Society of Clinical Oncology*, *22*(23), 4683–4690. http://doi.org/10.1200/JCO.2004.08.054.

Spigel, D. R., Ervin, T. J., Ramlau, R. A., Daniel, D. B., Goldschmidt, J. H., Blumenschein, G. R., … Peterson, A. C. (2013). Randomized phase II trial of Onartuzumab in combination with erlotinib in patients with advanced non-small-cell lung cancer. *Journal of Clinical Oncology: Official Journal of the American Society of Clinical Oncology*, *31*(32), 4105–4114. http://doi.org/10.1200/JCO.2012.47.4189.

Spina, A., De Pasquale, V., Cerulo, G., Cocchiaro, P., Della Morte, R., Avallone, L., & Pavone, L. (2015). HGF/c-MET axis in tumor microenvironment and metastasis formation. *Biomedicines*, *3*(1), 71–88. http://doi.org/10.3390/biomedicines3010071.

Starke, J., Maaser, K., Wehrle-Haller, B., & Friedl, P. (2013). Mechanotransduction of mesenchymal melanoma cell invasion into 3D collagen lattices: Filopod-mediated extension-relaxation cycles and force anisotropy. *Experimental Cell Research*, *319*(16), 2424–2433. http://doi.org/10.1016/j.yexcr.2013.04.003.

Stehn, J. R., Haass, N. K., Bonello, T., Desouza, M., Kottyan, G., Treutlein, H., … Gunning, P. W. (2013). A novel class of anticancer compounds targets the actin cytoskeleton in tumor cells. *Cancer Research*, *73*(16), 5169–5182. http://doi.org/10.1158/0008-5472.CAN-12-4501.

Stock, C., Gassner, B., Hauck, C. R., Arnold, H., Mally, S., Eble, J. A., … Schwab, A. (2005). Migration of human melanoma cells depends on extracellular pH and Na+/H+ exchange. *The Journal of Physiology*, *567*(Pt. 1), 225–238. http://doi.org/10.1113/jphysiol.2005.088344.

Stock, C., & Schwab, A. (2009). Protons make tumor cells move like clockwork. *Pflügers Archiv/European Journal of Physiology*, *458*(5), 981–992. http://doi.org/10.1007/s00424-009-0677-8.

Stransky, N., Egloff, A. M., Tward, A. D., Kostic, A. D., Cibulskis, K., Sivachenko, A., … Grandis, J. R. (2011). The mutational landscape of head and neck squamous cell carcinoma. *Science (New York, N.Y.)*, *333*(6046), 1157–1160. http://doi.org/10.1126/science.1208130.

Strobel, T., & Cannistra, S. A. (1999). Beta1-integrins partly mediate binding of ovarian cancer cells to peritoneal mesothelium in vitro. *Gynecologic Oncology*, *73*(3), 362–367. http://doi.org/10.1006/gyno.1999.5388.

Stupp, R., Hegi, M. E., Gorlia, T., Erridge, S. C., Perry, J., Hong, Y.-K., ... Weller, M. (2014). Cilengitide combined with standard treatment for patients with newly diagnosed glioblastoma with methylated MGMT promoter (CENTRIC EORTC 26071-22072 study): A multicentre, randomised, open-label, phase 3 trial. *The Lancet Oncology*, *15*(10), 1100–1108. http://dx.doi.org/10.1016/S1470-2045(14)70379-1.

Sugiyama, N., Varjosalo, M., Meller, P., Lohi, J., Chan, K. M., Zhou, Z., ... Lehti, K. (2010). FGF receptor-4 (FGFR4) polymorphism acts as an activity switch of a membrane type 1 matrix metalloproteinase-FGFR4 complex. *Proceedings of the National Academy of Sciences of the United States of America*, *107*(36), 15786–15791. http://doi.org/10.1073/pnas.0914459107.

Sulzmaier, F. J., Jean, C., & Schlaepfer, D. D. (2014). FAK in cancer: Mechanistic findings and clinical applications. *Nature Reviews. Cancer*, *14*(9), 598–610. http://doi.org/10.1038/nrc3792.

Taddei, M. L., Giannoni, E., Morandi, A., Ippolito, L., Ramazzotti, M., Callari, M., ... Chiarugi, P. (2014). Mesenchymal to amoeboid transition is associated with stem-like features of melanoma cells. *Cell Communication and Signaling: CCS*, *12*, 24. http://doi.org/10.1186/1478-811X-12-24.

Taddei, M. L., Parri, M., Angelucci, A., Bianchini, F., Marconi, C., Giannoni, E., ... Chiarugi, P. (2011). EphA2 induces metastatic growth regulating amoeboid motility and clonogenic potential in prostate carcinoma cells. *Molecular Cancer Research: MCR*, *9*(2), 149–160. http://doi.org/10.1158/1541-7786.MCR-10-0298.

Taherian, A., Li, X., Liu, Y., & Haas, T. A. (2011). Differences in integrin expression and signaling within human breast cancer cells. *BMC Cancer*, *11*, 293. http://doi.org/10.1186/1471-2407-11-293.

Tancioni, I., Uryu, S., Sulzmaier, F. J., Shah, N. R., Lawson, C., Miller, N. L. G., ... Schlaepfer, D. D. (2014). FAK Inhibition disrupts a β5 integrin signaling axis controlling anchorage-independent ovarian carcinoma growth. *Molecular Cancer Therapeutics*, *13*(8), 2050–2061. http://doi.org/10.1158/1535-7163.MCT-13-1063.

Taverna, D., Moher, H., Crowley, D., Borsig, L., Varki, A., & Hynes, R. O. (2004). Increased primary tumor growth in mice null for beta3- or beta3/beta5-integrins or selectins. *Proceedings of the National Academy of Sciences of the United States of America*, *101*(3), 763–768. http://doi.org/10.1073/pnas.0307289101.

Tavora, B., Batista, S., Reynolds, L. E., Jadeja, S., Robinson, S., Kostourou, V., ... Hodivala-Dilke, K. M. (2010). Endothelial FAK is required for tumour angiogenesis. *EMBO Molecular Medicine*, *2*(12), 516–528. http://doi.org/10.1002/emmm.201000106.

te Boekhorst, V., Preziosi, L., & Friedl, P. (2016). Plastiticy of cell migration in vivo and in silico. *Annual Review of Cell and Developmental Biology*. http://dx.doi.org/10.1146/annurev-cellbio-111315-125201.

Trendowski, M. (2014). Exploiting the inherent metastasis of leukemia to improve chemotherapeutic approaches. *Cell & Developmental Biology*, *03*(02). http://doi.org/10.4172/2168-9296.1000137.

Trevino, J. G., Summy, J. M., Lesslie, D. P., Parikh, N. U., Hong, D. S., Lee, F. Y., ... Gallick, G. E. (2006). Inhibition of SRC expression and activity inhibits tumor progression and metastasis of human pancreatic adenocarcinoma cells in an orthotopic nude mouse model. *The American Journal of Pathology*, *168*(3), 962–972. http://doi.org/10.2353/ajpath.2006.050570.

Trikha, M., Timar, J., Lundy, S. K., Szekeres, K., Cai, Y., Porter, A. T., & Honn, K. V. (1997). The high affinity alphaIIb beta3 integrin is involved in invasion of human melanoma cells. *Cancer Research*, *57*(12), 2522–2528. Retrieved from, http://www.ncbi.nlm.nih.gov/pubmed/9192835.

Trusolino, L., Bertotti, A., & Comoglio, P. M. (2010). MET signalling: Principles and functions in development, organ regeneration and cancer. *Nature Reviews. Molecular Cell Biology*, *11*(12), 834–848. http://doi.org/10.1038/nrm3012.

Ueda, J., Kajita, M., Suenaga, N., Fujii, K., & Seiki, M. (2003). Sequence-specific silencing of MT1-MMP expression suppresses tumor cell migration and invasion: Importance of MT1-MMP as a therapeutic target for invasive tumors. *Oncogene*, *22*(54), 8716–8722. http://doi.org/10.1038/sj.onc.1206962.

Valastyan, S., & Weinberg, R. A. (2011). Tumor metastasis: Molecular insights and evolving paradigms. *Cell*, *147*(2), 275–292. http://doi.org/10.1016/j.cell.2011.09.024.

van Helvert, S., & Friedl, P. (2016). Strain stiffening of fibrillar collagen during individual and collective cell migration identified by AFM nanoindentation. *ACS Applied Materials & Interfaces*. http://dx.doi.org/10.1021/acsami.6b01755.

Vaškovičová, K., Szabadosová, E., Čermák, V., Gandalovičová, A., Kasalová, L., Rösel, D., & Brábek, J. (2015). PKCα promotes the mesenchymal to amoeboid transition and increases cancer cell invasiveness. *BMC Cancer*, *15*(1), 326. http://doi.org/10.1186/s12885-015-1347-1.

Veracini, L., Grall, D., Schaub, S., Beghelli-de la Forest Divonne, S., Etienne-Grimaldi, M.-C., Milano, G., & Van Obberghen-Schilling, E. (2015). Elevated Src family kinase activity stabilizes E-cadherin-based junctions and collective movement of head and neck squamous cell carcinomas. *Oncotarget*, *6*(10), 7570–7583. http://doi.org/10.18632/oncotarget.3071.

Vermorken, J. B., Peyrade, F., Krauss, J., Mesía, R., Remenar, E., Gauler, T. C., … Clement, P. M. (2014). Cisplatin, 5-fluorouracil, and cetuximab (PFE) with or without cilengitide in recurrent/metastatic squamous cell carcinoma of the head and neck: Results of the randomized phase I/II ADVANTAGE trial (phase II part). *Annals of Oncology: Official Journal of the European Society for Medical Oncology/ESMO*, *25*(3), 682–688. http://doi.org/10.1093/annonc/mdu003.

Vitale, S., Avizienyte, E., Brunton, V. G., & Frame, M. C. (2008). Focal adhesion kinase is not required for Src-induced formation of invadopodia in KM12C colon cancer cells and can interfere with their assembly. *European Journal of Cell Biology*, *87*(8-9), 569–579. http://doi.org/10.1016/j.ejcb.2008.04.002.

Vuoriluoto, K., Haugen, H., Kiviluoto, S., Mpindi, J.-P., Nevo, J., Gjerdrum, C., … Ivaska, J. (2011). Vimentin regulates EMT induction by Slug and oncogenic H-Ras and migration by governing Axl expression in breast cancer. *Oncogene*, *30*(12), 1436–1448. http://doi.org/10.1038/onc.2010.509.

Wai Wong, C., Dye, D. E., & Coombe, D. R. (2012). The role of immunoglobulin superfamily cell adhesion molecules in cancer metastasis. *International Journal of Cell Biology*, *2012*, 340296. http://doi.org/10.1155/2012/340296.

Walsh, L. A., & Damjanovski, S. (2011). IGF-1 increases invasive potential of MCF 7 breast cancer cells and induces activation of latent TGF-β1 resulting in epithelial to mesenchymal transition. *Cell Communication and Signaling: CCS*, *9*(1), 10. http://doi.org/10.1186/1478-811X-9-10.

Walsh, C., Tanjoni, I., Uryu, S., Tomar, A., Nam, J.-O., Luo, H., … Schlaepfer, D. D. (2010). Oral delivery of PND-1186 FAK inhibitor decreases tumor growth and spontaneous breast to lung metastasis in pre-clinical models. *Cancer Biology & Therapy*, *9*(10), 778–790. Retrieved from, http://www.pubmedcentral.nih.gov/articlerender.fcgi?artid=2933309&tool=pmcentrez&rendertype=abstract.

Wan, L., Pantel, K., & Kang, Y. (2013). Tumor metastasis: Moving new biological insights into the clinic. *Nature Medicine*, *19*(11), 1450–1464. http://doi.org/10.1038/nm.3391.

Wang, Y., & McNiven, M. A. (2012). Invasive matrix degradation at focal adhesions occurs via protease recruitment by a FAK-p130Cas complex. *The Journal of Cell Biology, 196*(3), 375–385. http://doi.org/10.1083/jcb.201105153.

Wang, B., Wang, W., Niu, W., Liu, E., Liu, X., Wang, J., ... Niu, J. (2014). SDF-1/CXCR4 axis promotes directional migration of colorectal cancer cells through upregulation of integrin αvβ6. *Carcinogenesis, 35*(2), 282–291. http://doi.org/10.1093/carcin/bgt331.

Webb, B. A., Chimenti, M., Jacobson, M. P., & Barber, D. L. (2011). Dysregulated pH: A perfect storm for cancer progression. *Nature Reviews. Cancer, 11*(9), 671–677. http://doi.org/10.1083/jcb.201105153.

Weigelin, B., Bakker, G.-J., & Friedl, P. (2012). Intravital third harmonic generation microscopy of collective melanoma cell invasion: Principles of interface guidance and microvesicle dynamics. *IntraVital, 1*(1), 32–43. http://doi.org/10.4161/intv.21223.

Weigelin, B., Bakker, G.-J., & Friedl, P. (2016). Third harmonic generation microscopy of cells and tissue organization. *Journal of Cell Science, 129*(2), 245–255. http://doi.org/10.1242/jcs.152272.

Weijer, C. J. (2009). Collective cell migration in development. *Journal of Cell Science, 122*(Pt. 18), 3215–3223. http://doi.org/10.1242/jcs.036517.

Wells, A., Chao, Y. L., Grahovac, J., Wu, Q., & Lauffenburger, D. A. (2011). Epithelial and mesenchymal phenotypic switchings modulate cell motility in metastasis. *Frontiers in Bioscience (Landmark Edition), 16*, 815–837. Retrieved from, http://www.pubmedcentral.nih.gov/articlerender.fcgi?artid=4003907&tool=pmcentrez&rendertype=abstract.

Wendt, M. K., & Schiemann, W. P. (2009). Therapeutic targeting of the focal adhesion complex prevents oncogenic TGF-beta signaling and metastasis. *Breast Cancer Research: BCR, 11*(5), R68. http://doi.org/10.1186/bcr2360.

Wolf, K., Mazo, I., Leung, H., Engelke, K., von Andrian, U. H., Deryugina, E. I., ... Friedl, P. (2003). Compensation mechanism in tumor cell migration: Mesenchymal-amoeboid transition after blocking of pericellular proteolysis. *The Journal of Cell Biology, 160*(2), 267–277. http://doi.org/10.1083/jcb.200209006.

Wolf, K., Te Lindert, M., Krause, M., Alexander, S., Te Riet, J., Willis, A. L., ... Friedl, P. (2013). Physical limits of cell migration: Control by ECM space and nuclear deformation and tuning by proteolysis and traction force. *The Journal of Cell Biology, 201*(7), 1069–1084. http://doi.org/10.1083/jcb.201210152.

Wolf, K., Wu, Y. I., Liu, Y., Geiger, J., Tam, E., Overall, C., ... Friedl, P. (2007). Multi-step pericellular proteolysis controls the transition from individual to collective cancer cell invasion. *Nature Cell Biology, 9*(8), 893–904. http://doi.org/10.1038/ncb1616.

Wolfson, B., Eades, G., & Zhou, Q. (2015). Adipocyte activation of cancer stem cell signaling in breast cancer. *World Journal of Biological Chemistry, 6*(2), 39–47. http://doi.org/10.4331/wjbc.v6.i2.39.

Wyckoff, J. B., Pinner, S. E., Gschmeissner, S., Condeelis, J. S., & Sahai, E. (2006). ROCK- and myosin-dependent matrix deformation enables protease-independent tumor-cell invasion in vivo. *Current Biology: CB, 16*(15), 1515–1523. http://doi.org/10.1016/j.cub.2006.05.065.

Wyckoff, J. B., Wang, Y., Lin, E. Y., Li, J., Goswami, S., Stanley, E. R., ... Condeelis, J. (2007). Direct visualization of macrophage-assisted tumor cell intravasation in mammary tumors. *Cancer Research, 67*(6), 2649–2656. http://doi.org/10.1158/0008-5472.CAN-06-1823.

Wyckoff, J., Wang, W., Lin, E. Y., Wang, Y., Pixley, F., Stanley, E. R., ... Condeelis, J. (2004). A paracrine loop between tumor cells and macrophages is required for tumor cell migration in mammary tumors. *Cancer Research, 64*(19), 7022–7029. http://doi.org/10.1158/0008-5472.CAN-04-1449.

Xu, R., Boudreau, A., & Bissell, M. J. (2009). Tissue architecture and function: Dynamic reciprocity via extra- and intra-cellular matrices. *Cancer Metastasis Reviews*, *28*(1-2), 167–176. http://doi.org/10.1007/s10555-008-9178-z.

Xu, J., Lamouille, S., & Derynck, R. (2009). TGF-beta-induced epithelial to mesenchymal transition. *Cell Research*, *19*(2), 156–172. http://doi.org/10.1038/cr.2009.5.

Yadav, A., Kumar, B., Datta, J., Teknos, T. N., & Kumar, P. (2011). IL-6 promotes head and neck tumor metastasis by inducing epithelial-mesenchymal transition via the JAK-STAT3-SNAIL signaling pathway. *Molecular Cancer Research: MCR*, *9*(12), 1658–1667. http://doi.org/10.1158/1541-7786.MCR-11-0271.

Yamaguchi, N., Mizutani, T., Kawabata, K., & Haga, H. (2015). Leader cells regulate collective cell migration via Rac activation in the downstream signaling of integrin β1 and PI3K. *Scientific Reports*, *5*, 7656. http://doi.org/10.1038/srep07656.

Yap, T. A., Walton, M. I., Grimshaw, K. M., Te Poele, R. H., Eve, P. D., Valenti, M. R., … Garrett, M. D. (2012). AT13148 is a novel, oral multi-AGC kinase inhibitor with potent pharmacodynamic and antitumor activity. *Clinical Cancer Research: An Official Journal of the American Association for Cancer Research*, *18*(14), 3912–3923. http://doi.org/10.1158/1078-0432.CCR-11-3313.

Ying, H., Biroc, S. L., Li, W.-W., Alicke, B., Xuan, J.-A., Pagila, R., … Dinter, H. (2006). The Rho kinase inhibitor fasudil inhibits tumor progression in human and rat tumor models. *Molecular Cancer Therapeutics*, *5*(9), 2158–2164. http://doi.org/10.1158/1535-7163.MCT-05-0440.

Yoo, H. Y., Sung, M. K., Lee, S. H., Kim, S., Lee, H., Park, S., … Ko, Y. H. (2014). A recurrent inactivating mutation in RHOA GTPase in angioimmunoblastic T cell lymphoma. *Nature Genetics*, *46*(4), 371–375. http://doi.org/10.1038/ng.2916.

Yousif, N. G. (2014). Fibronectin promotes migration and invasion of ovarian cancer cells through up-regulation of FAK-PI3K/Akt pathway. *Cell Biology International*, *38*(1), 85–91. http://doi.org/10.1002/cbin.10184.

Zasadil, L. M., Andersen, K. A., Yeum, D., Rocque, G. B., Wilke, L. G., Tevaarwerk, A. J., … Weaver, B. A. (2014). Cytotoxicity of paclitaxel in breast cancer is due to chromosome missegregation on multipolar spindles. *Science Translational Medicine*, *6*(229), 229ra43. http://doi.org/10.1126/scitranslmed.3007965.

Zeisberg, M., & Neilson, E. G. (2009). Biomarkers for epithelial-mesenchymal transitions. *The Journal of Clinical Investigation*, *119*(6), 1429–1437. http://doi.org/10.1172/JCI36183.

Zhang, D., & Brodt, P. (2003). Type 1 insulin-like growth factor regulates MT1-MMP synthesis and tumor invasion via PI 3-kinase/Akt signaling. *Oncogene*, *22*(7), 974–982. http://doi.org/10.1038/sj.onc.1206197.

Zhang, B., Halder, S. K., Zhang, S., & Datta, P. K. (2009). Targeting transforming growth factor-beta signaling in liver metastasis of colon cancer. *Cancer Letters*, *277*(1), 114–120. http://doi.org/10.1016/j.canlet.2008.11.035.

Zhang, L., Huang, G., Li, X., Zhang, Y., Jiang, Y., Shen, J., … Qian, C. (2013). Hypoxia induces epithelial-mesenchymal transition via activation of SNAI1 by hypoxia-inducible factor -1α in hepatocellular carcinoma. *BMC Cancer*, *13*, 108. http://doi.org/10.1186/1471-2407-13-108.

Zhang, S., Huang, W.-C., Zhang, L., Zhang, C., Lowery, F. J., Ding, Z., … Yu, D. (2013). SRC family kinases as novel therapeutic targets to treat breast cancer brain metastases. *Cancer Research*, *73*(18), 5764–5774. http://doi.org/10.1158/0008-5472.CAN-12-1803.

Zhang, S., & Yu, D. (2012). Targeting Src family kinases in anti-cancer therapies: Turning promise into triumph. *Trends in Pharmacological Sciences*, *33*(3), 122–128. http://doi.org/10.1016/j.tips.2011.11.002.

Zhao, J., & Guan, J. L. (2009). Signal transduction by focal adhesion kinase in cancer. *Cancer Metastasis Reviews*, *28*(1–2), 35–49. http://dx.doi.org/10.1007/s10555-008-9165-4.

Zutter, M. M., Santoro, S. A., Staatz, W. D., & Tsung, Y. L. (1995). Re-expression of the alpha 2 beta 1 integrin abrogates the malignant phenotype of breast carcinoma cells. *Proceedings of the National Academy of Sciences of the United States of America*, *92*(16), 7411–7415. Retrieved from, http://www.pubmedcentral.nih.gov/articlerender.fcgi?artid=41349&tool=pmcentrez&rendertype=abstract.

CHAPTER EIGHT

Cytokine Regulation of Metastasis and Tumorigenicity

M. Yao, G. Brummer, D. Acevedo, N. Cheng[1]
University of Kansas Medical Center, Kansas City, KS, United States
[1]Corresponding author: e-mail address: ncheng@kumc.edu

Contents

1. Introduction 266
2. Interleukins 267
   2.1 Interleukin-1 (IL-1) 268
   2.2 Interleukin-6 (IL-6) 272
   2.3 Interleukin-17 (IL-17) 279
   2.4 Interleukins in Anticancer Therapy 281
3. Chemokines 284
   3.1 CCL2 289
   3.2 CCL5 299
   3.3 CXCL1 301
   3.4 CXCL8 303
   3.5 CXCL12 304
4. Chemokines in Therapy 307
5. Interferons 308
   5.1 IFN-α/IFN-β 313
   5.2 IFN-γ 315
   5.3 IFN-λ 317
   5.4 Exploiting IFNs in Anticancer Therapy 317
6. Tumor Necrosis Factor 321
   6.1 TNF-α: Signal Transduction and Expression Patterns 322
   6.2 TNF-α as a Tumor Suppressor 324
   6.3 TNF-α as a Tumor Promoter 329
   6.4 Exploiting the TNF-α Pathway in Anticancer Therapeutics 331
7. Closing Remarks 333
References 334

Abstract

The human body combats infection and promotes wound healing through the remarkable process of inflammation. Inflammation is characterized by the recruitment of stromal cell activity including recruitment of immune cells and induction of angiogenesis. These cellular processes are regulated by a class of soluble molecules called cytokines.

Advances in Cancer Research, Volume 132
ISSN 0065-230X
http://dx.doi.org/10.1016/bs.acr.2016.05.005

Based on function, cell target, and structure, cytokines are subdivided into several classes including: interleukins, chemokines, and lymphokines. While cytokines regulate normal physiological processes, chronic deregulation of cytokine expression and activity contributes to cancer in many ways. Gene polymorphisms of all types of cytokines are associated with risk of disease development. Deregulation RNA and protein expression of interleukins, chemokines, and lymphokines have been detected in many solid tumors and hematopoetic malignancies, correlating with poor patient prognosis. The current body of literature suggests that in some tumor types, interleukins and chemokines work against the human body by signaling to cancer cells and remodeling the local microenvironment to support the growth, survival, and invasion of primary tumors and enhance metastatic colonization. Some lymphokines are downregulated to suppress tumor progression by enhancing cytotoxic T cell activity and inhibiting tumor cell survival. In this review, we will describe the structure/function of several cytokine families and review our current understanding on the roles and mechanisms of cytokines in tumor progression. In addition, we will also discuss strategies for exploiting the expression and activity of cytokines in therapeutic intervention.

1. INTRODUCTION

The human body responds to biological stresses such as tissue injury or infection through the remarkable process of inflammation. Inflammation is characterized by the mobilization of immune cells, induction of angiogenesis, and alterations in the connective tissue, all of which result in tissue repair or clearance of the pathogen. The inflammatory process, which occurs in complex organisms such as mammals, birds, and reptiles (Montali, 1988), was first observed in injured tissues by London surgeon Dr. John Hunter who lived from 1728 until 1793 (Turk, 1994). Acute inflammation occurs during normal physiological functions such as wound healing of infection, and is defined as short term (Collins et al., 2014; Pullamsetti et al., 2011). Disease conditions such as allergic disorders, autoimmune diseases, and cancer are characterized by chronic inflammation resulting in the destruction of normal tissues (Izuhara & Harada, 1993; Jin, Scott, Vadas, & Burns, 1989; Konaka, Norcross, Maino, & Smith, 1981; Stahl et al., 1994). Cancer is often referred to as "wounds that do not heal" due to signs of chronic inflammation such as angiogenesis, recruitment of macrophages, and accumulation of fibroblasts (Dvorak, 2015).

Inflammatory responses are regulated by a broad class of soluble proteins termed cytokines (5–20 kDa). Based on function, cell target, and structure, cytokines are subdivided into several categories: interleukins, chemokines, and lymphokines. Interleukins are known for their ability to modulate immune cell activity, including proliferation, maturation, and migration (Skopinska & Ziembikiewicz, 1978). Chemokines are termed for their ability to stimulate directed cell migration (chemotaxis) (Deuel et al., 1981). Lymphokines are characterized by their secretion from lymphocytes and are subdivided into several molecular families that include: interferons, tumor necrosis factor (TNF), and transforming growth factors (Haber, Rosenau, & Goldberg, 1972; Kehrl et al., 1986; Williamson, Carswell, Rubin, Prendergast, & Old, 1983). Early identification of cytokines relied on experimentation of blood-derived factors in cell culture and in chick cam studies (Cantell, 1961; Isaacs, Burke, & Fadeeva, 1958; Lockart, Sreevalsan, & Horn, 1962). Genomics, proteomics, and bioinformatics technologies will continue to advance the discovery of new cytokines.

The expression and activity of cytokines are deregulated in many cancer types, contributing to chronic inflammation. Emerging studies indicate that interleukins, chemokines, and lymphokines play functionally redundant as well as distinct roles in order to sustain tumor growth, survival, and invasion. In the following sections, we will focus on the role of particular cytokines in the primary tumor and metastatic niche, highlighting advances in our understanding of how these cytokines modulate tumor progression. Furthermore, we will discuss the progress and challenges of utilizing our knowledge of cytokine biology to develop effective anticancer therapies. In this way, we hope this review will be informative to those who seek up to date information on the role of cytokines in tumorigenesis and metastasis.

2. INTERLEUKINS

Interleukins were initially discovered through studies on the pathogenesis of fever. They were described as secreted factors from leukocytes (lymphocytes), which regulated intercommunication among cells, thus giving rise to its current name. These early studies showed that interleukins regulated lymphocyte proliferation in response to antigenic stimuli (de Weck, Otz, Geczy, & Geczy, 1979; Farrar, Mizel, & Farrar, 1980). While interleukins were thought be primarily expressed by lymphocytes, interleukins are expressed by a host of immune cells and nonimmune cells. Interleukins are well-known regulators of inflammatory and immune responses caused by trauma or injuries that occur in the absence of microorganisms. Interleukins are generally conserved between mice and humans, and comprise a large family of cytokines, of which 17 subfamilies of interleukins have been identified, and possess varying structure and function (Dinarello, 1996; Dinarello et al., 2010; Dunna, Sims, Nicklin, & O'Neill, 2001; Garlanda, Dinarello, & Mantovani, 2013; Sims et al., 2001). There are many interleukins that play an important role in cancer, such as Il-4, IL-10, and IL-13, which can best be reviewed in Dennis, Blatner, Gounari, and Khazaie (2013), Geginat et al. (2016), Hallett, Venmar, and Fingleton (2012), and Suzuki, Leland, Joshi, and Puri (2015). Here, we will review some of the well-studied interleukins, IL-1 (IL-1α and IL-1β) and IL-6, and describe the emerging role of IL-17 in cancer.

2.1 Interleukin-1 (IL-1)

Due to amino acid sequence homology, structure, and receptor binding affinity, IL-1 represents a subfamily of cytokines, which regulate immune cell recruitment and the hypothalamus–pituitary–adrenal (HPA) axis, coordinating the fever response (Netea, Kullberg, & Van der Meer, 2000). The IL-1 family members are comprised of 11 cytokines: IL-1α, IL-1β, IL-18, IL-33, IL-1F5 through IL-1F10, and IL-1Ra, a receptor antagonist, which were classified based on their 12-stranded β-barrel structure (Dinarello, 1996; Thomas, Bazan, & Garcia, 2012). These interleukins bind to one of four cell surface receptor complexes. Each individual receptor subunit possesses an extracellular immunoglobulin domain and an intracellular Toll/IL-1 receptor (TIR) domain. Signaling is best demonstrated by IL-1α and IL-1β. IL-1α and IL-1β bind to a IL-1 receptor type I (IL-1RI), which recruits a second subunit, IL-1R acceptor protein (IL-1RAP); leading to recruitment of the adaptor proteins, toll interacting protein (TIR) and myeloid differentiation primary response gene 88 (MYD88), and activation of IL-1 receptor-associated kinase (IRAK). Phosphorylated IRAK4 complexes with TNF receptor-associated factor 6 (TRAF6), which leads to activation of several downstream pathways including NF-κB and p42/44MAPK (Apte et al., 2006; Dunne & O'Neill, 2003; Ninomiya-Tsuji et al., 1999). These pathways are summarized in Fig. 1. IL-1α and IL-1β exhibit some key differences signal transduction. Interestingly, a precursor form of IL-1α is important for intracellular signaling through translocation to the nucleus and regulation of gene expression (Maier, Statuto, & Ragnotti, 1994; Wessendorf, Garfinkel, Zhan, Brown, & Maciag, 1993). Although IL-1α and IL-1β signal through the same receptor, they exhibit different functions in noncancerous tissues. IL-1α is membrane associated and is thought to act locally in tissues to prime T cells and immunoglobulin production

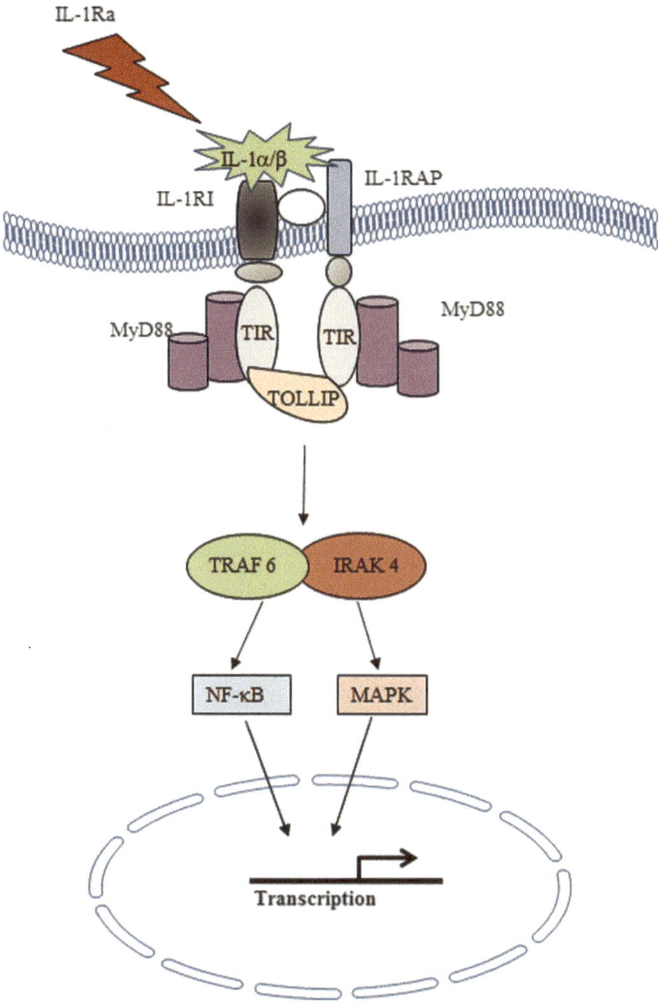

Fig. 1 IL-1 signal transduction pathway. IL-1α and IL-1β bind to IL-1 receptor type I (IL-1RI), which IL1R acceptor protein (IL-1RAP); leading to recruitment of the adaptor proteins, toll interacting protein (TIR) and myeloid differentiation primary response gene 88 (MYD88), activation of IL-1 receptor-associated kinase (IRAK), and tumor necrosis factor receptor-associated factor 6 (TRAF6). IL-1 signaling leads to activation NF-κB and p42/44MAPK pathways to modulate gene transcription.

during contact hypersensitivity (Kish, Gorbachev, & Fairchild, 2012; Kurt-Jones, Beller, Mizel, & Unanue, 1985; Nakae, Asano, Horai, Sakaguchi, & Iwakura, 2001). IL-1β exerts systemic effects, for example inducing fever (Kluger, Kozak, Leon, & Conn, 1998; Rathakrishnan et al., 2012). IL-1α and IL-1β signaling are regulated in part by IL-1 receptor antagonist (IL-1Ra), a soluble protein that binds to the IL-1 type I receptor subunit and blocks receptor activation by IL-1α or IL-1β (Greenfeder et al., 1995). IL-1RA thus plays an important role in the tight regulation of IL-1 signaling during normal tissue homeostasis.

In cancer, studies have reported deregulated expression of IL-1α, IL-1β, and IL-1Ra. Gene polymorphisms to IL-1α, IL-1β have been linked to various outcomes, such as increased risk for cancer development, and p53 gene mutations. A few studies have shown that increased expression of the IL-1Ra correlates with good prognosis (Table 1). Few studies on the prognostic significance of IL-1α expression have been conducted. However, increased IL-1β serum levels or increased protein expression correlate with poor prognosis in many carcinomas as well as glioblastoma (Table 2). Fewer studies have been performed to determine the prognostic significance of IL-1 receptor polymorphisms or expression patterns in cancer.

IL-1α and IL-1β signaling have been well studied in skin cancer. Treatment of mice with IL-1Ra inhibits melanoma metastasis and inhibits tumor angiogenesis (Chirivi, Garofalo, Padura, Mantovani, & Giavazzi, 1993; Voronov et al., 2003). IL-1β knockout mice show decreased tumor growth and lung metastasis; these phenotypes are associated with decreased tumor angiogenesis (Voronov et al., 2003). IL-1α deficiency also inhibits tumor angiogenesis in melanoma, but to a lesser extent than IL-1β. In IL-1β and IL-1α/IL-1β deficient mice, treatment with a skin carcinogen, 3-methylcholanthren results in decreased tumor incidence and slower tumor development, compared to wild-type mice. IL-1α/IL-1β deficient tumors are characterized by accumulation of fibroblasts and sparse infiltration of macrophages. Homozygous knockout of IL-1α is less effective than IL-1β or IL-1α/IL-1β at inhibiting tumor incidence and development. Tumors in IL-1α deficient mice are characterized by a late infiltration of macrophages, similar to tumors in wild-type mice (Krelin et al., 2007). Knockout of IL-1Ra enhances development of tumors, which are characterized by accumulation of neutrophils. These studies indicate that IL-1β may contribute to skin cancer progression differently than IL-1α, in part by remodeling the tumor microenvironment.

IL-1α and IL-1β may exert phenotypes similar to skin cancer. For example, IL-1β deficiency inhibits tumor angiogenesis more significantly than IL-1α, in prostate and mammary tumors (Voronov et al., 2003). Similar to melanoma, IL-1Ra inhibits growth of pancreatic cancers (Zhuang et al., 2015). In contrast to melanoma, IL-1α plays an important role in promoting the metastatic potential of pancreatic cancer cells, in part by inducing

Table 1 Single Nucleotide Polymorphisms of Genes Encoding Interleukin and Interleukin Receptors in Cancer

Interleukin	Polymorphism	Cancer Type	Clinical Relevance	References
IL-1α	rs1800587	Lung	Unknown	Bai et al. (2013)
IL-1β	rs1143634	Pancreatic	Associated with decreased survival	Barber et al. (2000)
	rs1143627	Breast	Associated with fatigue	Collado-Hidalgo, Bower, Ganz, Irwin, and Cole (2008)
		Lung	Associated with increased risk for development	Bai et al. (2013)
		HCC	Associated with Japanese patients	Wang et al. (2003)
		NSCLC	Associated with p53 gene mutation	Zienolddiny et al. (2004)
	rs16944	NSCLC	Associated with p53 gene mutation	Zienolddiny et al. (2004)
		Breast	Decreased risk for development	Ito et al. (2002)
IL-1Ra	rs1794068	Multiple myeloma	No association with risk	Zheng et al. (2000)
		Gastric	Associated with increased risk	Machado et al. (2001)
		Vulvar	Associated with decreased risk	Grimm et al. (2004)
IL-17A	rs2275913	Cervical, gastric, lung, and breast	Increased risk for development	He et al. (2015), Hou and Yang (2015), Sun, Wang, and Huang (2015), Wang, Jiang, et al. (2012), and Wang et al., 2014
		Papillary thyroid	Decreased risk for development	Lee et al. (2015)
		Colon	Associated with reduced efficacy for chemo- and radiotherapy	Omrane et al. (2015)

Continued

Table 1 Single Nucleotide Polymorphisms of Genes Encoding Interleukin and Interleukin Receptors in Cancer—cont'd

Interleukin	Polymorphism	Cancer Type	Clinical Relevance	References
IL-17F	rs763780	Cervical	Increased risk	Sun et al. (2015)
		Colon	Decreased risk for development and progression	Nemati, Golmoghaddam, Hosseini, Ghaderi, and Doroudchi (2015)
	rs12203582	Lung	Associated with increased risk for development	He et al. (2015)
IL-17RA	rs4819554	Papillary thyroid	Associated with decreased risk for development	Lee et al. (2015)
IL-17RB	rs1025689	Papillary thyroid	Associated with decreased risk for development	Lee et al. (2015)

NSCLC, Nonsmall cell lung cancer; *HCC*, hepatocellular carcinoma.

expression of HGF from adjacent stromal cells, which acts on cancer cells to enhance invasiveness (Xu et al., 2010). IL-1α derived from pancreatic and colon cancer cells are important for endothelial sprouting and angiogenesis (Matsuo, Sawai, Ma, et al., 2009; Matsuo, Sawai, Ochi, et al., 2009). IL-1α is important for regulating expression of prometastatic genes including CXCL8 and MMP3 (Chen et al., 1998; Nozaki, Sledge, & Nakshatri, 2000). These studies indicate that IL-1α is important for regulating the metastatic potential for multiple cancer types.

2.2 Interleukin-6 (IL-6)

The IL-6 family members are glycoproteins with a structure of four alpha helices and include: the original family member IL-6, as well as leukemia inhibitory factor (LIF), oncostatin M (OSM), ciliary neurotrophic factor (CNTF), IL-11, and cardiothrophin-1 (CT-1) (Hammacher et al., 1994; Nakashima & Taga, 1998). IL-6 is important for regulating multiple biological processes including: growth and differentiation of B and T cells, and colony formation of multipotential hematopoietic cells. IL-6 also regulates expression of hepatic acute phase proteins including: C-reactive protein,

Table 2 RNA and Protein Expression Patterns of Interleukins in Cancer

Interleukin	Cancer Type	Clinical Relevance	References
IL-1α	Prostate	Reduced stromal protein expression correlates with poor prognosis	Rodriguez-Berriguete et al. (2013)
IL-1β	Breast	Increased protein expression correlates with invasiveness, CD68+ macrophages	Jin et al. (1997) and Premkumar, Yuvaraj, Vijayasarathy, Gangadaran, and Sachdanandam (2007)
	Glioblastoma	Increased serum levels correlates with disease recurrence and decreased overall survival	Albulescu et al. (2013)
	Melanoma	Increased serum levels correlates with myeloid-derived suppressor cells, Tregs, tumor progression	Jiang et al. (2015)
	NSCLC	Increased serum levels associated with tumor progression, aggressiveness	Enewold et al. (2009)
	Prostate	Decreased serum expression levels correlates with poor prognosis in patients treated with tamoxifen, nutritional supplements	Rodriguez-Berriguete et al. (2013)
	Pancreatic	Increased protein expression, serum levels correlates with decreased overall and progression free survival, decreased efficacy to gemcitabine	Mitsunaga et al. (2013)
	Renal	Increased serum levels associated with tumor progression, poor prognosis	Xu et al. (2015)
IL-1Ra	Breast	Protein expressed in tumor cells correlates with ER levels	Miller et al. (2000)
	CML	No change between chronic phase and normal patients	Wetzler et al. (1994)

Continued

Table 2 RNA and Protein Expression Patterns of Interleukins in Cancer—cont'd

Interleukin	Cancer Type	Clinical Relevance	References
	Ovarian	Decreased protein concentration in ascites correlates with good prognosis	Mustea et al. (2008)
IL-6	Breast	Increased serum levels correlates with recurrence in Her2 tumors, decreased survival in hormone refractory metastatic breast cancer. Increased RNA expression correlates with good prognosis in breast cancer in general; RNA expression correlates with poor survival in triple negative breast cancer	Bachelot et al. (2003), Cho, Sung, Yeon, Ro, and Kim (2013), Hartman et al. (2013), and Karczewska, Nawrocki, Breborowicz, Filas, and Mackiewicz (2000)
	Colon cancer	Increased RNA expression correlates with disease recurrence	Olsen et al. (2015)
	Gastric	Increased serum levels correlated with increased disease progression, decreased patient survival	Ashizawa et al. (2005)
	HCC	Increased serum levels correlate with increased risk for development of HCC. Decreased serum levels in patients with hepatitis B virus-related hepatic carcinoma correlates with disease recurrence	Aleksandrova et al. (2014) and Cho et al. (2015)
	Leukemia	Increased protein expression correlates with stage	Levidou et al. (2014)
	Melanoma	Increased serum levels correlate with poor prognosis	Tas et al. (2005)
	NSCLC	Increased serum levels correlates with decreased overall survival	De Vita et al. (1998), Liao et al. (2014), and Wojciechowska-Lacka, Matecka-Nowak, Adamiak, Lacki, and Cerkaska-Gluszak (1996)

Table 2 RNA and Protein Expression Patterns of Interleukins in Cancer—cont'd

Interleukin	Cancer Type	Clinical Relevance	References
	Osteosarcoma	Increased serum levels correlate with decreased patient overall survival	Rutkowski, Kaminska, Kowalska, Ruka, and Steffen (2003)
	Ovarian	Protein expression in macrophages does not significant correlation with outcome. Protein expression in epithelium correlates with decreased survival	Isobe et al. (2015) and Wouters et al. (2014)
	Prostate	Increased serum levels correlates with disease recurrence	Domingo-Domenech et al. (2006)
	Pancreatic	Increased serum levels correlates with decreased survival	Mroczko, Groblewska, Gryko, Kedra, and Szmitkowski (2010)
	Renal	Increased protein expression correlates with decreased survival	Fu et al. (2015)
IL-17A	Bladder	Increased protein expression associated with lower tumor stage	Doroudchi et al. (2013)
	Breast	Increased protein expression correlates with FoxP3+ Tregs	Benevides et al. (2013)
	Colon	Increased RNA and protein expression correlates with FoxP3+ T regs	Wang et al. (2014)
IL-17F	Oral squamous cell	Increased serum protein levels correlates with overall stage and lymph node metastasis	Ding et al. (2015)
IL-17 (not specified)	NSCLC	Increased protein levels associated with smoking status, decreased survival, lymphangiogenesis	Chen et al. (2010)

CML, chronic myelocytic leukemia; *ER*, estrogen receptor; *HCC*, hepatocellular carcinoma; *NSCLC*, nonsmall cell lung cancer; *Tregs*, Tregulatory cells.

serum amyloid, haptoglobin, and fibrinogen, which are functional components of the complement system and coagulation during wound healing and infection (Bode, Albrecht, Haussinger, Heinrich, & Schaper, 2012; Koj, 1985). IL-6 carries out its functions by binding to an 80 kDa IL-6R subunit, which heterodimerizes with a polypeptide chain signal transducer, glycoprotein 130 (Gp130) (Hibi et al., 1990; Kishimoto, Akira, Narazaki, & Taga, 1995). Formation of this receptor complex leads to phosphorylation and activation of Janus kinase (JAK), which in turn phosphorylates a cytoplasmic portion of Gp130, leading to activation of several signaling pathways, including MAPK, PI3-kinase, and STAT1 and STAT3 pathways, which regulate gene transcription (Heinrich, Behrmann, Müller-Newen, Schaper, & Graeve, 1988). These pathways are summarized in Fig. 2.

Changes in IL-6 or its receptor IL-6R at the DNA, RNA, and protein levels are associated with cancer prognosis. IL-6 polymorphisms have been linked to increased risk for development of lung or breast cancer (Table 1). Increased serum levels of IL-6 or IL-6R are generally associated with poor prognosis for patients with carcinomas as well as sarcomas (Tables 2 and 3). Currently, the clinical relevance of IL-6R expression is less well understood than IL-6 expression in cancer.

Many functional studies using IL-6 knockout mice show an important role for IL-6 in hepatocyte cell growth (Yeoh et al., 2007), liver regeneration (Cressman et al., 1996), and regulation of hepatic acute phase proteins (Koj, 1985). IL-6 plays an important role in liver fibrosis and inflammation. Delivery of recombinant IL-6 to rats induces liver fibrosis and inflammation (Choi, Kang, Yang, & Pyun, 1994). However, other animal models show that IL-6 is associated with tissue repair and attenuation of fibrosis in toxin-induced liver injury, possibly indicating a context-dependent role for IL-6 in liver fibrosis (Kovalovich et al., 2000; Nasir et al., 2013). IL-6 serum levels are associated with chronic inflammatory liver disease and with the development of hepatocellular carcinoma (HCC) (Table 2). Much attention has been paid to the role of IL-6 in HCC. Treatment of mice with a chemical carcinogen, diethylnitrosamine (DEN) results in formation of HCC, associated with increased IL-6 production from resident macrophages (Naugler et al., 2007). Homozygous knockout of IL-6 significantly inhibits formation of HCC. Interestingly, IL-6 production is negatively regulated by estrogen (Naugler et al., 2007), which may be one factor contributing to the differences in HCC formation between men and women (Zhang, Ren, et al., 2015). Conditional knockout of GP130 in hepatocytes results in

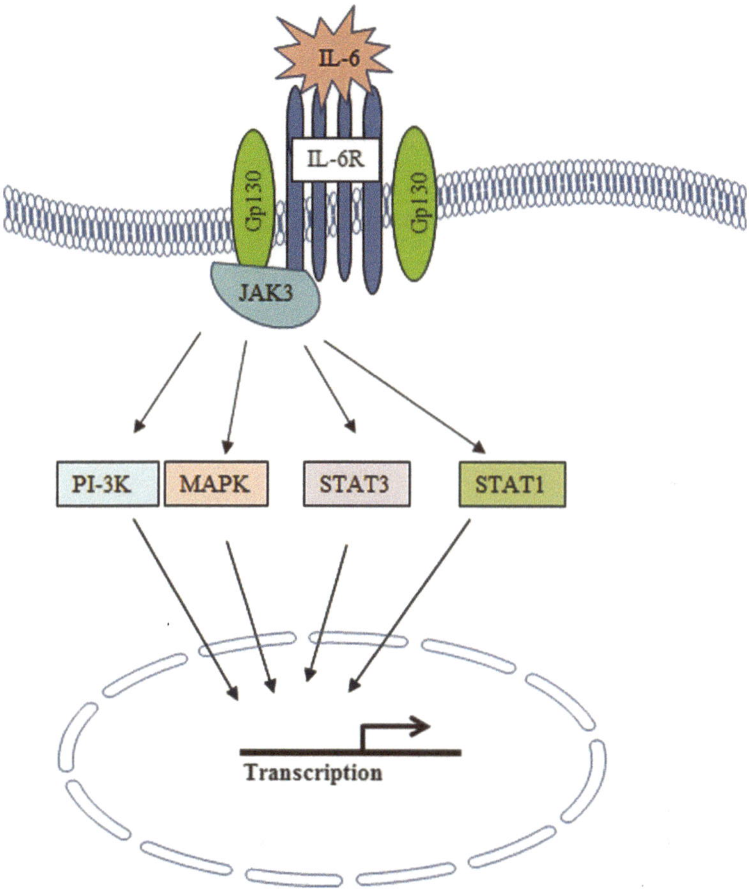

Fig. 2 IL-6 signal transduction. IL-6 binds to IL-6R, which heterodimerizes with glycoprotein 130 (Gp130). Formation of this receptor complex leads to phosphorylation and activation of Janus kinase (JAK), which in turn phosphorylates a cytoplasmic portion of Gp130, leading to activation of several signaling pathways, including MAPK, PI3-kinase, and STAT1 and STAT3 pathway to regulate gene transcription.

decreased recruitment of monocytes and peripheral mononuclear cells, and formation of tumors induced by DEN treatment (Hatting et al., 2015). IL-6 acts on macrophages to M2 macrophage polarization (Mauer et al., 2014), found proximal to HCC progenitor cells (Finkin et al., 2015). In coculture studies, IL-6 derived from macrophages promote expansion of HCC CD44+ dedifferentiated cells through Stat3-dependent mechanisms (Wan et al., 2014). In addition, autocrine IL-6 signaling in HCC progenitor cells is important for tumor growth and invasive progression (He et al., 2013).

Table 3 RNA and Protein Expression Patterns of IL-6 Receptors in Cancer

Cancer Type	Clinical Relevance	References
Breast	Increased RNA expression levels correlates with good prognosis	Karczewska et al. (2000)
Myeloma	Increased soluble receptor levels in serum correlates with decreased survival	Pulkki et al. (1996)
Osteosarcoma	Increased protein expression correlates with decreased survival	Rutkowski et al. (2003)
Ovarian	Increase protein expression in epithelium correlates with decreased progression free survival in one study; another shows correlation with increased survival	Isobe et al. (2015) and Wouters et al. (2014)
Pancreatic	Increased protein expression correlates with decreased survival	Denley et al. (2013)

These studies indicate that paracrine and autocrine IL-6 signaling are important for development and progression of HCC.

IL-6 has also been well studied in breast cancer. Of the molecular subtypes, triple negative (ER, PR, and Her2 negative) breast cancers express the highest levels of IL-6, which are important for anchorage-independent cell growth. Knockdown of IL-6 inhibits tumor formation and growth in animals (Hartman et al., 2013). Recent studies demonstrated that IL-6 receptor: JAK complexes interact with rearranged during transfection (RET) receptor tyrosine kinase:FAK complexes to modulate ER + breast cancer cell migration and invasion (Brocke-Heidrich et al., 2004). Fibroblast-derived IL-6 suppresses ERα levels and promotes tamoxifen resistance in luminal breast cancer cells (Sun et al., 2014). Furthermore, IL-6 derived from fibroblasts or mesenchymal stem cells (MSCs) enhance breast cancer cell growth, migration, and invasion through STAT3 signaling-dependent mechanisms (Di et al., 2014; Lieblein et al., 2008; Osuala et al., 2015; Studebaker et al., 2008). IL-6 modulation of Y-box binding protein-1 activity and regulation of Notch-3, Jagged-1, and carbonic anhydrase IX gene expression are important for these cellular processes (Castellana, Aasen, Moreno-Bueno, Dunn, & Ramon, 2015; Studebaker et al., 2008). These data demonstrate that IL-6 is important for progression of multiple breast cancer subtypes.

IL-6 acts on other cancer types. Antibody neutralization of IL-6 inhibits the growth of lung cancer xenografts (Song et al., 2014) and enhances chemosensitivity to melphalan in animal models of multiple myeloma (Hunsucker et al., 2011). In cell culture studies, IL-6 enhances survival and inhibits apoptosis of esophageal and multiple myeloma cells through Stat3- and MAPK-dependent mechanisms (Leu, Wong, Chang, Huang, & Hu, 2003; Loffler et al., 2007). In prostate cancer cells, IL-6 mediates survival of aggressive cancer cells via Mcl-1 expression (Cavarretta et al., 2008). These studies indicate an important role for IL-6 signaling in modulating cell survival across multiple cancer types.

IL-6 signaling may also regulate tumor progression via remodeling of the tumor microenvironment. In coculture studies, IL-6 derived from colon cancer cells enhance the phagocytic activity and migration of macrophages (Yeh, Wu, & Wu, 2016). In ovarian cancer, IL-6 induces leaky blood vessel formation by lack of pericyte coverage (Gopinathan et al., 2015; Nilsson, Langley, & Fidler, 2005). IL-6 also stimulates migration of Foxp3+CD4+ Tregulatory cells (Tregs) in a lung cancer model (Eikawa et al., 2010) and increases migration of MSCs to breast cancer cells (Rattigan, Hsu, Mishra, Glod, & Banerjee, 2010). These studies indicate that IL-6 has the capacity to significantly remodel the microenvironment by acting on multiple stromal cell types.

2.3 Interleukin-17 (IL-17)

IL-17 are secreted glycoproteins (35 kDa) with a conserved cysteine motif at the C-terminal region and function as homodimers (Hymowitz et al., 2001). There are six isoforms in the IL-17 family (A–F). The first isoform, IL-17A was originally cloned out from a rodent cDNA sequence derived from activated T cell hybridoma using subtractive hybridization (Rouvier, Luciani, Mattei, Denizot, & Golstein, 1993). Among the IL-17 isoforms, IL-17A is the most well studied in inflammation and cancer. Studies indicate that IL-17F share similar features and functions to IL-17A. Of the IL-17 family members, IL-17F shares the highest amino acid sequence homology to IL-17A (~50%) and the *IL-17F* gene is located on the same chromosome (6p12) as *IL-17A* (Kolls & Linden, 2004; Wang, Jiang, et al., 2012). Both IL-17A and IL-17F are expressed in: neutrophils, CD8+ cells and unique T cell subsets T cells such as Th17 CD4+ cells and γδ T cells (Ferretti, Bonneau, Dubois, Jones, & Trifilieff, 2003; Harrington et al., 2005; Infante-Duarte, Horton, Byrne, & Kamradt, 2000; Mills, 2008; Shin, Benbernou, Fekkar, Esnault, & Guenounou, 1998; Starnes et al., 2001). In contrast, the other IL-17 isoforms (B–D) are more commonly expressed in other CD4+ T cell subsets (Lee et al., 2001; Starnes, Broxmeyer, Robertson, & Hromas, 2002) and are expressed in various tissues including: pancreas, small intestine, prostate, skeletal muscle, and spinal cord (Li et al., 2000; Moore et al., 2002). IL-17A and IL-17F modulate allergic reactions and host defense against bacteria (Ishigame et al., 2009; Milner et al., 2008) by signaling to cell surface transmembrane receptors expressed on epithelial and stromal cells (Chang & Dong, 2011; Wright et al., 2008), and activation of NF-κB and p42/44MAPK pathways, thereby inducing production of other cytokines such as CXCL1, CXCL8, and CCL2 from epithelial tissues and stromal tissues. These cytokines in turn recruit myeloid cells to the site of inflammation (Awane, Andres, Li, & Reinecker, 1999; Shalom-Barak, Quach, & Lotz, 1998). Autocrine IL-17 signaling is important for regulating recruitment of γδT cells through a positive feedback loop (Sarkar, Cooney, & Fox, 2010).

Deregulated expression of IL-17A and IL-17F increase recruitment and activity of γδT cell subsets, which contribute to chronic inflammatory diseases (Li, Guo, et al., 2015; Lubberts, 2015; Luchtman, Ellwardt, Larochelle, & Zipp, 2014). Recent studies have revealed similar mechanisms through which IL-17 may promote cancer progression. In murine models for fibrosarcoma, γδT cells are the primary source for IL-17 production, and promote tumor growth and angiogenesis (Wakita et al., 2010). In breast cancer, IL-17 treatment of mammary tumor bearing mice increases tumor growth and angiogenesis (Du, Xu, Fang, & Qi, 2012). IL-17-mediated angiogenesis is associated with expression of CXCL8, MMP2, MMP9, and VEGF (Benevides et al., 2013). In HCC, IL-17A knockout mice show deceased tumor growth, while in vivo treatment with recombinant 17A enhance tumor growth. The tumor promoting effects of IL-17A are due in part to suppression of CD8+ T cell activity, increased expression of CXCL5 from tumor cells, and enhancing the recruitment of myeloid-derived suppressor cells (MDSCs), which suppress CD8+ cytotoxic T cell activity. MDSCs also enhance recruitment of γδT cells, demonstrating a complex cross-talk mechanism among γδT cells, CD8+ T cells, and MDSCs (Ma et al., 2014). This IL-17-mediated crosstalk among γδT cells and immature myeloid cells is important for enhancing resistance to anti-VEGF therapies in lymphoma, lung, and colon cancer (Chung et al., 2013). In ovarian cancer, γδT cells promote tumor growth by enhancing recruitment of macrophages expressing the IL-17RA, and soluble factors, which increase tumor angiogenesis and tumor growth (Rei et al., 2014).

These studies indicate that IL-17 functions in the primary tumor by suppressing cytotoxic T cell responses and enhancing angiogenesis and recruitment of macrophages and MDSCs.

Recent studies in breast cancer also indicate a role for IL-17 signaling in regulating metastasis. In a mouse model of lobular breast cancer, γδT cells express IL-17 which mediates recruitment of neutrophils, which suppress CD8+ T cell activity and enhance metastasis to the lymph node and lungs. Depletion of γδT cells or neutrophils inhibits metastasis to these sites (Coffelt et al., 2015). Mammary carcinoma cells injected into a model of IL-17A-mediated arthritis are prone to metastasis (Das Roy et al., 2009; Roy et al., 2011). Moreover, bone marrow stem cells expressing IL-17B interact with IL-17BR expressing breast cancer cells to promote bone metastasis (Goldstein, Reagan, Anderson, Kaplan, & Rosenblatt, 2010), indicating a possible role for IL-17 signaling in mediating the metastatic niche.

Emerging studies show that IL-17 also signals to cancer cells, which express IL-17RA. In a colorectal cancer model, IL-17RA promotes development of APC deficient colon cancer through p42/44MAPK, p38MAPK, and NF-κB-dependent pathways (Wang et al., 2014). In addition, IL-17A and IL-17E enhance the proliferation, survival, and chemoresistance of primary and transformed breast cancer cell lines via c-RAF/S6 kinase signaling (Mombelli et al., 2015). These studies indicate that multiple IL-17 isoforms signal to cancer cells to modulate cell, growth, and invasion.

The clinical relevance of IL-17 expression in cancer has remained unclear until recently. Genetic variants of IL-17A, IL-17B, IL-17RA, and IL-17RB have been associated with cancer risk in various tumors including papillary thyroid, cervical, and colon cancer (Table 1). Increased expression of IL-17A and IL-17F frequently associated with Tregs and poor patient prognosis in carcinomas (Table 2). Further studies are necessary to more clearly determine the clinical relevance of IL-17 and its receptors.

2.4 Interleukins in Anticancer Therapy

Current studies indicate that interleukins act on cancer cells and on multiple stromal cell types to promote tumor growth and metastasis (Fig. 3). Interleukins are currently a therapeutic target of interest in cancer. Several clinical trials targeting IL-6 have been conducted, with variable results (Table 4). While these clinical trials utilized neutralizing antibodies, another potential

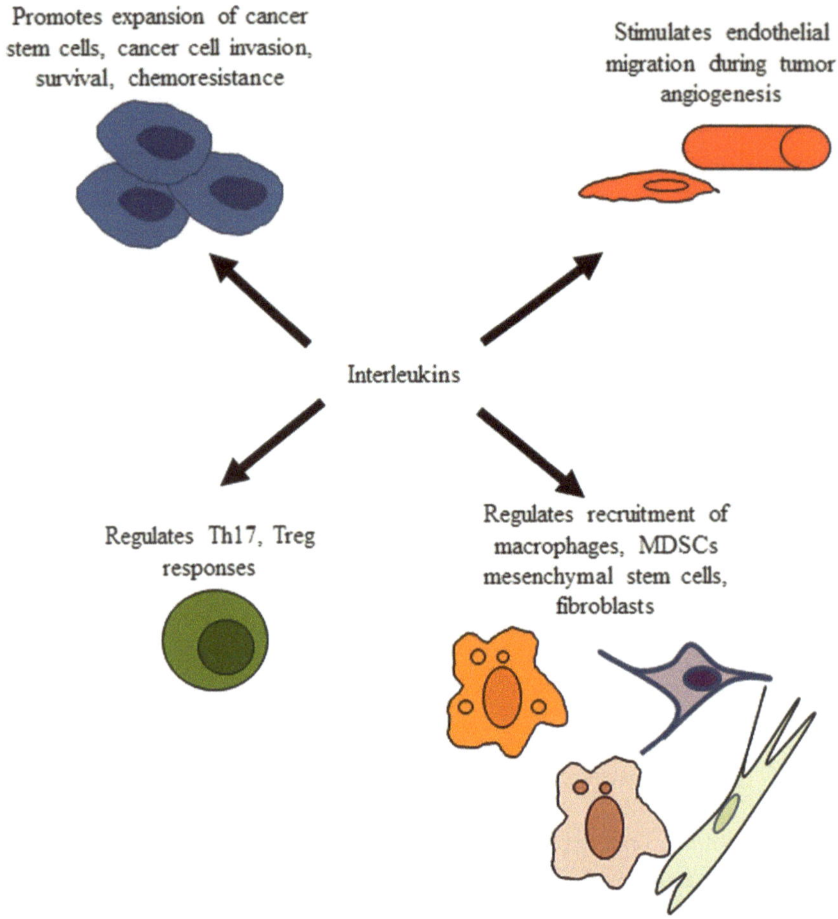

Fig. 3 Role of interleukins in cancer. Studies on IL-1a, IL-1b, and IL-6 indicate these interleukins promote cancer progression by signaling to cancer cells to promote stem cell expansion, cancer cell survival, invasion and chemoresistance, and by promoting tumor angiogenesis, recruiting myeloid cells and regulating T cell responses in the local microenvironment.

strategy is to administer IL-1RA. Since the concentration of circulating IL-1 in disease is relatively low, it is possible to administer soluble receptor antagonists at effective concentrations. The widely known recombinant form of IL-1Ra (Anakinra) works by blocking the IL-1 receptor (Dinarello & van der Meer, 2013). Anakinra was approved in 2001 to treat rheumatoid arthritis, and multiple clinical trials are in progress using Anakinra in combination therapy to treat advanced cancer (Table 4). However, a potential disadvantage of using soluble IL-1 receptor therapy is that these receptors might

Table 4 Targeting of Interleukin Pathways in Clinical Trials

Target	Drug	Sponsor	Disease	Status	References
IL-6	Siltuximab (CNTO 328) monoclonal antibody	Centocor, Inc.	Ovarian, pancreatic, colorectal, head and neck, and lung	Phase II completed	NCT00841191
	Siltuximab (CNTO 328) monoclonal antibody	Centocor, Inc.	Metastatic prostate cancer	Phase II terminated early, due to lack of efficacy	NCT00385827
	S0354, chimeric monoclonal antibody	Southwest Oncology Group, National Cancer Institute (NCI)	Castration resistant prostate cancer	Phase II, PSA response rate of 3.8% and a RECIST stable disease rate of 23%	NCT00433446 (Dorff et al., 2010)
IL-1	Anakinra (IL-Ra)	Warren Grant Magnuson Clinical Center	Unspecified adult solid tumor	Phase I, completed	NCT00072111
	Anakinra in combination with Afinitor	M.D. Anderson Cancer Center	Advanced cancer, unspecified	Phase I, in progress	NCT01624766
	Anakinra in combination with chemotherapy	Baylor Research Institute	Metastatic breast cancer	Phase I, in progress	NCT01802970

PSA, prostate serum antigen; *RECIST*, response evaluation criteria in solid tumors.
Unless specified, clinical trials are occurring, or have occurred in the United States. Clinicaltrials.gov trial number is referenced, unless otherwise specified.

prolong the clearance of IL-1 and affect the delayed response to IL-1 (Dinarello, 1996). A clear understanding on the pharmacokinetics and toxicology of soluble interleukin receptor antagonists will be necessary to achieve therapeutic success.

3. CHEMOKINES

Chemokines were first discovered as a class of proteins that direct the migration of neutrophils and monocytes through the formation of concentration gradients (Locati et al., 1994; Sozzani et al., 1991). Since then, chemokines are known to recruit other immune cell types including T cells and natural killer (NK) cells, and also act on mesenchymal cells to promote angiogenesis during inflammation. Chemokines encompass a large family of proteins (5–10 kDa) in which over 40 ligands and 16 receptors have currently been identified, and have been classified into several groups, depending on the composition of a conserved cysteine motif at the NH2 terminus. Current classes include CC, CXC, and CXC3C, in which the X is a noncysteine amino acid residue. Within the C–X–C class, chemokines are further divided depending on the presence of a Glu-Leu-Arg (ELR) motif. This motif is important for chemokine: receptor binding and mediating neutrophil activity and angiogenesis (Clark-Lewis, Dewald, Geiser, Moser, & Baggiolini, 1993; Hebert, Vitangcol, & Baker, 1991; Strieter et al., 1995). While chemokines are capable of binding multiple receptors, chemokine ligands may exhibit a strong affinity to specific receptors. For example, CXCL1 is capable of binding to CXCR1 and CXCR2, but exhibits a stronger affinity to CXCR2 (Lowman et al., 1996). Furthermore, in animal studies, CCL2 and CCR2 knockout mice show defects in macrophage recruitment, without compensatory upregulation of other chemokine ligands or receptors (Boring et al., 1997; Huang, Wang, Kivisakk, Rollins, & Ransohoff, 2001; Kurihara, Warr, Loy, & Bravo, 1997). These studies indicate a unique role for CCL2/CCR2 signaling in regulating macrophage recruitment. Thus, despite the potential for extensive functional redundancy, certain chemokine signaling can exert unique biological functions.

Chemokines transduce signals by binding to seven transmembrane G protein-coupled receptors. The exceptions are Duffy and D6 receptors, which bind to and sequester several chemokine ligands (Hansell, Hurson, & Nibbs, 2011; Horuk, 2015). Ligand binding at the NH_2 terminus of the chemokine receptor leads to phosphorylation of serine/threonine residues in the intracellular receptor region, conformational changes, and activation of a heterotrimeric small G protein complex comprised of G_α, G_β, and G_γ subunits. Activation of the G protein complex to a GTP bound state leads to disassociation of the G_α subunit form the G_β and G_γ partners, and activation of downstream pathways including: PI-3 kinase (PI-3K), the Rho family of GTPases, and p42/44MAPK (Fang et al., 2012; Jimenez-Sainz, Fast, Mayor, & Aragay, 2003). Examples of C–C and C–X–C chemokine signaling are shown in Figs. 4 and 5. Chemokines also activate G protein-independent pathways such as p38MAPK (Mellado et al., 1998; Vlahakis

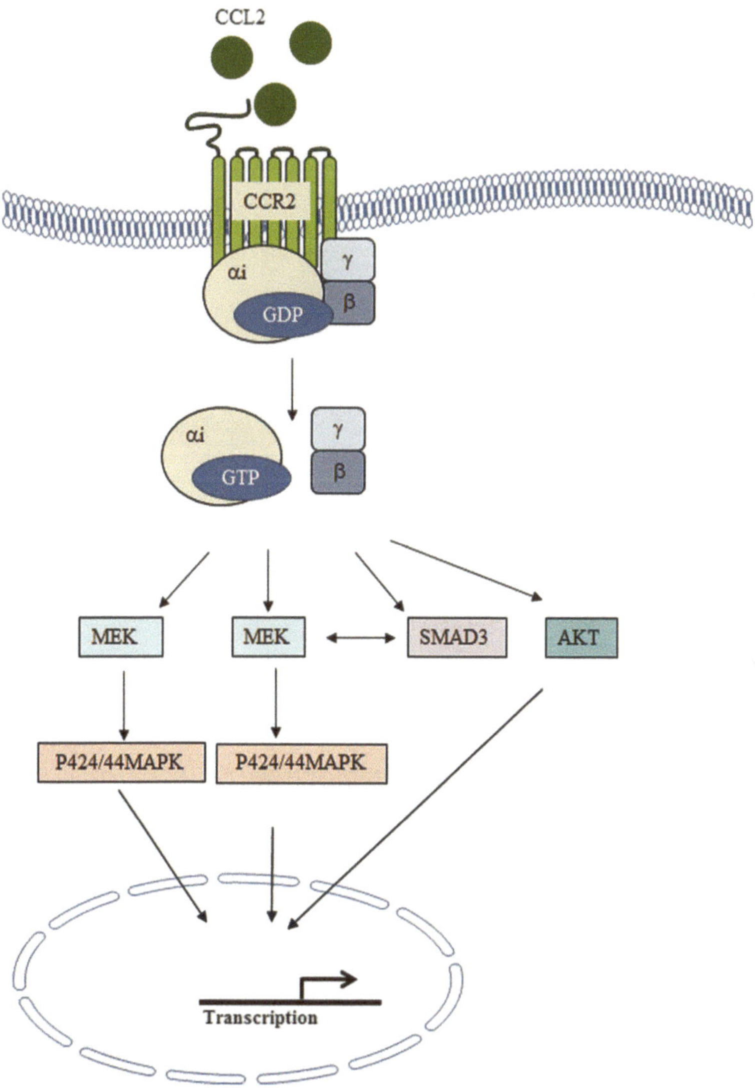

Fig. 4 CCL2 signal transduction pathway. As shown for cancer cells, a G protein complex, comprised of Gαi, γ and β subunits are bound to CCR2 in a GDP inactive bound state. CCL2 binding to CCR2 activates that G protein complex through GTP binding to the Gαi subunit, lead to diassociation from β/γ subunits, and activation of p42/44MAPK, Smad3 and AKT pathways to modulate gene transcription.

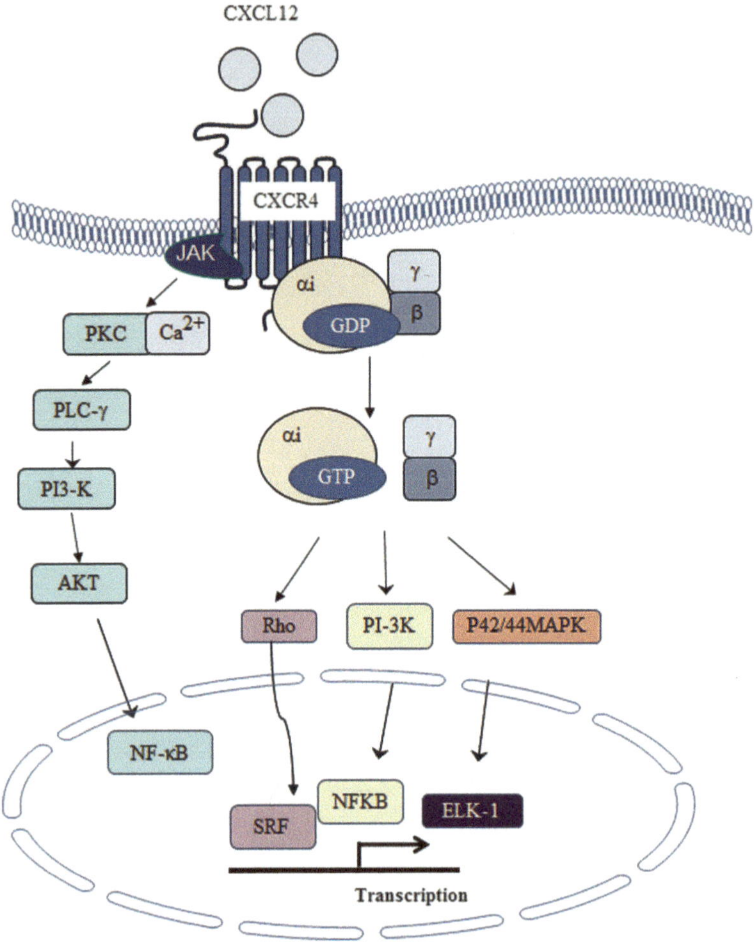

Fig. 5 CXCL12 chemokine signal transduction pathway. As shown for cancer cells, a G protein complex, comprised of Gαi, γ and β subunits are bound to CCR2 in a GDP inactive bound state CXCL12 binding to CXCR4 activates that G protein complex through GTP binding to the Gαi subunit, lead to diassociation from β/γ subunits, and activation of Rho, PI-3 kinase, and p42/44MAPK pathways. These pathways modulate activity of transcription factors, SRF, NF-kB, and ELK-1. CXCL12/CXCR4 signaling also activates G protein-independent pathways to modulate gene transcription, such as PKC signaling via JAK-dependent mechanisms.

et al., 2002). These pathways are important in immune cell recruitment, endothelial cell migration, and proliferation of adipocytes during normal physiologic processes including: tissue development, wound healing, and immunity (Burger & Kipps, 2006; Cotton & Claing, 2009).

To maintain normal cellular homeostasis, chemokine signaling is tightly regulated on multiple levels. Chemokine levels are kept at low levels in the resting adult. During acute inflammation, chemokine expression is temporarily induced by other cytokines including TNF-κ and IL-1β (Biswas et al., 1998; Hacke et al., 2010). In normal tissues, prolonged chemokine ligand/receptor binding eventually leads to receptor internalization and downregulation of downstream signaling. In cancer, expression of chemokines and receptors in the C–C and C–X–C families are chronically upregulated and are often associated with poor patient prognosis. For the purposes of brevity, we will focus on a few members from the C–C and C–X–C families. These chemokines have acquired various names due to discovery from multiple research groups over time. For purposes of clarification, these names, their formal name (Zlotnik & Yoshie, 2000), and their binding receptors are summarized (Table 5). Gene variants of chemokines and their binding receptors are associated with risk of cancer development (Tables 6 and 7). Increased RNA and protein expression of chemokines or the corresponding receptors are upregulated in many tumor types (Tables 8–10). These findings demonstrate a clinical relevance for chemokine expression at the genomic, RNA, and protein levels. As demonstrated for the following chemokines, deregulation of these chemokine signaling pathways may have important implications on existing and future anticancer therapies.

Table 5 Chemokine Ligands and Their Binding Receptors Discussed in This Review

Alternative Names (Murine)	Alternative Names (Human)	Approved Name	Receptor
JE, Scya2	SCYA2, MCP1, MCP-1, MCAF, SMC-CF, GDCF-2, HC11, MGC9434	CCL2	CCR2, CCR4, CCR5, Duffy, D6
Scya5	D17S136E, SCYA5, RANTES, SISd, TCP228, MGC17164	CCL5	CCR5, Duffy, D6
Gro1, Mgsa, KC, Scyb1	MGSA, GRO1, FSP, SCYB1, GROα, MGSA-α, NAP-3	CXCL1	CXCR2, CXCR1, Duffy
Unknown	IL-8	CXCL8	CXCR2, CXCR1
Scyb12	SDF1A, SDF1B, SDF1, SCYB12, SDF-1a, SDF-1b, PBSF, TLSF-a, TLSF-b, TPAR1	CXCL12	CXCR4

Binding receptors are color coded.

Table 6 Single Nucleotide Polymorphisms of Genes Encoding Chemokine Ligands in Cancer

Chemokine	Polymorphism	Cancer Type	Role	References
CCL2	rs1024611	Bladder, breast, colon, endometrial, gastric, melanoma, oral squamous carcinoma Prostate, renal	Increased risk of development	Attar et al. (2010), Kucukgergin et al. (2012), Liu et al. (2013), Rodero et al. (2007), Bektas-Kayhan, Unur, Boy-Metin, and Cakmakoglu (2012), Ghilardi, Biondi, La Torre, Battaglioli, and Scorza (2005), Gu et al. (2011), and Sun et al. (2011)
	rs3760399	Prostate	Associated with disease progression in patients with prostectomy	Sun et al. (2011)
	rs2857654, rs2530797	Prostate	Increased risk of development	Sun et al. (2011)
	rs3760396	NSCLC	Decreased risk of development	Ma et al. (2011)
CCL5	rs2107538	Oral cancer	Increased risk	Weng et al. (2010)
CXCL8	rs4073	Breast	Increased risk	Snoussi et al. (2010)
		Prostate	No risk	Yang et al. (2006)
		Ovarian	Increased therapeutic responsiveness	Schultheis et al. (2008)
CXCL12	rs2297630	Renal cell	Increased risk	Kwon et al. (2011)
	rs1801157	Breast cancer	Associated with increased risk of development	de Oliveira et al. (2011) and Razmkhah, Talei, Doroudchi, Khalili-Azad, and Ghaderi (2005)

Table 6 Single Nucleotide Polymorphisms of Genes Encoding Chemokine Ligands in Cancer—cont'd

Chemokine	Polymorphism	Cancer Type	Role	References
		Colon	Associated with increased risk of development	Chang et al. (2009)
		Esophago-gastric	Associated with disease progression	Schimanski et al. (2011)
		HCC	Predicts lymph node metastasis in stage 3, T3 cancer	Chang et al. (2009)
		Myeloid leukemia	Not associated with risk	El-Ghany, El-Saadany, Bahaa, Ibrahim, and Hussien (2014)
		Nasopharangeal	Associated with decreased DMFS and progression free survival	Chen et al. (2015)
		Prostate	Variations in risk	Hirata et al. (2007) and Isman et al. (2012)
		Renal cell	Associated with decreased survival	Cai et al. (2013)
	rs1804429	NSCLC	Associated with decreased survival	Ma et al. (2011)

3.1 CCL2

CCL2/CCR2 signaling is best known for its role in regulating macrophage recruitment and polarization during inflammation. CCL2 regulates cellular adhesion and chemotaxis of macrophages through activation of β1 integrins and p38MAPK signaling pathways (Ashida, Arai, Yamasaki, & Kita, 2001).

Table 7 Single Nucleotide Polymorphisms of Chemokine Receptor Genes in Cancer

Chemokine Receptor	Polymorphism	Cancer Type	Role	References
CCR2	rs1799864	Her2 + breast cancer, bladder, endometrial, and renal	Increased risk for development	Attar et al. (2010), Banin-Hirata et al. (2016), Kucukgergin et al. (2012), and Liu et al. (2013)
		Prostate	Decreased risk for development	Zambra, Biolchi, Brum, and Chies (2013)
CCR5	rs333	Breast, prostate	No risk found	Zambra et al. (2013)
		Melanoma	Unfavorable prognosis for patients receiving immunotherapy	Ugurel et al. (2008)
		Cervical, gallbladder	Increased risk	Srivastava, Pandey, Choudhuri, and Mittal (2008) and Singh, Sachan, Jain, and Mittal (2008)
CXCR1	rs559029	Oral	Variable risk	Weng et al. (2010)
	rs2230054	Prostate	No risk found	Yang et al. (2006)
	rs2234671	Colon	Chemotherapeutic responsiveness	Gerger et al. (2011)
	rs4073	Breast	Increased risk of developing invasive breast cancer	Kamali-Sarvestani, Aliparasti, and Atefi (2007)
CXCR2	rs1801032	Breast	Increased risk	Snoussi et al. (2010)
			Associated with disease progression	Kamali-Sarvestani et al. (2007)
	rs2230054	Ovarian cancer	Associated with progression free survival	Schultheis et al. (2008)

Table 7 Single Nucleotide Polymorphisms of Chemokine Receptor Genes in Cancer—cont'd

Chemokine Receptor	Polymorphism	Cancer Type	Role	References
	rs1226580	Prostate cancer	No risk found	Yang et al. (2006)
	rs1799939	Pancreatic	Increased risk	Donahue and Hines (2009)
CXCR4	rs2228014	Breast, HCC	No risk found	Okuyama Kishima et al. (2015) and Chang et al. (2009)
		Renal cell	Correlates with decreased survival	Cai et al. (2013)

Prolonged signaling in macrophages leads to: activation of β-arrestin, receptor internalization, and downregulation of signaling (Aragay et al., 1998). These mechanisms prolong inflammation in normal tissues. In many cancer types, overexpression of CCL2 or presence of gene variants is associated with macrophage recruitment and poor patient prognosis. Yet, in ovarian, pancreatic and nonsmall cell lung cancer, CCL2 protein expression, and macrophage recruitment correlate with favorable survival (Table 8). While CCL2 binds promiscuously to CCR1–5, it binds with a particularly high affinity to CCR2 (Kurihara & Bravo, 1996; Monteclaro & Charo, 1996; Sarau et al., 1997; Wang, Hishinuma, Oppenheim, & Matsushima, 1993), whose prognostic significance has been less well studied (Table 10). There are a few common patterns of expression. For example, CCL2 and CCR2 polymorphisms have been detected in Her2+ breast cancers, prostate, and renal cancers, correlating with increased risk of cancer development (Tables 6 and 7). These studies indicate an important prognostic significance for CCL2 and CCR2 coexpression in cancer.

Animal studies indicate a tumor promoting role for CCL2 signaling in certain types of cancers. CCL2 knockout or treatment with CCR2 antagonists in animal models inhibits progression of HCC (Li, Yao, et al., 2015). In prostate cancer, CCL2 neutralizing antibodies inhibit growth and progression of xenografts and reduce macrophage recruitment to the primary tumor (Zhang, Lu, & Pienta, 2010). In breast cancer, Her2/neu transgenic mice deficient in CCL2 expression show a longer latency to tumor

Table 8 RNA and Protein Expression Patterns of C–C Chemokines in Cancer

Chemokine	Cancer Type	Clinical Relevance	References
CCL2	Breast	Increased protein expression in stroma, epithelium, correlates with increased disease recurrence, and macrophage recruitment	Fang et al. (2012), Fujimoto et al. (2009), Saji et al. (2001), and Valkovic, Lucin, Krstulja, Dobi-Babic, and Jonjic (1998)
	Colon	Increased protein expression in adenoma	Tanaka et al. (2006)
	Leukemia	Increased RNA levels	de Vasconcellos et al. (2011)
	Melanoma	Increased protein expression in primary tumor and metastatic tissues	Koga et al. (2008)
	Neuroblastoma	Increased RNA levels associated with natural killer cell recruitment	Metelitsa et al. (2004)
	NSCLC	Increased protein expression correlates with increased overall survival	Zhang, Qin, et al. (2013)
	Ovarian	Increased serum levels correlates with tumor grade, age, and prior to treatment. Decreased RNA levels in tumors. Increased protein expression in tumors correlates with improved chemoresponsiveness, patient survival	Hefler et al. (1999), Arnold, Huggard, Cummings, Ramm, and Chenevix-Trench (2005), and Fader et al. (2010)
	Pancreatic	Increased protein, RNA levels in tumor, increased serum levels	Monti et al. (2003)
	Papillary thyroid	Increased protein expression in tumors correlates with lymph notes metastasis, tumor recurrence	Tanaka et al. (2009)

Table 8 RNA and Protein Expression Patterns of C–C Chemokines in Cancer—cont'd

Chemokine	Cancer Type	Clinical Relevance	References
	Prostate cancer	Varying serum levels. Decreased RNA levels in carcinoma vs. benign hyperplasia negatively correlates with tumor grade	Agarwal, He, Siddiqui, Wei, and Macoska (2013), Tsaur et al. (2015), and Mazzucchelli et al. (1996)
CCL5	Breast	Protein expression correlates with tumor stage, CD44+ stem cell in luminal breast cancer. Protein expression associated with poor outcome in stage II ER-breast cancer	Berghuis et al. (2011), Yaal-Hahoshen et al. (2006), Zhang, Ren, et al. (2015), Zhang et al. (2009), and Zumwalt, Arnold, Goel, and Boland (2015)
	Cervical	Increased serum levels correlates with progressive disease, poor prognosis	Niwa et al. (2001)
	Colon	Increased RNA, protein expression correlates with CD8+ T cells and increased survival	Zumwalt et al. (2015)
	Ewing's sarcoma	Increased RNA, protein expression correlates with CD8+ T cells and increased survival	Berghuis et al. (2011)
	Gastric	Increased serum levels correlates with progressive disease, poor prognosis	Sugasawa et al. (2008) and Sima et al. (2014)
	HCC	Increased serum levels correlates with progressive disease, poor prognosis	Sadeghi et al. (2015)
	NSCLC	Decreased levels in serum in patients treated with EGFR TKIs correlates with increased survival	Umekawa et al. (2013)
	Prostate	No significant changes in serum levels	Agarwal et al. (2013)

NSCLC, nonsmall cell lung cancer; *HCC*, hepatocellular carcinoma; *EGFR TKIs*, epidermal growth factor receptor tyrosine kinase inhibitors.

Table 9 RNA and Protein Expression Patterns of C–X–C Chemokines in Cancer

Chemokine	Cancer Type	Clinical Relevance	References
CXCL1	Bladder	Increased protein expression correlates with tumor stage	Miyake, Lawton, et al. (2013)
	Breast cancer	Increased RNA, protein expression in stroma and epithelium associated with disease recurrence, tumor stage	Bieche et al. (2007), Zou et al. (2014), and Razmkhah et al. (2012)
	Colon	Increased protein expression correlates with tumor grade, unfavorable relapse free survival	Oladipo et al. (2011)
	Gastric	Increased RNA, protein expression in tumor correlates with disease progression	Cheng et al. (2011)
	Pancreatic	Increased protein expression associated with neutrophil recruitment, poor prognosis	Li, Xu, et al. (2015)
	Prostate	No significant changes in serum levels. Increased protein expression in tumor correlates with tumor grade	Agarwal et al. (2013) and Miyake, Lawton, Goodison, Urquidi, and Rosser (2014)
CXCL8	Breast	Increased RNA, protein expression levels in tumor and tumor stroma, increased serum levels correlates with tumor stage, associated with metastasis	Razmkhah et al. (2012), Bieche et al. (2007), and Metelitsa et al. (2004)
	Colon	Increased protein expression in tumors associated with poor prognosis	Xiao et al. (2015) and Cheng et al. (2014)
	Pancreatic	Increased protein expression in tumor correlates with malignant disease, unfavorable prognosis	Frick et al. (2008) and Metelitsa et al. (2004)
	Prostate	No significant changes in serum levels	Agarwal et al. (2013)
CXCL12	B cell chronic lymphocytic leukemia	Decreased protein expression does not correlate with stage or bone marrow infiltration	Barretina et al. (2003)

Table 9 RNA and Protein Expression Patterns of C–X–C Chemokines in Cancer—cont'd

Chemokine	Cancer Type	Clinical Relevance	References
	Breast cancer	Increased RNA, protein expression correlates with overall survival	Mirisola et al. (2009) and Razis et al. (2012)
	Colon cancer	Increased protein expression correlates with decreased patient survival, increased lymph node metastasis	Akishima-Fukasawa et al. (2009)
	Glioma	Increased protein expression in tumor correlates with disease progression	Salmaggi et al. (2005)
	Head and neck	Decreased RNA levels correlates with decreased metastasis free, disease free, overall survival	Clatot et al. (2011)
	Head and neck cancer	Positive protein expression in tumor does not associate with prognosis	Fukushima, Sugita, Niino, Mihashi, and Ohshima (2012) and Almofti et al. (2004)
	Osteo-sarcoma	Increased protein expression in tumor associates with favorable outcome, fewer metastases	Baumhoer et al. (2012)
	Pancreatic	Increased RNA, protein levels correlates with decreased overall survival, and decreased relapse free survival for patients with stage II cancer	Koshiba et al. (2000) and Liang et al. (2010)
	Prostate	No significant changes in serum levels	Agarwal et al. (2013)
	Renal	Increased RNA, protein expression in tumor, increased serum levels correlates with tumor stage, lymph node metastasis	Tsaur et al. (2011), Wang, Jiang, et al. (2012), and Wang et al., 2014
	Urothelial	Increased protein expression correlates with grade and stage	Batsi et al. (2014)

NSCLC, nonsmall cell lung cancer; *HCC*, hepatocellular carcinoma; *EGFR TKIs*, epidermal growth factor receptor tyrosine kinase inhibitors.

Table 10 RNA and Protein Expression Patterns of Chemokine Receptors in Cancer

Chemokine Receptor	Cancer Type	Clinical Relevance	References
CCR2	Breast	Increased protein expression epithelium	Fang et al. (2012)
	Prostate	Increased RNA levels in tumor negatively correlates with Gleason score and grade	Tsaur et al. (2015)
	NSCLC	Positive protein expression in tumor does not associate with clinicopathologic variables	Zhang, Qin, et al. (2013)
CCR5	Breast	Increased protein expression in tumors associated with lymph node metastasis	Metelitsa et al. (2004)
	Colon	Increased protein expression in CD8+ T cells associates with favorable prognosis	Musha et al. (2005)
	Leukemia	Increased RNA expression correlates with favorable prognosis	Bigildeev, Shipounova, Svinareva, and Drize (2011)
	Ovarian	Increased protein expression immune infiltrating cells, weak expression on tumor cells is a weak predictor of outcome	Dong et al. (2006)
	Renal	Expression in infiltrating Th1 T cells correlates with favorable prognosis	Kondo et al. (2006)
CXCR1	Endometrial	Increased RNA, protein correlates with tumor grade, decreased survival	Ewington et al. (2012)
	HCC	Increased protein expression correlates with neutrophil recruitment, poor prognosis	Oladipo et al. (2011)

Table 10 RNA and Protein Expression Patterns of Chemokine Receptors in Cancer—cont'd

Chemokine Receptor	Cancer Type	Clinical Relevance	References
CXCR2	Colon	Increased protein expression in tumor epithelium	Oladipo et al. (2011)
	Endometrial	Increased RNA, protein expression associated with tumor grade, decreased survival	Ewington et al. (2012)
	Esophageal	Increased protein expression correlates with lymph node metastasis	Sui et al. (2014)
	HCC	Increased protein expression in peritumoral stroma correlates with poor prognosis	Li, Xu, et al. (2015)
	Laryngeal squamous cell	Increased RNA, protein expression correlates with overall stage, lymph node metastasis, associated with decreased survival	Han et al. (2012)
	Lung	Increased protein expression associated with smoking lifestyle, poor patient prognosis	Saintigny et al. (2013)
	Renal	Increased protein expression in tumor correlates with decreased overall survival and relapse free survival	An et al. (2015)
CXCR4	B cell chronic lymphocytic leukemia	Increased protein expression correlates with leukocyte count, variable association with stage	Barretina et al. (2003) and Ghobrial et al. (2004)
	Breast cancer	Increased expression in tumors associated with reduced overall survival, relapse free survival	Yasuoka et al. (2008)
	Cervical	Increased protein expression correlates with tumor grade	Huang, Zhang, Cui, Zhao, and Zheng (2013)

Continued

Table 10 RNA and Protein Expression Patterns of Chemokine Receptors in Cancer—cont'd

Chemokine Receptor	Cancer Type	Clinical Relevance	References
	Colon	Increased protein expression associated with advanced disease, lymph-vascular invasion, decreased patient survival	Zhang et al. (2012) and Wang et al. (2010)
	Multiple myeloma	Increased protein expression correlates with good survival	Bao et al. (2013)
	Pancreatic	Increased protein expression; variable associations	Koshiba et al. (2000), Krieg, Riemer, Telan, Gabbert, and Knoefel (2015), and Wang et al. (2013)
	Prostate	Increased protein expression correlates with increased lymph node, bone metastasis, and poor prognosis	Chen and Zhong (2015)
	Head and neck	Increased RNA, protein expression does not associate with prognosis	Fukushima et al. (2012) and Clatot et al. (2011)
	Oral squamous cell	Increased protein expression correlates with lymph node metastasis and recurrence	Almofti et al. (2004)
	Urothelial	Increased protein expression correlates with grade and stage	Batsi et al. (2014)

NSCLC, nonsmall cell lung cancer; *HCC*, hepatocellular carcinoma.

development (Conti, Dube, & Rollins, 2004). In mice bearing breast tumor xenografts, treatment with CCL2 neutralizing antibodies decrease tumor growth and metastasis, associated with decreased angiogenesis and M2 macrophage recruitment (Fujimoto et al., 2009; Hembruff, Jokar, Yang, & Cheng, 2010; Qian et al., 2011). Monocyte recruitment and M2 polarization are regulated by CCL2/CCR2 signaling through MAPK pathways (Roca et al., 2009; Sierra-Filardi et al., 2014). CCL2 may also function with

CCL3 and CCR1 to regulate macrophage recruitment during breast metastasis (Kitamura et al., 2015). In melanoma and pancreatic mouse models, siRNA knockdown CCL2 or antibody neutralizations inhibit recruitment of dendritic cells and Tregs and decreased tumor growth and metastasis (Kudo-Saito, Shirako, Ohike, Tsukamoto, & Kawakami, 2013). These studies indicate that CCL2 promotes tumor progression through recruitment and activation of multiple immune cell types.

While CCL2 recruitment of macrophages is a well-established mechanism for regulating tumor development and progression, emerging studies indicate that CCL2 signals to cancer cells. In cell culture studies, treatment with CCL2 recombinant proteins promotes prostate cancer cell proliferation and inhibits autophagic cell death through AKT signaling, which enhances expression of survivin proteins (Zhang et al., 2010). CCL2 signaling in breast cancer cells does not activate AKT, but activates p42/44MAPK and Smad3 pathways through G protein-dependent mechanisms, resulting in increased RhoA expression (Fang et al., 2012). In addition, CCL2 enhances mammosphere formation in certain breast cancer cell lines, indicating a role of regulating cancer stem cell renewal. In addition to apoptosis, CCL2 expression is important for breast cancer cell survival by inhibiting necrosis and autophagy (Fang et al., 2015), indicating that CCL2 regulates survival through modulation of different forms of programmed cell death. These studies indicate that CCL2 signaling modulates cancer cell survival, growth, and invasion. A summary of known CCL2 pathways in cancer cells is shown in Fig. 4.

In some instances, CCL2 may also suppressive tumor progression. CCL2 overexpression in colon cancer cells or rat gliosarcoma cells inhibits tumor development in immunocompetent mice and is associated with recruitment of M1 macrophages at the site of injection (Tsuchiyama, Nakamoto, Sakai, Mukaida, & Kaneko, 2008; Yamashiro et al., 1994). CCL2 is also associated with M1 macrophage recruitment in certain animal models of HCC (Tsuchiyama et al., 2008). In a B16 melanoma model, CCL2/CCR2 signaling mediates recruitment of γδ T cells, which express IFN-γ and are cytotoxic to cancer cells (Lanca et al., 2013). In breast cancer, one study has shown that neutrophils are activated by CCL2 in the primary tumor, and become cytotoxic to metastatic cells in the lung, thereby inhibiting seeding (Granot et al., 2011). These studies indicate that CCL2 suppresses tumor progression through recruitment of immune cells in a context- and tissue-dependent manner.

3.2 CCL5

CCL5 and one of its cognate receptors, CCR5 are best known for its role in HIV, facilitating viral entry into cells (Watson, Jenkinson, Kazmierski, &

Kenakin, 2005). Emerging studies show that changes to CCL5 or CCR5 at the DNA, RNA, and/or protein level may be associated with development or progression of cancer. A CCL5 gene variant is associated with development of oral cancer, while a CCR5 polymorphism is associated with increased risk for gallbladder and cervical cancer (Tables 6 and 7). Increased RNA and protein expression of CCL5 or CCR5 are associated with poor prognosis for gallbladder, cervical, and breast cancer. Other studies have shown that CCL5 or CCR5 expression may be markers for favorable prognosis as shown for Ewing's sarcoma, colon cancer, and renal cancer. In these cases, CCL5 or CCR5 expression is associated with accumulation of CD8+ infiltrating T cells (Tables 8 and 10). However, more work remains to be done to determine whether a functional relationship exists between CCL5/CCR5 and cytotoxic T cells.

Multiple in vivo and in vitro mechanistic studies indicate that CCL5 promotes tumor progression by remodeling the tumor microenvironment. In pancreatic cancer, Treg cells express high levels of CCR5, which are recruited to CCL5 overexpressing tumors. Knockdown of CCL5 or pharmacologic inhibition of CCL5 inhibits pancreatic tumor growth (Tan et al., 2009). In addition, CCL5 may also modulate activity of MDSCs from the bone marrow, and suppress activity cytotoxic T cells, as demonstrated in a model of triple negative breast cancer (Zhang, Lv, et al., 2013; Zhang, Qin, et al., 2013). CCL5 signaling through CCR3 regulate Th2 (IL4(+)CD4(+) T) cellular responses to promote metastasis of luminal breast cancers (Yasuhara et al., 2015). In addition to modulating immune cell recruitment and activity, CCL5 promotes VEGF-dependent angiogenesis in tumors, as demonstrated for chondrosarcoma and osteosarcoma (Liu et al., 2014; Wang et al., 2015). These studies indicate that within the primary tumor, CCL5 suppresses cytotoxic T cell activity, increases recruitment of Tregs, promotes Th2 responses, and promotes tumor angiogenesis.

CCL5 signals directly on cancer cells to promote survival, invasion, and stem cell renewal. In breast cancer, CCL5 expressed by MSCs act on breast cancer cells to promote invasion and metastasis (Karnoub et al., 2007). Furthermore, CCL5 enhances growth and invasion of CD44+/CD24− cells, a population of stem cell cells. These effects are partially inhibited by CCR5 antibody neutralization (Zhang et al., 2009). CCL5/CCR5 modulation of breast cancer cell invasion and stem cell renewal may be regulated by β-catenin signaling (Yasuhara et al., 2015). CCL5 promotes cancer cell survival and chemoresistance through STAT3-dependent mechanisms (Yi et al., 2013). These studies indicate that CCL5 signaling regulates invasion and survival of breast cancer cells, in part, through paracrine interactions between MSCs and cancer cells. In ovarian cancer, CD133+ cells exhibit cancer stem cell-like characteristics and secrete high levels of CCL5, which induce epithelial to mesenchymal transition (EMT) of CD133– cells. Autocrine CCL5 signaling in CD133+ cells is important for migration and invasion through NF-κB-mediated MMP9 expression (Long et al., 2012). Furthermore, CCL5 stimulates EMT and in vitro invasion of CD133– cells through mechanisms dependent on CCR1, CCR3, and CCR5. Furthermore, CCL5 signaling enhances metastasis of CD133– cells to the bowel and liver in animal models (Long et al., 2015). These studies indicate that CCL5 regulates interactions between cancer stems and noncancer stem cell populations to promote tumor progression by signaling to multiple receptors expressed on cancer cells. In colon cancer, CCL5 derived from MDSCs enhances migration and invasion of cancer cells through MALAT1- and Snail-dependent mechanisms (Kan et al., 2015). These studies indicate that CCL5 regulates interactions between MDSCs and cancer cells.

Studies reveal that CCL5 regulates migration, invasion, and survival through many other different pathways. In oral cancer, CCL5 signals exclusively through CCR5 to enhance migration through mechanisms dependent on POLC, PKC, and NF-kB signaling and increased MMP9 production (Chuang et al., 2009). In prostate cancer, CCL5 signals through CCR1 to promote invasion of chemoresistant PC3 prostate cancer cells, through p42/44MAPK and Rac signaling and MMP2- and MMP9-dependent mechanisms (Kato et al., 2013). In human HCC cells, CCL5 regulates migration through Syndecans 1 and 4 and PI-3K/AKT pathways (Bai et al., 2014; Charni et al., 2009). It remains to be determined whether CCL5 regulation of these pathways are common to all cell types or whether these pathways are distinct for each cell type.

3.3 CXCL1

As an ELR+ chemokine, CXCL1 signals through CXCR2 to promote angiogenesis (Miyake, Goodison, Urquidi, Gomes Giacoia, & Rosser, 2013) and regulates recruitment of neutrophils and basophils during inflammation (Chen et al., 2001; Geiser, Dewald, Ehrengruber, Clark-Lewis, & Baggiolini, 1993; Moser, Schumacher, von Tscharner, Clark-Lewis, & Baggiolini, 1991). In cancer, genetic variants of CXCR2 but not CXCL1 have been associated with increased cancer risk (Tables 6 and 7). Increased

RNA and protein expression of CXCL1 or CXCR2 are frequently associated with neutrophil recruitment, disease recurrence, and disease progression (Tables 9 and 10).

The role of CXCL1 has been best characterized in melanoma, where it was originally as identified as a melanoma growth stimulatory activity protein (Richmond & Thomas, 1988; Thomas & Richmond, 1988). Overexpression of CXCL1 transforms immortalized murine melanocytes through Ras-dependent mechanisms, enabling tumor formation in nude and SCID mice (Balentien, Mufson, Shattuck, Derynck, & Richmond, 1991; Dhawan & Richmond, 2002b; Owen et al., 1997). CXCL1 antibody neutralization or siRNA knockdown of CXCL1 in melanocytes enhances tumor cell apoptosis and inhibits tumor growth and invasion (Haghnegahdar et al., 2000; Luan et al., 1997; Singh, Sadanandam, Varney, Nannuru, & Singh, 2010). Transgenic overexpression of CXCL1 coupled with treatment of 7,12-dimethyl-benz(a) anthracene promotes tumor development in mice, compared to CXCL1 overexpression alone. Furthermore, homozygous knockout of the tumor suppressor gene INK-4a/ARF in CXCL1 overexpressing melanocytes increases tumor development to 85% of mice when these cells are transplanted in nude mice (Dhawan & Richmond, 2002b). These studies indicate that CXCL1 cooperates with oncogenic drivers or loss of tumor suppressors to promote tumor development.

Studies have since demonstrated important roles for CXCL1/CXCR2 signaling in regulating the primary tumor and metastatic niche. In breast cancer, knockdown of CXCL1 in PyVmT or LM2–4175 breast cancer cells inhibits primary tumor growth in mammary orthotopic injection models (Acharyya et al., 2012). CXCL1 is part of a gene signature associated with recurrent lung metastasis in mammary tumor models (Minn et al., 2007, 2005). CXCL1 increases invasiveness of circulating tumor cells (Kim et al., 2009). Knockdown of CXCL1 in breast cancer cells inhibits lung metastasis of mammary carcinoma cells in tail vein injection models (Acharyya et al., 2012). In coculture studies, bone-derived MSCs increase PyVmT mammary carcinoma cell migration, which is blocked by antibody neutralization of CXCL1 or CXCR2 (Halpern, Kilbarger, & Lynch, 2011). These studies indicate that CXCL1/CXCR2 signaling in mammary carcinoma cells functions to regulate tumor growth and invasion to the lung and bone. In vitro, CXCL1 signals through CXCR2 to enhance the growth, motility, and invasion of various cancer cell lines including: breast, melanoma, ovarian cancer, lung cancer cells, and esophageal cells. These cellular activities are regulated through a common set of pathways including:

NF-κB, Ras, MAPK, and AKT (Dhawan & Richmond, 2002a; Dong, Kabir, Lee, & Son, 2013; Saintigny et al., 2013; Wang, Hendricks, Wamunyokoli, & Parker, 2006; Wang, Khachigian, et al., 2009). In summary, CXCL1/CXCR2 signaling in cancer cells modulates tumor progression by enhancing cell growth, motility, and invasion.

CXCL1 also signals to CXCR2+ stromal cells to regulate cancer progression. CXCR2–/– mice transplanted with metastatic melanoma or mammary carcinoma cells exhibit reduced tumor growth and metastasis, due in part to decreased angiogenesis and neutrophil recruitment (Sharma, Nannuru, Varney, & Singh, 2015; Singh, Varney, & Singh, 2009). In transgenic and xenograft models of breast cancer, doxorubicin, and cyclophosphamide activate a CXCL1 signaling mechanism whereby CXCL1 recruited CXCR2 expressing myeloid cells (Cd11b+/Gr1+) to the primary tumor. These myeloid cells promote cancer cell survival and invasion through S100A8- and S100A9-dependent mechanisms. Treatment of tumor bearing mice with CXCR2 antagonists enhance chemosensitivity and inhibit tumor metastasis (Acharyya et al., 2012). These studies indicate that CXCL1/CXCR2 signaling to endothelial and bone marrow cells are important mechanisms to tumor progression.

3.4 CXCL8

As an ELR+ chemokine, CXCL8 (also known as IL-8) shares many functions with CXCL1 in inflammation and cancer. While polymorphisms of CXCL8 or CXCR1, a binding receptor shows variable associations with prognosis (Tables 6 and 7), increased RNA, and protein expression of CXCL8 or CXCR1 frequently correlates with unfavorable cancer prognosis (Tables 9 and 10). CXCL8 stimulates promotes angiogenesis in corneal models (Koch et al., 1992), and tumor angiogenesis in animal models of: pancreatic, glioblastoma, lung carcinoma, prostate, ovarian, and colon cancer (Arenberg et al., 1996; Brat, Bellail, & Van Meir, 2005; Devapatla, Sharma, & Woo, 2015; Inoue et al., 2000; Matsuo, Ochi, et al., 2009; Ning et al., 2011). CXCL8 also stimulates chemotaxis of neutrophils and basophils (Geiser et al., 1993). Similarly to CXCL1, CXCL8 overexpression in cancer cell lines enhances tumor cell growth and invasion in in vitro and in vivo in melanoma models (Schadendorf et al., 1993; Singh, Gutman, Reich, & Bar-Eli, 1995). Yet, there are also key differences between CXCL1 and CXCL8. In contrast to CXCL1 and CXCR2, CXCL8, and CXCLR1 are expressed in humans, but not in mice (Mestas & Hughes,

2004; Zlotnik & Yoshie, 2000). CXCL8 binds preferentially to CXCR1 than CXCR2 (Nasser et al., 2009). Pharmacologic blockade of CXCR1 or CXCL8 selective antagonists inhibit CXCL8-mediated tumor growth and invasion as demonstrated in breast and lung cancer models (Ginestier et al., 2010; Khan, Wang, et al., 2015). Emerging studies have also revealed unique functions and mechanisms for CXCL8 signaling in cancer progression.

CXCL8 plays an important role in cancer stem cell renewal and survival. In breast cancer, CXCL8 enhances self-renewal of ALDH1 + cells and promotes cancer cell survival of breast cancer cell through CXCR1-dependent mechanisms (Charafe-Jauffret et al., 2009). Treatment of breast cancer cells with CXCR1 neutralizing antibodies or the small molecule inhibitor repertaxin decrease Aldeflour activity and enhance cellular apoptosis, which are mediated through FasL, AKT, FAK, and Fox03A signaling mechanisms (Ginestier et al., 2010). In pancreatic cancer, CXCL8 stimulates sphere formation and self-renewal of CD44+/CD24– cells, which are inhibited by pharmacologic or antibody neutralization of CXCR1 (Chen et al., 2014; Maxwell, Neisen, Messenger, & Waugh, 2014). In nasopharyngeal carcinoma, CXCL8 stimulated growth of tumor spheroids in vitro through a PI-3K/AKT- and CXCR2-dependent mechanism (Lo et al., 2013). CXCL8 promotes survival of prostate cancer cells to 5-FU through CXCR2- and Bcl2-dependent mechanisms (Wilson et al., 2012). These studies indicate that CXCL8 regulates cancer progression by signaling to cancer cells to mediate stem cell renewal and survival, with important implications on chemoresistance.

3.5 CXCL12

As an ELR-chemokine, CXCL12 is normally expressed throughout the body including pancreas, heart, spleen, and brain (Abe et al., 2015; Yu et al., 2006) and in circulating platelets (Berahovich et al., 2014; Chatterjee et al., 2015). CXCL12 signaling through CXCR4 is important to immunity by modulating recruitment of T cells and monocytes (Chatterjee et al., 2015; Inngjerdingen, Torgersen, & Maghazachi, 2002) and promoting survival and growth of pre-B cells (Nagasawa, Kikutani, & Kishimoto, 1994; Wang, Fairhurst, et al., 2009). CXCL12 contributes to angiogenesis and vasculogenesis by promoting endothelial migration and integration into maturing blood vessels (Jin, Zhao, & Yuan, 2013; Newey et al., 2014). CXCL12 also mobilizes a number of hematopoietic stem and progenitor cells, and MSCs to the blood or connective tissues

(Aiuti, Webb, Bleul, Springer, & Gutierrez-Ramos, 1997; Christensen, Wright, Wagers, & Weissman, 2004; Hu et al., 2013). These studies indicate an important role for CXCL12/CXCR4 signaling in maintaining immunity and the vasculature throughout the body.

Within the chemokine family, CXCL12 and CXCR4 are among the most well-studied molecules in cancer. While the associations between CXCL12 and CXCR4 polymorphisms and cancer risk are variable among cancer type (Tables 6 and 7), numerous studies have reported in carcinomas, gliomas, and leukemia. Frequently, increased expression of either CXCL12 or CXCR4 correlates with unfavorable prognosis (Tables 9 and 10). CXCL12/CXCR4 signaling promotes progression of the primary tumor in part by signaling to cancer cells. In ovarian cancer, CXCR4 overexpression in SKOV3 cancer cells leads to increased tumor and metastasis when these cells are transplanted in nude mice. In vitro CXCL12 signaling through CXCR4 signaling in ovarian cancer cells promotes cell proliferation, migration, and invasion in vitro (Guo et al., 2015). CXCL12 mediates tumor growth and metastasis of HeLA cells in vivo through mTORC1/Raptor-dependent mechanisms (Dillenburg-Pilla et al., 2015). These studies indicate that CXCL12 signaling regulates invasion through multiple pathways. Stromal cells in the primary tumor are an important source of CXCL12. CXCL12 is highly expressed in CAFs (Allinen et al., 2004; Izumi et al., 2015). Cotransplantation of carcinoma-associated fibroblasts with breast cancer cells increase primary tumor growth, which are abrogated by antibody neutralization of CXCL12 (Orimo et al., 2005). In coculture studies, CXCL12 derived from fibroblasts enhance gastric cancer cell invasion through integrin-β1 clustering (Izumi et al., 2015), and increase breast cancer cell proliferation and invasion (Allinen et al., 2004). The vascular endothelium is another important source of CXCL12 expression. In a model of T cell acute lymphoblastic leukemia (T-ALL), cre-lox-mediated conditional knockout of CXCL12 in vascular endothelial cells, or knockout of CXCR4 in T-ALL cells significantly inhibit T-ALL growth and progression in mice (Pitt et al., 2015). These studies indicate that CXCL12 regulates interactions between cancer cells and fibroblasts or with endothelial cells to promote progression of the primary tumor.

The effects of CXCL12 on stromal remodeling in the primary tumor are well documented. CXCL12 expression in primary basal cell carcinomas correlates with tumor angiogenesis (Chu et al., 2009). Delivery of recombinant CXCL12 to animal models increases vascular density in animal models of basal cell carcinoma (Chu et al., 2009) and synergizes with VEGF to promote angiogenesis in ovarian cancer in vivo (Kryczek et al., 2005). In breast cancer, fibroblast-derived CXCL12 promotes tumor growth and angiogenesis in part by signaling to endothelial progenitors (Orimo et al., 2005). In a MMTV-Wnt1 transgenic model of mammary tumorigenesis, CXCL12 enhance tumor growth by increasing tumor angiogenesis and recruitment of CD11b+/Gr1+ cells (Liu et al., 2010). These studies indicate that CXCL12 promotion of tumor angiogenesis and recruitment of myeloid cells are additional mechanisms to enhancing tumor progression.

CXCL12/CXCR4 signaling contributes to metastatic progression through several ways.

In breast cancer, CXCR4+ cancer cells have been detected in the bone marrow of patients and increased CXCR4 in the primary tumor may predict bone metastasis (Cabioglu et al., 2005). CXCR4 overexpression in breast cancer cells increases metastases to the brain, lungs, lymph nodes, and bone when delivered to mice (Wurth et al., 2015). In vitro, CXCL12 treatment of breast cancer cells enhances cell survival and increases vascular permeability, facilitating transendothelial migration (Lee, Lee, Avraham, & Avraham, 2004; Wurth et al., 2015). In prostate cancer, overexpression of CXCR4 in prostate cancer cells increases metastatic growth in bone tissue and increased osteolysis (Chinni et al., 2008). Bone marrow-derived stromal cells express CXCL12 signal to prostate cancer cells that express CXCR4, resulting in increased prostate cancer cell migration through an endothelial layer (Taichman et al., 2002). These studies indicate an important role for CXCL12/CXCR4 signaling in facilitating metastasis, particularly to the bone.

The signal transduction pathways of CXCL12/CXCR4 signaling have been mapped out extensively through analysis of multiple cancer cell types including: breast, prostate and pancreatic cancer cells, leukemia cells, and metastatic lymphangioleiomyomatosis and angiomyolipoma cells (Fig. 5). CXCL12 binding to CXCR2 triggers G protein-dependent pathways including: PI3-kinase, p42/44MAPK, and Rho (Clements, Markwick, Puri, & Johnson, 2010; Evelyn et al., 2007; Fernandis, Prasad, Band, Klosel, & Ganju, 2004). CXCL12 also regulate PKC signaling through JAK-dependent mechanisms (Mills et al., 2016). These pathways in turn activate SRF, NF-κB, and Elk-mediated transcription (Begley, MacDonald, Day, & Macoska, 2007; Chinni et al., 2006; Singh et al., 2012). These signaling pathways regulate cell growth, migration, and invasion, and in cooperation with WNT/beta-catenin and Her2 (Chinni et al., 2006, 2008; Garcia-Irigoyen et al., 2015; Song, Gao, Chu, Han, & Qu, 2015).

Through a transgenic mammary tumor model, CXCL12 has also been shown to regulate in vivo invasion of carcinoma cells through EGF- and CSF-1-dependent mechanisms (Hernandez et al., 2009). These studies indicate that CXCL12 modulates cancer cell invasion in cooperation with multiple oncogenes.

Recent studies have shown that CXCL12 may be expressed as several different isoforms (CXCL12-α, -β, and -γ), adding to the complexity to the mechanisms of CXCL12/CXCR4 signaling. In breast cancer, these isoforms have been implicated in primary tumor growth and are expressed in metastatic pleural effusions from patients. CXCL12-γ expressing fibroblasts cografted with CXCR4+ breast cancer cells enhance metastasis to the bone. Interestingly, these fibroblasts are also detected in metastatic sites including bone and lung, indicating that fibroblasts may travel from the primary tumor to secondary sites to modulate metastasis, through CXCL12-dependent mechanisms (Ray et al., 2015). However, more work needs to be done to identify the contributions of specific CXCL12 isoforms to cancer progression.

4. CHEMOKINES IN THERAPY

Current literature indicates that C–C and C–X–C chemokines modulate the primary tumor and metastatic microenvironments by signaling to cancer cells, and recruitment of stromal cells including endothelial cells, bone marrow-derived cells, and Th2 cells (Fig. 6). Given the importance of chemokine signaling in cancer progression, there is a great deal of interest in developing new therapies targeting chemokine ligands or receptors. A number of chemokine ligand and receptor antagonists are currently being tested in clinical trials, as monotherapies or in combination with conventional chemotherapies (Table 11). There are several challenges to developing effective chemokine-based therapies. It is unclear how to avoid targeting tumor suppressive components of the immune system, such as NK cells or CD8+ cytotoxic T cells. As chemokines and chemokine receptors are ubiquitously expressed in the human body, it would also be important to develop delivery strategies to maximize drug uptake to tumor tissues. Even when successfully addressing these issues, targeting of chemokine pathways may still lead to unexpected side effects. Recent studies have shown that when treatment of mammary tumor bearing mice with CCL2 neutralizing antibodies is interrupted, tumor growth recurs accompanied by increased malignancy and tumor angiogenesis (Bonapace et al., 2014). A design of

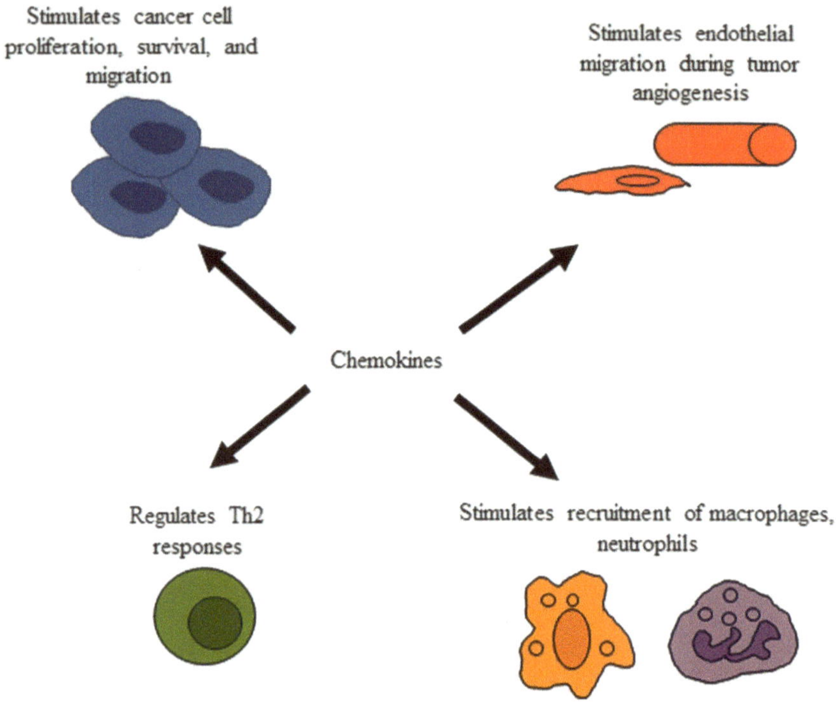

Fig. 6 Role of chemokines in cancer. Chemokines act on cancer cells to promote cancer cell growth, survival. In addition, chemokines stimulate Th2 responses, tumor angiogenesis, and recruitment of bone marrow cells to the local tumor microenvironment.

effective strategies will come from clearly understanding the molecular and cellular mechanisms of chemokine signaling, and how they cooperate with other oncogenes to modulate tumor progression.

5. INTERFERONS

Interferons (IFNs) are homodimeric soluble proteins in the cytokine class that were originally named for their ability to interfere with viral replication inside host cells. There are three types of IFNs: I, II, and III. Type I IFNs consist of interferon alpha (IFN-α) and interferon beta (IFN-β). Type II IFN consists of IFN-γ. The most recently discovered type of IFNs is type III IFN (IFN-λ). Upon recognition of a pathogen—be it viral, bacterial, fungal, or tumor cells—various host cells will release IFNs to signal to their adjacent cells to produce antiviral machinery and to increase their production of major histocompatibility complex (MHC) proteins. IFNs upregulate the presentation of antigens within possibly infected cells, allowing removal

Table 11 Chemokine Receptor Antagonists in Clinical Trials

Target	Drug	Sponsor	Disease	Status	References
CCR4	KW-0761 (mogamulizuma)	Amgen	Adult T cell leukemia, T cell lymphoma	Phase II, ongoing	Ogura et al. (2014), Tanba et al. (2016), and Ueda (2015)
CXCR4	Plerixafor	Genzyme	Advanced pancreatic, ovarian, colorectal	Phase I (recruiting)	NCT02179970
CXCR4	BMS-936564 in combination with Lenalidomide/Dexamethasone or Bortezomib/Dexamethasone	Bristol-Myers Squibb	Multiple myeloma	Phase Ib, ongoing	NCT01359657
CCR2	MLN1202	Southwest Oncology Group	Bone metastasis	Phase II completed, 14/43 patients show urinary N-telopeptide response	NCT01015560
CCR5	Maraviroc	National Center for Tumor Diseases, Heidelberg	Metastatic colorectal cancer	Phase I completed, no results posted	NCT01736813
CXCR1	Reparixin	Dompé Farmaceutici S.p.A	Early breast cancer	Phase I terminated, enrollment target not reached	NCT01861054
CXCR1	Reparixin in combination with paclitaxel	Dompé Farmaceutici S.p.A	Her2-metastatic breast cancer	Phase Ib completed, no results posted	NCT01861054

Unless specified, clinical trials are occurring, or have occurred in the United States. Clinicaltrials.gov trial number is referenced, unless otherwise specified.

of infected or tumorigenic cells by resident lymphocytes and macrophages. Increased levels of IFN result in fever, muscle aches, and other flu-like symptoms.

The signal transduction pathways of type I and II IFNs have been well characterized in immune cells. The biological effects of type I and II IFNs are mediated through the homodimeric IFN receptor. Type I IFNs signaling through the IFN alpha receptor (IFNAR), type II IFNs signaling through the IFN gamma receptor (IFNGR). Both IFN-α/β and IFN-γ signal through JAK/STAT mechanisms, but there are also key signaling differences between type I and type II IFN signaling. In type I IFN signaling, homodimeric IFN-α or IFN-β binds to preaggregated proteins IFNAR1 and IFNAR2. Binding causes autophosphorylation of TYK2, which induces phosphorylation of JAK1 (Gauzzi et al., 1996). Cytosolic STAT1 and STAT2 are subsequently phosphorylated and transported to the nucleus via the nuclear importer adapter importin alpha 5 (Holloway, Dang, Jans, & Coulson, 2014; Leung, Qureshi, Kerr, Darnell, & Stark, 1995; Qureshi, Salditt-Georgieff, & Darnell, 1995). Phosphorylated STAT1/STAT2 complex associates with IRF9 to form ISGF3, which can then act as a transcriptional activator of downstream effects of IFN-α (Au-Yeung, Mandhana, & Horvath, 2013). These classical type I IFN pathways are summarized in Fig. 7. In type II IFN signaling, IFN-γ binds to IFNGR1 and IFNGR2, JAK2 phosphorylates JAK1 which then phosphorylates only STAT1 (Ahmed & Johnson, 2006). Phospho-STAT1 dimerizes and is localized to the nucleus (Wang, Tyring, Townsend, & Evers, 1998). Alternatively, it has been shown that IFN-λ can be internalized on its own, whereupon it can interact with IFNGR1 and phosphorylated STAT1 and translocate to the nucleus via a classical importin-dependent pathway (Ahmed & Johnson, 2006). Once inside the nucleus, phosphorylated STAT1 (and to a lesser degree, NF-kB) can induce IRF-1. IRF-1 associates with the pSTAT1 dimer and causes transcriptional activation of downstream effects (Connett, Hunt, Hickerson, Wu, & Doherty, 2003). These classical type II IFN pathways are summarized in Fig. 8. It should be noted that IFN-α,β, and γ are capable of activating pathways including p42/44MAPK, STAT3, and STAT5 (Giannopoulou et al., 2006; Tanabe et al., 2005; Wang et al., 1998) to modulate T cell proliferation and activity.

Having been discovered in 2003, type III IFNs are less well understood than type I or type II IFNs. Currently, four ligands in the type III IFN class have been identified (IFN-λ1, IFN-λ2, IFN-λ3, and IFN-λ4), which bind to a unique receptor, IFN-λR, which is comprised of IFNLR1 and IL10Rβ subunits (Lazear, Nice, & Diamond, 2015). Type III IFNs activate signaling

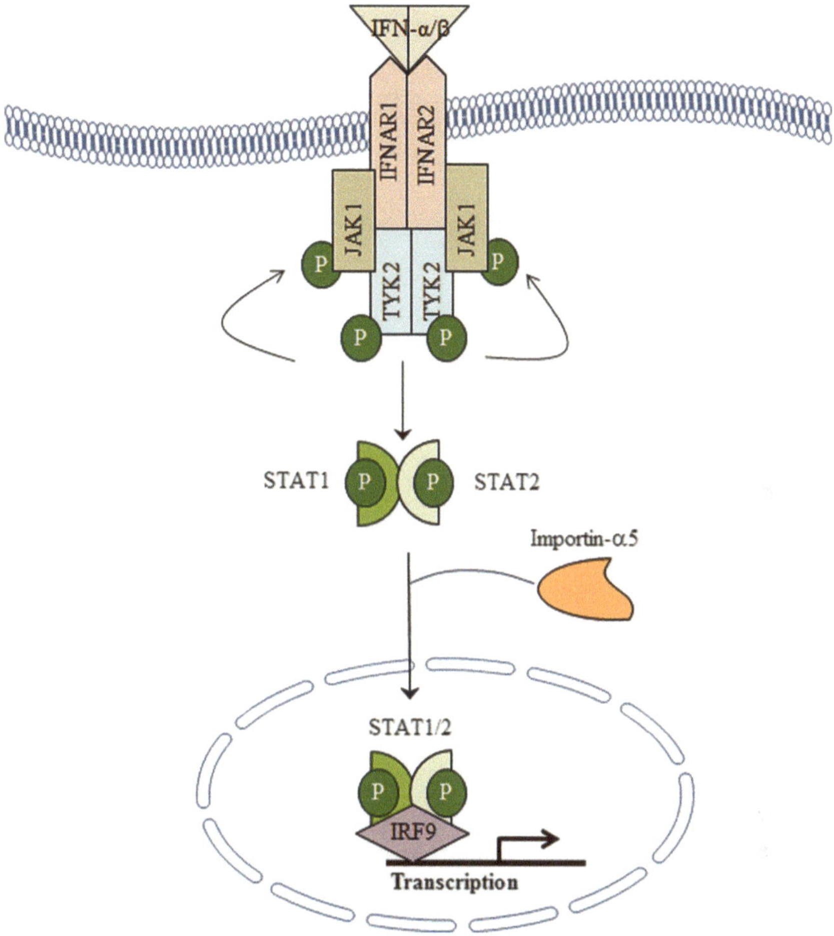

Fig. 7 Type I IFN signal transduction pathway. IFN-α or IFN-β binds to IFNAR1 and IFNAR2, leading to autophosphorylation of TYK2, which induces phosphorylation of JAK1. Cytsolic STAT1 and STAT2 is subsequently phosphorylated and transported to the nucleus via the nuclear importer adapter importin alpha 5. Phosphorylated STAT1/STAT2 complex associates with IRF9 to bind DNA and modulate gene transcription.

proteins common to type I IFNs, such as Jak1 and Tyk2, which phosphorylate Stat1 and Stat2 proteins (Dumoutier et al., 2004). Activation of the JAK/Stat pathways result in transcription of IFN stimulated genes, such as IRF1 and IRF7, which regulate gene expression of IFNs (Onoguchi et al., 2007; Osterlund, Pietila, Veckman, Kotenko, & Julkunen, 2007; Thomson et al., 2009). In contrast to IFN-α, IFN-λ also activates Jak2, indicating some differences between type I and type III IFN signaling (Odendall et al., 2014).

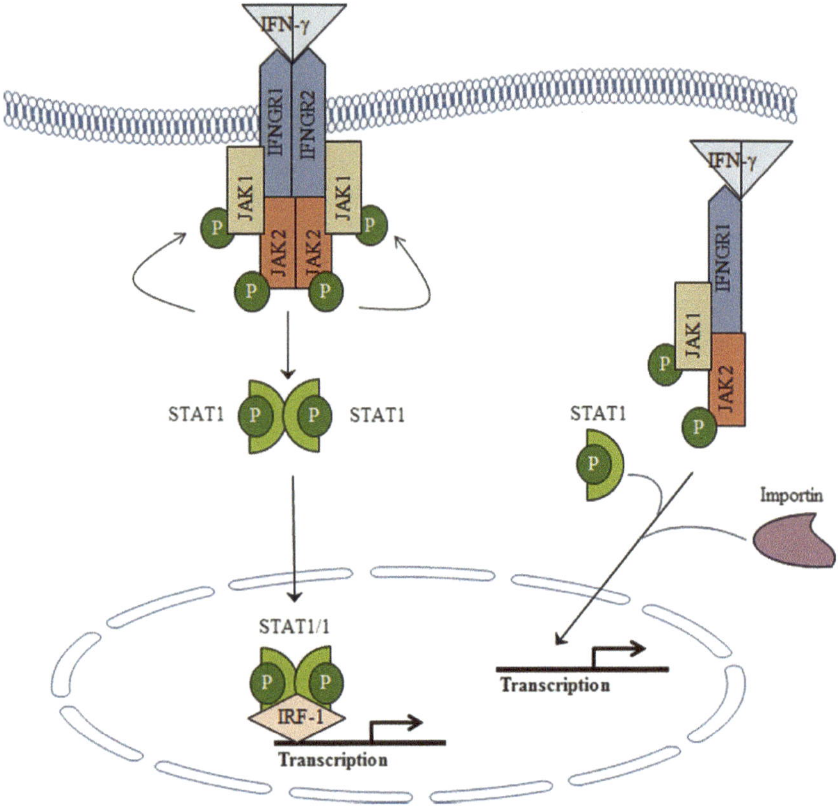

Fig. 8 Type II IFN signal transduction pathway. IFN-γ binds to IFNGR1 and IFNGR2, JAK2 phosphorylates JAK1 which then phosphorylates only STAT1. Phospho-STAT1 dimerizes and is localized to the nucleus. Alternatively, it has been shown that IFN-λ can be internalized on its own, whereupon it can interact with IFNGR1 and phosphorylated STAT1 and translocate to the nucleus via a classical importin-dependent pathway. Once inside the nucleus, phosphorylated STAT1 dimers associate with induce IRF-1, bind DNA, and modulate gene transcription.

IFN signaling is modulated through a variety of mechanisms. One mechanism involves soluble IFN receptors (IFNAR2), which bind type IFNs, thereby antagonizing cell signaling. Interestingly, through ligand binding, soluble IFNAR2 can exert antiproliferative signals in thymocytes, indicating that soluble IFN receptors can exert cellular effects (de Weerd, Samarajiwa, & Hertzog, 2007; Hardy et al., 2001). Another mechanism relates to regulation of IFN expression. A variety of immune cells produce IFN in response to viral and microbial molecules, through mechanisms dependent on pattern-recognition receptors, which include Toll-like receptors (TLRs) (Ivashkiv & Donlin, 2014). A mechanism that distinguishes type I and III IFN induction is the NF-κB pathway, which appears to be more important in TLR9-mediated production of type III IFN (Iversen, Ank, Melchjorsen, & Paludan, 2010). As we will see, deregulated expression and activity of IFNs significantly affect the development and progression of many types of cancers.

5.1 IFN-α/IFN-β

The clinical relevance on the expression of type I IFN and their receptors in cancer remain poorly understood. Some gene polymorphisms of type I IFN-related genes are associated with lower IFN gene expression and favorable patient prognosis for systemic sclerosis, an autoimmune disease (Wu & Assassi, 2013). However, the significance of gene polymorphisms in cancer remains unclear. A few clinical studies have reported changes in expression of IFN receptors in tumor tissues. In patients, increased serum levels of soluble IFNAR2 are associated with increased malignancy in many carcinomas including: bladder, breast, colon, and HCC (Table 12). Further studies at the genomic, RNA, and protein levels are necessary to more clearly determine the clinical relevance on the expression of type I IFNs in cancer.

Functional studies indicate that type I IFN signaling is important for eliminating cancer cells and abolishing metastatic niches. This is best demonstrated in animal models of sarcomas. IFNAR−/− knockout mice show increase susceptibility to methylcholanthrene (MCA)-induced sarcomas (Dunn et al., 2005). Sarcomas developed from STAT1 knockout mice are tumorigenic and metastatic. These features are significantly diminished by reconstituting STAT1 expression (Huang, Bucana, Van Arsdall, & Fidler, 2002). These studies indicate that decreased type I IFN-STAT1 signaling prevents immune rejection of tumor cells in mice, enabling tumor establishment and progression.

Type I IFNs signal to multiple immune cell types to suppress tumor progression. One study shows that exogenous IFN-β downregulates the expression of CXCR2 ligands (CXCL1, CXCL2, and CXCL5), whose expression gradient causes extravasation and recruitment of neutrophils to the tumor microenvironment. Neutrophils increase angiogenesis and support tumor growth. Therefore, IFN-β appears to be important for preventing neutrophil-mediated angiogenesis and growth of tumors (Jablonska, Wu, Andzinski, Leschner, & Weiss, 2014). The role of type I IFN and MDSCs in clearing metastatic cells is illustrated in a study using a mouse with a

Table 12 RNA and Protein expression Patterns of Interferons and Their Receptors in Cancer

Interferon	Cancer Type	Clinical Relevance	References
IFNAR2	Bladder	Increased serum levels associated with malignancy	Ambrus et al. (2003)
	Breast	Increased serum levels associated with malignancy	Ambrus et al. (2003)
	Colon	Increased serum levels associated with malignancy	Ambrus et al. (2003)
	HCC	Increased protein associated with carcinoma; no association to increased response to IFN-α	Kondo, Yukinaka, Nomura, Nakaya, and Ito (2000) and Yano et al. (1999)
	Lung	Increased serum levels associated with malignancy	Ambrus et al. (2003)
	Ovarian	Increased serum levels associated with malignancy	Ambrus et al. (2003)
	Prostate	Increased serum levels associated with decreased response to IFN-α. Increased serum levels associated with malignancy	Ambrus et al. (2003) and Booy, van Eijck, Dogan, van Koetsveld, and Hofland (2014)
	Renal	Increased tumor mRNA; increased serum mRNA associated with progression of RCC and shorter overall survival; increased overall survival	Furuya et al. (2011) and Kamai et al. (2007)
	Uterine	Increased serum levels associated with malignancy	Ambrus et al. (2003)
IFN-γ	Cervical	Decreased intratumoral mRNA correlates with poor prognosis	Tartour et al. (1998)
	HCC	Decreased serum levels correlate with disease recurrence	Lee et al. (2013)
	Ovarian	Increased RNA correlated with malignant but not normal tissue	Pisa et al. (1992)
IFN-γR	Breast	Decreased protein expression correlates with invasiveness	Garcia-Tunon et al. (2007)
	Basal cell	Decreased protein expression correlates with malignancy	Kooy, Tank, Vuzevski, van Joost, and Prens (1998)

HCC, hepatocellular carcinoma.

selective knockout of IFNAR in the MDSC cell population. Mice lacking MDSCs with functional IFNAR are unable to clear highly immunogenic cancer cell transplants that WT mice are capable of clearing, suggesting a strong role for type I IFNs in MDSC-mediated cancer cell elimination (Diamond et al., 2011). Type I IFNs may also regulate activity of tumor-associated macrophages (TAMs), which create a hospitable microenvironment for tumor growth and angiogenesis. In IFNAR–/– mice, tumors are associated with fewer TAMs via IHC and flow cytometry than wild-type mice and exhibited slower growth and decreased angiogenesis as a result (U'Ren, Guth, Kamstock, & Dow, 2010). IFN-α also appears important in tumor-bearing mice at suppressing development of MDSCs (Zoglmeier et al., 2011). All of these studies illustrate that type I IFNs suppress tumor growth and establishment in part by suppressing the activity of bone marrow cells.

Type I IFNs also suppress tumor progression by signaling to cancer cells. One study has found that IFN-β increases the MHC class I expression on the surface of breast tumor cells through an IFNAR-dependent manner, thus increasing their antigenicity to cytotoxic T cells (Wan et al., 2012). In addition, IFN signaling within cancer cells increases the expression levels of p53 in those cells, increasing their sensitivity to stresses, and predisposing them to apoptosis (Juang et al., 2004; Takaoka et al., 2003). These studies indicate that type I IFNs act directly on tumor cells expressing IFNAR to both directly cause apoptosis and to increase their antigenicity to cytotoxic T cells.

5.2 IFN-γ

Decreased expression and activity of type II IFNs are associated with increased tumor development. In noncancerous tissues, the expression of type II IFN is limited to a subset of antigen presenting cells, including (but not limited to) T-cells, NK cells, and NKT cells (Jameson & Grossberg, 1979). During tumor development, γδ T cells may be an important source of IFN-γ. This is supported by the finding that γδ T-cell deficient mice develop more tumors when challenged with the carcinogen MCA or B16 malignant melanoma cells. Selective knockout of IFN-γ in this mouse model also increases tumor development following challenges with MCA and B16 cells, further implicating the role γδ T cells have in early expression of IFN-γ to suppress tumor formation (Gao et al., 2003). In human tissues, changes in RNA and protein expression levels of IFN-γ and IFN-γR have been detected in various cancer types. In many carcinomas, decreased RNA and protein expression levels correlate with increased tumor malignancy and disease recurrence (Table 12). Gene polymorphisms of type II IFNs may predict treatment response of hepatitis patients (Indolfi, Azzari, & Resti, 2014; Noureddin et al., 2015). However, the significance of gene polymorphisms of type I IFNs in cancer remains unclear. Further clinical studies of type II IFNs at the genomic, RNA, and protein levels are necessary to more clearly understand the relevance of type II IFNs in cancer.

The importance of IFN-γ in tumor surveillance is well established through multiple animal studies. Mice injected with fibrosarcoma cells expressing dominant-negative IFNGR show increased tumor growth, suggesting that endogenously produced IFN-γ acts directly on cancer cells to inhibit tumor establishment and growth (Dighe, Richards, Old, & Schreiber, 1994). These findings have been confirmed in additional carcinogenesis studies (Betts et al., 2007; Wakita et al., 2009). Further studies indicate that IFN-γ mediates the tumor suppressive activities of IL-12. Administration of IL-12 to mice induces regression of sarcomas. This regression is diminished by delivery of neutralization antibodies to IFN-γ, but not anti-TNF-α (Nastala et al., 1994). These studies demonstrate that importance of IFN-γ in the primary tumor.

Similar to type I IFNs, IFN-γ has direct cytotoxic effects on cancer cells. One study has shown that IFN-γ administration induces cell death in multiple ovarian cancer cell lines and primary cell lines derived from ascites samples (Wall, Burke, Barton, Smyth, & Balkwill, 2003). Another study has shown that IFN-γ induces cell cycle arrest and caspase-dependent apoptosis in ovarian cancer cells (Barton, Davies, Balkwill, & Burke, 2005). In an indirect manner, cells are made more susceptible to lysis by cytotoxic T lymphocytes as a result of MHC class I and II upregulation upon administration of exogenous IFN-γ (Propper et al., 2003). These studies indicate that type II IFN signaling in cancer cells inhibits cell survival through immune cell mediated and independent mechanisms.

Dysregulation of IFN-γ signaling appears to be important in the establishment of the metastatic niche. One study has shown that IFN-λ and an IFN-λ-induced transcription factor, IRF-1, are essential to NK-cell-mediated destruction of metastatic cells in a lung-metastasis model. The study shows that IFN-λ administration significantly reduces the number of metastatic lung nodules, and that this effect is abolished by either knocking out IRF-1 or by selectively diminishing the NK cell population in the mice (Ksienzyk et al., 2011). IFN-γ also prevents metastasis is by disrupting access to blood vessels by decreasing angiogenesis through Th1 helper T cell responses. This mechanism is demonstrated in one study, which has examined the effects of IFN-γ has on Th1-mediated tumor ablation. Th1 lymphocytes are shown to decrease both tumor burden and angiogenesis in a mouse model of pancreatic adenomas. Interestingly, these same cells increase carcinogenesis and drive angiogenesis when anti-IFN-γ antibodies are delivered. These data illustrate that IFN-λ not only differentiates lymphocytes toward a Th1 phenotype, but also synergistically allows them to eliminate tumor cells and decrease malignant access to the vasculature (Muller-Hermelink et al., 2008).

Interestingly, IFN-γ is beneficial for tumor development and dissemination in some studies. One such study has shown that PD-L1 is upregulated on ovarian cancer cells upon administration of IFN-γ—or conversely, downregulated upon knockout of IFNGR. PD-L1 normally suppresses T lymphocytes through binding their PD1 receptors, so increased expression of PD-L1 resulted in less infiltrating CD8 T cells, more peritoneal invasion in a mouse model, and shorter survival (Abiko et al., 2015). It is possible that these tumor promoting effects are in part dependent on the tumor type, microenvironment, and/or stage of tumor progression (Zaidi & Merlino, 2011).

5.3 IFN-λ

Animal studies indicate that type III IFNs function primarily as a tumor suppressor. Overexpression of IFN-λ2 in melanoma and colon carcinoma cells results in increased cell cycle arrest and apoptosis, and delay in tumor formation and tumor progression in mice (Sato, Ohtsuki, Hata, Kobayashi, & Murakami, 2006). Similarly, overexpression of IFN-λ2 in fibrosarcoma cells and hepatoma cells inhibits tumor growth and metastasis in mice (Abushahba et al., 2010; Numasaki et al., 2007). In addition to acting on cancer cells, IFN-l may act on multiple stromal cell types to suppress tumor growth and progression. IFN-λ inhibits tumor angiogenesis and positively regulates activity of NK cells, CD8+ T cells, and neutrophils (Abushahba et al., 2010; Lasfar et al., 2006; Numasaki et al., 2007). Interestingly, IFN-α, but not IFN-L inhibits the numbers of CD4+CD25+Foxp3+ Tregs in hepatomas (Abushahba et al., 2010), indicating different mechanisms for IFN-α and IFN-λ-mediated tumor suppression. The role of type III IFNs in cancer progression remains under heavy investigation.

5.4 Exploiting IFNs in Anticancer Therapy

In contrast to interleukins and chemokines, current studies on IFN indicate that these cytokines primarily suppress tumor progression by inhibiting cancer cell growth and survival, and suppressing stromal cell reactivity (Fig. 9).

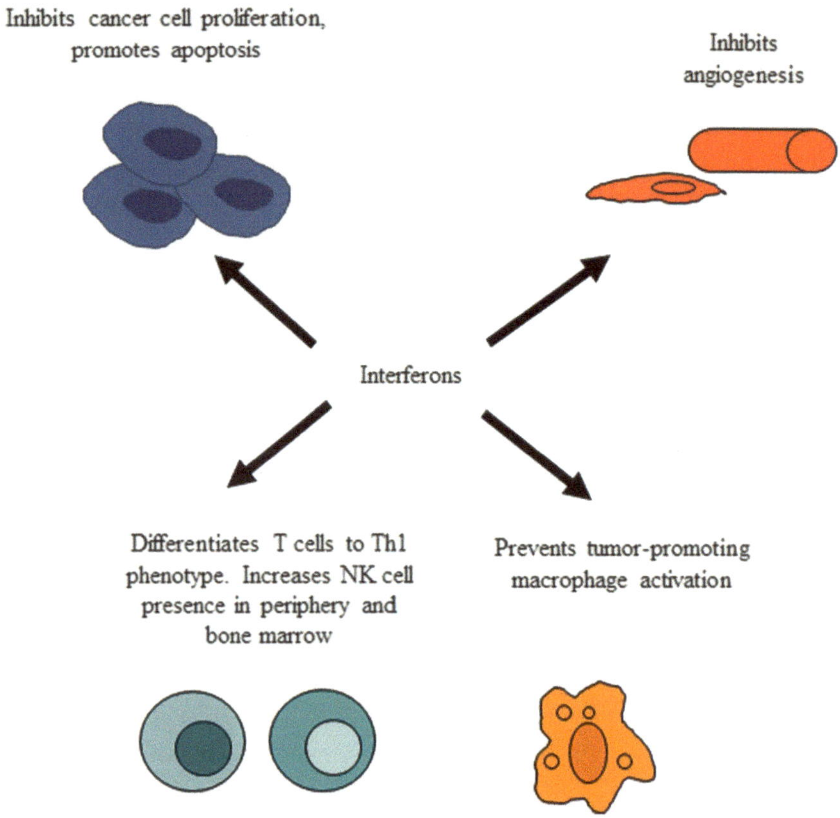

Fig. 9 Role of interferons in cancer. IFNs act on cancer cells to inhibit cell growth and enhance apoptosis, and remodel the microenvironment by inhibiting angiogenesis and macrophage polarization, and increasing the presence of cytotoxic T cells.

IFNs are used as therapeutic agents in a growing list of human diseases, including viral hepatitis and multiple sclerosis (Khan, Tanasescu, & Constantinescu, 2015; Zhang, Zhang, & Cui, 2015). IFN delivery has also shown promising results in aggressive cancers, with a subset of patients showing a partial response or stable disease. These cancers include: metastatic renal cancer, metastatic colon cancer, and recurrent lung cancer. However, adverse events have been reported in phase II clinical trials for the treatment of pancreatic cancer, resulting in trial termination. A summary of human trials can be found in Table 13. Interestingly, anthracycline chemotherapeutics induce expression of IFN-α in cancer cells through TLR3-dependent mechanisms, suggesting a role between IFN signaling and chemoresistance (Sistigu et al., 2014). Recent mouse studies have investigated the effectiveness of delivering IFN overexpressing MSCs or human umbilical cord

Table 13 Interferon-Based Anticancer Therapies in Clinical Trials

Drug	Sponsor	Disease	Status	References
IFN-α	M.D. Anderson Cancer Center	Bladder cancer, urothelial cancer	Phase I completed. No results posted	NCT00082719
IFN-α in combination with 13-cis-retinoic acid, paclitaxel	Eastern Cooperative Oncology Group	Recurrent NSCLC	Phase II, 3/19 patients showed partial response, 5/19 show stable disease	NCT00062010
IFN-α in combination with radiation therapy, 5-Fluorouracil	M.D. Anderson Cancer Center	Pancreatic cancer	Phase II completed, median overall survival: 42 months	NCT00068575
IFN-α2b in combination with celecoxib	Case Comprehensive Cancer Center	Metastatic kidney cancer	Phase II, completed, 3/17 patients show partial response; 5/17 show stable disease	NCT01158534
IFN-α in combination with tumor cell vaccine, Aldesleukin	Dartmouth-Hitchcock Medical Center	Metastatic renal carcinoma	Phase II, 9/18 show clinical response as measured by RECIST	NCT00085436
IFN-α in combination with cisplatin, 5-Fluorouracil, radiation therapy	Masonic Cancer Center, University of Minnesota	Pancreatic cancer	Phase II, terminated, 7/7 patients show adverse event	NCT00262951
Pegylated IFN-α with adjuvant therapy	National Cancer Institute (NCI)	Diffuse Intrinsic Pontine Glioma	Phase II completed, 32 patients analyzed, 2-year survival = 14.29 months	NCT00036569

Continued

Table 13 Interferon-Based Anticancer Therapies in Clinical Trials—cont'd

Drug	Sponsor	Disease	Status	References
IFN-γ in combination with carboplatin, GM-CSF	M.D. Anderson Cancer Center	Ovarian cancer, fallopian tube cancer, peritoneal cancer	Phase II completed, 9/54 show complete response, 21/54 show partial response	NCT00501644
IFN-γ in combination with 5-Fluorouracil, Leukovorin, Bevacizumab	Accelerated Community Oncology Research Network	Metastatic colon cancer	Phase II completed, No prior chemotherapy: 6/20 show partial response, 7/20 patients show stable disease. Prior chemotherapy: 3/28 show partial response, 15/28 show stable disease	NCT00786643

Due to the large volume of trials conducted, the trials listed are within the last 10 years and are completed or terminated. Unless specified, clinical trials are occurring, or have occurred in the United States. Clinicaltrials.gov trial number is referenced, unless otherwise specified. *RECIST*, response evaluation criteria in solid tumors.

matrix cells, with varying degrees of therapeutic effectiveness (Rachakatla et al., 2008; Ren, Kumar, Chanda, Chen, et al., 2008; Ren, Kumar, Chanda, Kallman, et al., 2008). Although the use of IFNs has been successful in treating some diseases and malignancies, it is important to consider the possible negative consequences of such treatments. Perhaps the most important reason for this caution is due to immunoediting, or the ability of immunological processes to select for more resistant cancers, which has been shown to be true of interferon signaling (reviewed in Dunn, Koebel, & Schreiber, 2006). A more clear understanding on the role of the immune system and IFN signaling may contribute to more effective IFN-based strategies to treat cancer.

6. TUMOR NECROSIS FACTOR

The cytokine TNF is notable for its long history in cancer research. About 100 years ago, the New York surgeon William Coley developed a cancer treatment with a mixture of bacteria products called "Coley's toxin," which stimulated patient responses (Aggarwal, Gupta, & Kim, 2012). Subsequent efforts led to the discovery of a factor that was induced in patients with bacterial infections (Carswell et al., 1975). Furthermore, this factor induced tumor necrosis when injected into several animal models, ultimately leading to the name "Tumor Necrosis Factor" (Balkwill et al., 1986; Brouckaert, Leroux-Roels, Guisez, Tavernier, & Fiers, 1986; Pennica et al., 1984). While showing initial promise as an anticancer therapeutic, TNF-α administration has been shown to induce severe side effects, limiting its clinical application (Brouckaert et al., 1986; Havell, Fiers, & North, 1988; Kettelhut, Fiers, & Goldberg, 1987). TNF plays important roles in tumor progression (Balkwill, 2009). Continuing efforts are still working to utilize both the pro- and anticancer functions of TNF in tumor targeting remedies.

The human TNF superfamily contains 19 ligands that bind to 29 receptors (Aggarwal et al., 2012; Locksley, Killeen, & Lenardo, 2001). The TNF superfamily ligands are type II proteins, containing both membrane-bound and cleaved soluble forms. Both forms assemble into active trimers during signaling. A few notable family members include: TNF-α (TNFSF2), TNF-β (TNFSF1), FASL (TNFSF6), and RANKL (TNFSF11). The TNF superfamily receptors are divided into two subgroups based on whether they contain an intracellular death domain, which is involved in protein interaction with cell death pathway components. The TNF superfamily plays diverse roles, such as inflammatory regulation, cell apoptosis, development of hematopoietic cell lineages, and tissue morphogenesis (Aggarwal et al., 2012; Locksley et al., 2001). TNF-α is unique in the superfamily due to its critical role in proinflammation and is implicated in multiple diseases including cancer. Of all of the TNF ligands, TNF-α has been the most extensively studied and will be the focus in this review.

6.1 TNF-α: Signal Transduction and Expression Patterns

TNF-α is produced as a type II transmembrane protein consisting of homotrimers. It can be cleaved into soluble trimer form (sTNF), consisting of three 17 kDa protomers (Wajant, Pfizenmaier, & Scheurich, 2003). TNF binds to two receptors, TNFR1(p55/TNFRSF1A) and TNFR2(p75/TNFRSF1B), which are also bound by the family member TNFβ. While both membrane and soluble TNF signal to TNFR1, only the membrane form of TNF activates strong signals via TNFR2. The TNF-α receptors can also be cleaved into soluble forms, and function as nonsignaling inhibitors (Wajant et al., 2003). The classical downstream TNF-α signaling involves the activation of the NF-κB signaling pathway and JNK pathways (Schwabe & Brenner, 2006; Wajant et al., 2003), which are summarized in Fig. 10. TNF-α binding to TNFRI recruits the adaptor protein TRADD, which brings together TRAF2 and kinase RIP. TRAF2 and RIP phosphorylates Nemo/IKKα/IKKβ complex, leading to subsequent phosphorylation, and degradation of IkB. NF-κB translocates to the nucleus to regulate gene transcription (Wajant et al., 2003). Activation of NF-kB by TNF-α signaling plays an important role in induction of proinflammatory responses, such as secretion of cytokines and chemokines, and recruitment of neutrophils, macrophage, and dendritic cells (Lawrence, 2009). TNFR2 lacks a death domain, but can direct bind to TRAF2 after ligand binding and activate NF-κB signaling. Furthermore, TNF-α-mediated recruitment of TRADD, TRAF2, and RIP adaptor proteins leads to phosphorylation and activation of apoptosis signal-regulating kinase (ASK1), which mediates JNK signaling (Tobiume et al., 2001). JNK signaling enhances activity of the transcription factor c-Jun and Itch, an E3 ligase that ubiquitinates and degrades the caspase8 inhibitor c-FLIP, thereby enhancing programmed cell death (Chang et al., 2006; Diehl et al., 1994; Westwick, Weitzel, Minden, Karin, & Brenner, 1994). TNF-α signaling can also activate other pathways including: p42/44MAPK, p38MAPK, FADD, mTOR pathways to regulate apoptosis, and cell survival (Cabal-Hierro et al., 2014; Kataoka, 2009; Ruspi et al., 2014; Wajant et al., 2003).

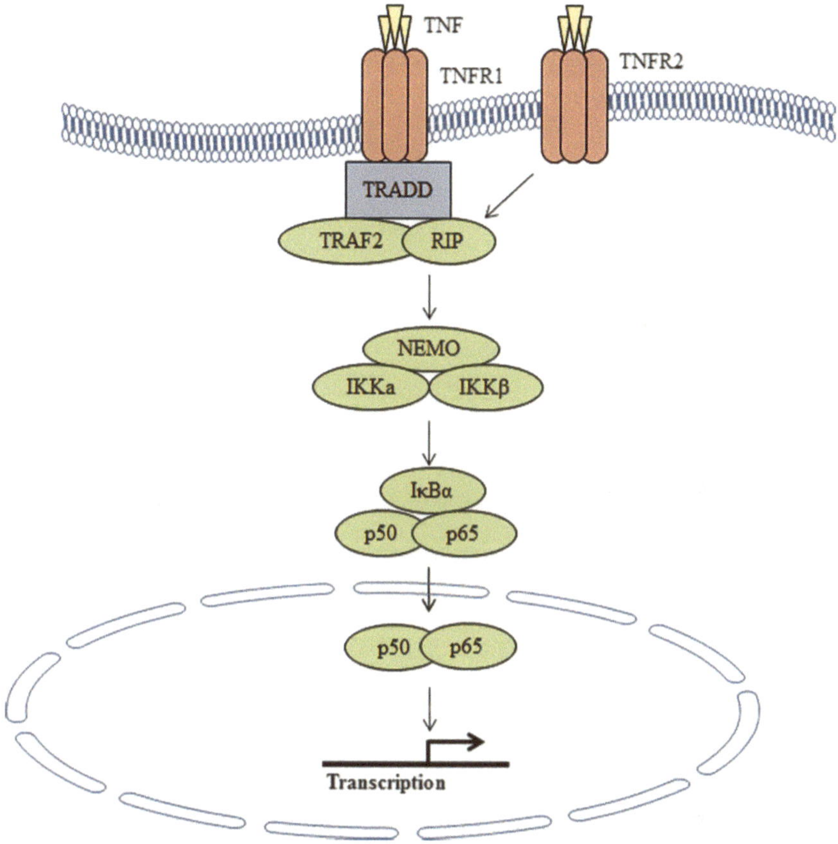

Fig. 10 TNF signal transduction pathway. TNF-α binding to TNFR1 recruits the adaptor protein TRADD via its DD domain, which brings together TRAF2 and kinase RIP. TRAP2 and RIP activates the Nemo/IKKα/IKKβ complex, leading to subsequent phosphorylation and degradation of IkB. NF-κB then translocates to the nucleus to modulate gene transcription. TNFR2 directly binds to TRAF2 after ligand binding and also activates NF-κB signaling.

The expression patterns of TNF-α and its receptors vary among cell types. TNF-α is expressed in macrophages, NK cells, B cells, and T cells. TNFR1 is ubiquitously expressed in most nucleated cells, while TNFR2 expression is mostly restricted to immune cells (Wajant et al., 2003). Knockout mice of TNF-α and TNFR1 have revealed their functional importance in the immune system. TNF knockout mice are viable and fertile, but exhibit defects in splenic B cell follicular formation and humoral immune response (Pasparakis, Alexopoulou, Episkopou, & Kollias, 1996). Both TNF and TNFR1 mutant mice are susceptible to microbial infection, indicating they play important roles in innate immune response (Pasparakis et al.,

Table 14 Clinical Relevance of TNF-α Single Nucleotide Polymorphisms in Cancer

Polymorphism	Cancer Type	Clinical Relevance	References
rs1800629	Breast	Increased risk of metastasis in triple negative breast	Li, Yao, et al. (2015)
	Oropharynx	Associated with reduced recurrence free survival	Zhang et al. (2014)
rs1800630	Oropharynx	Associated with reduced recurrence free survival	Zhang et al. (2014)
	HCC	Increased risk of development	Yang, Qiu, Yu, Zeng, and Bei (2012)

HCC, hepatocellular carcinoma.

1996; Rothe et al., 1993). Knockout of TNF-β results in defects in lymphoid organ development (Banks et al., 1995). Overexpression of TNF-α has been linked to autoimmune diseases and cancer. Neutralizing antibodies to TNF-α or TNF receptors have been approved to treat several autoimmune diseases (Croft, Benedict, & Ware, 2013; Feldmann & Maini, 2003).

Changes in expression of TNF-α have been detected in various cancers. At the genomic level, polymorphisms of TNF-α are associated with increased risk of cancer develop or poor patient prognosis (Table 14). Increased RNA and protein expression levels of TNF-α in tumors frequently associate with disease progression and unfavorable outcome (Table 15). TNF-α is mainly expressed in cancer cells and TAMs. Serum soluble TNF-α has been detected in cancer patients and predicts worse outcome. Serum levels of TNFR1 and TNFR2 have been detected in lymphoma patients correlating with worse outcome (Table 15).

6.2 TNF-α as a Tumor Suppressor

Historically, TNF-α has been shown to promote tumor necrosis. Injections of high concentrations of recombinant human TNF-α induce tumor necrosis in synergistic or xenograft tumor models (Balkwill et al., 1986; Brouckaert et al., 1986; Havell et al., 1988; Kettelhut et al., 1987). The necrosis is rapidly induced within the tumor center, characterized by tumor vascular hemorrhage, leading to the hypothesis that TNF-α functioned as antitumor factor (Nawroth et al., 1988). Injections of high dose recombinant TNF-α into mouse models and in patients induce severe toxicity, including cytokine storm and septic shock, thus limiting its clinical usage (Havell et al., 1988; Kettelhut et al., 1987). This toxicity is mainly due to enhanced TNFR2 signaling. Flaws in experimental design may have also contributed

Table 15 RNA and Protein Expression of TNF Ligands and Receptors in Cancer

Ligand/Receptor	Cancer Type	Clinical Relevance	References
TNF-α	Ovarian	Increased RNA, protein expression in tumors, in macrophages, and cancer cells. Positive expression correlates with tumor grade in serous subtype	Naylor, Malik, Stamp, Jobling, and Balkwill (1990) and Naylor, Stamp, Foulkes, Eccles, and Balkwill (1993)
	Esophagus	Protein expression in cancer cells, correlates with disease progression	Tselepis et al. (2002)
	Colon	RNA expressed in tumor associated macrophage	Naylor, Stamp, and Balkwill (1990)
	Prostate	Serum TNF correlates with cachexia and mortality	Nakashima et al. (1998)
		High serum TNF correlates with disease progression and reduced survival	Michalaki, Syrigos, Charles, and Waxman (2004)
	Breast	Increased protein expression in tumors, expressed in macrophages and cancer cells, correlates with lymph node metastasis	Leek et al. (1998)
		Serum TNF correlates with worse recurrence free survival	Bozcuk et al. (2004)
	Pancreas	Serum TNF associated with metastatic cancer, cachexia	Karayiannakis et al. (2001)
	Kidney	Serum TNF correlates with stage and grade	Yoshida et al. (2002)
	Lymphoma	Serum TNF correlates with disease progression and worse outcome	Warzocha et al. (1997, 1998)

Continued

Table 15 RNA and Protein Expression of TNF Ligands and Receptors in Cancer—cont'd

Ligand/ Receptor	Cancer Type	Clinical Relevance	References
TNFR1	Ovarian	Increased RNA, protein expression in cancer cells	Naylor et al. (1993)
	Leukemia	Serum TNFR1 High expression correlates with worse outcome in AML	Vinante et al. (1998)
	Lymphoma	Serum TNFR1 correlates with disease progression and worse outcome	Warzocha et al. (1997, 1998)
TNFR2	Ovarian	Increased RNA, protein expression in infiltrating cells	Naylor et al. (1993)
	Leukemia	Increased serum levels in AML and ALL	Vinante et al. (1998)
	Lymphoma	Increased serum correlates with disease progression and worse outcome	Warzocha et al. (1997, 1998)

AML, acute myelocytic leukemia; *ALL*, acute lymphocytic leukemia.

to faulty conclusions about the safety of using recombinant TNF-α in patients. Human TNF-α has a lower binding affinity than mouse TNF-α to mouse TNFR2. It is possible that human TNF-α delivered into preclinical models would not induce the level of toxicity seen in patients during clinical trials (Lewis et al., 1991). To overcome the toxicity, local/isolated infusion, and targeted delivery approaches have been developed and are currently being tested in the clinic (Table 16). Of note, recombinant TNF-α is approved in Europe for treatment of sarcoma in the limbs using isolated limb perfusion, in combination with chemotherapy drug melphalan (Verhoef et al., 2007).

TNF-α in induces tumor necrosis through induction of tumor vascular clotting and hemorrhage (Nawroth et al., 1988). TNF-α signals to endothelial cells lining tumor vascular through TNFR1, and induces fibrin deposition in tumor microvessels and leads to coagulant formation (Nawroth et al., 1988; Zhang et al., 1996). It is interesting to note that TNF-α sensitizes tumor cell lines, such as Meth-A fibrosarcoma, secretes the endothelial

Table 16 Delivery of TNF-α as an Anticancer Therapeutic in Clinical Trials

Drug	Sponsor	Disease	Status	References
L19TNF-α in combination with L19-IL2	University Hospital of Sienna, National Tumor Institute (Italy)	Advanced melanoma	Phase II completed, 1/20 complete response, 10/20 partial response; 5/20 disease stable	Danielli et al. (2015)
L19TNF-α in combination with doxorubicin	Philogen S.p.A	Advanced solid tumors	Phase I (recruiting)	NCT02076620
TNF-α bound colloidal gold	National Cancer Institute, CytImmune Sciences	Advanced solid tumors	Phase I completed, reduced toxicity, no tumor response	Libutti et al. (2010)
NGR-TNF	Istituto Clinico Humanitas, Istituto Scientifico San Raffaele, European Institute of Oncology (Italy)	HCC	Phase II completed, 1/27 partial response, 6/27 stable disease	Santoro, Pressiani, et al. (2010)
NGR-TNF	Istituto Scientifico San Raffaele, European Institute of Oncology, RCCS Fonda-zione Istituto Nazionale dei Tumori, Milan, Istituto Nazionale per la Ricerca sul Cancro, Istituto Clinico Humanitas (Italy)	Pleural mesothelioma	Phase II completed, 1/43 partial response 18/43 stable disease	Gregorc et al. (2010)
NGR-TNF	Istituto Clinico Humanitas, Ospedale San Martino, Istituto Scientifico San Raffaele, Università Vita-Salute San Raffaele	CRC	Phase II completed, 1/33 partial response, 12/33 stable disease	Santoro, Rimassa, et al. (2010)

Continued

Table 16 Delivery of TNF-α as an Anticancer Therapeutic in Clinical Trials—cont'd

Drug	Sponsor	Disease	Status	References
NGR-TNF in combination with oxaliplatin	National Cancer Institute, The Lombardi Comprehensive Cancer Center	Metastatic colorectal cancer	Phase II completed low dose, 1/12 patient partial response, 5/12 stable disease	Mammoliti et al. (2011)
NGR-TNF in combination with doxorubin	Catholic University of Rome San Raffaele Institute, MolMed, Università Vita-Salute San Raffaele (Italy)	Advanced ovarian cancer	Phase II completed, 8/35 partial response, 15/35 stable disease	Lorusso et al. (2012)
NGR-TNF in combination with standard chemotherapy	MolMed S.p.A	Advanced NSCLC	Phase II, in progress	NCT00994097
NGR-TNF in combination with doxorubicin	MolMed S.p.A	Sarcomas	Phase II, in progress	NCT00484341
TNFerade in combination with radiation and 5-Fluoruracil	Sidney Kimmel Comprehensive Cancer Center, Johns Hopkins Hospital, University of California Irvine Medical Center, Lee Moffitt Cancer Center, University of Colorado School of Medicine	Pancreatic cancer	Phase II completed, not effective in prolonging survival	Herman et al. (2013)

HCC, hepatocellular carcinoma; *NSCLC*, nonsmall cell lung cancer.
Unless specified, clinical trials are occurring, or have occurred in the United States. Clinicaltrials.gov trial number is referenced, unless otherwise specified.

monocyte activating peptide II (EMAP II). EMAP II enhances the effect of TNF-α on tumor vasculature by upregulating TNFR1 expression in endothelial cells (Kayton & Libutti, 2001). TNF-α also induces blood vascular permeability and enhances efficacy of chemotherapies (Seynhaeve et al., 2007).

TNF-α induction of tumor necrosis is also mediated by the immune system. The tumor suppressive effects TNF-α are reduced in T cell deficient mice, and TNF-α administration induces memory of host immune cells (Havell et al., 1988; Palladino et al., 1987). TNF-α plays multiple roles in innate and adaptive immunity. TNF-α signaling is important for T cell survival and activation (Chatzidakis, Fousteri, Tsoukatou, Kollias, & Mamalaki, 2007; Kim & Teh, 2001), dendritic cell maturation (Calzascia et al., 2007; Ding et al., 2011), and NK cell activation (Kashii, Giorda, Herberman, Whiteside, & Vujanovic, 1999). All of these mechanisms are important for immune surveillance during tumor development and progression.

TNF-α inhibits tumor progression by signaling to cancer cells. In a transgenic islet cancer model, TNF-α promotes senescence of cancer cells by upregulating expression of the cell cycle inhibitor p16 and p19 (Braumuller et al., 2013). In addition, TNF-α signaling in cancer cells increases expression of the antigen presenting complex, thereby increasing recognition by immune cells for elimination (Johnson & Pober, 1991; van den Elsen, 2011). In addition, TNF-α signaling through TNFR1 induces apoptosis in a variety of cancer cell lines (Colotta, Peri, Villa, & Mantovani, 1984; Parrington, 1979). However, its ability to induce apoptosis is much weaker compared to other TNF family ligands such as FASL and TRAIL (Aggarwal et al., 2012), and the effect is only prominent when cancer cells are under metabolic stress (Balkwill, 2009). Continuing insights from studying the role of TNF-α in autoimmune disease (Croft et al., 2012; Feldmann & Maini, 2003) could provide further clarity into the role of TNF-α in cancer.

6.3 TNF-α as a Tumor Promoter

TNF-α may also function to promote tumor progression. Knockdown of TNF-α or TNFR1 in mice delay the progression of skin cancer (Bertrand et al., 2015; Moore et al., 1999; Suganuma et al., 1999) and liver cancer (Nakagawa et al., 2014; Park et al., 2010). Chronic inflammation associated with TNF-α increase susceptibility to tumor formation (Cooks et al., 2013; Nakagawa et al., 2014; Park et al., 2010). For example, mice fed on a high fat diet exhibit liver inflammation, and are prone to spontaneous tumor formation, associated with increased TNF-α expression. TNF-α antibody neutralization, TNFR1 knockout, or NF-κB inhibition reduces tumor development (Nakagawa et al., 2014; Park et al., 2010).

These studies indicate that TNF-α promote development and progression of some carcinomas.

TNF-α signaling promotes tumor progression by regulating activity of multiple stromal cell types. TNF-α increases expression of other cytokines including: CCL2 and IL-6 (Lawrence, 2009), which promote tumor progression through recruitment of myeloid cells. TNF-α modulates expression of CCL7 and VEGF, which function as proangiogenic factors (Ferrara, 2002; Qian et al., 2010). TNF-α signaling through TNFR1 induces cell death in activated CD8+ T cells and inhibits T cell-mediated tumor rejection (Bertrand et al., 2015; Zheng et al., 1995). TNF-α signaling through TNFR2 mediates protumor activities of TNF-α through different mechanisms. For one, TNF-α/TNFR2 signaling mediates expansion of Tregs in tumors (Chopra et al., 2013; Okubo, Mera, Wang, & Faustman, 2013). TNF-α/TNFR2 signaling also supports the survival and immunosuppressive functions of MDSCs in cancer models (Sade-Feldman et al., 2013; Zhao et al., 2012). These studies indicate that TNFR1 and TNFR2 signaling are important for regulating the tumor promoting activities of TNF-α by remodeling the tumor microenvironment.

TNF-α signaling also promotes tumor progression by signaling to cancer cells. TNF-α induces cancer cells to undergo EMT, by enhancing NF-kB signaling, and increasing expression of the EMT transcriptional factors Zeb (Chua et al., 2007), Snail (Wu et al., 2009), and Twist (Li et al., 2012). In breast cancer, TNF-α signaling increases Snail expression by inhibiting its ubiquitination-mediated protein degradation (Wu et al., 2009). Decreased Snail or Twist1 expression in cancer cells reduces metastasis and inflammation in mouse models, indicating an important role for TNF-α in EMT-directed metastasis (Li et al., 2012; Wu et al., 2009). TNF-α activation of p38MAPK promotes breast cancer stem cell expansion under hypoxic conditions (Wu et al., 2015). In melanoma, TNF-α signaling increases cell survival during radiation therapy. Autocrine TNF-α signaling in TAMs lead to increased VEGF production and increased melanoma outgrowth after radiation in mice (Meng et al., 2010). In glioblastoma, TNF-α activation of NF-kB signaling in cancer cells induces differentiation into a mesenchymal state and increases resistance to radiation treatment (Bhat et al., 2013). These studies indicate that TNF-α promotes tumor progression by enhancing EMT, tumor cell survival, and cancer stem cell expansion.

In summary, studies indicate that TNF-α plays a dual role in cancer. As a tumor suppressor, TNF-α remodels the tumor microenvironment by increasing activity of cytotoxic T cells, promoting maturation of dendritic cell, and inhibiting tumor angiogenesis. TNF-α also signals directly to

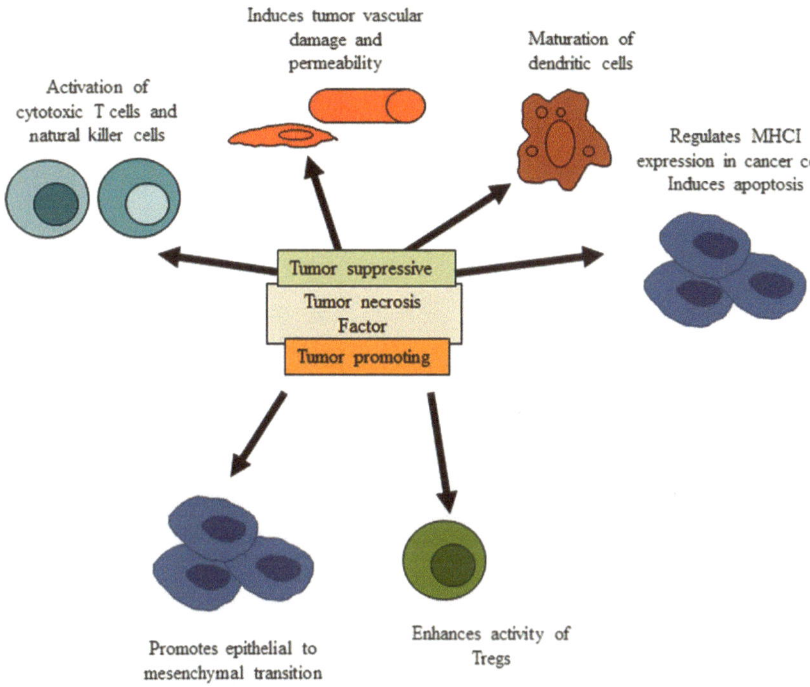

Fig. 11 Role of TNF-α in cancer. TNF plays a dual role in tumor progression. As a tumor suppressor, TNF promotes activation of cytoxic T cells and natural killer cells, inhibits tumor angiogenesis, promotes maturation of myeloid cells, signals to cancer cells to induce apoptosis, and regulate MHC I expression. As a tumor promoter, TNF promotes EMT of cancer cells and enhances activity of Tregs.

cancer cells to promote apoptosis and alter expression of MHC proteins, promoting recognition to T cells. As a tumor promoter, TNF-α increases EMT of cancer cells and increases activity of Tregs. The dual role of TNF-α as a tumor promoter and tumor suppressor is summarized in Fig. 11.

6.4 Exploiting the TNF-α Pathway in Anticancer Therapeutics

Delivery of TNF-α has been extensively tested in clinical trials as an anticancer therapeutic (Table 16). To circumvent toxic side effects, trials have been designed to deliver TNF-α locally to tissues. Recombinant human TNF-α (Tasonamin) has been approved for treatment in sarcoma in combination with melphalan via limb perfusion (Verhoef et al., 2007). TNF-α has also been explored as a nanotherapy (TNF-bound colloidal gold) or through gene delivery (TNFerade), but clinical trials have revealed no significant

Table 17 Delivery of TNF-α Inhibitors as Anticancer Therapeutics in Clinical Trials

Drug	Sponsor	Disease	Status	References
Etanercept	University of Oxford, Churchill Hospital, and Cancer Research UK Translational Oncology Laboratory, Queen Mary's School of Medicine and Dentistry (United Kingdom)	Recurrent ovarian cancer	Phase I completed, 6/30 disease stabilization	Madhusudan et al. (2005)
Etanercept	Cancer Research United Kingdom Medical Oncology Unit, University of Oxford, The Churchill, Oxford Radcliffe Hospitals, Cancer Research United Kingdom Translational Oncology Laboratory, Queen Mary's School of Medicine and Dentistry (United Kingdom)	Metastatic breast cancer	Phase II completed, 1/16 disease stabilization	Madhusudan et al. (2004)
Etanercept in combination with gemcitabine	Ohio State University Comprehensive Cancer Center	Advanced pancreatic cancer	Phase II completed, no significant enhancement of gemicitabine alone	Wu et al. (2013)
Infliximab in combination with Sorafenib	The Royal Marsden Hospital NHS Foundation Trust, Ortho Biotech Oncology Research & Development (United Kingdom)	Renal cell carcinoma	No additional benefit, increased side effects	Larkin et al. (2010)
Infliximab	Royal Marsden Hospital (United Kingdom)	Renal cell carcinoma	Phase II completed, low dosage, 3/19 partial response, 3/19 stable disease; high dosage 11/19 stable disease	Larkin et al. (2010)

Unless specified, clinical trials are occurring, or have occurred in the United States. Clinicaltrials.gov trial number is referenced, unless otherwise specified.

antitumor benefits. Other strategies have been developed to increase specific targeting of TNF-α, such as L19-TNF and NGR-TNF. The L19-TNF is a fusion of TNF-α and an L19 antibody fragment, which recognizes fibronectin, a protein highly expressed in the tumor microenvironment (Borsi et al., 2003; Halin et al., 2003). The NGR-TNF is a fusion of TNF-α and a cyclic peptide Cys-Asn-Gly-Arg-Cys (CNGRC) (Curnis et al., 2000). Both of these designs increase tumor vascular targeting and reduce toxicity. Ongoing clinical trials are using the targeted TNF-α proteins in combination with chemotherapy drugs.

Clinical trials have been conducted to evaluate the effects of targeting TNF-α in cancer, using TNF inhibitors approved in autoimmune diseases (Table 17). In clinical trials with Etanercept (a soluble TNFR2 antibody) or Infliximab (anti-TNF-α antibody), limited patient responses have been observed, with no follow-up clinical studies. It is possible that patient responsiveness is dependent on tissue type and stage of the cancer. To target such a pleiotropic factor, a more refined therapeutic strategy may be necessary. One potential approach is to target TNFR1 or TNFR2, as each receptor appears to display different functions during tumor progression. As TNF-α signaling is involved in immunosuppression during tumor progression, it is highly possibly that targeting TNF-α may be useful for enhancement of immunotherapies.

7. CLOSING REMARKS

This review has focused on multiple families of cytokines including: interleukins, chemokines, interferons, and TNF. These cytokines appear to play distinct and overlapping roles in regulating tumor progression, by signaling to cancer cells and remodeling the tumor microenvironment. Ongoing clinical trials demonstrate mixed results for delivering tumor suppressive cytokines such as interferons and targeting tumor promoting cytokines such as chemokines. Studies indicate that exploiting cytokine pathways in cancer are challenging for multiple reasons. For one, some cytokines play dual tumor suppressive and tumor promoting roles. Cytokines also cooperate with each other to regulate tumor progression. In addition, the role of these cytokines may depend on the tissue type and context, such as stage of disease or exposure to other anticancer therapies. A clear understanding on the roles of cytokines in tumor progression may lead to a more successful design of cytokine-based therapies to treat cancer.

REFERENCES

Abe, P., Molnar, Z., Tzeng, Y. S., Lai, D. M., Arnold, S. J., & Stumm, R. (2015). Intermediate progenitors facilitate intracortical progression of thalamocortical axons and interneurons through CXCL12 chemokine signaling. *The Journal of Neuroscience*, *35*, 13053–13063.

Abiko, K., Matsumura, N., Hamanishi, J., Horikawa, N., Murakami, R., Yamaguchi, K., et al. (2015). IFN-gamma from lymphocytes induces PD-L1 expression and promotes progression of ovarian cancer. *British Journal of Cancer*, *112*, 1501–1509.

Abushahba, W., Balan, M., Castaneda, I., Yuan, Y., Reuhl, K., Raveche, E., et al. (2010). Antitumor activity of type I and type III interferons in BNL hepatoma model. *Cancer Immunology, Immunotherapy*, *59*, 1059–1071.

Acharyya, S., Oskarsson, T., Vanharanta, S., Malladi, S., Kim, J., Morris, P. G., et al. (2012). A CXCL1 paracrine network links cancer chemoresistance and metastasis. *Cell*, *150*, 165–178.

Agarwal, M., He, C., Siddiqui, J., Wei, J. T., & Macoska, J. A. (2013). CCL11 (eotaxin-1): A new diagnostic serum marker for prostate cancer. *Prostate*, *73*, 573–581.

Aggarwal, B. B., Gupta, S. C., & Kim, J. H. (2012). Historical perspectives on tumor necrosis factor and its superfamily: 25 years later, a golden journey. *Blood*, *119*, 651–665.

Ahmed, C. M., & Johnson, H. M. (2006). IFN-gamma and its receptor subunit IFNGR1 are recruited to the IFN-gamma-activated sequence element at the promoter site of IFN-gamma-activated genes: Evidence of transactivational activity in IFNGR1. *Journal of Immunology*, *177*, 315–321.

Aiuti, A., Webb, I. J., Bleul, C., Springer, T., & Gutierrez-Ramos, J. C. (1997). The chemokine SDF-1 is a chemoattractant for human CD34+ hematopoietic progenitor cells and provides a new mechanism to explain the mobilization of CD34+ progenitors to peripheral blood. *The Journal of Experimental Medicine*, *185*, 111–120.

Akishima-Fukasawa, Y., Nakanishi, Y., Ino, Y., Moriya, Y., Kanai, Y., & Hirohashi, S. (2009). Prognostic significance of CXCL12 expression in patients with colorectal carcinoma. *American Journal of Clinical Pathology*, *132*, 202–210, quiz 307.

Albulescu, R., Codrici, E., Popescu, I. D., Mihai, S., Necula, L. G., Petrescu, D., et al. (2013). Cytokine patterns in brain tumour progression. *Mediators of Inflammation*, *2013*, 979748.

Aleksandrova, K., Boeing, H., Nothlings, U., Jenab, M., Fedirko, V., Kaaks, R., et al. (2014). Inflammatory and metabolic biomarkers and risk of liver and biliary tract cancer. *Hepatology*, *60*, 858–871.

Allinen, M., Beroukhim, R., Cai, L., Brennan, C., Lahti-Domenici, J., Huang, H., et al. (2004). Molecular characterization of the tumor microenvironment in breast cancer. *Cancer Cell*, *6*, 17–32.

Almofti, A., Uchida, D., Begum, N. M., Tomizuka, Y., Iga, H., Yoshida, H., et al. (2004). The clinicopathological significance of the expression of CXCR4 protein in oral squamous cell carcinoma. *International Journal of Oncology*, *25*, 65–71.

Ambrus, J. L., Sr., Dembinski, W., Ambrus, J. L., Jr., Sykes, D. E., Akhter, S., Kulaylat, M. N., et al. (2003). Free interferon-alpha/beta receptors in the circulation of patients with adenocarcinoma. *Cancer*, *98*, 2730–2733.

An, H., Xu, L., Chang, Y., Zhu, Y., Yang, Y., Chen, L., et al. (2015). CXC chemokine receptor 2 is associated with postoperative recurrence and survival of patients with non-metastatic clear-cell renal cell carcinoma. *European Journal of Cancer*, *51*, 1953–1961.

Apte, R. N., Krelin, Y., Song, X., Dotan, S., Recih, E., Elkabets, M., et al. (2006). Effects of micro-environment- and malignant cell-derived interleukin-1 in carcinogenesis, tumour invasiveness and tumour–host interactions. *European Journal of Cancer*, *42*, 751–759.

Aragay, A. M., Mellado, M., Frade, J. M., Martin, A. M., Jimenez-Sainz, M. C., Martinez, A. C., et al. (1998). Monocyte chemoattractant protein-1-induced CCR2B receptor desensitization mediated by the G protein-coupled receptor kinase 2. *Proceedings of the National Academy of Sciences of the United States of America*, *95*, 2985–2990.

Arenberg, D. A., Kunkel, S. L., Polverini, P. J., Glass, M., Burdick, M. D., & Strieter, R. M. (1996). Inhibition of interleukin-8 reduces tumorigenesis of human non-small cell lung cancer in SCID mice. *The Journal of Clinical Investigation*, *97*, 2792–2802.

Arnold, J. M., Huggard, P. R., Cummings, M., Ramm, G. A., & Chenevix-Trench, G. (2005). Reduced expression of chemokine (C-C motif) ligand-2 (CCL2) in ovarian adenocarcinoma. *British Journal of Cancer*, *92*, 2024–2031.

Ashida, N., Arai, H., Yamasaki, M., & Kita, T. (2001). Distinct signaling pathways for MCP-1-dependent integrin activation and chemotaxis. *The Journal of Biological Chemistry*, *276*, 16555–16560.

Ashizawa, T., Okada, R., Suzuki, Y., Takagi, M., Yamazaki, T., Sumi, T., et al. (2005). Clinical significance of interleukin-6 (IL-6) in the spread of gastric cancer: Role of IL-6 as a prognostic factor. *Gastric Cancer*, *8*, 124–131.

Attar, R., Agachan, B., Kuran, S. B., Cacina, C., Sozen, S., Yurdum, L. M., et al. (2010). Association of CCL2 and CCR2 gene variants with endometrial cancer in Turkish women. *In Vivo*, *24*, 243–248.

Au-Yeung, N., Mandhana, R., & Horvath, C. M. (2013). Transcriptional regulation by STAT1 and STAT2 in the interferon JAK-STAT pathway. *JAK-STAT*, *2*, 23931–23940.

Awane, M., Andres, P. G., Li, D. J., & Reinecker, H. C. (1999). NF-kappa B-inducing kinase is a common mediator of IL-17-, TNF-alpha-, and IL-1 beta-induced chemokine promoter activation in intestinal epithelial cells. *Journal of Immunology*, *162*, 5337–5344.

Bachelot, T., Ray-Coquard, I., Menetrier-Caux, C., Rastkha, M., Duc, A., & Blay, J. Y. (2003). Prognostic value of serum levels of interleukin 6 and of serum and plasma levels of vascular endothelial growth factor in hormone-refractory metastatic breast cancer patients. *British Journal of Cancer*, *88*, 1721–1726.

Bai, H., Weng, Y., Bai, S., Jiang, Y., Li, B., He, F., et al. (2014). CCL5 secreted from bone marrow stromal cells stimulates the migration and invasion of Huh7 hepatocellular carcinoma cells via the PI3K-Akt pathway. *International Journal of Oncology*, *45*, 333–343.

Bai, L., Yu, H., Wang, H., Su, H., Zhao, J., & Zhao, Y. (2013). Genetic single-nucleotide polymorphisms of inflammation-related factors associated with risk of lung cancer. *Medical Oncology*, *30*, 414–419.

Balentien, E., Mufson, B. E., Shattuck, R. L., Derynck, R., & Richmond, A. (1991). Effects of MGSA/GRO alpha on melanocyte transformation. *Oncogene*, *6*, 1115–1124.

Balkwill, F. (2009). Tumour necrosis factor and cancer. *Nature Reviews. Cancer*, *9*, 361–371.

Balkwill, F. R., Lee, A., Aldam, G., Moodie, E., Thomas, J. A., Tavernier, J., et al. (1986). Human tumor xenografts treated with recombinant human tumor necrosis factor alone or in combination with interferons. *Cancer Research*, *46*, 3990–3993.

Banin-Hirata, B. K., Losi-Guembarovski, R., Oda, J. M., de Oliveira, C. E., Campos, C. Z., Mazzuco, T. L., et al. (2016). CCR2-V64I genetic polymorphism: A possible involvement in HER2+ breast cancer. *Clinical and Experimental Medicine*, *16*, 139–145.

Banks, T. A., Rouse, B. T., Kerley, M. K., Blair, P. J., Godfrey, V. L., Kuklin, N. A., et al. (1995). Lymphotoxin-alpha-deficient mice. Effects on secondary lymphoid organ development and humoral immune responsiveness. *Journal of Immunology*, *155*, 1685–1693.

Bao, L., Lai, Y., Liu, Y., Qin, Y., Zhao, X., Lu, X., et al. (2013). CXCR4 is a good survival prognostic indicator in multiple myeloma patients. *Leukemia Research*, *37*, 1083–1088.

Barber, M. D., Powell, J. J., Lynch, S. F., Fearon, K. C. H., & Ross, J. A. (2000). A polymorphism of the interleukin-1 β gene influences survival in pancreatic cancer. *British Journal of Cancer*, *83*, 1443–1447.

Barretina, J., Junca, J., Llano, A., Gutierrez, A., Flores, A., Blanco, J., et al. (2003). CXCR4 and SDF-1 expression in B-cell chronic lymphocytic leukemia and stage of the disease. *Annals of Hematology, 82*, 500–505.

Barton, C., Davies, D., Balkwill, F., & Burke, F. (2005). Involvement of both intrinsic and extrinsic pathways in IFN-gamma-induced apoptosis that are enhanced with cisplatin. *European Journal of Cancer, 41*, 1474–1486.

Batsi, O., Giannopoulou, I., Nesseris, I., Valavanis, C., Gakiopoulou, H., Patsouris, E. S., et al. (2014). Immunohistochemical evaluation of CXCL12-CXCR4 axis and VEGFR3 expression in primary urothelial cancer and its recurrence. *Anticancer Research, 34*, 3537–3542.

Baumhoer, D., Smida, J., Zillmer, S., Rosemann, M., Atkinson, M. J., Nelson, P. J., et al. (2012). Strong expression of CXCL12 is associated with a favorable outcome in osteosarcoma. *Modern Pathology, 25*, 522–528.

Begley, L. A., MacDonald, J. W., Day, M. L., & Macoska, J. A. (2007). CXCL12 activates a robust transcriptional response in human prostate epithelial cells. *The Journal of Biological Chemistry, 282*, 26767–26774.

Bektas-Kayhan, K., Unur, M., Boy-Metin, Z., & Cakmakoglu, B. (2012). MCP-1 and CCR2 gene variants in oral squamous cell carcinoma. *Oral Diseases, 18*, 55–59.

Benevides, L., Cardoso, C. R., Tiezzi, D. G., Marana, H. R., Andrade, J. M., & Silva, J. S. (2013). Enrichment of regulatory T cells in invasive breast tumor correlates with the upregulation of IL-17A expression and invasiveness of the tumor. *European Journal of Immunology, 43*, 1518–1528.

Berahovich, R. D., Zabel, B. A., Lewen, S., Walters, M. J., Ebsworth, K., Wang, Y., et al. (2014). Endothelial expression of CXCR7 and the regulation of systemic CXCL12 levels. *Immunology, 141*, 111–122.

Berghuis, D., Santos, S. J., Baelde, H. J., Taminiau, A. H., Egeler, R. M., Schilham, M. W., et al. (2011). Pro-inflammatory chemokine-chemokine receptor interactions within the Ewing sarcoma microenvironment determine CD8(+) T-lymphocyte infiltration and affect tumour progression. *The Journal of Pathology, 223*, 347–357.

Bertrand, F., Rochotte, J., Colacios, C., Montfort, A., Tilkin-Mariame, A. F., Touriol, C., et al. (2015). Blocking tumor necrosis factor alpha enhances CD8 T-cell-dependent immunity in experimental melanoma. *Cancer Research, 75*, 2619–2628.

Betts, G., Twohig, J., Van den Broek, M., Sierro, S., Godkin, A., & Gallimore, A. (2007). The impact of regulatory T cells on carcinogen-induced sarcogenesis. *British Journal of Cancer, 96*, 1849–1854.

Bhat, K. P., Balasubramaniyan, V., Vaillant, B., Ezhilarasan, R., Hummelink, K., Hollingsworth, F., et al. (2013). Mesenchymal differentiation mediated by NF-kappaB promotes radiation resistance in glioblastoma. *Cancer Cell, 24*, 331–346.

Bieche, I., Chavey, C., Andrieu, C., Busson, M., Vacher, S., Le Corre, L., et al. (2007). CXC chemokines located in the 4q21 region are up-regulated in breast cancer. *Endocrine-Related Cancer, 14*, 1039–1052.

Bigildeev, A. E., Shipounova, I. N., Svinareva, D. A., & Drize, N. J. (2011). Leukemia cells invading the liver express liver chemokine receptors and possess characteristics of leukemia stem cells in mice with MPD-like myeloid leukemia. *Experimental Hematology, 39*, 187–194.

Biswas, P., Delfanti, F., Bernasconi, S., Mengozzi, M., Cota, M., Polentarutti, N., et al. (1998). Interleukin-6 induces monocyte chemotactic protein-1 in peripheral blood mononuclear cells and in the U937 cell line. *Blood, 91*, 258–265.

Bode, J. G., Albrecht, U., Haussinger, D., Heinrich, P. C., & Schaper, F. (2012). Hepatic acute phase proteins—Regulation by IL-6- and IL-1-type cytokines involving STAT3 and its crosstalk with NF-kappaB-dependent signaling. *European Journal of Cell Biology, 91*, 496–505.

Bonapace, L., Coissieux, M. M., Wyckoff, J., Mertz, K. D., Varga, Z., Junt, T., et al. (2014). Cessation of CCL2 inhibition accelerates breast cancer metastasis by promoting angiogenesis. *Nature, 515*, 130–133.

Booy, S., van Eijck, C. H., Dogan, F., van Koetsveld, P. M., & Hofland, L. J. (2014). Influence of type-I Interferon receptor expression level on the response to type-I Interferons in human pancreatic cancer cells. *Journal of Cellular and Molecular Medicine, 18*, 492–502.

Boring, L., Gosling, J., Chensue, S. W., Kunkel, S. L., Farese, R. V., Jr., Broxmeyer, H. E., et al. (1997). Impaired monocyte migration and reduced type 1 (Th1) cytokine responses in C-C chemokine receptor 2 knockout mice. *The Journal of Clinical Investigation, 100*, 2552–2561.

Borsi, L., Balza, E., Carnemolla, B., Sassi, F., Castellani, P., Berndt, A., et al. (2003). Selective targeted delivery of TNFalpha to tumor blood vessels. *Blood, 102*, 4384–4392.

Bozcuk, H., Uslu, G., Samur, M., Yildiz, M., Ozben, T., Ozdogan, M., et al. (2004). Tumour necrosis factor-alpha, interleukin-6, and fasting serum insulin correlate with clinical outcome in metastatic breast cancer patients treated with chemotherapy. *Cytokine, 27*, 58–65.

Brat, D. J., Bellail, A. C., & Van Meir, E. G. (2005). The role of interleukin-8 and its receptors in gliomagenesis and tumoral angiogenesis. *Neuro-Oncology*, 7, 122–133.

Braumuller, H., Wieder, T., Brenner, E., Assmann, S., Hahn, M., Alkhaled, M., et al. (2013). T-helper-1-cell cytokines drive cancer into senescence. *Nature, 494*, 361–365.

Brocke-Heidrich, K., Kretzschmar, A. K., Pfeifer, G., Henze, C., Loffler, D., Koczan, D., et al. (2004). Interleukin-6-dependent gene expression profiles in multiple myeloma INA-6 cells reveal a Bcl-2 family-independent survival pathway closely associated with Stat3 activation. *Blood, 103*, 242–251.

Brouckaert, P. G., Leroux-Roels, G. G., Guisez, Y., Tavernier, J., & Fiers, W. (1986). In vivo anti-tumour activity of recombinant human and murine TNF, alone and in combination with murine IFN-gamma, on a syngeneic murine melanoma. *International Journal of Cancer, 38*, 763–769.

Burger, J. A., & Kipps, T. J. (2006). CXCR4: A key receptor in the crosstalk between tumor cells and their microenvironment. *Blood, 107*, 1761–1767.

Cabal-Hierro, L., Rodriguez, M., Artime, N., Iglesias, J., Ugarte, L., Prado, M. A., et al. (2014). TRAF-mediated modulation of NF-kB AND JNK activation by TNFR2. *Cellular Signalling, 26*, 2658–2666.

Cabioglu, N., Sahin, A., Doucet, M., Yavuz, E., Igci, A., Yildirim, E. O., et al. (2005). Chemokine receptor CXCR4 expression in breast cancer as a potential predictive marker of isolated tumor cells in bone marrow. *Clinical & Experimental Metastasis, 22*, 39–46.

Cai, C., Wang, L. H., Dong, Q., Wu, Z. J., Li, M. Y., & Sun, Y. H. (2013). Association of CXCL12 and CXCR4 gene polymorphisms with the susceptibility and prognosis of renal cell carcinoma. *Tissue Antigens, 82*, 165–170.

Calzascia, T., Pellegrini, M., Hall, H., Sabbagh, L., Ono, N., Elford, A. R., et al. (2007). TNF-alpha is critical for antitumor but not antiviral T cell immunity in mice. *The Journal of Clinical Investigation, 117*, 3833–3845.

Cantell, K. (1961). Production and action of interferon in HeLa cells. *Archiv für die Gesamte Virusforschung, 10*, 510–521.

Carswell, E. A., Old, L. J., Kassel, R. L., Green, S., Fiore, N., & Williamson, B. (1975). An endotoxin-induced serum factor that causes necrosis of tumors. *Proceedings of the National Academy of Sciences of the United States of America, 72*, 3666–3670.

Castellana, B., Aasen, T., Moreno-Bueno, G., Dunn, S. E., & Ramon, Y. C. S. (2015). Interplay between YB-1 and IL-6 promotes the metastatic phenotype in breast cancer cells. *Oncotarget, 6*, 38239–38256.

Cavarretta, I. T., Neuwirt, H., Zaki, M. H., Steiner, H., Hobisch, A., Nemeth, J. A., et al. (2008). Mcl-1 is regulated by IL-6 and mediates the survival activity of the cytokine in a model of late stage prostate carcinoma. *Advances in Experimental Medicine and Biology, 617*, 547–555.

Chang, S. H., & Dong, C. (2011). Signaling of interleukin-17 family cytokines in immunity and inflammation. *Cellular Signalling*, *23*, 1069–1075.

Chang, L., Kamata, H., Solinas, G., Luo, J. L., Maeda, S., Venuprasad, K., et al. (2006). The E3 ubiquitin ligase itch couples JNK activation to TNFalpha-induced cell death by inducing c-FLIP(L) turnover. *Cell*, *124*, 601–613.

Chang, S. C., Lin, P. C., Yang, S. H., Wang, H. S., Li, A. F., & Lin, J. K. (2009). SDF-1alpha G801A polymorphism predicts lymph node metastasis in stage T3 colorectal cancer. *Annals of Surgical Oncology*, *16*, 2323–2330.

Charafe-Jauffret, E., Ginestier, C., Iovino, F., Wicinski, J., Cervera, N., Finetti, P., et al. (2009). Breast cancer cell lines contain functional cancer stem cells with metastatic capacity and a distinct molecular signature. *Cancer Research*, *69*, 1302–1313.

Charni, F., Friand, V., Haddad, O., Hlawaty, H., Martin, L., Vassy, R., et al. (2009). Syndecan-1 and syndecan-4 are involved in RANTES/CCL5-induced migration and invasion of human hepatoma cells. *Biochimica et Biophysica Acta*, *1790*, 1314–1326.

Chatterjee, M., von Ungern-Sternberg, S. N., Seizer, P., Schlegel, F., Buttcher, M., Sindhu, N. A., et al. (2015). Platelet-derived CXCL12 regulates monocyte function, survival, differentiation into macrophages and foam cells through differential involvement of CXCR4-CXCR7. *Cell Death & Disease*, *6*, e1989.

Chatzidakis, I., Fousteri, G., Tsoukatou, D., Kollias, G., & Mamalaki, C. (2007). An essential role for TNF in modulating thresholds for survival, activation, and tolerance of CD8+ T cells. *Journal of Immunology*, *178*, 6735–6745.

Chen, Z., Colon, I., Ortiz, N., Callister, M., Dong, G., Pegram, M. Y., et al. (1998). Effects of interleukin-1alpha, interleukin-1 receptor antagonist, and neutralizing antibody on proinflammatory cytokine expression by human squamous cell carcinoma lines. *Cancer Research*, *58*, 3668–3676.

Chen, L., Fan, J., Chen, H., Meng, Z., Chen, Z., Wang, P., et al. (2014). The IL-8/CXCR1 axis is associated with cancer stem cell-like properties and correlates with clinical prognosis in human pancreatic cancer cases. *Scientific Reports*, *4*, 5911.

Chen, S. C., Mehrad, B., Deng, J. C., Vassileva, G., Manfra, D. J., Cook, D. N., et al. (2001). Impaired pulmonary host defense in mice lacking expression of the CXC chemokine lungkine. *The Journal of Immunology*, *166*, 3362–3368.

Chen, X., Wan, J., Liu, J., Xie, W., Diao, X., Xu, J., et al. (2010). Increased IL-17-producing cells correlate with poor survival and lymphangiogenesis in NSCLC patients. *Lung Cancer*, *69*, 348–354.

Chen, R., Xu, Y., Du, X., Liu, N., Li, Y., He, Q., et al. (2015). CXCL12 genetic variants as prognostic markers in nasopharyngeal carcinoma. *OncoTargets and Therapy*, *8*, 2835–2842.

Chen, Q., & Zhong, T. (2015). The association of CXCR4 expression with clinicopathological significance and potential drug target in prostate cancer: A meta-analysis and literature review. *Drug Design, Development and Therapy*, *9*, 5115–5122.

Cheng, X. S., Li, Y. F., Tan, J., Sun, B., Xiao, Y. C., Fang, X. B., et al. (2014). CCL20 and CXCL8 synergize to promote progression and poor survival outcome in patients with colorectal cancer by collaborative induction of the epithelial-mesenchymal transition. *Cancer Letters*, *348*, 77–87.

Cheng, W. L., Wang, C. S., Huang, Y. H., Tsai, M. M., Liang, Y., & Lin, K. H. (2011). Overexpression of CXCL1 and its receptor CXCR2 promote tumor invasion in gastric cancer. *Annals of Oncology*, *22*, 2267–2276.

Chinni, S. R., Sivalogan, S., Dong, Z., Filho, J. C., Deng, X., Bonfil, R. D., et al. (2006). CXCL12/CXCR4 signaling activates Akt-1 and MMP-9 expression in prostate cancer cells: The role of bone microenvironment-associated CXCL12. *Prostate*, *66*, 32–48.

Chinni, S. R., Yamamoto, H., Dong, Z., Sabbota, A., Bonfil, R. D., & Cher, M. L. (2008). CXCL12/CXCR4 transactivates HER2 in lipid rafts of prostate cancer cells and promotes growth of metastatic deposits in bone. *Molecular Cancer Research*, *6*, 446–457.

Chirivi, R. G., Garofalo, A., Padura, I. M., Mantovani, A., & Giavazzi, R. (1993). Interleukin 1 receptor antagonist inhibits the augmentation of metastasis induced by interleukin 1 or lipopolysaccharide in a human melanoma/nude mouse system. *Cancer Research*, *53*, 5051–5054.

Cho, H. J., Kim, S. S., Ahn, S. J., Park, S. Y., Park, J. H., Kim, J. K., et al. (2015). Low serum interleukin-6 levels as a predictive marker of recurrence in patients with hepatitis B virus related hepatocellular carcinoma who underwent curative treatment. *Cytokine*, *73*, 245–252.

Cho, Y. A., Sung, M. K., Yeon, J. Y., Ro, J., & Kim, J. (2013). Prognostic role of interleukin-6, interleukin-8, and leptin levels according to breast cancer subtype. *Cancer Research and Treatment*, *45*, 210–219.

Choi, I., Kang, H. S., Yang, Y., & Pyun, K. H. (1994). IL-6 induces hepatic inflammation and collagen synthesis in vivo. *Clinical & Experimental Immunology*, *95*, 530–535.

Chopra, M., Riedel, S. S., Biehl, M., Krieger, S., von Krosigk, V., Bauerlein, C. A., et al. (2013). Tumor necrosis factor receptor 2-dependent homeostasis of regulatory T cells as a player in TNF-induced experimental metastasis. *Carcinogenesis*, *34*, 1296–1303.

Christensen, J. L., Wright, D. E., Wagers, A. J., & Weissman, I. L. (2004). Circulation and chemotaxis of fetal hematopoietic stem cells. *PLoS Biology*, *2*, E75.

Chu, C. Y., Cha, S. T., Lin, W. C., Lu, P. H., Tan, C. T., Chang, C. C., et al. (2009). Stromal cell-derived factor-1alpha (SDF-1alpha/CXCL12)-enhanced angiogenesis of human basal cell carcinoma cells involves ERK1/2-NF-kappaB/interleukin-6 pathway. *Carcinogenesis*, *30*, 205–213.

Chua, H. L., Bhat-Nakshatri, P., Clare, S. E., Morimiya, A., Badve, S., & Nakshatri, H. (2007). NF-kappaB represses E-cadherin expression and enhances epithelial to mesenchymal transition of mammary epithelial cells: Potential involvement of ZEB-1 and ZEB-2. *Oncogene*, *26*, 711–724.

Chuang, J. Y., Yang, W. H., Chen, H. T., Huang, C. Y., Tan, T. W., Lin, Y. T., et al. (2009). CCL5/CCR5 axis promotes the motility of human oral cancer cells. *Journal of Cellular Physiology*, *220*, 418–426.

Chung, A. S., Wu, X., Zhuang, G., Ngu, H., Kasman, I., Zhang, J., et al. (2013). An interleukin-17-mediated paracrine network promotes tumor resistance to anti-angiogenic therapy. *Nature Medicine*, *19*, 1114–1123.

Clark-Lewis, I., Dewald, B., Geiser, T., Moser, B., & Baggiolini, M. (1993). Platelet factor 4 binds to interleukin 8 receptors and activates neutrophils when its N terminus is modified with Glu-Leu-Arg. *Proceedings of the National Academy of Sciences of the United States of America*, *90*, 3574–3577.

Clatot, F., Picquenot, J. M., Choussy, O., Gouerant, S., Moldovan, C., Schultheis, D., et al. (2011). Intratumoural level of SDF-1 correlates with survival in head and neck squamous cell carcinoma. *Oral Oncology*, *47*, 1062–1068.

Clements, D., Markwick, L. J., Puri, N., & Johnson, S. R. (2010). Role of the CXCR4/CXCL12 axis in lymphangioleiomyomatosis and angiomyolipoma. *The Journal of Immunology*, *185*, 1812–1821.

Coffelt, S. B., Kersten, K., Doornebal, C. W., Weiden, J., Vrijland, K., Hau, C. S., et al. (2015). IL-17-producing gammadelta T cells and neutrophils conspire to promote breast cancer metastasis. *Nature*, *522*, 345–348.

Collado-Hidalgo, A., Bower, J. E., Ganz, P. A., Irwin, M. R., & Cole, S. W. (2008). Cytokine gene polymorphisms and fatigue in breast cancer survivors: Early findings. *Brain, Behavior, and Immunity*, *22*, 1197–1200.

Collins, J. W., Keeney, K. M., Crepin, V. F., Rathinam, V. A., Fitzgerald, K. A., Finlay, B. B., et al. (2014). Citrobacter rodentium: Infection, inflammation and the microbiota. *Nature Reviews. Microbiology*, *12*, 612–623.

Colotta, F., Peri, G., Villa, A., & Mantovani, A. (1984). Rapid killing of actinomycin D-treated tumor cells by human mononuclear cells. I. Effectors belong to the monocyte-macrophage lineage. *Journal of Immunology, 132*, 936–944.

Connett, J. M., Hunt, S. R., Hickerson, S. M., Wu, S. J., & Doherty, G. M. (2003). Localization of IFN-gamma-activated Stat1 and IFN regulatory factors 1 and 2 in breast cancer cells. *Journal of Interferon & Cytokine Research, 23*, 621–630.

Conti, I., Dube, C., & Rollins, B. J. (2004). Chemokine-based pathogenetic mechanisms in cancer. In D. J. Chadwick & J. Goode (Eds.), *Cancer and inflammation* (pp. 29–39): Novartis Foundation. http://dx.doi.org/10.1002/0470856734.ch3.

Cooks, T., Pateras, I. S., Tarcic, O., Solomon, H., Schetter, A. J., Wilder, S., et al. (2013). Mutant p53 prolongs NF-kappaB activation and promotes chronic inflammation and inflammation-associated colorectal cancer. *Cancer Cell, 23*, 634–646.

Cotton, M., & Claing, A. (2009). G protein-coupled receptors stimulation and the control of cell migration. *Cellular Signalling, 21*(7), 1045–1053.

Cressman, D. E., Greenbaum, L. E., DeAngelis, R. A., Ciliberto, G., Furth, E. E., Poli, V., et al. (1996). Liver failure and defective hepatocyte regeneration in interleukin-6-deficient mice. *Science, 274*, 1379–1383.

Croft, M., Benedict, C. A., & Ware, C. F. (2013). Clinical targeting of the TNF and TNFR superfamilies. *Nature Reviews. Drug Discovery, 12*, 147–168.

Croft, M., Duan, W., Choi, H., Eun, S. Y., Madireddi, S., & Mehta, A. (2012). TNF superfamily in inflammatory disease: Translating basic insights. *Trends in Immunology, 33*, 144–152.

Curnis, F., Sacchi, A., Borgna, L., Magni, F., Gasparri, A., & Corti, A. (2000). Enhancement of tumor necrosis factor alpha antitumor immunotherapeutic properties by targeted delivery to aminopeptidase N (CD13). *Nature Biotechnology, 18*, 1185–1190.

Danielli, R., Patuzzo, R., Di Giacomo, A. M., Gallino, G., Maurichi, A., Di Florio, A., et al. (2015). Intralesional administration of L19-IL2/L19-TNF in stage III or stage IVM1a melanoma patients: Results of a phase II study. *Cancer Immunology, Immunotherapy, 64*, 999–1009.

Das Roy, L., Pathangey, L. B., Tinder, T. L., Schettini, J. L., Gruber, H. E., & Mukherjee, P. (2009). Breast-cancer-associated metastasis is significantly increased in a model of autoimmune arthritis. *Breast Cancer Research, 11*, R56.

Denley, S. M., Jamieson, N. B., McCall, P., Oien, K. A., Morton, J. P., Carter, C. R., et al. (2013). Activation of the IL-6R/Jak/stat pathway is associated with a poor outcome in resected pancreatic ductal adenocarcinoma. *Journal of Gastrointestinal Surgery, 17*, 887–898.

Dennis, K. L., Blatner, N. R., Gounari, F., & Khazaie, K. (2013). Current status of interleukin-10 and regulatory T-cells in cancer. *Current Opinion in Oncology, 25*, 637–645.

de Oliveira, K. B., Guembarovski, R. L., Oda, J. M., Mantovani, M. S., Carrera, C. M., Reiche, E. M., et al. (2011). CXCL12 rs1801157 polymorphism and expression in peripheral blood from breast cancer patients. *Cytokine, 55*, 260–265.

Deuel, T. F., Senior, R. M., Chang, D., Griffin, G. L., Heinrikson, R. L., & Kaiser, E. T. (1981). Platelet factor 4 is chemotactic for neutrophils and monocytes. *Proceedings of the National Academy of Sciences of the United States of America, 78*, 4584–4587.

Devapatla, B., Sharma, A., & Woo, S. (2015). CXCR2 inhibition combined with sorafenib improved antitumor and antiangiogenic response in preclinical models of ovarian cancer. *PLoS One, 10*, e0139237.

de Vasconcellos, J. F., Laranjeira, A. B., Zanchin, N. I., Otubo, R., Vaz, T. H., Cardoso, A. A., et al. (2011). Increased CCL2 and IL-8 in the bone marrow microenvironment in acute lymphoblastic leukemia. *Pediatric Blood & Cancer, 56*, 568–577.

De Vita, F., Orditura, M., Auriemma, A., Infusino, S., Roscigno, A., & Catalano, G. (1998). Serum levels of interleukin-6 as a prognostic factor in advanced non-small cell lung cancer. *Oncology Reports*, *5*, 649–652.

de Weck, A. L., Otz, U., Geczy, C., & Geczy, A. F. (1979). Activity and molecular properties of mitogenic factors produced by mixed lymphocyte reactions (MLC-MF) in humans and guinea pigs. *Transplantation Proceedings*, *11*, 1387–1389.

de Weerd, N. A., Samarajiwa, S. A., & Hertzog, P. J. (2007). Type I interferon receptors: Biochemistry and biological functions. *The Journal of Biological Chemistry*, *282*, 20053–20057.

Dhawan, P., & Richmond, A. (2002a). A novel NF-kappa B-inducing kinase-MAPK signaling pathway up-regulates NF-kappa B activity in melanoma cells. *The Journal of Biological Chemistry*, *277*, 7920–7928.

Dhawan, P., & Richmond, A. (2002b). Role of CXCL1 in tumorigenesis of melanoma. *Journal of Leukocyte Biology*, *72*, 9–18.

Di, G. H., Liu, Y., Lu, Y., Liu, J., Wu, C., & Duan, H. F. (2014). IL-6 secreted from senescent mesenchymal stem cells promotes proliferation and migration of breast cancer cells. *PLoS One*, *9*, e113572.

Diamond, M. S., Kinder, M., Matsushita, H., Mashayekhi, M., Dunn, G. P., Archambault, J. M., et al. (2011). Type I interferon is selectively required by dendritic cells for immune rejection of tumors. *The Journal of Experimental Medicine*, *208*, 1989–2003.

Diehl, A. M., Yin, M., Fleckenstein, J., Yang, S. Q., Lin, H. Z., Brenner, D. A., et al. (1994). Tumor necrosis factor-alpha induces c-jun during the regenerative response to liver injury. *The American Journal of Physiology*, *267*, G552–G561.

Dighe, A. S., Richards, E., Old, L. J., & Schreiber, R. D. (1994). Enhanced in vivo growth and resistance to rejection of tumor cells expressing dominant negative IFN gamma receptors. *Immunity*, *1*, 447–456.

Dillenburg-Pilla, P., Patel, V., Mikelis, C. M., Zarate-Blades, C. R., Doci, C. L., Amornphimoltham, P., et al. (2015). SDF-1/CXCL12 induces directional cell migration and spontaneous metastasis via a CXCR4/Galphai/mTORC1 axis. *FASEB Journal*, *29*, 1056–1068.

Dinarello, C. A. (1996). Biologic basis for interleukin-1 in disease. *Blood*, *87*, 2095–2147.

Dinarello, C., Arend, W., Sims, J., Smith, D., Blumberg, H., O'Neill, L., et al. (2010). IL-1 family nomenclature. *Nature Immunology*, *11*, 973.

Dinarello, C. A., & van der Meer, J. W. (2013). Treating inflammation by blocking interleukin-1 in humans. *Seminars in Immunology*, *25*, 469–484.

Ding, L., Hu, E. L., Xu, Y. J., Huang, X. F., Zhang, D. Y., Li, B., et al. (2015). Serum IL-17 F combined with VEGF as potential diagnostic biomarkers for oral squamous cell carcinoma. *Tumour Biology*, *36*, 2523–2529.

Ding, X., Yang, W., Shi, X., Du, P., Su, L., Qin, Z., et al. (2011). TNF receptor 1 mediates dendritic cell maturation and CD8 T cell response through two distinct mechanisms. *Journal of Immunology*, *187*, 1184–1191.

Domingo-Domenech, J., Oliva, C., Rovira, A., Codony-Servat, J., Bosch, M., Filella, X., et al. (2006). Interleukin 6, a nuclear factor-kappaB target, predicts resistance to docetaxel in hormone-independent prostate cancer and nuclear factor-kappaB inhibition by PS-1145 enhances docetaxel antitumor activity. *Clinical Cancer Research*, *12*, 5578–5586.

Donahue, T. R., & Hines, O. J. (2009). CXCR2 and RET single nucleotide polymorphisms in pancreatic cancer. *World Journal of Surgery*, *33*, 710–715.

Dong, H. P., Elstrand, M. B., Holth, A., Silins, I., Berner, A., Trope, C. G., et al. (2006). NK- and B-cell infiltration correlates with worse outcome in metastatic ovarian carcinoma. *American Journal of Clinical Pathology*, *125*, 451–458.

Dong, Y. L., Kabir, S. M., Lee, E. S., & Son, D. S. (2013). CXCR2-driven ovarian cancer progression involves upregulation of proinflammatory chemokines by potentiating NF-kappaB activation via EGFR-transactivated Akt signaling. *PLoS One*, *8*, e83789.

Dorff, T. B., Goldman, B., Pinski, J. K., Mack, P. C., Lara, P. N., Jr., Van Veldhuizen, P. J., Jr., et al. (2010). Clinical and correlative results of SWOG S0354: A phase II trial of CNTO328 (siltuximab), a monoclonal antibody against interleukin-6, in chemotherapy-pretreated patients with castration-resistant prostate cancer. *Clinical Cancer Research*, *16*, 3028–3034.

Doroudchi, M., Saidi, M., Malekzadeh, M., Golmoghaddam, H., Khezri, A., & Ghaderi, A. (2013). Elevated IL-17A levels in early stages of bladder cancer regardless of smoking status. *Future Oncology*, *9*, 295–304.

Du, J. W., Xu, K. Y., Fang, L. Y., & Qi, X. L. (2012). Interleukin-17, produced by lymphocytes, promotes tumor growth and angiogenesis in a mouse model of breast cancer. *Molecular Medicine Reports*, *6*, 1099–1102.

Dumoutier, L., Tounsi, A., Michiels, T., Sommereyns, C., Kotenko, S. V., & Renauld, J. C. (2004). Role of the interleukin (IL)-28 receptor tyrosine residues for antiviral and antiproliferative activity of IL-29/interferon-lambda 1: Similarities with type I interferon signaling. *The Journal of Biological Chemistry*, *279*, 32269–32274.

Dunn, G. P., Bruce, A. T., Sheehan, K. C., Shankaran, V., Uppaluri, R., Bui, J. D., et al. (2005). A critical function for type I interferons in cancer immunoediting. *Nature Immunology*, *6*, 722–729.

Dunn, G. P., Koebel, C. M., & Schreiber, R. D. (2006). Interferons, immunity and cancer immunoediting. *Nature Reviews. Immunology*, *6*, 836–848.

Dunna, E., Sims, J. E., Nicklin, M. J., & O'Neill, L. A. (2001). Annotating genes with potential roles in the immune system: Six new members of the IL-1 family. *Trends in Immunology*, *22*, 533–536.

Dunne, A., & O'Neill, L. A. (2003). The interleukin-1 receptor/Toll-like receptor superfamily: Signal transduction during inflammation and host defense. *Science STKE*, *2003*, re3.

Dvorak, H. F. (2015). Tumors: Wounds that do not heal-redux. *Cancer Immunology Research*, *3*, 1–11.

Eikawa, S., Ohue, Y., Kitaoka, K., Aji, T., Uenaka, A., Oka, M., et al. (2010). Enrichment of Foxp3 + CD4 regulatory T cells in migrated T cells to IL-6- and IL-8-expressing tumors through predominant induction of CXCR1 by IL-6. *The Journal of Immunology*, *185*, 6734–6740.

El-Ghany, H. M., El-Saadany, Z. A., Bahaa, N. M., Ibrahim, N. Y., & Hussien, S. M. (2014). Stromal cell derived factor-1 (CXCL12) chemokine gene variant in myeloid leukemias. *Clinical Laboratory*, *60*, 735–741.

Enewold, L., Mechanic, L. E., Bowman, E. D., Zheng, Y. L., Yu, Z., Trivers, G., et al. (2009). Serum concentrations of cytokines and lung cancer survival in African Americans and Caucasians. *Cancer Epidemiology, Biomarkers & Prevention*, *18*, 215–222.

Evelyn, C. R., Wade, S. M., Wang, Q., Wu, M., Iniguez-Lluhi, J. A., Merajver, S. D., et al. (2007). CCG-1423: A small-molecule inhibitor of RhoA transcriptional signaling. *Molecular Cancer Therapeutics*, *6*, 2249–2260.

Ewington, L., Taylor, A., Sriraksa, R., Horimoto, Y., Lam, E. W., & El-Bahrawy, M. A. (2012). The expression of interleukin-8 and interleukin-8 receptors in endometrial carcinoma. *Cytokine*, *59*, 417–422.

Fader, A. N., Rasool, N., Vaziri, S. A., Kozuki, T., Faber, P. W., Elson, P., et al. (2010). CCL2 expression in primary ovarian carcinoma is correlated with chemotherapy response and survival outcomes. *Anticancer Research*, *30*, 4791–4798.

Fang, W. B., Jokar, I., Zou, A., Lambert, D., Dendukuri, P., & Cheng, N. (2012). CCL2/CCR2 chemokine signaling coordinates survival and motility of breast cancer cells through Smad3 protein- and p42/44 mitogen-activated protein kinase (MAPK)-dependent mechanisms. *The Journal of Biological Chemistry*, *287*, 36593–36608.

Fang, W. B., Yao, M., Jokar, I., Alhakamy, N., Berkland, C., Chen, J., et al. (2015). The CCL2 chemokine is a negative regulator of autophagy and necrosis in luminal B breast cancer cells. *Breast Cancer Research and Treatment*, *150*, 309–320.

Farrar, W. L., Mizel, S. B., & Farrar, J. J. (1980). Participation of lymphocyte activating factor (Interleukin 1) in the induction of cytotoxic T cell responses. *The Journal of Immunology*, *124*, 1371–1377.

Feldmann, M., & Maini, R. N. (2003). Lasker Clinical Medical Research Award. TNF defined as a therapeutic target for rheumatoid arthritis and other autoimmune diseases. *Nature Medicine*, *9*, 1245–1250.

Fernandis, A. Z., Prasad, A., Band, H., Klosel, R., & Ganju, R. K. (2004). Regulation of CXCR4-mediated chemotaxis and chemoinvasion of breast cancer cells. *Oncogene*, *23*, 157–167.

Ferrara, N. (2002). VEGF and the quest for tumour angiogenesis factors. *Nature Reviews Cancer*, *2*, 795–803.

Ferretti, S., Bonneau, O., Dubois, G. R., Jones, C. E., & Trifilieff, A. (2003). IL-17, produced by lymphocytes and neutrophils, is necessary for lipopolysaccharide-induced airway neutrophilia: IL-15 as a possible trigger. *Journal of Immunology*, *170*, 2106–2112.

Finkin, S., Yuan, D., Stein, I., Taniguchi, K., Weber, A., Unger, K., et al. (2015). Ectopic lymphoid structures function as microniches for tumor progenitor cells in hepatocellular carcinoma. *Nature Immunology*, *16*, 1235–1244.

Frick, V. O., Rubie, C., Wagner, M., Graeber, S., Grimm, H., Kopp, B., et al. (2008). Enhanced ENA-78 and IL-8 expression in patients with malignant pancreatic diseases. *Pancreatology*, *8*, 488–497.

Fu, Q., Chang, Y., An, H., Fu, H., Zhu, Y., Xu, L., et al. (2015). Prognostic value of interleukin-6 and interleukin-6 receptor in organ-confined clear-cell renal cell carcinoma: A 5-year conditional cancer-specific survival analysis. *British Journal of Cancer*, *113*, 1581–1589.

Fujimoto, H., Sangai, T., Ishii, G., Ikehara, A., Nagashima, T., Miyazaki, M., et al. (2009). Stromal MCP-1 in mammary tumors induces tumor-associated macrophage infiltration and contributes to tumor progression. *International Journal of Cancer*, *125*, 1276–1284.

Fukushima, S., Sugita, Y., Niino, D., Mihashi, H., & Ohshima, K. (2012). Clinicopathological analysis of olfactory neuroblastoma. *Brain Tumor Pathology*, *29*, 207–215.

Furuya, N., Kamai, T., Shirataki, H., Yanai, Y., Fukuda, T., Mizuno, T., et al. (2011). Serum interferon alpha receptor 2 mRNA may predict efficacy of interferon alpha with/without low-dose sorafenib for metastatic clear cell renal cell carcinoma. *Cancer Immunology, Immunotherapy*, *60*, 793–808.

Gao, Y., Yang, W., Pan, M., Scully, E., Girardi, M., Augenlicht, L. H., et al. (2003). Gamma delta T cells provide an early source of interferon gamma in tumor immunity. *The Journal of Experimental Medicine*, *198*, 433–442.

Garcia-Irigoyen, O., Latasa, M. U., Carotti, S., Uriarte, I., Elizalde, M., Urtasun, R., et al. (2015). Matrix metalloproteinase 10 contributes to hepatocarcinogenesis in a novel crosstalk with the stromal derived factor 1/C-X-C chemokine receptor 4 axis. *Hepatology*, *62*, 166–178.

Garcia-Tunon, I., Ricote, M., Ruiz, A. A., Fraile, B., Paniagua, R., & Royuela, M. (2007). Influence of IFN-gamma and its receptors in human breast cancer. *BMC Cancer*, *7*, 158.

Garlanda, C., Dinarello, C. A., & Mantovani, A. (2013). The interlukin-1 family: Back to the future. *Immunity*, *39*, 1003–1018.

Gauzzi, M. C., Velazquez, L., McKendry, R., Mogensen, K. E., Fellous, M., & Pellegrini, S. (1996). Interferon-alpha-dependent activation of Tyk2 requires phosphorylation of positive regulatory tyrosines by another kinase. *The Journal of Biological Chemistry*, *271*, 20494–20500.

Geginat, J., Larghi, P., Paroni, M., Nizzoli, G., Penatti, A., Pagani, M., et al. (2016). The light and the dark sides of Interleukin-10 in immune-mediated diseases and cancer. *Cytokine & Growth Factor Reviews*. http://dx.doi.org/10.1016/j.cytogfr.2016.02.003.

Geiser, T., Dewald, B., Ehrengruber, M. U., Clark-Lewis, I., & Baggiolini, M. (1993). The interleukin-8-related chemotactic cytokines GRO alpha, GRO beta, and GRO gamma activate human neutrophil and basophil leukocytes. *The Journal of Biological Chemistry*, *268*, 15419–15424.

Gerger, A., El-Khoueiry, A., Zhang, W., Yang, D., Singh, H., Bohanes, P., et al. (2011). Pharmacogenetic angiogenesis profiling for first-line Bevacizumab plus oxaliplatin-based chemotherapy in patients with metastatic colorectal cancer. *Clinical Cancer Research*, *17*, 5783–5792.

Ghilardi, G., Biondi, M. L., La Torre, A., Battaglioli, L., & Scorza, R. (2005). Breast cancer progression and host polymorphisms in the chemokine system: Role of the macrophage chemoattractant protein-1 (MCP-1) -2518 G allele. *Clinical Chemistry*, *51*, 452–455.

Ghobrial, I. M., Bone, N. D., Stenson, M. J., Novak, A., Hedin, K. E., Kay, N. E., et al. (2004). Expression of the chemokine receptors CXCR4 and CCR7 and disease progression in B-cell chronic lymphocytic leukemia/ small lymphocytic lymphoma. *Mayo Clinic Proceedings*, *79*, 318–325.

Giannopoulou, M., Iszkula, S. C., Dai, C., Tan, X., Yang, J., Michalopoulos, G. K., et al. (2006). Distinctive role of Stat3 and Erk-1/2 activation in mediating interferon-gamma inhibition of TGF-beta1 action. *American Journal of Physiology. Renal Physiology*, *290*, F1234–F1240.

Ginestier, C., Liu, S., Diebel, M. E., Korkaya, H., Luo, M., Brown, M., et al. (2010). CXCR1 blockade selectively targets human breast cancer stem cells in vitro and in xenografts. *The Journal of Clinical Investigation*, *120*, 485–497.

Goldstein, R. H., Reagan, M. R., Anderson, K., Kaplan, D. L., & Rosenblatt, M. (2010). Human bone marrow-derived MSCs can home to orthotopic breast cancer tumors and promote bone metastasis. *Cancer Research*, *70*, 10044–10050.

Gopinathan, G., Milagre, C., Pearce, O. M., Reynolds, L. E., Hodivala-Dilke, K., Leinster, D. A., et al. (2015). Interleukin-6 stimulates defective angiogenesis. *Cancer Research*, *75*, 3098–3107.

Granot, Z., Henke, E., Comen, E. A., King, T. A., Norton, L., & Benezra, R. (2011). Tumor entrained neutrophils inhibit seeding in the premetastatic lung. *Cancer Cell*, *20*, 300–314.

Greenfeder, S. A., Nunes, P., Kwee, L., Labow, M., Chizzonite, R. A., & Ju, G. (1995). Molecular cloning and characterization of a second subunit of the interleukin 1 receptor complex. *The Journal of Biological Chemistry*, *270*, 13757–13765.

Gregorc, V., Zucali, P. A., Santoro, A., Ceresoli, G. L., Citterio, G., De Pas, T. M., et al. (2010). Phase II study of asparagine-glycine-arginine-human tumor necrosis factor alpha, a selective vascular targeting agent, in previously treated patients with malignant pleural mesothelioma. *Journal of Clinical Oncology*, *28*, 2604–2611.

Grimm, C., Berger, I., Tomovski, C., Zeillinger, R., Concin, N., Leodolter, S., et al. (2004). A polymorphism of the interleukin-1 receptor antagonist plays a prominent role within the interleukin-1 gene cluster in vulvar carcinogenesis. *Gynecologic Oncology*, *92*, 936–940.

Gu, H., Ni, M., Guo, X., Feng, P., Xu, Y., Gu, X., et al. (2011). The functional polymorphism in monocyte chemoattractant protein-1 gene increases susceptibility to gastric cancer. *Medical Oncology*, *28*(Suppl. 1), S280–S285.

Guo, Q., Gao, B. L., Zhang, X. J., Liu, G. C., Xu, F., Fan, Q. Y., et al. (2015). CXCL12-CXCR4 axis promotes proliferation, migration, invasion, and metastasis of ovarian cancer. *Oncology Research*, *22*, 247–258.

Haber, J., Rosenau, W., & Goldberg, M. (1972). Separate factors in phytohaemagglutinin induce lymphotoxin, interferon, and nucleic acid synthesis. *Nature: New Biology*, *238*, 60–61.

Hacke, K., Rincon-Orozco, B., Buchwalter, G., Siehler, S. Y., Wasylyk, B., Wiesmuller, L., et al. (2010). Regulation of MCP-1 chemokine transcription by p53. *Molecular Cancer*, *9*, 82.

Haghnegahdar, H., Du, J., Wang, D., Strieter, R. M., Burdick, M. D., Nanney, L. B., et al. (2000). The tumorigenic and angiogenic effects of MGSA/GRO proteins in melanoma. *Journal of Leukocyte Biology*, *67*, 53–62.

Halin, C., Gafner, V., Villani, M. E., Borsi, L., Berndt, A., Kosmehl, H., et al. (2003). Synergistic therapeutic effects of a tumor targeting antibody fragment, fused to interleukin 12 and to tumor necrosis factor alpha. *Cancer Research*, *63*, 3202–3210.

Hallett, M. A., Venmar, K. T., & Fingleton, B. (2012). Cytokine stimulation of epithelial cancer cells: The similar and divergent functions of IL-4 and IL-13. *Cancer Research*, *72*, 6338–6343.

Halpern, J. L., Kilbarger, A., & Lynch, C. C. (2011). Mesenchymal stem cells promote mammary cancer cell migration in vitro via the CXCR2 receptor. *Cancer Letters*, *308*, 91–99.

Hammacher, A., Ward, L. D., Weinstock, J., Treutlein, H., Yasukawa, K., & Simpson, R. J. (1994). Structure–function analysis of human IL-6: Identification of two distinct regions that are important for receptor binding. *Protein Science*, *3*, 2280–2293.

Han, L., Jiang, B., Wu, H., Wang, X., Tang, X., Huang, J., et al. (2012). High expression of CXCR2 is associated with tumorigenesis, progression, and prognosis of laryngeal squamous cell carcinoma. *Medical Oncology*, *29*, 2466–2472.

Hansell, C. A., Hurson, C. E., & Nibbs, R. J. (2011). DARC and D6: Silent partners in chemokine regulation? *Immunology and Cell Biology*, *89*, 197–206.

Hardy, M. P., Owczarek, C. M., Trajanovska, S., Liu, X., Kola, I., & Hertzog, P. J. (2001). The soluble murine type I interferon receptor Ifnar-2 is present in serum, is independently regulated, and has both agonistic and antagonistic properties. *Blood*, *97*, 473–482.

Harrington, L. E., Hatton, R. D., Mangan, P. R., Turner, H., Murphy, T. L., Murphy, K. M., et al. (2005). Interleukin 17-producing CD4+ effector T cells develop via a lineage distinct from the T helper type 1 and 2 lineages. *Nature Immunology*, *6*, 1123–1132.

Hartman, Z. C., Poage, G. M., den Hollander, P., Tsimelzon, A., Hill, J., Panupinthu, N., et al. (2013). Growth of triple-negative breast cancer cells relies upon coordinate autocrine expression of the proinflammatory cytokines IL-6 and IL-8. *Cancer Research*, *73*, 3470–3480.

Hatting, M., Spannbauer, M., Peng, J., Al Masaoudi, M., Sellge, G., Nevzorova, Y. A., et al. (2015). Lack of gp130 expression in hepatocytes attenuates tumor progression in the DEN model. *Cell Death & Disease*, *6*, e1667.

Havell, E. A., Fiers, W., & North, R. J. (1988). The antitumor function of tumor necrosis factor (TNF), I. Therapeutic action of TNF against an established murine sarcoma is indirect, immunologically dependent, and limited by severe toxicity. *The Journal of Experimental Medicine*, *167*, 1067–1085.

He, G., Dhar, D., Nakagawa, H., Font-Burgada, J., Ogata, H., Jiang, Y., et al. (2013). Identification of liver cancer progenitors whose malignant progression depends on autocrine IL-6 signaling. *Cell*, *155*, 384–396.

He, Y., Du, Y., Wei, S., Shi, J., Mei, Z., Qian, L., et al. (2015). IL-17A and IL-17 F single nucleotide polymorphisms associated with lung cancer in Chinese population. *The Clinical Respiratory Journal*.

Hebert, C. A., Vitangcol, R. V., & Baker, J. B. (1991). Scanning mutagenesis of interleukin-8 identifies a cluster of residues required for receptor binding. *The Journal of Biological Chemistry, 266*, 18989–18994.

Hefler, L., Tempfer, C., Heinze, G., Mayerhofer, K., Breitenecker, G., Leodolter, S., et al. (1999). Monocyte chemoattractant protein-1 serum levels in ovarian cancer patients. *British Journal of Cancer, 81*, 855–859.

Heinrich, P. C., Behrmann, I., Müller-Newen, G., Schaper, F., & Graeve, L. (1988). Interleukin-6-type cytokine signalling through the gp130/Jak/STAT pathway. *The Biochemical Journal, 334*, 297–314.

Hembruff, S. L., Jokar, I., Yang, L., & Cheng, N. (2010). Loss of transforming growth factor-beta signaling in mammary fibroblasts enhances CCL2 secretion to promote mammary tumor progression through macrophage-dependent and -independent mechanisms. *Neoplasia, 12*, 425–433.

Herman, J. M., Wild, A. T., Wang, H., Tran, P. T., Chang, K. J., Taylor, G. E., et al. (2013). Randomized phase III multi-institutional study of TNFerade biologic with fluorouracil and radiotherapy for locally advanced pancreatic cancer: Final results. *Journal of Clinical Oncology, 31*, 886–894.

Hernandez, L., Smirnova, T., Kedrin, D., Wyckoff, J., Zhu, L., Stanley, E. R., et al. (2009). The EGF/CSF-1 paracrine invasion loop can be triggered by heregulin beta1 and CXCL12. *Cancer Research, 69*, 3221–3227.

Hibi, M., Murakami, M., Saito, M., Hirano, T., Taga, T., & Kishimoto, T. (1990). Molecular cloning and expression of an IL-6 signal transducer, gp130. *Cell, 63*, 1149–1157.

Hirata, H., Hinoda, Y., Kikuno, N., Kawamoto, K., Dahiya, A. V., Suehiro, Y., et al. (2007). CXCL12 G801A polymorphism is a risk factor for sporadic prostate cancer susceptibility. *Clinical Cancer Research, 13*, 5056–5062.

Holloway, G., Dang, V. T., Jans, D. A., & Coulson, B. S. (2014). Rotavirus inhibits IFN-induced STAT nuclear translocation by a mechanism that acts after STAT binding to importin-alpha. *The Journal of General Virology, 95*, 1723–1733.

Horuk, R. (2015). The duffy antigen receptor for chemokines DARC/ACKR1. *Frontiers in Immunology, 6*, 279.

Hou, C., & Yang, F. (2015). Interleukin-17A gene polymorphism is associated with susceptibility to gastric cancer. *International Journal of Clinical and Experimental Pathology, 8*, 7378–7384.

Hu, C., Yong, X., Li, C., Lu, M., Liu, D., Chen, L., et al. (2013). CXCL12/CXCR4 axis promotes mesenchymal stem cell mobilization to burn wounds and contributes to wound repair. *The Journal of Surgical Research, 183*, 427–434.

Huang, S., Bucana, C. D., Van Arsdall, M., & Fidler, I. J. (2002). Stat1 negatively regulates angiogenesis, tumorigenicity and metastasis of tumor cells. *Oncogene, 21*, 2504–2512.

Huang, D. R., Wang, J., Kivisakk, P., Rollins, B. J., & Ransohoff, R. M. (2001). Absence of monocyte chemoattractant protein 1 in mice leads to decreased local macrophage recruitment and antigen-specific T helper cell type 1 immune response in experimental autoimmune encephalomyelitis. *The Journal of Experimental Medicine, 193*, 713–726.

Huang, Y., Zhang, J., Cui, Z. M., Zhao, J., & Zheng, Y. (2013). Expression of the CXCL12/CXCR4 and CXCL16/CXCR6 axes in cervical intraepithelial neoplasia and cervical cancer. *Chinese Journal of Cancer, 32*, 289–296.

Hunsucker, S. A., Magarotto, V., Kuhn, D. J., Kornblau, S. M., Wang, M., Weber, D. M., et al. (2011). Blockade of interleukin-6 signalling with siltuximab enhances melphalan cytotoxicity in preclinical models of multiple myeloma. *British Journal of Haematology, 152*, 579–592.

Hymowitz, S. G., Filvaroff, E. H., Yin, J. P., Lee, J., Cai, L., Risser, P., et al. (2001). IL-17 s adopt a cystine knot fold: Structure and activity of a novel cytokine, IL-17F, and implications for receptor binding. *EMBO Journal, 20*, 5332–5341.

Indolfi, G., Azzari, C., & Resti, M. (2014). Polymorphisms in the IFNL3/IL28B gene and hepatitis C: From adults to children. *World Journal of Gastroenterology*, *20*, 9245–9252.
Infante-Duarte, C., Horton, H. F., Byrne, M. C., & Kamradt, T. (2000). Microbial lipopeptides induce the production of IL-17 in Th cells. *The Journal of Immunology*, *165*, 6107–6115.
Inngjerdingen, M., Torgersen, K. M., & Maghazachi, A. A. (2002). Lck is required for stromal cell-derived factor 1 alpha (CXCL12)-induced lymphoid cell chemotaxis. *Blood*, *99*, 4318–4325.
Inoue, K., Slaton, J. W., Eve, B. Y., Kim, S. J., Perrotte, P., Balbay, M. D., et al. (2000). Interleukin 8 expression regulates tumorigenicity and metastases in androgen-independent prostate cancer. *Clinical Cancer Research*, *6*, 2104–2119.
Isaacs, A., Burke, D. C., & Fadeeva, L. (1958). Effect of interferon on the growth of viruses on the chick chorion. *British Journal of Experimental Pathology*, *39*, 447–451.
Ishigame, H., Kakuta, S., Nagai, T., Kadoki, M., Nambu, A., Komiyama, Y., et al. (2009). Differential roles of interleukin-17A and -17F in host defense against mucoepithelial bacterial infection and allergic responses. *Immunity*, *30*, 108–119.
Isman, F. K., Kucukgergin, C., Dasdemir, S., Cakmakoglu, B., Sanli, O., & Seckin, S. (2012). Association between SDF1-3'A or CXCR4 gene polymorphisms with predisposition to and clinicopathological characteristics of prostate cancer with or without metastases. *Molecular Biology Reports*, *39*, 11073–11079.
Isobe, A., Sawada, K., Kinose, Y., Ohyagi-Hara, C., Nakatsuka, E., Makino, H., et al. (2015). Interleukin 6 receptor is an independent prognostic factor and a potential therapeutic target of ovarian cancer. *PLoS One*, *10*, e0118080.
Ito, L. S., Iwata, H., Hamajima, N., Saito, T., Matsuo, K., Mizutani, M., et al. (2002). Significant reduction in breast cancer risk for Japanese women with interleukin 1B-31 CT/TT relative to CC genotype. *Japanese Journal of Clinical Oncology*, *32*, 398–402.
Ivashkiv, L. B., & Donlin, L. T. (2014). Regulation of type I interferon responses. *Nature Reviews. Immunology*, *14*, 36–49.
Iversen, M. B., Ank, N., Melchjorsen, J., & Paludan, S. R. (2010). Expression of type III interferon (IFN) in the vaginal mucosa is mediated primarily by dendritic cells and displays stronger dependence on NF-kappaB than type I IFNs. *Journal of Virology*, *84*, 4579–4586.
Izuhara, K., & Harada, N. (1993). Interleukin-4 (IL-4) induces protein tyrosine phosphorylation of the IL-4 receptor and association of phosphatidylinositol 3-kinase to the IL-4 receptor in a mouse T cell line, HT2. *The Journal of Biological Chemistry*, *268*, 13097–13102.
Izumi, D., Ishimoto, T., Miyake, K., Sugihara, H., Eto, K., Sawayama, H., et al. (2015). CXCL12/CXCR4 activation by cancer-associated fibroblasts promotes integrin beta1 clustering and invasiveness in gastric cancer. *International Journal of Cancer*, *138*(5), 1207–1219.
Jablonska, J., Wu, C. F., Andzinski, L., Leschner, S., & Weiss, S. (2014). CXCR2-mediated tumor-associated neutrophil recruitment is regulated by IFN-beta. *International Journal of Cancer*, *134*, 1346–1358.
Jameson, P., & Grossberg, S. E. (1979). Production of interferon by human tumor cell lines. *Archives of Virology*, *62*, 209–219.
Jiang, H., Gebhardt, C., Umansky, L., Beckhove, P., Schulze, T. J., Utikal, J., et al. (2015). Elevated chronic inflammatory factors and myeloid-derived suppressor cells indicate poor prognosis in advanced melanoma patients. *International Journal of Cancer*, *136*, 2352–2360.
Jimenez-Sainz, M. C., Fast, B., Mayor, F., Jr., & Aragay, A. M. (2003). Signaling pathways for monocyte chemoattractant protein 1-mediated extracellular signal-regulated kinase activation. *Molecular Pharmacology*, *64*, 773–782.

Jin, B., Scott, J. L., Vadas, M. A., & Burns, G. F. (1989). TGF beta down-regulates TLiSA1 expression and inhibits the differentiation of precursor lymphocytes into CTL and LAK cells. *Immunology, 66*, 570–576.

Jin, L., Yuan, R. Q., Fuchs, A., Yao, Y., Joseph, A., Schwall, R., et al. (1997). Expression of interleukin-1beta in human breast carcinoma. *Cancer, 80*, 421–434.

Jin, J., Zhao, W. C., & Yuan, F. (2013). CXCR7/CXCR4/CXCL12 axis regulates the proliferation, migration, survival and tube formation of choroid-retinal endothelial cells. *Ophthalmic Research, 50*, 6–12.

Johnson, D. R., & Pober, J. S. (1991). Tumor necrosis factor regulation of major histocompatibility complex gene expression. *Immunologic Research, 10*, 141–155.

Juang, S. H., Wei, S. J., Hung, Y. M., Hsu, C. Y., Yang, D. M., Liu, K. J., et al. (2004). IFN-beta induces caspase-mediated apoptosis by disrupting mitochondria in human advanced stage colon cancer cell lines. *Journal of Interferon & Cytokine Research, 24*, 231–243.

Kamai, T., Yanai, Y., Arai, K., Abe, H., Yamanishi, T., Kurimoto, M., et al. (2007). Increased interferon alpha receptor 2 mRNA levels is associated with renal cell carcinoma metastasis. *BMC Cancer, 7*, 159.

Kamali-Sarvestani, E., Aliparasti, M. R., & Atefi, S. (2007). Association of interleukin-8 (IL-8 or CXCL8) -251T/A and CXCR2 +1208C/T gene polymorphisms with breast cancer. *Neoplasma, 54*, 484–489.

Kan, J. Y., Wu, D. C., Yu, F. J., Wu, C. Y., Ho, Y. W., Chiu, Y. J., et al. (2015). Chemokine (C-C Motif) ligand 5 is involved in tumor-associated dendritic cell-mediated colon cancer progression through non-coding RNA MALAT-1. *Journal of Cellular Physiology, 230*, 1883–1894.

Karayiannakis, A. J., Syrigos, K. N., Polychronidis, A., Pitiakoudis, M., Bounovas, A., & Simopoulos, K. (2001). Serum levels of tumor necrosis factor-alpha and nutritional status in pancreatic cancer patients. *Anticancer Research, 21*, 1355–1358.

Karczewska, A., Nawrocki, S., Breborowicz, D., Filas, V., & Mackiewicz, A. (2000). Expression of interleukin-6, interleukin-6 receptor, and glycoprotein 130 correlates with good prognoses for patients with breast carcinoma. *Cancer, 88*, 2061–2071.

Karnoub, A. E., Dash, A. B., Vo, A. P., Sullivan, A., Brooks, M. W., Bell, G. W., et al. (2007). Mesenchymal stem cells within tumour stroma promote breast cancer metastasis. *Nature, 449*, 557–563.

Kashii, Y., Giorda, R., Herberman, R. B., Whiteside, T. L., & Vujanovic, N. L. (1999). Constitutive expression and role of the TNF family ligands in apoptotic killing of tumor cells by human NK cells. *The Journal of Immunology, 163*, 5358–5366.

Kataoka, T. (2009). Chemical biology of inflammatory cytokine signaling. *The Journal of Antibiotics (Tokyo), 62*, 655–667.

Kato, T., Fujita, Y., Nakane, K., Mizutani, K., Terazawa, R., Ehara, H., et al. (2013). CCR1/CCL5 interaction promotes invasion of taxane-resistant PC3 prostate cancer cells by increasing secretion of MMPs 2/9 and by activating ERK and Rac signaling. *Cytokine, 64*, 251–257.

Kayton, M. L., & Libutti, S. K. (2001). Endothelial monocyte activating polypeptide II (EMAP II) enhances the effect of TNF on tumor-associated vasculature. *Current Opinion in Investigational Drugs, 2*, 136–138.

Kehrl, J. H., Wakefield, L. M., Roberts, A. B., Jakowlew, S., Alvarez-Mon, M., Derynck, R., et al. (1986). Production of transforming growth factor beta by human T lymphocytes and its potential role in the regulation of T cell growth. *The Journal of Experimental Medicine, 163*, 1037–1050.

Kettelhut, I. C., Fiers, W., & Goldberg, A. L. (1987). The toxic effects of tumor necrosis factor in vivo and their prevention by cyclooxygenase inhibitors. *Proceedings of the National Academy of Sciences of the United States of America, 84*, 4273–4277.

Khan, U. T., Tanasescu, R., & Constantinescu, C. S. (2015). PEGylated IFNbeta-1a in the treatment of multiple sclerosis. *Expert Opinion on Biological Therapy*, *15*, 1077–1084.

Khan, M. N., Wang, B., Wei, J., Zhang, Y., Li, Q., Luan, X., et al. (2015). CXCR1/2 antagonism with CXCL8/Interleukin-8 analogue CXCL8(3-72)K11R/G31P restricts lung cancer growth by inhibiting tumor cell proliferation and suppressing angiogenesis. *Oncotarget*, *6*, 21315–21327.

Kim, M. Y., Oskarsson, T., Acharyya, S., Nguyen, D. X., Zhang, X. H., Norton, L., et al. (2009). Tumor self-seeding by circulating cancer cells. *Cell*, *139*, 1315–1326.

Kim, E. Y., & Teh, H. S. (2001). TNF type 2 receptor (p75) lowers the threshold of T cell activation. *Journal of Immunology*, *167*, 6812–6820.

Kish, D. D., Gorbachev, A. V., & Fairchild, R. L. (2012). IL-1 receptor signaling is required at multiple stages of sensitization and elicitation of the contact hypersensitivity response. *Journal of Immunology*, *188*, 1761–1771.

Kishimoto, T., Akira, S., Narazaki, M., & Taga, T. (1995). Interleukin-6 family of cytokines and gp130. *Blood*, *86*, 1243–1254.

Kitamura, T., Qian, B. Z., Soong, D., Cassetta, L., Noy, R., Sugano, G., et al. (2015). CCL2-induced chemokine cascade promotes breast cancer metastasis by enhancing retention of metastasis-associated macrophages. *The Journal of Experimental Medicine*, *212*, 1043–1059.

Kluger, M. J., Kozak, W., Leon, L. R., & Conn, C. A. (1998). The use of knockout mice to understand the role of cytokines in fever. *Clinical and Experimental Pharmacology & Physiology*, *25*, 141–144.

Koch, A. E., Polverini, P. J., Kunkel, S. L., Harlow, L. A., DiPietro, L. A., Elner, V. M., et al. (1992). Interleukin-8 as a macrophage-derived mediator of angiogenesis. *Science*, *258*, 1798–1801.

Koga, M., Kai, H., Egami, K., Murohara, T., Ikeda, A., Yasuoka, S., et al. (2008). Mutant MCP-1 therapy inhibits tumor angiogenesis and growth of malignant melanoma in mice. *Biochemical and Biophysical Research Communications*, *365*, 279–284.

Koj, A. (1985). Cytokines regulating acute inflammation and synthesis of acute phase proteins. *Blut*, *51*, 267–274.

Kolls, J. K., & Linden, A. (2004). Interleukin-17 family members and inflammation. *Immunity*, *21*, 467–476.

Konaka, Y., Norcross, M. A., Maino, V. C., & Smith, R. T. (1981). Anti-Thy-1-mediated T cell activation. Role of soluble factors and expression of interleukin 2 receptors on T cells. *European Journal of Immunology*, *11*, 445–450.

Kondo, T., Nakazawa, H., Ito, F., Hashimoto, Y., Osaka, Y., Futatsuyama, K., et al. (2006). Favorable prognosis of renal cell carcinoma with increased expression of chemokines associated with a Th1-type immune response. *Cancer Science*, *97*, 780–786.

Kondo, Y., Yukinaka, M., Nomura, M., Nakaya, Y., & Ito, S. (2000). Early diagnosis of interferon-induced myocardial disorder in patients with chronic hepatitis C: Evaluation by myocardial imaging with 123I-BMIPP. *Journal of Gastroenterology*, *35*, 127–135.

Kooy, A. J., Tank, B., Vuzevski, V. D., van Joost, T., & Prens, E. P. (1998). Expression of interferon-gamma receptors and interferon-gamma-induced up-regulation of intercellular adhesion molecule-1 in basal cell carcinoma; decreased expression of IFN-gamma R and shedding of ICAM-1 as a means to escape immune surveillance. *The Journal of Pathology*, *184*, 169–176.

Koshiba, T., Hosotani, R., Miyamoto, Y., Ida, J., Tsuji, S., Nakajima, S., et al. (2000). Expression of stromal cell-derived factor 1 and CXCR4 ligand receptor system in pancreatic cancer: A possible role for tumor progression. *Clinical Cancer Research*, *6*, 3530–3535.

Kovalovich, K., DeAngelis, R. A., Li, W., Furth, E. E., Ciliberto, G., & Taub, R. (2000). Increased toxin-induced liver injury and fibrosis in interleukin-6-deficient mice. *Hepatology*, *31*, 149–159.

Krelin, Y., Voronov, E., Dotan, S., Elkabets, M., Reich, E., Fogel, M., et al. (2007). Interleukin-1beta-driven inflammation promotes the development and invasiveness of chemical carcinogen-induced tumors. *Cancer Research*, *67*, 1062–1071.

Krieg, A., Riemer, J. C., Telan, L. A., Gabbert, H. E., & Knoefel, W. T. (2015). CXCR4-A prognostic and clinicopathological biomarker for pancreatic ductal adenocarcinoma: A meta-analysis. *PLoS One*, *10*, e0130192.

Kryczek, I., Lange, A., Mottram, P., Alvarez, X., Cheng, P., Hogan, M., et al. (2005). CXCL12 and vascular endothelial growth factor synergistically induce neoangiogenesis in human ovarian cancers. *Cancer Research*, *65*, 465–472.

Ksienzyk, A., Neumann, B., Nandakumar, R., Finsterbusch, K., Grashoff, M., Zawatzky, R., et al. (2011). IRF-1 expression is essential for natural killer cells to suppress metastasis. *Cancer Research*, *71*, 6410–6418.

Kucukgergin, C., Isman, F. K., Dasdemir, S., Cakmakoglu, B., Sanli, O., Gokkusu, C., et al. (2012). The role of chemokine and chemokine receptor gene variants on the susceptibility and clinicopathological characteristics of bladder cancer. *Gene*, *511*, 7–11.

Kudo-Saito, C., Shirako, H., Ohike, M., Tsukamoto, N., & Kawakami, Y. (2013). CCL2 is critical for immunosuppression to promote cancer metastasis. *Clinical & Experimental Metastasis*, *30*, 393–405.

Kurihara, T., & Bravo, R. (1996). Cloning and functional expression of mCCR2, a murine receptor for the C-C chemokines JE and FIC. *The Journal of Biological Chemistry*, *271*, 11603–11607.

Kurihara, T., Warr, G., Loy, J., & Bravo, R. (1997). Defects in macrophage recruitment and host defense in mice lacking the CCR2 chemokine receptor. *The Journal of Experimental Medicine*, *186*, 1757–1762.

Kurt-Jones, E. A., Beller, D. I., Mizel, S. B., & Unanue, E. R. (1985). Identification of a membrane-associated interleukin 1 in macrophages. *Proceedings of the National Academy of Sciences of the United States of America*, *82*, 1204–1208.

Kwon, E. M., Salinas, C. A., Kolb, S., Fu, R., Feng, Z., Stanford, J. L., et al. (2011). Genetic polymorphisms in inflammation pathway genes and prostate cancer risk. *Cancer Epidemiology, Biomarkers & Prevention*, *20*, 923–933.

Lanca, T., Costa, M. F., Goncalves-Sousa, N., Rei, M., Grosso, A. R., Penido, C., et al. (2013). Protective role of the inflammatory CCR2/CCL2 chemokine pathway through recruitment of type 1 cytotoxic gammadelta T lymphocytes to tumor beds. *The Journal of Immunology*, *190*, 6673–6680.

Larkin, J. M., Ferguson, T. R., Pickering, L. M., Edmonds, K., James, M. G., Thomas, K., et al. (2010). A phase I/II trial of sorafenib and infliximab in advanced renal cell carcinoma. *British Journal of Cancer*, *103*, 1149–1153.

Lasfar, A., Lewis-Antes, A., Smirnov, S. V., Anantha, S., Abushahba, W., Tian, B., et al. (2006). Characterization of the mouse IFN-lambda ligand-receptor system: IFN-lambdas exhibit antitumor activity against B16 melanoma. *Cancer Research*, *66*, 4468–4477.

Lawrence, T. (2009). The nuclear factor NF-kappaB pathway in inflammation. *Cold Spring Harbor Perspectives in Biology*, *1*, a001651.

Lazear, H. M., Nice, T. J., & Diamond, M. S. (2015). Interferon-lambda: Immune functions at barrier surfaces and beyond. *Immunity*, *43*, 15–28.

Lee, Y. C., Chung, J. H., Kim, S. K., Rhee, S. Y., Chon, S., Oh, S. J., et al. (2015). Association between interleukin 17/interleukin 17 receptor gene polymorphisms and papillary thyroid cancer in Korean population. *Cytokine*, *71*, 283–288.

Lee, J., Ho, W. H., Maruoka, M., Corpuz, R. T., Baldwin, D. T., Foster, J. S., et al. (2001). IL-17E, a novel proinflammatory ligand for the IL-17 receptor homolog IL-17Rh1. *The Journal of Biological Chemistry*, *276*, 1660–1664.
Lee, I. C., Huang, Y. H., Chau, G. Y., Huo, T. I., Su, C. W., Wu, J. C., et al. (2013). Serum interferon gamma level predicts recurrence in hepatocellular carcinoma patients after curative treatments. *International Journal of Cancer*, *133*, 2895–2902.
Lee, B. C., Lee, T. H., Avraham, S., & Avraham, H. K. (2004). Involvement of the chemokine receptor CXCR4 and its ligand stromal cell-derived factor 1alpha in breast cancer cell migration through human brain microvascular endothelial cells. *Molecular Cancer Research*, *2*, 327–338.
Leek, R. D., Landers, R., Fox, S. B., Ng, F., Harris, A. L., & Lewis, C. E. (1998). Association of tumour necrosis factor alpha and its receptors with thymidine phosphorylase expression in invasive breast carcinoma. *British Journal of Cancer*, *77*, 2246–2251.
Leu, C. M., Wong, F. H., Chang, C., Huang, S. F., & Hu, C. P. (2003). Interleukin-6 acts as an antiapoptotic factor in human esophageal carcinoma cells through the activation of both STAT3 and mitogen-activated protein kinase pathways. *Oncogene*, *22*, 7809–7818.
Leung, S., Qureshi, S. A., Kerr, I. M., Darnell, J. E., Jr., & Stark, G. R. (1995). Role of STAT2 in the alpha interferon signaling pathway. *Molecular and Cellular Biology*, *15*, 1312–1317.
Levidou, G., Sachanas, S., Pangalis, G. A., Kalpadakis, C., Yiakoumis, X., Moschogiannis, M., et al. (2014). Immunohistochemical analysis of IL-6, IL-8/CXCR2 axis, Tyr p-STAT-3, and SOCS-3 in lymph nodes from patients with chronic lymphocytic leukemia: Correlation between microvascular characteristics and prognostic significance. *BioMed Research International*, *2014*, 251479.
Lewis, M., Tartaglia, L. A., Lee, A., Bennett, G. L., Rice, G. C., Wong, G. H., et al. (1991). Cloning and expression of cDNAs for two distinct murine tumor necrosis factor receptors demonstrate one receptor is species specific. *Proceedings of the National Academy of Sciences of the United States of America*, *88*, 2830–2834.
Li, H., Chen, J., Huang, A., Stinson, J., Heldens, S., Foster, J., et al. (2000). Cloning and characterization of IL-17B and IL-17C, two new members of the IL-17 cytokine family. *Proceedings of the National Academy of Sciences of the United States of America*, *97*, 773–778.
Li, D., Guo, B., Wu, H., Tan, L., Chang, C., & Lu, Q. (2015). Interleukin-17 in systemic lupus erythematosus: A comprehensive review. *Autoimmunity*, *48*, 353–361.
Li, C. W., Xia, W., Huo, L., Lim, S. O., Wu, Y., Hsu, J. L., et al. (2012). Epithelial–mesenchymal transition induced by TNF-alpha requires NF-kappaB-mediated transcriptional upregulation of Twist1. *Cancer Research*, *72*, 1290–1300.
Li, L., Xu, L., Yan, J., Zhen, Z. J., Ji, Y., Liu, C. Q., et al. (2015). CXCR2-CXCL1 axis is correlated with neutrophil infiltration and predicts a poor prognosis in hepatocellular carcinoma. *Journal of Experimental & Clinical Cancer Research*, *34*, 129.
Li, X., Yao, W., Yuan, Y., Chen, P., Li, B., Li, J., et al. (2015). Targeting of tumour-infiltrating macrophages via CCL2/CCR2 signalling as a therapeutic strategy against hepatocellular carcinoma. *Gut*, 1–11.
Liang, J. J., Zhu, S., Bruggeman, R., Zaino, R. J., Evans, D. B., Fleming, J. B., et al. (2010). High levels of expression of human stromal cell-derived factor-1 are associated with worse prognosis in patients with stage II pancreatic ductal adenocarcinoma. *Cancer Epidemiology, Biomarkers & Prevention*, *19*, 2598–2604.
Liao, C., Yu, Z., Guo, W., Liu, Q., Wu, Y., Li, Y., et al. (2014). Prognostic value of circulating inflammatory factors in non-small cell lung cancer: A systematic review and meta-analysis. *Cancer Biomarkers*, *14*, 469–481.
Libutti, S. K., Paciotti, G. F., Byrnes, A. A., Alexander, H. R., Jr., Gannon, W. E., Walker, M., et al. (2010). Phase I and pharmacokinetic studies of CYT-6091, a novel PEGylated colloidal gold-rhTNF nanomedicine. *Clinical Cancer Research*, *16*, 6139–6149.

Lieblein, J. C., Ball, S., Hutzen, B., Sasser, A. K., Lin, H. J., Huang, T. H., et al. (2008). STAT3 can be activated through paracrine signaling in breast epithelial cells. *BMC Cancer*, *8*, 302.

Liu, G. T., Chen, H. T., Tsou, H. K., Tan, T. W., Fong, Y. C., Chen, P. C., et al. (2014). CCL5 promotes VEGF-dependent angiogenesis by down-regulating miR-200b through PI3K/Akt signaling pathway in human chondrosarcoma cells. *Oncotarget*, *5*, 10718–10731.

Liu, B. Y., Soloviev, I., Chang, P., Lee, J., Huang, X., Zhong, C., et al. (2010). Stromal cell-derived factor-1/CXCL12 contributes to MMTV-Wnt1 tumor growth involving Gr1+CD11b+ cells. *PLoS One*, *5*, e8611.

Liu, G. X., Zhang, X., Li, S., Koiiche, R. D., Sindsceii, J. H., & Song, H. (2013). Monocyte chemotactic protein-1 and CC chemokine receptor 2 polymorphisms and prognosis of renal cell carcinoma. *Tumour Biology*, *34*, 2741–2746.

Lo, M. C., Yip, T. C., Ngan, K. C., Cheng, W. W., Law, C. K., Chan, P. S., et al. (2013). Role of MIF/CXCL8/CXCR2 signaling in the growth of nasopharyngeal carcinoma tumor spheres. *Cancer Letters*, *335*, 81–92.

Locati, M., Zhou, D., Luini, W., Evangelista, V., Mantovani, A., & Sozzani, S. (1994). Rapid induction of arachidonic acid release by monocyte chemotactic protein-1 and related chemokines. Role of Ca2+ influx, synergism with platelet-activating factor and significance for chemotaxis. *The Journal of Biological Chemistry*, *269*, 4746–4753.

Lockart, R. Z., Jr., Sreevalsan, T., & Horn, B. (1962). Inhibition of viral RNA synthesis by interferon. *Virology*, *18*, 493–494.

Locksley, R. M., Killeen, N., & Lenardo, M. J. (2001). The TNF and TNF receptor superfamilies: Integrating mammalian biology. *Cell*, *104*, 487–501.

Loffler, D., Brocke-Heidrich, K., Pfeifer, G., Stocsits, C., Hackermuller, J., Kretzschmar, A. K., et al. (2007). Interleukin-6 dependent survival of multiple myeloma cells involves the Stat3-mediated induction of microRNA-21 through a highly conserved enhancer. *Blood*, *110*, 1330–1333.

Long, H., Xiang, T., Qi, W., Huang, J., Chen, J., He, L., et al. (2015). CD133+ ovarian cancer stem-like cells promote non-stem cancer cell metastasis via CCL5 induced epithelial–mesenchymal transition. *Oncotarget*, *6*, 5846–5859.

Long, H., Xie, R., Xiang, T., Zhao, Z., Lin, S., Liang, Z., et al. (2012). Autocrine CCL5 signaling promotes invasion and migration of CD133+ ovarian cancer stem-like cells via NF-kappaB-mediated MMP-9 upregulation. *Stem Cells*, *30*, 2309–2319.

Lorusso, D., Scambia, G., Amadio, G., di Legge, A., Pietragalla, A., De Vincenzo, R., et al. (2012). Phase II study of NGR-hTNF in combination with doxorubicin in relapsed ovarian cancer patients. *British Journal of Cancer*, *107*, 37–42.

Lowman, H. B., Slagle, P. H., DeForge, L. E., Wirth, C. M., Gillece-Castro, B. L., Bourell, J. H., et al. (1996). Exchanging interleukin-8 and melanoma growth-stimulating activity receptor binding specificities. *The Journal of Biological Chemistry*, *271*, 14344–14352.

Luan, J., Shattuck-Brandt, R., Haghnegahdar, H., Owen, J. D., Strieter, R., Burdick, M., et al. (1997). Mechanism and biological significance of constitutive expression of MGSA/GRO chemokines in malignant melanoma tumor progression. *Journal of Leukocyte Biology*, *62*, 588–597.

Lubberts, E. (2015). The IL-23-IL-17 axis in inflammatory arthritis. *Nature Reviews Rheumatology*, *11*, 415–429.

Luchtman, D. W., Ellwardt, E., Larochelle, C., & Zipp, F. (2014). IL-17 and related cytokines involved in the pathology and immunotherapy of multiple sclerosis: Current and future developments. *Cytokine & Growth Factor Reviews*, *25*, 403–413.

Ma, S., Cheng, Q., Cai, Y., Gong, H., Wu, Y., Yu, X., et al. (2014). IL-17A produced by gammadelta T cells promotes tumor growth in hepatocellular carcinoma. *Cancer Research*, *74*, 1969–1982.

Ma, H., Shu, Y., Pan, S., Chen, J., Dai, J., Jin, G., et al. (2011). Polymorphisms of key chemokine genes and survival of non-small cell lung cancer in Chinese. *Lung Cancer, 74*, 164–169.

Machado, J. C., Pharoah, P., Sousa, S., Carvalho, R., Oliveira, C., Figueiredo, C., et al. (2001). Interleukin 1B and interleukin 1RN polymorphisms are associated with increased risk of gastric carcinoma. *Gastroenterology, 121*, 823–829.

Madhusudan, S., Foster, M., Muthuramalingam, S. R., Braybrooke, J. P., Wilner, S., Kaur, K., et al. (2004). A phase II study of etanercept (Enbrel), a tumor necrosis factor alpha inhibitor in patients with metastatic breast cancer. *Clinical Cancer Research, 10*, 6528–6534.

Madhusudan, S., Muthuramalingam, S. R., Braybrooke, J. P., Wilner, S., Kaur, K., Han, C., et al. (2005). Study of etanercept, a tumor necrosis factor-alpha inhibitor, in recurrent ovarian cancer. *Journal of Clinical Oncology, 23*, 5950–5959.

Maier, J. A., Statuto, M., & Ragnotti, G. (1994). Endogenous interleukin 1 alpha must be transported to the nucleus to exert its activity in human endothelial cells. *Molecular and Cellular Biology, 14*, 1845–1851.

Mammoliti, S., Andretta, V., Bennicelli, E., Caprioni, F., Comandini, D., Fornarini, G., et al. (2011). Two doses of NGR-hTNF in combination with capecitabine plus oxaliplatin in colorectal cancer patients failing standard therapies. *Annals of Oncology, 22*, 973–978.

Matsuo, Y., Ochi, N., Sawai, H., Yasuda, A., Takahashi, H., Funahashi, H., et al. (2009). CXCL8/IL-8 and CXCL12/SDF-1alpha co-operatively promote invasiveness and angiogenesis in pancreatic cancer. *International Journal of Cancer, 124*, 853–861.

Matsuo, Y., Sawai, H., Ma, J., Xu, D., Ochi, N., Yasuda, A., et al. (2009a). IL-1alpha secreted by colon cancer cells enhances angiogenesis: The relationship between IL-1alpha release and tumor cells' potential for liver metastasis. *Journal of Surgical Oncology, 99*, 361–367.

Matsuo, Y., Sawai, H., Ochi, N., Yasuda, A., Takahashi, H., Funahashi, H., et al. (2009b). Interleukin-1alpha secreted by pancreatic cancer cells promotes angiogenesis and its therapeutic implications. *The Journal of Surgical Research, 153*, 274–281.

Mauer, J., Chaurasia, B., Goldau, J., Vogt, M. C., Ruud, J., Nguyen, K. D., et al. (2014). Signaling by IL-6 promotes alternative activation of macrophages to limit endotoxemia and obesity-associated resistance to insulin. *Nature Immunology, 15*, 423–430.

Maxwell, P. J., Neisen, J., Messenger, J., & Waugh, D. J. (2014). Tumor-derived CXCL8 signaling augments stroma-derived CCL2-promoted proliferation and CXCL12-mediated invasion of PTEN-deficient prostate cancer cells. *Oncotarget, 5*, 4895–4908.

Mazzucchelli, L., Loetscher, P., Kappeler, A., Uguccioni, M., Baggiolini, M., Laissue, J. A., et al. (1996). Monocyte chemoattractant protein-1 gene expression in prostatic hyperplasia and prostate adenocarcinoma. *The American Journal of Pathology, 149*, 501–509.

Mellado, M., Rodriguez-Frade, J. M., Aragay, A., del Real, G., Martin, A. M., Vila-Coro, A. J., et al. (1998). The chemokine monocyte chemotactic protein 1 triggers Janus kinase 2 activation and tyrosine phosphorylation of the CCR2B receptor. *The Journal of Immunology, 161*, 805–813.

Meng, Y., Beckett, M. A., Liang, H., Mauceri, H. J., van Rooijen, N., Cohen, K. S., et al. (2010). Blockade of tumor necrosis factor alpha signaling in tumor-associated macrophages as a radiosensitizing strategy. *Cancer Research, 70*, 1534–1543.

Mestas, J., & Hughes, C. C. (2004). Of mice and not men: Differences between mouse and human immunology. *The Journal of Immunology, 172*, 2731–2738.

Metelitsa, L. S., Wu, H. W., Wang, H., Yang, Y., Warsi, Z., Asgharzadeh, S., et al. (2004). Natural killer T cells infiltrate neuroblastomas expressing the chemokine CCL2. *The Journal of Experimental Medicine, 199*, 1213–1221.

Michalaki, V., Syrigos, K., Charles, P., & Waxman, J. (2004). Serum levels of IL-6 and TNF-alpha correlate with clinicopathological features and patient survival in patients with prostate cancer. *British Journal of Cancer, 90*, 2312–2316.

Miller, L. J., Kurtzman, S. H., Anderson, K., Wang, Y., Stankus, M., Renna, M., et al. (2000). Interleukin-1 family expression in human breast cancer: Interleukin-1 receptor antagonist. *Cancer Investigation, 18*, 293–302.

Mills, K. H. (2008). Induction, function and regulation of IL-17-producing T cells. *European Journal of Immunology, 38*, 2636–2649.

Mills, S. C., Goh, P. H., Kudatsih, J., Ncube, S., Gurung, R., Maxwell, W., et al. (2016). Cell migration towards CXCL12 in leukemic cells compared to breast cancer cells. *Cellular Signalling, 28*, 316–324.

Milner, J. D., Brenchley, J. M., Laurence, A., Freeman, A. F., Hill, B. J., Elias, K. M., et al. (2008). Impaired T(H)17 cell differentiation in subjects with autosomal dominant hyper-IgE syndrome. *Nature, 452*, 773–776.

Minn, A. J., Gupta, G. P., Padua, D., Bos, P., Nguyen, D. X., Nuyten, D., et al. (2007). Lung metastasis genes couple breast tumor size and metastatic spread. *Proceedings of the National Academy of Sciences of the United States of America, 104*, 6740–6745.

Minn, A. J., Gupta, G. P., Siegel, P. M., Bos, P. D., Shu, W., Giri, D. D., et al. (2005). Genes that mediate breast cancer metastasis to lung. *Nature, 436*, 518–524.

Mirisola, V., Zuccarino, A., Bachmeier, B. E., Sormani, M. P., Falter, J., Nerlich, A., et al. (2009). CXCL12/SDF1 expression by breast cancers is an independent prognostic marker of disease-free and overall survival. *European Journal of Cancer, 45*, 2579–2587.

Mitsunaga, S., Ikeda, M., Shimizu, S., Ohno, I., Furuse, J., Inagaki, M., et al. (2013). Serum levels of IL-6 and IL-1beta can predict the efficacy of gemcitabine in patients with advanced pancreatic cancer. *British Journal of Cancer, 108*, 2063–2069.

Miyake, M., Goodison, S., Urquidi, V., Gomes Giacoia, E., & Rosser, C. J. (2013). Expression of CXCL1 in human endothelial cells induces angiogenesis through the CXCR2 receptor and the ERK1/2 and EGF pathways. *Laboratory Investigation, 93*, 768–778.

Miyake, M., Lawton, A., Goodison, S., Urquidi, V., Gomes-Giacoia, E., Zhang, G., et al. (2013). Chemokine (C-X-C) ligand 1 (CXCL1) protein expression is increased in aggressive bladder cancers. *BMC Cancer, 13*, 322.

Miyake, M., Lawton, A., Goodison, S., Urquidi, V., & Rosser, C. J. (2014). Chemokine (C-X-C motif) ligand 1 (CXCL1) protein expression is increased in high-grade prostate cancer. *Pathology, Research and Practice, 210*, 74–78.

Mombelli, S., Cochaud, S., Merrouche, Y., Garbar, C., Antonicelli, F., Laprevotte, E., et al. (2015). IL-17A and its homologs IL-25/IL-17E recruit the c-RAF/S6 kinase pathway and the generation of pro-oncogenic LMW-E in breast cancer cells. *Scientific Reports, 5*, 11874.

Montali, R. J. (1988). Comparative pathology of inflammation in the higher vertebrates (reptiles, birds and mammals). *Journal of Comparative Pathology, 99*, 1–26.

Monteclaro, F. S., & Charo, I. F. (1996). The amino-terminal extracellular domain of the MCP-1 receptor, but not the RANTES/MIP-1alpha receptor, confers chemokine selectivity. Evidence for a two-step mechanism for MCP-1 receptor activation. *The Journal of Biological Chemistry, 271*, 19084–19092.

Monti, P., Leone, B. E., Marchesi, F., Balzano, G., Zerbi, A., Scaltrini, F., et al. (2003). The CC chemokine MCP-1/CCL2 in pancreatic cancer progression: Regulation of expression and potential mechanisms of antimalignant activity. *Cancer Research, 63*, 7451–7461.

Moore, R. J., Owens, D. M., Stamp, G., Arnott, C., Burke, F., East, N., et al. (1999). Mice deficient in tumor necrosis factor-alpha are resistant to skin carcinogenesis. *Nature Medicine, 5*, 828–831.

Moore, E. E., Presnell, S., Garrigues, U., Guilbot, A., LeGuern, E., Smith, D., et al. (2002). Expression of IL-17B in neurons and evaluation of its possible role in the chromosome 5q-linked form of Charcot–Marie–Tooth disease. *Neuromuscular Disorders, 12*, 141–150.

Moser, B., Schumacher, C., von Tscharner, V., Clark-Lewis, I., & Baggiolini, M. (1991). Neutrophil-activating peptide 2 and gro/melanoma growth-stimulatory activity interact with neutrophil-activating peptide 1/interleukin 8 receptors on human neutrophils. *The Journal of Biological Chemistry*, *266*, 10666–10671.

Mroczko, B., Groblewska, M., Gryko, M., Kedra, B., & Szmitkowski, M. (2010). Diagnostic usefulness of serum interleukin 6 (IL-6) and C-reactive protein (CRP) in the differentiation between pancreatic cancer and chronic pancreatitis. *Journal of Clinical Laboratory Analysis*, *24*, 256–261.

Muller-Hermelink, N., Braumuller, H., Pichler, B., Wieder, T., Mailhammer, R., Schaak, K., et al. (2008). TNFR1 signaling and IFN-gamma signaling determine whether T cells induce tumor dormancy or promote multistage carcinogenesis. *Cancer Cell*, *13*, 507–518.

Musha, H., Ohtani, H., Mizoi, T., Kinouchi, M., Nakayama, T., Shiiba, K., et al. (2005). Selective infiltration of CCR5(+)CXCR3(+) T lymphocytes in human colorectal carcinoma. *International Journal of Cancer*, *116*, 949–956.

Mustea, A., Pirvulescu, C., Konsgen, D., Braicu, E. I., Yuan, S., Sun, P., et al. (2008). Decreased IL-1 RA concentration in ascites is associated with a significant improvement in overall survival in ovarian cancer. *Cytokine*, *42*, 77–84.

Nagasawa, T., Kikutani, H., & Kishimoto, T. (1994). Molecular cloning and structure of a pre-B-cell growth-stimulating factor. *Proceedings of the National Academy of Sciences of the United States of America*, *91*, 2305–2309.

Nakae, S., Asano, M., Horai, R., Sakaguchi, N., & Iwakura, Y. (2001). IL-1 enhances T cell-dependent antibody production through induction of CD40 ligand and OX40 on T cells. *The Journal of Immunology*, *167*, 90–97.

Nakagawa, H., Umemura, A., Taniguchi, K., Font-Burgada, J., Dhar, D., Ogata, H., et al. (2014). ER stress cooperates with hypernutrition to trigger TNF-dependent spontaneous HCC development. *Cancer Cell*, *26*, 331–343.

Nakashima, J., Tachibana, M., Ueno, M., Miyajima, A., Baba, S., & Murai, M. (1998). Association between tumor necrosis factor in serum and cachexia in patients with prostate cancer. *Clinical Cancer Research*, *4*, 1743–1748.

Nakashima, K., & Taga, T. (1998). gp130 and the IL-6 family of cytokines: Signaling mechanisms and thrombopoietic activities. *Seminars in Hematology*, *35*, 210–221.

Nasir, G. A., Mohsin, S., Khan, M., Shams, S., Ali, G., Khan, S. N., et al. (2013). Mesenchymal stem cells and Interleukin-6 attenuate liver fibrosis in mice. *Journal of Translational Medicine*, *11*, 78.

Nasser, M. W., Raghuwanshi, S. K., Grant, D. J., Jala, V. R., Rajarathnam, K., & Richardson, R. M. (2009). Differential activation and regulation of CXCR1 and CXCR2 by CXCL8 monomer and dimer. *The Journal of Immunology*, *183*, 3425–3432.

Nastala, C. L., Edington, H. D., McKinney, T. G., Tahara, H., Nalesnik, M. A., Brunda, M. J., et al. (1994). Recombinant IL-12 administration induces tumor regression in association with IFN-gamma production. *The Journal of Immunology*, *153*, 1697–1706.

Naugler, W. E., Sakurai, T., Kim, S., Maeda, S., Kim, K., Elsharkawy, A. M., et al. (2007). Gender disparity in liver cancer due to sex differences in MyD88-dependent IL-6 production. *Science*, *317*, 121–124.

Nawroth, P., Handley, D., Matsueda, G., De Waal, R., Gerlach, H., Blohm, D., et al. (1988). Tumor necrosis factor/cachectin-induced intravascular fibrin formation in meth A fibrosarcomas. *The Journal of Experimental Medicine*, *168*, 637–647.

Naylor, M. S., Malik, S. T., Stamp, G. W., Jobling, T., & Balkwill, F. R. (1990). In situ detection of tumour necrosis factor in human ovarian cancer specimens. *European Journal of Cancer*, *26*, 1027–1030.

Naylor, M. S., Stamp, G. W., & Balkwill, F. R. (1990). Investigation of cytokine gene expression in human colorectal cancer. *Cancer Research*, *50*, 4436–4440.

Naylor, M. S., Stamp, G. W., Foulkes, W. D., Eccles, D., & Balkwill, F. R. (1993). Tumor necrosis factor and its receptors in human ovarian cancer. Potential role in disease progression. *The Journal of Clinical Investigation, 91*, 2194–2206.

Nemati, K., Golmoghaddam, H., Hosseini, S. V., Ghaderi, A., & Doroudchi, M. (2015). Interleukin-17FT7488 allele is associated with a decreased risk of colorectal cancer and tumor progression. *Gene, 561*, 88–94.

Netea, M. G., Kullberg, B. J., & Van der Meer, J. W. (2000). Circulating cytokines as mediators of fever. *Clinical Infectious Diseases, 31*(Suppl. 5), S178–S184.

Newey, S. E., Tsaknakis, G., Khoo, C. P., Athanassopoulos, T., Camicia, R., Zhang, Y., et al. (2014). The hematopoietic chemokine CXCL12 promotes integration of human endothelial colony forming cell-derived cells into immature vessel networks. *Stem Cells and Development, 23*, 2730–2743.

Nilsson, M. B., Langley, R. R., & Fidler, I. J. (2005). Interleukin-6, secreted by human ovarian carcinoma cells, is a potent proangiogenic cytokine. *Cancer Research, 65*, 10794–10800.

Ning, Y., Manegold, P. C., Hong, Y. K., Zhang, W., Pohl, A., Lurje, G., et al. (2011). Interleukin-8 is associated with proliferation, migration, angiogenesis and chemosensitivity in vitro and in vivo in colon cancer cell line models. *International Journal of Cancer, 128*, 2038–2049.

Ninomiya-Tsuji, J., Kishimoto, K., Hiyama, A., Inoue, J., Cao, Z., & Matsumoto, K. (1999). The kinase TAK1 can activate the NIK-I kappaB as well as the MAP kinase cascade in the IL-1 signalling pathway. *Nature, 398*, 252–256.

Niwa, Y., Akamatsu, H., Niwa, H., Sumi, H., Ozaki, Y., & Abe, A. (2001). Correlation of tissue and plasma RANTES levels with disease course in patients with breast or cervical cancer. *Clinical Cancer Research*, 7, 285–289.

Noureddin, M., Rotman, Y., Zhang, F., Park, H., Rehermann, B., Thomas, E., et al. (2015). Hepatic expression levels of interferons and interferon-stimulated genes in patients with chronic hepatitis C: A phenotype-genotype correlation study. *Genes and Immunity, 16*, 321–329.

Nozaki, S., Sledge, G. W., Jr., & Nakshatri, H. (2000). Cancer cell-derived interleukin 1alpha contributes to autocrine and paracrine induction of pro-metastatic genes in breast cancer. *Biochemical and Biophysical Research Communications, 275*, 60–62.

Numasaki, M., Tagawa, M., Iwata, F., Suzuki, T., Nakamura, A., Okada, M., et al. (2007). IL-28 elicits antitumor responses against murine fibrosarcoma. *The Journal of Immunology, 178*, 5086–5098.

Odendall, C., Dixit, E., Stavru, F., Bierne, H., Franz, K. M., Durbin, A. F., et al. (2014). Diverse intracellular pathogens activate type III interferon expression from peroxisomes. *Nature Immunology, 15*, 717–726.

Ogura, M., Ishida, T., Hatake, K., Taniwaki, M., Ando, K., Tobinai, K., et al. (2014). Multicenter phase II study of mogamulizumab (KW-0761), a defucosylated anti-cc chemokine receptor 4 antibody, in patients with relapsed peripheral T-cell lymphoma and cutaneous T-cell lymphoma. *Journal of Clinical Oncology, 32*, 1157–1163.

Okubo, Y., Mera, T., Wang, L., & Faustman, D. L. (2013). Homogeneous expansion of human T-regulatory cells via tumor necrosis factor receptor 2. *Scientific Reports, 3*, 3153.

Okuyama Kishima, M., Brajao de Oliveira, K., Ariza, C. B., de Oliveira, C. E., Losi Guembarovski, R., Banin Hirata, B. K., et al. (2015). Genetic polymorphism and expression of CXCR4 in breast cancer. *Analytical Cellular Pathology (Amsterdam), 2015*, 289510.

Oladipo, O., Conlon, S., O'Grady, A., Purcell, C., Wilson, C., Maxwell, P. J., et al. (2011). The expression and prognostic impact of CXC-chemokines in stage II and III colorectal cancer epithelial and stromal tissue. *British Journal of Cancer, 104*, 480–487.

Olsen, J., Kirkeby, L. T., Olsen, J., Eiholm, S., Jess, P., Gogenur, I., et al. (2015). High interleukin-6 mRNA expression is a predictor of relapse in colon cancer. *Anticancer Research, 35*, 2235–2240.

Omrane, I., Medimegh, I., Baroudi, O., Ayari, H., Bedhiafi, W., Stambouli, N., et al. (2015). Involvement of IL17A, IL17F and IL23R polymorphisms in colorectal cancer therapy. *PLoS One*, *10*, e0128911.

Onoguchi, K., Yoneyama, M., Takemura, A., Akira, S., Taniguchi, T., Namiki, H., et al. (2007). Viral infections activate types I and III interferon genes through a common mechanism. *The Journal of Biological Chemistry*, *282*, 7576–7581.

Orimo, A., Gupta, P. B., Sgroi, D. C., Arenzana-Seisdedos, F., Delaunay, T., Naeem, R., et al. (2005). Stromal fibroblasts present in invasive human breast carcinomas promote tumor growth and angiogenesis through elevated SDF-1/CXCL12 secretion. *Cell*, *121*, 335–348.

Osterlund, P. I., Pietila, T. E., Veckman, V., Kotenko, S. V., & Julkunen, I. (2007). IFN regulatory factor family members differentially regulate the expression of type III IFN (IFN-lambda) genes. *The Journal of Immunology*, *179*, 3434–3442.

Osuala, K. O., Sameni, M., Shah, S., Aggarwal, N., Simonait, M. L., Franco, O. E., et al. (2015). Il-6 signaling between ductal carcinoma in situ cells and carcinoma-associated fibroblasts mediates tumor cell growth and migration. *BMC Cancer*, *15*, 584.

Owen, J. D., Strieter, R., Burdick, M., Haghnegahdar, H., Nanney, L., Shattuck-Brandt, R., et al. (1997). Enhanced tumor-forming capacity for immortalized melanocytes expressing melanoma growth stimulatory activity/growth-regulated cytokine beta and gamma proteins. *International Journal of Cancer*, *73*, 94–103.

Palladino, M. A., Jr., Shalaby, M. R., Kramer, S. M., Ferraiolo, B. L., Baughman, R. A., Deleo, A. B., et al. (1987). Characterization of the antitumor activities of human tumor necrosis factor-alpha and the comparison with other cytokines: Induction of tumor-specific immunity. *The Journal of Immunology*, *138*, 4023–4032.

Park, E. J., Lee, J. H., Yu, G. Y., He, G., Ali, S. R., Holzer, R. G., et al. (2010). Dietary and genetic obesity promote liver inflammation and tumorigenesis by enhancing IL-6 and TNF expression. *Cell*, *140*, 197–208.

Parrington, F. R. (1979). The evolution of the mammalian middle and outer ears: A personal review. *Biological Reviews of the Cambridge Philosophical Society*, *54*, 369–387.

Pasparakis, M., Alexopoulou, L., Episkopou, V., & Kollias, G. (1996). Immune and inflammatory responses in TNF alpha-deficient mice: A critical requirement for TNF alpha in the formation of primary B cell follicles, follicular dendritic cell networks and germinal centers, and in the maturation of the humoral immune response. *The Journal of Experimental Medicine*, *184*, 1397–1411.

Pennica, D., Nedwin, G. E., Hayflick, J. S., Seeburg, P. H., Derynck, R., Palladino, M. A., et al. (1984). Human tumour necrosis factor: Precursor structure, expression and homology to lymphotoxin. *Nature*, *312*, 724–729.

Pisa, P., Halapi, E., Pisa, E. K., Gerdin, E., Hising, C., Bucht, A., et al. (1992). Selective expression of interleukin 10, interferon gamma, and granulocyte-macrophage colony-stimulating factor in ovarian cancer biopsies. *Proceedings of the National Academy of Sciences of the United States of America*, *89*, 7708–7712.

Pitt, L. A., Tikhonova, A. N., Hu, H., Trimarchi, T., King, B., Gong, Y., et al. (2015). CXCL12-producing vascular endothelial niches control acute T cell leukemia maintenance. *Cancer Cell*, *27*, 755–768.

Premkumar, V. G., Yuvaraj, S., Vijayasarathy, K., Gangadaran, S. G., & Sachdanandam, P. (2007). Serum cytokine levels of interleukin-1beta, -6, -8, tumour necrosis factor-alpha and vascular endothelial growth factor in breast cancer patients treated with tamoxifen and supplemented with co-enzyme Q(10), riboflavin and niacin. *Basic & Clinical Pharmacology & Toxicology*, *100*, 387–391.

Propper, D. J., Chao, D., Braybrooke, J. P., Bahl, P., Thavasu, P., Balkwill, F., et al. (2003). Low-dose IFN-gamma induces tumor MHC expression in metastatic malignant melanoma. *Clinical Cancer Research*, *9*, 84–92.

Pulkki, K., Pelliniemi, T. T., Rajamaki, A., Tienhaara, A., Laakso, M., & Lahtinen, R. (1996). Soluble interleukin-6 receptor as a prognostic factor in multiple myeloma. *Finnish Leukaemia Group. British Journal of Haematology, 92*, 370–374.

Pullamsetti, S. S., Savai, R., Janssen, W., Dahal, B. K., Seeger, W., Grimminger, F., et al. (2011). Inflammation, immunological reaction and role of infection in pulmonary hypertension. *Clinical Microbiology and Infection, 17*, 7–14.

Qian, B. Z., Li, J., Zhang, H., Kitamura, T., Zhang, J., Campion, L. R., et al. (2011). CCL2 recruits inflammatory monocytes to facilitate breast-tumour metastasis. *Nature, 475*, 222–225.

Qian, D. Z., Rademacher, B. L., Pittsenbarger, J., Huang, C. Y., Myrthue, A., Higano, C. S., et al. (2010). CCL2 is induced by chemotherapy and protects prostate cancer cells from docetaxel-induced cytotoxicity. *Prostate, 70*, 433–442.

Qureshi, S. A., Salditt-Georgieff, M., & Darnell, J. E., Jr. (1995). Tyrosine-phosphorylated Stat1 and Stat2 plus a 48-kDa protein all contact DNA in forming interferon-stimulated-gene factor 3. *Proceedings of the National Academy of Sciences of the United States of America, 92*, 3829–3833.

Rachakatla, R. S., Pyle, M. M., Ayuzawa, R., Edwards, S. M., Marini, F. C., Weiss, M. L., et al. (2008). Combination treatment of human umbilical cord matrix stem cell-based interferon-beta gene therapy and 5-fluorouracil significantly reduces growth of metastatic human breast cancer in SCID mouse lungs. *Cancer Investigation, 26*, 662–670.

Rathakrishnan, A., Wang, S. M., Hu, Y., Khan, A. M., Ponnampalavanar, S., Lum, L. C., et al. (2012). Cytokine expression profile of dengue patients at different phases of illness. *PLoS One*, 7, e52215.

Rattigan, Y., Hsu, J. M., Mishra, P. J., Glod, J., & Banerjee, D. (2010). Interleukin 6 mediated recruitment of mesenchymal stem cells to the hypoxic tumor milieu. *Experimental Cell Research, 316*, 3417–3424.

Ray, P., Stacer, A. C., Fenner, J., Cavnar, S. P., Meguiar, K., Brown, M., et al. (2015). CXCL12-gamma in primary tumors drives breast cancer metastasis. *Oncogene, 34*, 2043–2051.

Razis, E., Kalogeras, K. T., Kotoula, V., Eleftheraki, A. G., Nikitas, N., Kronenwett, R., et al. (2012). Improved outcome of high-risk early HER2 positive breast cancer with high CXCL13-CXCR5 messenger RNA expression. *Clinical Breast Cancer, 12*, 183–193.

Razmkhah, M., Jaberipour, M., Safaei, A., Talei, A. R., Erfani, N., & Ghaderi, A. (2012). Chemokine and chemokine receptors: A comparative study between metastatic and nonmetastatic lymph nodes in breast cancer patients. *European Cytokine Network, 23*, 72–77.

Razmkhah, M., Talei, A. R., Doroudchi, M., Khalili-Azad, T., & Ghaderi, A. (2005). Stromal cell-derived factor-1 (SDF-1) alleles and susceptibility to breast carcinoma. *Cancer Letters, 225*, 261–266.

Rei, M., Goncalves-Sousa, N., Lanca, T., Thompson, R. G., Mensurado, S., Balkwill, F. R., et al. (2014). Murine CD27(-) Vgamma6(+) gammadelta T cells producing IL-17A promote ovarian cancer growth via mobilization of protumor small peritoneal macrophages. *Proceedings of the National Academy of Sciences of the United States of America, 111*, E3562–E3570.

Ren, C., Kumar, S., Chanda, D., Chen, J., Mountz, J. D., & Ponnazhagan, S. (2008a). Therapeutic potential of mesenchymal stem cells producing interferon-alpha in a mouse melanoma lung metastasis model. *Stem Cells, 26*, 2332–2338.

Ren, C., Kumar, S., Chanda, D., Kallman, L., Chen, J., Mountz, J. D., et al. (2008b). Cancer gene therapy using mesenchymal stem cells expressing interferon-beta in a mouse prostate cancer lung metastasis model. *Gene Therapy, 15*, 1446–1453.

Richmond, A., & Thomas, H. G. (1988). Melanoma growth stimulatory activity: Isolation from human melanoma tumors and characterization of tissue distribution. *Journal of Cellular Biochemistry*, *36*, 185–198.

Roca, H., Varsos, Z. S., Sud, S., Craig, M. J., Ying, C., & Pienta, K. J. (2009). CCL2 and interleukin-6 promote survival of human CD11b+ peripheral blood mononuclear cells and induce M2-type macrophage polarization. *The Journal of Biological Chemistry*, *284*, 34342–34354.

Rodero, M., Rodero, P., Descamps, V., Lebbe, C., Wolkenstein, P., Aegerter, P., et al. (2007). Melanoma susceptibility and progression: Association study between polymorphisms of the chemokine (CCL2) and chemokine receptors (CX3CR1, CCR5). *Journal of Dermatological Science*, *46*, 72–76.

Rodriguez-Berriguete, G., Sanchez-Espiridion, B., Cansino, J. R., Olmedilla, G., Martinez-Onsurbe, P., Sanchez-Chapado, M., et al. (2013). Clinical significance of both tumor and stromal expression of components of the IL-1 and TNF-alpha signaling pathways in prostate cancer. *Cytokine*, *64*, 555–563.

Rothe, J., Lesslauer, W., Lotscher, H., Lang, Y., Koebel, P., Kontgen, F., et al. (1993). Mice lacking the tumour necrosis factor receptor 1 are resistant to TNF-mediated toxicity but highly susceptible to infection by Listeria monocytogenes. *Nature*, *364*, 798–802.

Rouvier, E., Luciani, M. F., Mattei, M. G., Denizot, F., & Golstein, P. (1993). CTLA-8, cloned from an activated T cell, bearing AU-rich messenger RNA instability sequences, and homologous to a herpesvirus saimiri gene. *The Journal of Immunology*, *150*, 5445–5456.

Roy, L. D., Ghosh, S., Pathangey, L. B., Tinder, T. L., Gruber, H. E., & Mukherjee, P. (2011). Collagen induced arthritis increases secondary metastasis in MMTV-PyV MT mouse model of mammary cancer. *BMC Cancer*, *11*, 365.

Ruspi, G., Schmidt, E. M., McCann, F., Feldmann, M., Williams, R. O., Stoop, A. A., et al. (2014). TNFR2 increases the sensitivity of ligand-induced activation of the p38 MAPK and NF-kappaB pathways and signals TRAF2 protein degradation in macrophages. *Cell Signal*, *26*, 683–690.

Rutkowski, P., Kaminska, J., Kowalska, M., Ruka, W., & Steffen, J. (2003). Cytokine and cytokine receptor serum levels in adult bone sarcoma patients: Correlations with local tumor extent and prognosis. *Journal of Surgical Oncology*, *84*, 151–159.

Sade-Feldman, M., Kanterman, J., Ish-Shalom, E., Elnekave, M., Horwitz, E., & Baniyash, M. (2013). Tumor necrosis factor-alpha blocks differentiation and enhances suppressive activity of immature myeloid cells during chronic inflammation. *Immunity*, *38*, 541–554.

Sadeghi, M., Lahdou, I., Oweira, H., Daniel, V., Terness, P., Schmidt, J., et al. (2015). Serum levels of chemokines CCL4 and CCL5 in cirrhotic patients indicate the presence of hepatocellular carcinoma. *British Journal of Cancer*, *113*, 756–762.

Saintigny, P., Massarelli, E., Lin, S., Ahn, Y. H., Chen, Y., Goswami, S., et al. (2013). CXCR2 expression in tumor cells is a poor prognostic factor and promotes invasion and metastasis in lung adenocarcinoma. *Cancer Research*, *73*, 571–582.

Saji, H., Koike, M., Yamori, T., Saji, S., Seiki, M., Matsushima, K., et al. (2001). Significant correlation of monocyte chemoattractant protein-1 expression with neovascularization and progression of breast carcinoma. *Cancer*, *92*, 1085–1091.

Salmaggi, A., Gelati, M., Pollo, B., Marras, C., Silvani, A., Balestrini, M. R., et al. (2005). CXCL12 expression is predictive of a shorter time to tumor progression in low-grade glioma: A single-institution study in 50 patients. *Journal of Neuro-Oncology*, *74*, 287–293.

Santoro, A., Pressiani, T., Citterio, G., Rossoni, G., Donadoni, G., Pozzi, F., et al. (2010). Activity and safety of NGR-hTNF, a selective vascular-targeting agent, in previously treated patients with advanced hepatocellular carcinoma. *British Journal of Cancer*, *103*, 837–844.

Santoro, A., Rimassa, L., Sobrero, A. F., Citterio, G., Sclafani, F., Carnaghi, C., et al. (2010). Phase II study of NGR-hTNF, a selective vascular targeting agent, in patients with metastatic colorectal cancer after failure of standard therapy. *European Journal of Cancer, 46*, 2746–2752.

Sarau, H. M., Rush, J. A., Foley, J. J., Brawner, M. E., Schmidt, D. B., White, J. R., et al. (1997). Characterization of functional chemokine receptors (CCR1 and CCR2) on EoL-3 cells: A model system to examine the role of chemokines in cell function. *The Journal of Pharmacology and Experimental Therapeutics, 283*, 411–418.

Sarkar, S., Cooney, L. A., & Fox, D. A. (2010). The role of T helper type 17 cells in inflammatory arthritis. *Clinical and Experimental Immunology, 159*, 225–237.

Sato, A., Ohtsuki, M., Hata, M., Kobayashi, E., & Murakami, T. (2006). Antitumor activity of IFN-lambda in murine tumor models. *The Journal of Immunology, 176*, 7686–7694.

Schadendorf, D., Moller, A., Algermissen, B., Worm, M., Sticherling, M., & Czarnetzki, B. M. (1993). IL-8 produced by human malignant melanoma cells in vitro is an essential autocrine growth factor. *The Journal of Immunology, 151*, 2667–2675.

Schimanski, C. C., Jordan, M., Schlaegel, F., Schmidtmann, I., Lang, H., Galle, P. R., Moehler, M., & Gockel, I. (2011). SNP rs1801157 significantly correlates with distant metastasis in CXCL12 expressing esophagogastric cancer. *International Journal of Oncology, 39*, 515–520.

Schultheis, A. M., Lurje, G., Rhodes, K. E., Zhang, W., Yang, D., Garcia, A. A., et al. (2008). Polymorphisms and clinical outcome in recurrent ovarian cancer treated with cyclophosphamide and bevacizumab. *Clinical Cancer Research, 14*, 7554–7563.

Schwabe, R. F., & Brenner, D. A. (2006). Mechanisms of liver injury. I. TNF-alpha-induced liver injury: Role of IKK, JNK, and ROS pathways. *American Journal of Physiology Gastrointestinal and Liver Physiology, 290*, G583–G589.

Seynhaeve, A. L., Hoving, S., Schipper, D., Vermeulen, C. E., de Wiel-Ambagtsheer, G., van Tiel, S. T., et al. (2007). Tumor necrosis factor alpha mediates homogeneous distribution of liposomes in murine melanoma that contributes to a better tumor response. *Cancer Research, 67*, 9455–9462.

Shalom-Barak, T., Quach, J., & Lotz, M. (1998). Interleukin-17-induced gene expression in articular chondrocytes is associated with activation of mitogen-activated protein kinases and NF-kappaB. *The Journal of Biological Chemistry, 273*, 27467–27473.

Sharma, B., Nannuru, K. C., Varney, M. L., & Singh, R. K. (2015). Host Cxcr2-dependent regulation of mammary tumor growth and metastasis. *Clinical & Experimental Metastasis, 32*, 65–72.

Shin, H. C., Benbernou, N., Fekkar, H., Esnault, S., & Guenounou, M. (1998). Regulation of IL-17, IFN-gamma and IL-10 in human CD8(+) T cells by cyclic AMP-dependent signal transduction pathway. *Cytokine, 10*, 841–850.

Sierra-Filardi, E., Nieto, C., Dominguez-Soto, A., Barroso, R., Sanchez-Mateos, P., Puig-Kroger, A., et al. (2014). CCL2 shapes macrophage polarization by GM-CSF and M-CSF: Identification of CCL2/CCR2-dependent gene expression profile. *The Journal of Immunology, 192*, 3858–3867.

Sima, A. R., Sima, H. R., Rafatpanah, H., Hosseinnezhad, H., Ghaffarzadehgan, K., Valizadeh, N., et al. (2014). Serum chemokine ligand 5 (CCL5/RANTES) level might be utilized as a predictive marker of tumor behavior and disease prognosis in patients with gastric adenocarcinoma. *Journal of Gastrointestinal Cancer, 45*, 476–480.

Sims, J. E., Nicklin, M. J., Bazan, J. F., Barton, J. L., Busfield, S. J., Ford, J. E., et al. (2001). A new nomenclature for IL-1 family genes. *Trends in Immunology, 22*, 536–537.

Singh, A. P., Arora, S., Bhardwaj, A., Srivastava, S. K., Kadakia, M. P., Wang, B., et al. (2012). CXCL12/CXCR4 protein signaling axis induces sonic hedgehog expression in pancreatic cancer cells via extracellular regulated kinase- and Akt kinase-mediated activation of nuclear factor kappaB: Implications for bidirectional tumor-stromal interactions. *The Journal of Biological Chemistry, 287*, 39115–39124.

Singh, R. K., Gutman, M., Reich, R., & Bar-Eli, M. (1995). Ultraviolet B irradiation promotes tumorigenic and metastatic properties in primary cutaneous melanoma via induction of interleukin 8. *Cancer Research*, *55*, 3669–3674.

Singh, H., Sachan, R., Jain, M., & Mittal, B. (2008). CCR5-Delta32 polymorphism and susceptibility to cervical cancer: Association with early stage of cervical cancer. *Oncology Research*, *17*, 87–91.

Singh, S., Sadanandam, A., Varney, M. L., Nannuru, K. C., & Singh, R. K. (2010). Small interfering RNA-mediated CXCR1 or CXCR2 knock-down inhibits melanoma tumor growth and invasion. *International Journal of Cancer*, *126*, 328–336.

Singh, S., Varney, M., & Singh, R. K. (2009). Host CXCR2-dependent regulation of melanoma growth, angiogenesis, and experimental lung metastasis. *Cancer Research*, *69*, 411–415.

Sistigu, A., Yamazaki, T., Vacchelli, E., Chaba, K., Enot, D. P., Adam, J., et al. (2014). Cancer cell-autonomous contribution of type I interferon signaling to the efficacy of chemotherapy. *Nature Medicine*, *20*, 1301–1309.

Skopinska, E., & Ziembikiewicz, A. (1978). The appearance of factors influencing migration of lymphocytes (LyMIF and LyMSF) in rabbit mixed lymphocyte cultures. *Archivum Immunologiae et Therapiae Experimentalis*, *26*, 389–392.

Snoussi, K., Mahfoudh, W., Bouaouina, N., Fekih, M., Khairi, H., Helal, A. N., et al. (2010). Combined effects of IL-8 and CXCR2 gene polymorphisms on breast cancer susceptibility and aggressiveness. *BMC Cancer*, *10*, 283.

Song, Z. Y., Gao, Z. H., Chu, J. H., Han, X. Z., & Qu, X. J. (2015). Downregulation of the CXCR4/CXCL12 axis blocks the activation of the Wnt/beta-catenin pathway in human colon cancer cells. *Biomedicine & Pharmacotherapy*, *71*, 46–52.

Song, L., Smith, M. A., Doshi, P., Sasser, K., Fulp, W., Altiok, S., et al. (2014). Antitumor efficacy of the anti-interleukin-6 (IL-6) antibody siltuximab in mouse xenograft models of lung cancer. *Journal of Thoracic Oncology*, *9*, 974–982.

Sozzani, S., Luini, W., Molino, M., Jilek, P., Bottazzi, B., Cerletti, C., et al. (1991). The signal transduction pathway involved in the migration induced by a monocyte chemotactic cytokine. *The Journal of Immunology*, *147*, 2215–2221.

Srivastava, A., Pandey, S. N., Choudhuri, G., & Mittal, B. (2008). CCR5 Delta32 polymorphism: Associated with gallbladder cancer susceptibility. *Scandinavian Journal of Immunology*, *67*, 516–522.

Stahl, N., Boulton, T. G., Farruggella, T., Ip, N. Y., Davis, S., Witthuhn, B. A., et al. (1994). Association and activation of Jak-Tyk kinases by CNTF-LIF-OSM-IL-6 beta receptor components. *Science*, *263*, 92–95.

Starnes, T., Broxmeyer, H. E., Robertson, M. J., & Hromas, R. (2002). Cutting edge: IL-17D, a novel member of the IL-17 family, stimulates cytokine production and inhibits hemopoiesis. *The Journal of Immunology*, *169*, 642–646.

Starnes, T., Robertson, M. J., Sledge, G., Kelich, S., Nakshatri, H., Broxmeyer, H. E., et al. (2001). Cutting edge: IL-17F, a novel cytokine selectively expressed in activated T cells and monocytes, regulates angiogenesis and endothelial cell cytokine production. *The Journal of Immunology*, *167*, 4137–4140.

Strieter, R. M., Polverini, P. J., Kunkel, S. L., Arenberg, D. A., Burdick, M. D., Kasper, J., et al. (1995). The functional role of the ELR motif in CXC chemokine-mediated angiogenesis. *The Journal of Biological Chemistry*, *270*, 27348–27357.

Studebaker, A. W., Storci, G., Werbeck, J. L., Sansone, P., Sasser, A. K., Tavolari, S., et al. (2008). Fibroblasts isolated from common sites of breast cancer metastasis enhance cancer cell growth rates and invasiveness in an interleukin-6-dependent manner. *Cancer Research*, *68*, 9087–9095.

Suganuma, M., Okabe, S., Marino, M. W., Sakai, A., Sueoka, E., & Fujiki, H. (1999). Essential role of tumor necrosis factor alpha (TNF-alpha) in tumor promotion as revealed by TNF-alpha-deficient mice. *Cancer Research*, *59*, 4516–4518.

Sugasawa, H., Ichikura, T., Tsujimoto, H., Kinoshita, M., Morita, D., Ono, S., et al. (2008). Prognostic significance of expression of CCL5/RANTES receptors in patients with gastric cancer. *Journal of Surgical Oncology*, *97*, 445–450.

Sui, P., Hu, P., Zhang, T., Zhang, X., Liu, Q., & Du, J. (2014). High expression of CXCR-2 correlates with lymph node metastasis and predicts unfavorable prognosis in resected esophageal carcinoma. *Medical Oncology*, *31*, 809.

Sun, X., Mao, Y., Wang, J., Zu, L., Hao, M., Cheng, G., et al. (2014). IL-6 secreted by cancer-associated fibroblasts induces tamoxifen resistance in luminal breast cancer. *Oncogene*, *33*(35), 4450.

Sun, T., Mary, L. G., Oh, W. K., Freedman, M. L., Pomerantz, M., Pienta, K. J., et al. (2011). Inherited variants in the chemokine CCL2 gene and prostate cancer aggressiveness in a Caucasian cohort. *Clinical Cancer Research*, *17*, 1546–1552.

Sun, L. X., Wang, X. B., & Huang, X. J. (2015). Association analysis of rs2275913G>A and rs763780T>C interleukin 17 polymorphisms in Chinese women with cervical cancer. *Genetics and Molecular Research*, *14*, 13612–13617.

Suzuki, A., Leland, P., Joshi, B. H., & Puri, R. K. (2015). Targeting of IL-4 and IL-13 receptors for cancer therapy. *Cytokine*, *75*, 79–88.

Taichman, R. S., Cooper, C., Keller, E. T., Pienta, K. J., Taichman, N. S., & McCauley, L. K. (2002). Use of the stromal cell-derived factor-1/CXCR4 pathway in prostate cancer metastasis to bone. *Cancer Research*, *62*, 1832–1837.

Takaoka, A., Hayakawa, S., Yanai, H., Stoiber, D., Negishi, H., Kikuchi, H., et al. (2003). Integration of interferon-alpha/beta signalling to p53 responses in tumour suppression and antiviral defence. *Nature*, *424*, 516–523.

Tan, M. C., Goedegebuure, P. S., Belt, B. A., Flaherty, B., Sankpal, N., Gillanders, W. E., et al. (2009). Disruption of CCR5-dependent homing of regulatory T cells inhibits tumor growth in a murine model of pancreatic cancer. *The Journal of Immunology*, *182*, 1746–1755.

Tanabe, Y., Nishibori, T., Su, L., Arduini, R. M., Baker, D. P., & David, M. (2005). Cutting edge: Role of STAT1, STAT3, and STAT5 in IFN-alpha beta responses in T lymphocytes. *The Journal of Immunology*, *174*, 609–613.

Tanaka, K., Kurebayashi, J., Sohda, M., Nomura, T., Prabhakar, U., Yan, L., et al. (2009). The expression of monocyte chemotactic protein-1 in papillary thyroid carcinoma is correlated with lymph node metastasis and tumor recurrence. *Thyroid*, *19*, 21–25.

Tanaka, S., Tatsuguchi, A., Futagami, S., Gudis, K., Wada, K., Seo, T., et al. (2006). Monocyte chemoattractant protein 1 and macrophage cyclooxygenase 2 expression in colonic adenoma. *Gut*, *55*, 54–61.

Tanba, K., Uoshima, N., Uchiyama, H., Kawata, E., Isa, R., Yamaguchi, J., et al. (2016). Toxic epidermal necrolysis in adult T cell leukemia/lymphoma treated with mogamulizumab. *Annals of Hematology*, *95*, 661–662.

Tartour, E., Gey, A., Sastre-Garau, X., Lombard Surin, I., Mosseri, V., & Fridman, W. H. (1998). Prognostic value of intratumoral interferon gamma messenger RNA expression in invasive cervical carcinomas. *Journal of the National Cancer Institute*, *90*, 287–294.

Tas, F., Oguz, H., Argon, A., Duranyildiz, D., Camlica, H., Yasasever, V., et al. (2005). The value of serum levels of IL-6, TNF-alpha, and erythropoietin in metastatic malignant melanoma: Serum IL-6 level is a valuable prognostic factor at least as serum LDH in advanced melanoma. *Medical Oncology*, *22*, 241–246.

Thomas, C., Bazan, J. F., & Garcia, K. C. (2012). Structure of the activating IL-1 receptor signaling complex. *Nature Structural & Molecular Biology*, *19*, 455–457.

Thomas, H. G., & Richmond, A. (1988). Immunoaffinity purification of melanoma growth stimulatory activity. *Archives of Biochemistry and Biophysics*, *260*, 719–724.

Thomson, S. J., Goh, F. G., Banks, H., Krausgruber, T., Kotenko, S. V., Foxwell, B. M., et al. (2009). The role of transposable elements in the regulation of IFN-lambda1 gene

expression. *Proceedings of the National Academy of Sciences of the United States of America*, *106*, 11564–11569.
Tobiume, K., Matsuzawa, A., Takahashi, T., Nishitoh, H., Morita, K., Takeda, K., et al. (2001). ASK1 is required for sustained activations of JNK/p38 MAP kinases and apoptosis. *EMBO Reports*, *2*, 222–228.
Tsaur, I., Noack, A., Makarevic, J., Oppermann, E., Waaga-Gasser, A. M., Gasser, M., et al. (2015). CCL2 Chemokine as a potential biomarker for prostate cancer: A pilot study. *Cancer Research and Treatment*, *47*, 306–312.
Tselepis, C., Perry, I., Dawson, C., Hardy, R., Darnton, S. J., McConkey, C., et al. (2002). Tumour necrosis factor-alpha in Barrett's oesophagus: A potential novel mechanism of action. *Oncogene*, *21*, 6071–6081.
Tsuchiyama, T., Nakamoto, Y., Sakai, Y., Mukaida, N., & Kaneko, S. (2008). Optimal amount of monocyte chemoattractant protein-1 enhances antitumor effects of suicide gene therapy against hepatocellular carcinoma by M1 macrophage activation. *Cancer Science*, *99*, 2075–2082.
Turk, J. L. (1994). Inflammation: John Hunter's "A treatise on the blood, inflammation and gun-shot wounds". *International Journal of Experimental Pathology*, *75*, 385–395.
Ueda, R. (2015). Clinical application of anti-CCR4 monoclonal antibody. *Oncology*, *89*(Suppl. 1), 16–21.
Ugurel, S., Schrama, D., Keller, G., Schadendorf, D., Brocker, E. B., Houben, R., et al. (2008). Impact of the CCR5 gene polymorphism on the survival of metastatic melanoma patients receiving immunotherapy. *Cancer Immunology, Immunotherapy*, *57*, 685–691.
Umekawa, K., Kimura, T., Kudoh, S., Suzumura, T., Oka, T., Nagata, M., et al. (2013). Plasma RANTES, IL-10, and IL-8 levels in non-small-cell lung cancer patients treated with EGFR-TKIs. *BMC Research Notes*, *6*, 139.
U'Ren, L., Guth, A., Kamstock, D., & Dow, S. (2010). Type I interferons inhibit the generation of tumor-associated macrophages. *Cancer Immunology, Immunotherapy*, *59*, 587–598.
Valkovic, T., Lucin, K., Krstulja, M., Dobi-Babic, R., & Jonjic, N. (1998). Expression of monocyte chemotactic protein-1 in human invasive ductal breast cancer. *Pathology, Research and Practice*, *194*, 335–340.
van den Elsen, P. J. (2011). Expression regulation of major histocompatibility complex class I and class II encoding genes. *Frontiers in Immunology*, *2*, 48.
Verhoef, C., de Wilt, J. H., Grunhagen, D. J., van Geel, A. N., ten Hagen, T. L., & Eggermont, A. M. (2007). Isolated limb perfusion with melphalan and TNF-alpha in the treatment of extremity sarcoma. *Current Treatment Options in Oncology*, *8*, 417–427.
Vinante, F., Rigo, A., Tecchio, C., Morosato, L., Nadali, G., Ricetti, M. M., et al. (1998). Serum levels of p55 and p75 soluble TNF receptors in adult acute leukaemia at diagnosis: Correlation with clinical and biological features and outcome. *British Journal of Haematology*, *102*, 1025–1034.
Vlahakis, S. R., Villasis-Keever, A., Gomez, T., Vanegas, M., Vlahakis, N., & Paya, C. V. (2002). G protein-coupled chemokine receptors induce both survival and apoptotic signaling pathways. *The Journal of Immunology*, *169*, 5546–5554.
Voronov, E., Shouval, D. S., Krelin, Y., Cagnano, E., Benharroch, D., Iwakura, Y., et al. (2003). IL-1 is required for tumor invasiveness and angiogenesis. *Proceedings of the National Academy of Sciences of the United States of America*, *100*, 2645–2650.
Wajant, H., Pfizenmaier, K., & Scheurich, P. (2003). Tumor necrosis factor signaling. *Cell Death and Differentiation*, *10*, 45–65.
Wakita, D., Chamoto, K., Ohkuri, T., Narita, Y., Ashino, S., Sumida, K., et al. (2009). IFN-gamma-dependent type 1 immunity is crucial for immunosurveillance against squamous cell carcinoma in a novel mouse carcinogenesis model. *Carcinogenesis*, *30*, 1408–1415.
Wakita, D., Sumida, K., Iwakura, Y., Nishikawa, H., Ohkuri, T., Chamoto, K., et al. (2010). Tumor-infiltrating IL-17-producing gammadelta T cells support the progression of tumor by promoting angiogenesis. *European Journal of Immunology*, *40*, 1927–1937.

Wall, L., Burke, F., Barton, C., Smyth, J., & Balkwill, F. (2003). IFN-gamma induces apoptosis in ovarian cancer cells in vivo and in vitro. *Clinical Cancer Research, 9*, 2487–2496.

Wan, S., Pestka, S., Jubin, R. G., Lyu, Y. L., Tsai, Y. C., & Liu, L. F. (2012). Chemotherapeutics and radiation stimulate MHC class I expression through elevated interferon-beta signaling in breast cancer cells. *PLoS One*, 7, e32542.

Wan, S., Zhao, E., Kryczek, I., Vatan, L., Sadovskaya, A., Ludema, G., et al. (2014). Tumor-associated macrophages produce interleukin 6 and signal via STAT3 to promote expansion of human hepatocellular carcinoma stem cells. *Gastroenterology, 147*, 1393–1404.

Wang, A., Fairhurst, A. M., Tus, K., Subramanian, S., Liu, Y., Lin, F., et al. (2009). CXCR4/CXCL12 hyperexpression plays a pivotal role in the pathogenesis of lupus. *Journal of Immunology, 182*, 4448–4458.

Wang, B., Hendricks, D. T., Wamunyokoli, F., & Parker, M. I. (2006). A growth-related oncogene/CXC chemokine receptor 2 autocrine loop contributes to cellular proliferation in esophageal cancer. *Cancer Research, 66*, 3071–3077.

Wang, J. M., Hishinuma, A., Oppenheim, J. J., & Matsushima, K. (1993). Studies of binding and internalization of human recombinant monocyte chemotactic and activating factor (MCAF) by monocytic cells. *Cytokine, 5*, 264–275.

Wang, L., Jiang, Y., Zhang, Y., Wang, Y., Huang, S., Wang, Z., et al. (2012). Association analysis of IL-17A and IL-17F polymorphisms in Chinese Han women with breast cancer. *PLoS One*, 7, e34400.

Wang, Y., Kato, N., Hoshida, Y., Yoshida, H., Taniguchi, H., Goto, T., et al. (2003). Interleukin-1beta gene polymorphisms associated with hepatocellular carcinoma in hepatitis C virus infection. *Hepathology, 37*, 65–71.

Wang, B., Khachigian, L. M., Esau, L., Birrer, M. J., Zhao, X., Parker, M. I., et al. (2009). A key role for early growth response-1 and nuclear factor-kappaB in mediating and maintaining GRO/CXCR2 proliferative signaling in esophageal cancer. *Molecular Cancer Research*, 7, 755–764.

Wang, K., Kim, M. K., Di Caro, G., Wong, J., Shalapour, S., Wan, J., Zhang, W., Zhong, Z., Sanchez-Lopez, E., Wu, L. W., Taniguchi, K., Feng, Y., Fearon, E., Grivennikov, S. I., & Karin, M. (2014). Interleukin-17 receptor a signaling in transformed enterocytes promotes early colorectal tumorigenesis. *Immunity, 41*, 1052–1063.

Wang, S. C., Lin, J. K., Wang, H. S., Yang, S. H., Li, A. F., & Chang, S. C. (2010). Nuclear expression of CXCR4 is associated with advanced colorectal cancer. *International Journal of Colorectal Disease, 25*, 1185–1191.

Wang, S. W., Liu, S. C., Sun, H. L., Huang, T. Y., Chan, C. H., Yang, C. Y., et al. (2015). CCL5/CCR5 axis induces vascular endothelial growth factor-mediated tumor angiogenesis in human osteosarcoma microenvironment. *Carcinogenesis, 36*, 104–114.

Wang, Z., Ma, Q., Li, P., Sha, H., Li, X., & Xu, J. (2013). Aberrant expression of CXCR4 and beta-catenin in pancreatic cancer. *Anticancer Research, 33*, 4103–4110.

Wang, S., Tyring, S. K., Townsend, C. M., Jr., & Evers, B. M. (1998). Interferon-mediated activation of the STAT signaling pathway in a human carcinoid tumor. *Annals of Surgical Oncology, 5*, 642–649.

Warzocha, K., Bienvenu, J., Ribeiro, P., Moullet, I., Dumontet, C., Neidhardt-Berard, E. M., et al. (1998). Plasma levels of tumour necrosis factor and its soluble receptors correlate with clinical features and outcome of Hodgkin's disease patients. *British Journal of Cancer*, 77, 2357–2362.

Warzocha, K., Salles, G., Bienvenu, J., Bastion, Y., Dumontet, C., Renard, N., et al. (1997). Tumor necrosis factor ligand-receptor system can predict treatment outcome in lymphoma patients. *Journal of Clinical Oncology, 15*, 499–508.

Watson, C., Jenkinson, S., Kazmierski, W., & Kenakin, T. (2005). The CCR5 receptor-based mechanism of action of 873140, a potent allosteric noncompetitive HIV entry inhibitor. *Molecular Pharmacology, 67*, 1268–1282.

Weng, C. J., Chien, M. H., Lin, C. W., Chung, T. T., Zavras, A. I., Tsai, C. M., et al. (2010). Effect of CC chemokine ligand 5 and CC chemokine receptor 5 genes polymorphisms on the risk and clinicopathological development of oral cancer. *Oral Oncology*, *46*, 767–772.

Wessendorf, J. H., Garfinkel, S., Zhan, X., Brown, S., & Maciag, T. (1993). Identification of a nuclear localization sequence within the structure of the human interleukin-1 alpha precursor. *The Journal of Biological Chemistry*, *268*, 22100–22104.

Westwick, J. K., Weitzel, C., Minden, A., Karin, M., & Brenner, D. A. (1994). Tumor necrosis factor alpha stimulates AP-1 activity through prolonged activation of the c-Jun kinase. *The Journal of Biological Chemistry*, *269*, 26396–26401.

Wetzler, M., Kurzrock, R., Estrov, Z., Kantarjian, H., Gisslinger, H., Underbrink, M. P., et al. (1994). Altered levels of interleukin-1 beta and interleukin-1 receptor antagonist in chronic myelogenous leukemia: Clinical and prognostic correlates. *Blood*, *84*, 3142–3147.

Williamson, B. D., Carswell, E. A., Rubin, B. Y., Prendergast, J. S., & Old, L. J. (1983). Human tumor necrosis factor produced by human B-cell lines: Synergistic cytotoxic interaction with human interferon. *Proceedings of the National Academy of Sciences of the United States of America*, *80*, 5397–5401.

Wilson, C., Maxwell, P. J., Longley, D. B., Wilson, R. H., Johnston, P. G., & Waugh, D. J. (2012). Constitutive and treatment-induced CXCL8-signalling selectively modulates the efficacy of anti-metabolite therapeutics in metastatic prostate cancer. *PLoS One*, *7*, e36545.

Wojciechowska-Lacka, A., Matecka-Nowak, M., Adamiak, E., Lacki, J. K., & Cerkaska-Gluszak, B. (1996). Serum levels of interleukin-10 and interleukin-6 in patients with lung cancer. *Neoplasma*, *43*, 155–158.

Wouters, M., Dijkgraaf, E. M., Kuijjer, M. L., Jordanova, E. S., Hollema, H., Welters, M., et al. (2014). Interleukin-6 receptor and its ligand interleukin-6 are opposite markers for survival and infiltration with mature myeloid cells in ovarian cancer. *Oncoimmunology*, *3*, e962397.

Wright, J. F., Bennett, F., Li, B., Brooks, J., Luxenberg, D. P., Whitters, M. J., et al. (2008). The human IL-17F/IL-17A heterodimeric cytokine signals through the IL-17RA/IL-17RC receptor complex. *The Journal of Immunology*, *181*, 2799–2805.

Wu, M., & Assassi, S. (2013). The role of type 1 interferon in systemic sclerosis. *Frontiers in Immunology*, *4*, 266.

Wu, M. Z., Chen, S. F., Nieh, S., Benner, C., Ger, L. P., Jan, C. I., et al. (2015). Hypoxia drives breast tumor malignancy through a TET-TNFalpha-p38-MAPK signaling axis. *Cancer Research*, *75*, 3912–3924.

Wu, Y., Deng, J., Rychahou, P. G., Qiu, S., Evers, B. M., & Zhou, B. P. (2009). Stabilization of snail by NF-kappaB is required for inflammation-induced cell migration and invasion. *Cancer Cell*, *15*, 416–428.

Wu, C., Fernandez, S. A., Criswell, T., Chidiac, T. A., Guttridge, D., Villalona-Calero, M., et al. (2013). Disrupting cytokine signaling in pancreatic cancer: A phase I/II study of etanercept in combination with gemcitabine in patients with advanced disease. *Pancreas*, *42*, 813–818.

Wurth, R., Tarn, K., Jernigan, D., Fernandez, S. V., Cristofanilli, M., Fatatis, A., et al. (2015). A preclinical model of inflammatory breast cancer to study the involvement of CXCR4 and ACKR3 in the metastatic process. *Translational Oncology*, *8*, 358–367.

Xiao, Y. C., Yang, Z. B., Cheng, X. S., Fang, X. B., Shen, T., Xia, C. F., et al. (2015). CXCL8, overexpressed in colorectal cancer, enhances the resistance of colorectal cancer cells to anoikis. *Cancer Letters*, *361*, 22–32.

Xu, D., Matsuo, Y., Ma, J., Koide, S., Ochi, N., Yasuda, A., et al. (2010). Cancer cell-derived IL-1alpha promotes HGF secretion by stromal cells and enhances metastatic potential in pancreatic cancer cells. *Journal of Surgical Oncology*, *102*, 469–477.

Xu, L., Zhu, Y., An, H., Liu, Y., Lin, Z., Wang, G., et al. (2015). Clinical significance of tumor-derived IL-1beta and IL-18 in localized renal cell carcinoma: Associations with recurrence and survival. *Urologic Oncology*, *33*, 68.e9–68.e16.

Yaal-Hahoshen, N., Shina, S., Leider-Trejo, L., Barnea, I., Shabtai, E. L., Azenshtein, E., et al. (2006). The chemokine CCL5 as a potential prognostic factor predicting disease progression in stage II breast cancer patients. *Clinical Cancer Research, 12*, 4474–4480.

Yamashiro, S., Takeya, M., Nishi, T., Kuratsu, J., Yoshimura, T., Ushio, Y., et al. (1994). Tumor-derived monocyte chemoattractant protein-1 induces intratumoral infiltration of monocyte-derived macrophage subpopulation in transplanted rat tumors. *The American Journal of Pathology, 145*, 856–867.

Yang, Y., Qiu, X. Q., Yu, H. P., Zeng, X. Y., & Bei, C. H. (2012). TNF-alpha -863 polymorphisms and the risk of hepatocellular carcinoma. *Experimental and Therapeutic Medicine, 3*, 513–518.

Yang, H. P., Woodson, K., Taylor, P. R., Pietinen, P., Albanes, D., Virtamo, J., et al. (2006). Genetic variation in interleukin 8 and its receptor genes and its influence on the risk and prognosis of prostate cancer among Finnish men in a large cancer prevention trial. *European Journal of Cancer Prevention, 15*, 249–253.

Yano, H., Iemura, A., Haramaki, M., Ogasawara, S., Takayama, A., Akiba, J., et al. (1999). Interferon alfa receptor expression and growth inhibition by interferon alfa in human liver cancer cell lines. *Hepatology, 29*, 1708–1717.

Yasuhara, R., Irie, T., Suzuki, K., Sawada, T., Miwa, N., Sasaki, A., et al. (2015). The beta-catenin signaling pathway induces aggressive potential in breast cancer by up-regulating the chemokine CCL5. *Experimental Cell Research, 338*, 22–31.

Yasuoka, H., Tsujimoto, M., Yoshidome, K., Nakahara, M., Kodama, R., Sanke, T., et al. (2008). Cytoplasmic CXCR4 expression in breast cancer: Induction by nitric oxide and correlation with lymph node metastasis and poor prognosis. *BMC Cancer, 8*, 340.

Yeh, K. Y., Wu, T. H., & Wu, T. L. (2016). Colorectal cancer cell-derived interleukin-6 enhances the phagocytic capacity and migration of THP-1 cells. *Cytokine, 79*, 82–89.

Yeoh, G. C., Ernst, M., Rose-John, S., Akhurst, B., Payne, C., Long, S., et al. (2007). Opposing roles of gp130-mediated STAT-3 and ERK-1/2 signaling in liver progenitor cell migration and proliferation. *Hepatology, 45*, 486–494.

Yi, E. H., Lee, C. S., Lee, J. K., Lee, Y. J., Shin, M. K., Cho, C. H., et al. (2013). STAT3-RANTES autocrine signaling is essential for tamoxifen resistance in human breast cancer cells. *Molecular Cancer Research, 11*, 31–42.

Yoshida, N., Ikemoto, S., Narita, K., Sugimura, K., Wada, S., Yasumoto, R., et al. (2002). Interleukin-6, tumour necrosis factor alpha and interleukin-1beta in patients with renal cell carcinoma. *British Journal of Cancer, 86*, 1396–1400.

Yu, L., Cecil, J., Peng, S. B., Schrementi, J., Kovacevic, S., Paul, D., et al. (2006). Identification and expression of novel isoforms of human stromal cell-derived factor 1. *Gene, 374*, 174–179.

Zaidi, M. R., & Merlino, G. (2011). The two faces of interferon-gamma in cancer. *Clinical Cancer Research, 17*, 6118–6124.

Zambra, F. M., Biolchi, V., Brum, I. S., & Chies, J. A. (2013). CCR2 and CCR5 genes polymorphisms in benign prostatic hyperplasia and prostate cancer. *Human Immunology, 74*, 1003–1008.

Zhang, Y., Deng, Y., Wendt, T., Liliensiek, B., Bierhaus, A., Greten, J., et al. (1996). Intravenous somatic gene transfer with antisense tissue factor restores blood flow by reducing tumor necrosis factor-induced tissue factor expression and fibrin deposition in mouse meth-A sarcoma. *The Journal of Clinical Investigation, 97*, 2213–2224.

Zhang, S. S., Han, Z. P., Jing, Y. Y., Tao, S. F., Li, T. J., Wang, H., et al. (2012). CD133 (+)CXCR4(+) colon cancer cells exhibit metastatic potential and predict poor prognosis of patients. *BMC Medicine, 10*, 85.

Zhang, J., Lu, Y., & Pienta, K. J. (2010). Multiple roles of chemokine (C-C motif) ligand 2 in promoting prostate cancer growth. *Journal of the National Cancer Institute, 102*, 522–528.

Zhang, Y., Lv, D., Kim, H. J., Kurt, R. A., Bu, W., Li, Y., et al. (2013). A novel role of hematopoietic CCL5 in promoting triple-negative mammary tumor progression by regulating generation of myeloid-derived suppressor cells. *Cell Research*, *23*, 394–408.

Zhang, X. W., Qin, X., Qin, C. Y., Yin, Y. L., Chen, Y., & Zhu, H. L. (2013). Expression of monocyte chemoattractant protein-1 and CC chemokine receptor 2 in non-small cell lung cancer and its significance. *Cancer Immunology, Immunotherapy*, *62*, 563–570.

Zhang, Y., Ren, J. S., Shi, J. F., Li, N., Wang, Y. T., Qu, C., et al. (2015). International trends in primary liver cancer incidence from 1973 to 2007. *BMC Cancer*, *15*, 94.

Zhang, C., Sturgis, E. M., Zheng, H., Song, X., Wei, P., Jin, L., et al. (2014). Genetic variants in TNF-alpha promoter are predictors of recurrence in patients with squamous cell carcinoma of oropharynx after definitive radiotherapy. *International Journal of Cancer*, *134*, 1907–1915.

Zhang, Y., Yao, F., Yao, X., Yi, C., Tan, C., Wei, L., et al. (2009). Role of CCL5 in invasion, proliferation and proportion of CD44+/CD24- phenotype of MCF-7 cells and correlation of CCL5 and CCR5 expression with breast cancer progression. *Oncology Reports*, *21*, 1113–1121.

Zhang, Y. L., Zhang, J., & Cui, L. Y. (2015). Interferon-alpha combined with lamivudine versus lamivudine monotherapy for the emergence of YMDD mutations in chronic hepatitis B infection: A meta-analysis of randomized controlled trials. *Hepato-Gastroenterology*, *62*, 133–139.

Zhao, X., Rong, L., Zhao, X., Li, X., Liu, X., Deng, J., et al. (2012). TNF signaling drives myeloid-derived suppressor cell accumulation. *The Journal of Clinical Investigation*, *122*, 4094–4104.

Zheng, L., Fisher, G., Miller, R. E., Peschon, J., Lynch, D. H., & Lenardo, M. J. (1995). Induction of apoptosis in mature T cells by tumour necrosis factor. *Nature*, *377*, 348–351.

Zheng, C., Huang, D. R., Bergenbrant, S., Sundblad, A., Osterborg, A., Björkholm, M., et al. (2000). Interleukin 6, tumour necrosis factor alpha, interleukin 1beta and interleukin 1 receptor antagonist promoter or coding gene polymorphisms in multiple myeloma. *British Journal of Haematology*, *109*, 39–45.

Zhuang, Z., Ju, H. Q., Aguilar, M., Gocho, T., Li, H., Iida, T., et al. (2015). IL1 receptor antagonist inhibits pancreatic cancer growth by abrogating NF-kappaB activation. *Clinical Cancer Research*, *22*(6), 1432–1444.

Zienolddiny, S., Ryberg, D., Maggini, V., Skaug, V., Canzian, F., & Haugen, A. (2004). Polymorphisms of the interleukin-1 beta gene are associated with increased risk of non-small cell lung cancer. *International Journal of Cancer*, *109*, 353–356.

Zlotnik, A., & Yoshie, O. (2000). Chemokines: A new classification system and their role in immunity. *Immunity*, *12*, 121–127.

Zoglmeier, C., Bauer, H., Norenberg, D., Wedekind, G., Bittner, P., Sandholzer, N., et al. (2011). CpG blocks immunosuppression by myeloid-derived suppressor cells in tumor-bearing mice. *Clinical Cancer Research*, *17*, 1765–1775.

Zou, A., Lambert, D., Yeh, H., Yasukawa, K., Behbod, F., Fan, F., et al. (2014). Elevated CXCL1 expression in breast cancer stroma predicts poor prognosis and is inversely associated with expression of TGF-beta signaling proteins. *BMC Cancer*, *14*, 781.

Zumwalt, T. J., Arnold, M., Goel, A., & Boland, C. R. (2015). Active secretion of CXCL10 and CCL5 from colorectal cancer microenvironments associates with GranzymeB+ CD8+ T-cell infiltration. *Oncotarget*, *6*, 2981–2991.

INDEX

Note: Page numbers followed by "*f*" indicate figures, and "*t*" indicate tables.

A

Actin dynamics, 231–232
Acute inflammation, 266
Adaptive immune cells, 61–62
Adhesion system
  focal adhesion kinase (FAK), 235–236
  integrins, 233–235
  Src family kinases (SFKs), 236–237
ADT. *See* Androgen deprivation therapy (ADT)
All-*trans* retinoic acid (atRA), 52–53
Amoeboid migration, 213–215, 214*f*
Amoeboid-to-collective transition (ACT), 214*f*
Amoeboid-to-mesenchymal (AMT) migration, 214*f*
Androgen deprivation therapy (ADT), 93–94
Antagomir, 198–199
Anticancer therapy
  implications for, 229–240
  interferons (IFNs) in, 317–321
  interleukins (IL) in, 281–283
  tumor necrosis factor-α (TNF-α) pathway in, 331–333
Antimigration therapy, 239–240
Apoptosis signal-regulating kinase (ASK1), 322
Autocrine TNF-α signaling, 330

B

Basement membrane (BM), 211–212
B-cell chronic lymphocytic leukemia (CLL), 176
Bioluminescence imaging (BLI), 10–11
Biomarkers
  definition, 17–18
  for metastasis, 17–20
Bladder cancer, 115*t*
Blebby amoeboid migration, 214*f*
BLI. *See* Bioluminescence imaging (BLI)
Blood-based tumor-activatable microcircle approach, 10
Bone marrow, 53–54
Bone microenvironment, extrinsic signals induce dormancy within, 53–54
BRAF mutation, 23–25
Breast cancer, 115*t*, 280–281
  hereditary modifiers of metastasis in, 82–93, 85*t*
  metastasis in mice, 189*f*
Breast cancer metastasis suppressor 1 (BRMS1), 114–118, 115*t*
  metastasis inhibited by, 117*t*
  role in human cancer, 126–128*t*
Bromodomain containing 4 (BRD4), 88

C

Cancer-associated fibroblasts (CAFs), 223–226
Cancer cell dormancy
  endosteal and perivascular niches support, 54–56
  reactivation from dormancy, 56–58
Cancer cell migration, 210–213
  amoeboid migration, 213–215
  collective cell migration, 216–218
  collective plasticity, 222
  collective-to-single-cell transition, 220–222
  and cytokines, 226–229
  growth factor modulation of, 226–229, 227–228*t*
  mechanisms, modes and plasticity of, 214*f*
  mesenchymal migration, 215–216
  metabolic control, 229, 230*t*
  modes by ECM organization, 223–226, 224–225*t*
  plasticity of single, 219–220
  therapeutic targeting of, 229–240
Cancer Genetic Markers of Susceptibility (CGEMS), 95–96
Cancer stem cells (CSCs), 195

Carbohydrate antigen 15.3 (CA15.3), 18–19
Carcinoembryonic antigen related cell adhesion molecule 1 (CEACAM-1), 9
Castration resistant prostate cancer (CRPC), 23–25
CAT. *See* Collective-to-amoeboid transition (CAT)
Cathepsin, 146
CCL2. *See* Chemokine C–C-motif ligand 2 (CCL2)
CCL5, 299–301
CEACAM-1. *See* Carcinoembryonic antigen related cell adhesion molecule 1 (CEACAM-1)
Cell Adhesion Molecule 1 (Cadm1), 90–91
Cell–cell adhesion, 212, 214*f*, 217, 221
Cell–cell interaction, 5
Cell–ECM adhesion, 219–220
Cell migration, cancer, 210–213
Cell subset, 240
Cellular dormancy, 46–47, 62–63, 62*f*
Cellular homeostasis, 287
ceRNAs. *See* Competing endogenous RNAs (ceRNAs)
CGEMS. *See* Cancer Genetic Markers of Susceptibility (CGEMS)
Chemokine C–C-motif ligand 2 (CCL2), 59, 285*f*, 289–299
Chemokines, 266–267, 284–307, 287*t*
  ligands in cancer, 288–289*t*
  receptor antagonist in clinical trial, 309*t*
  receptor genes in cancer, 290–291*t*
  RNA and protein expression patterns, 296–298*t*
    of C–C, 292–293*t*
    of C–X–C, 294–295*t*
  in therapy, 307–308, 308*f*
  transduce signals, 284–286
Chemotherapy, 148, 154–155
Chromatin, 112–113
  immunoprecipitation assay, 124
  remodeling, 112–113, 118–119, 122–124
Chronic inflammation, 329–330
Chronic lymphocytic leukemia (CLL), 176
Circulating tumor cells (CTCs), 3–4, 16–17, 23–26, 218–219
CLL. *See* Chronic lymphocytic leukemia (CLL)
cMET, 239
CoCo, 57–58
Coley's toxin, 321
Collective cell migration, 216–218
Collective-to-amoeboid transition (CAT), 214*f*
Collective-to-single-cell transition, 220–222
Competing endogenousRNAs (ceRNAs), 185
Cortical actin network contraction, 213–215
CRPC. *See* Castration resistant prostate cancer (CRPC)
CTCs. *See* Circulating tumor cells (CTCs)
CXCL1, 301–303
CXCL8, 303–304
CXCL12, 286*f*, 304–307
Cytokines, 226–229
CytoSelect™ Tumor-Endothelium Adhesion assay, 5
Cytoskeletal organization and function, 231–233

D

Diasporin pathway, 88–89
Disseminated tumor cells (DTCs), 46–47, 49, 52–54
  dormancy marker expression in, 48*f*
  dormant, 46–47, 49–50, 54–57
  extrinsic signals inducing dormancy in, 55*f*
  solitary, 49, 59–61
Distant metastasis-free survival (DMFS), 91
Dormancy. *See also* Cancer cell dormancy; Tumor cell dormancy
  cellular, 46–47, 62–63, 62*f*
  immune system and, 58–65
  macrophages on, effects of, 60*f*
  NK cells on, effects of, 60*f*
  population-based, 54–56, 62–63, 62*f*
  T-cells on, effects of, 62–63, 62*f*
Dormancy-inducing signaling pathway, 48*f*
Dormant DTCs, 46–47, 49–50, 54–57
Doxorubicin-loaded nanobubbles, 27
DTCs. *See* Disseminated tumor cells (DTCs)

E

E-cadherin, 190–192, 191*f*
ECM. *See* Extracellular matrix (ECM)
Ectopic growth
  macrophages, 151–152
  natural killer (NK) cells, 154
  neutrophils, 152–153
  T lymphocytes, 153–154
EGFR. *See* Epidermal growth factor receptor (EGFR)
EMAP II. *See* Endothelial monocyte activating peptide II (EMAP II)
EMT. *See* Epithelial–mesenchymal transition (EMT)
Endothelial cells, 4–5
Endothelial monocyte activating peptide II (EMAP II), 326–329
Epidermal growth factor (EGF), 145–146
Epidermal growth factor receptor (EGFR), 19–20
Epigenetic regulation, 112
Epigenetics, 113, 118–119
Epithelial–mesenchymal transition (EMT), 197, 199–200, 214*f*, 215, 220–221
  induction of, 186–192
  role of, 196–197
Exonic miRNA genes, 170–171
Exportin-5, 169–170
Expression QTL (eQTL) mapping, 87–88
Extracellular matrix (ECM), 146, 214*f*, 215–216
  gene expression, 87–88
  organization, 223–226
  proteins, 5

F

Filopodal amoeboid migration, 214*f*
Fluorescence resonance energy transfer (FRET), 122–123
Focal adhesion kinase (FAK), 235–236
  inhibitors, 235–236

G

Gap junctional intercellular communication (GJIC), 118
Gas filled spheres, 27
G-CSF expression, 147–148
Gene expression, metastatic, 74–75
Gene polymorphism, 270
Genes
  metastasis-associated, 112
  metastasis-promoting, 113
Genetic analysis human cohorts, 80–82
Genome-wide association studies (GWAS), 75–76, 81–82
Germline polymorphism, 75–76, 77*f*, 81–82, 87–88, 94–95
GJIC. *See* Gap junctional intercellular communication (GJIC)
Glioblastoma, 330
Green fluorescent protein (GFP), 181
GWAS. *See* Genome-wide association studies (GWAS)

H

HCC. *See* Hepatocellular carcinoma (HCC)
Hematopoietic stem cells (HSCs), 64, 181
Hepatocellular carcinoma (HCC), 276–278, 280–281
Hepatocyte growth factor (HGF), 221
Histone acetyltransferases (HAT), 112, 123–124
Histone deacetylases (HDAC), 112–113, 123–124
HSCs. *See* Hematopoietic stem cells (HSCs)
Human cohorts, genetic analysis, 80–82
Human melanoma xenograft mouse model, 9
Human patient-derived xenograft mouse model, 8
Human sodium–iodide symporter (hNIS), 15–16

I

IFN alpha receptor (IFNAR), 310
IFN gamma receptor (IFNGR), 310
IFN receptors (IFNAR2), 312–313
IFNs. *See* Interferons (IFNs)
IHC. *See* Immunohistochemical analysis (IHC)
IL. *See* Interleukin (IL)
IL1R acceptor protein (IL-1RAP), 269*f*
IL-1 receptor antagonist (IL-1Ra), 268–270
IL-1 receptor-associated kinase (IRAK), 268–270, 269*f*

Immune editing hypothesis, 63–64
Immune regulation, of metastatic cascade, 142–143*f*
Immune system and dormancy, 58–65
Immunohistochemical analysis (IHC), 9
Inducible nitric oxide synthase (iNOS), 152–153
Inflammation, 266
  acute, 266
Inflammatory process, 266
Inflammatory responses, 266–267
Inhibitors
  focal adhesion kinase (FAK), 235–236
  microtubules, 232
iNOS. *See* Inducible nitric oxide synthase (iNOS)
Insulin-like growth factor receptor (IGFR) signaling, 226
Integrins, 233–235
Interferons (IFNs), 308–321
  in anticancer therapy, 317–321
  IFN-α, 313–315
  IFN-β, 313–315
  IFN-γ, 315–317
  IFN-λ, 317
  RNA and protein expression patterns, 314*t*
  signal transduction pathway, 311–312*f*
Intergenic miRNA genes, 170–171
Interleukin (IL), 267–283
  in anticancer therapy, 281–283, 318*f*, 319–320*t*
  pathways in clinical trials, 283*t*
  receptor in cancer, 271–272*t*
  RNA and protein expression patterns of, 273–275*t*
  role in cancer, 282*f*
Interleukin-1 (IL-1), 268–272
  signal transduction pathway, 269*f*
Interleukin-6 (IL-6), 272–279
  RNA and protein expression patterns, 278*t*
  signal transduction, 277*f*
Interleukin-17 (IL-17), 279–281
Invasion, 113
  of tumor cells, 142–143*f*
In vitro models, 4–6
In vivo metastasis, 49
In vivo models, 6–8
IRAK. *See* IL-1 receptor-associated kinase (IRAK)
Iron oxide nanoparticles (IONPs), 28

J

Janus kinase (JAK), 272–276, 277*f*

K

Kaplan–Meier plots, 51

L

Latency competent cancer cells, 61
Leukemia inhibitory factor receptor (LIFR), 190, 191*f*
lin-4, molecular hallmarks of, 169*f*
Liver metastatic lesion, 10–11
lncRNAs. *See* Long noncoding RNAs (lncRNAs)
Locked nucleic acids (LNAs), 198–199
Long noncoding RNAs (lncRNAs), 166
Luciferin, 10–11

M

Macrophage-derived matrix metalloproteinases (MMPs), 146, 237
  inhibition therapy, 238
  stromal-derived, 142–143*f*
Macrophages, 141–143
  on dormancy, 60*f*
  ectopic growth, 151–152
  invasion and intravasation, 145–147
Magnetic resonance imaging (MRI), 11–12
Major histocompatibility complex (MHC) proteins, 308–310
Mammary carcinoma cells, 281
MAMs. *See* Metastasis-associated macrophages (MAMs)
MDSCs. *See* Myeloid derived suppressor cells (MDSCs)
Melanoma, 115*t*
  hereditary modifiers of metastasis in, 96–99
Mesenchymal migration, 215–216
Mesenchymal stem cells (MSCs), 278
Mesenchymal-to-epithelial transition (MET), 186–187, 214*f*

Messenger RNAs (mRNAs), 166
  degradation, 172
Metabolic control, cancer cell migration, 229, 230*t*
Metastasis, 2–3, 3*f*, 59, 112, 140
  biomarkers for, 17–20
  definition, 74
  epidemiological analysis, 76–80
  formation, 46
  hereditary modifiers of
    breast cancer, 82–93, 85*t*
    melanoma, 96–99
    prostate cancer, 93–96, 97–98*t*
  inhibition by *BRMS1*, 117*t*
  microRNAs in, 188*t*, 193–194*f*
  suppression, 114–116
  susceptibility loci, 83*f*
  systemic, 211–212
  technical monitoring of, 239–240
Metastasis-associated genes, 112
Metastasis-associated macrophages (MAMs), 59–61
Metastasis-associated microRNAs, 124–125
Metastasis-associated protein 1 (MTA1), 112–113
Metastasis models, 6–7
Metastasis-promoting genes, 113
Metastasis-suppressing miRNAs, 195–197
Metastasis suppression, 114–116, 118, 121, 123–124
Metastatic cascade, immune regulation of, 142–143*f*
Metastatic evasion, diversity of, 218–219
Metastatic gene expression, 74–75
Metastatic lesion, 211
Metastatic lesions detection
  biomarkers for metastasis, 17–20
  challenges, 29–30
  circulating tumor cells, 16–17, 23–26
  combining imaging with therapy (theranostics), 26–29
  imaging procedures, 20–23
  lab tests/histopathology, 9
  molecular-genetic imaging (promoter-based protocols), 12–16, 13*f*
  noninvasive blood tests, 9–10
  small animal imaging, 10–12
Microbubbles, 27
Microenvironment, 112, 118, 124–125
MicroRNAs (miRNAs), 91–92, 166, 168–169
  biogenesis and mechanisms of action, 169–173
  canonical pathway, 170*f*
  discovery of, 166–169, 167–168*t*
  expression, 174
    deregulated, 176–180, 178–180*t*
  function, 173–176
  inhibitors, 198–199
  in metastasis, 188*t*, 193–194*f*
  metastasis-associated, 124–125
  metastasis-promoting, 187–195
  metastasis-suppressing, 195–197
  mimics, 197–198
  as oncogenes/tumor suppressors, 181–184
  regulation of metastasis, 186–197
  regulatory networks in cancer, 184–185
  as therapeutic targets, 197–200
Microtubule (MT), 232–233
  destabilization, 232–233
  inhibitors, 232
Migration
  amoeboid, 213–215
  inhibition by molecular targeted therapy, 238–239
Minimal residual disease, 64–65
mir-17–19b cluster, 182*f*
Mitochondrial DNA (mtDNA), 92–93
MMPs. *See* Macrophage-derived matrix metalloproteinases (MMPs)
Molecular-genetic approach, 28–29
Molecular targeted therapy, migration inhibition by, 238–239
Monocytes, 59–61, 60*f*, 141
  survival and extravasation, 148–151
Mouse mammary tumor virus (MMTV) promoter, 7–8, 84
Mouse model
  PyMT, 84–86
  transgenic, 83*f*, 84, 99
mRNAs. *See* Messenger RNAs (mRNAs)
MSCs. *See* Mesenchymal stem cells (MSCs)
mtDNA. *See* Mitochondrial DNA (mtDNA)
Mucins, 18–19
Myeloid cells, 59, 282*f*

Myeloid derived suppressor cells (MDSCs), 280–281
Myosin II activity, 213–215

N

Nanobubbles, doxorubicin-loaded, 27
Nanoparticles, 28
Natural killer (NK) cells, 145
on dormancy, 60*f*
ectopic growth, 154
survival and extravasation, 151
Necdin (Ndn), 88–89
NED. *See* No evidence of disease (NED)
Neoplastic cells, 111
Neutrophils, 144
ectopic growth, 152–153
invasion and intravasation, 147–148
survival and extravasation, 149–150
NFkB, 124
NK cells. *See* Natural killer (NK) cells
No evidence of disease (NED), 46–47
Noncoding RNAs, 166–169
Noninvasive blood test, 9–10
Nonsmall cell lung cancer (NSCLC), 19–20, 115*t*
Nonsynonymous polymorphism, 87–88
NSCLC. *See* Nonsmall cell lung cancer (NSCLC)
Nucleosome remodeling and histone deacetylase (NuRD), 112–113

O

Oncogene, microRNAs as, 181–184
Oncomine, 190
Oncomir, 183, 192
Ovarian cancer, 115*t*

P

Perforin, 154
Pericellular proteolysis, 214*f*, 222
Phosphorylation, 121
Plasticity, of single cancer cell migration, 219–220
Polymorphism
gene, 270
germline, 75–76, 94–95
nonsynonymous, 87–88
single nucleotide, 271–272*t*
Polyoma virus middle T antigen (PyMT), 7–8
Prader–Willi syndrome, 88–89
Pre-miRNAs, 169–170
Primary tumor cells, 2–3, 211
Progression model, 74–75
Prostate cancer, hereditary modifiers of metastasis in, 93–96, 97–98*t*
Prostate, Lung, Colorectal, and Ovarian (PLCO) cancer screening trial, 95–96
Prostate-specific antigen (PSA), 18
promoter/enhancer, 13–14
Proteases, 237–238
Protein expression
patterns of C–C chemokine, 292–293*t*
patterns of C–X–C chemokine, 294–295*t*
Proteolysis, 214*f*, 216, 222
PSA. *See* Prostate-specific antigen (PSA)
PyMT. *See* Polyoma virus middle T antigen (PyMT)
Pyruvate kinase M2 (PKM2), 194

Q

Quantitative trait locus (QTL) mapping, 84–86, 95–96
Quantum dots, 26–27

R

Radionuclides, 11–12
Recombinant human TNF-α, 331–333
Rectal cancer, 115*t*
RhoA via rho kinase (ROCK) signaling, 213–215
Rho GTPases, 231
Ribosomal RNA Processing 1 Homolog B (Rrp1b), 87–88
Ribosomal RNAs (rRNAs), 166
RNA-induced silencing complex (RISC), 170*f*
RNA interference (RNAi), 166
RNAs, noncoding, 166–169
ROCK inhibition, 231–232
rRNAs. *See* Ribosomal RNAs (rRNAs)

S

Scratch–wound assay, 5
Secreted embryonic alkaline phosphatase (SEAP), 10, 14

SFKs. *See* Src family kinases (SFKs)
Signal-Induced Proliferation-Associated gene 1 (Sipa1), 84–86
Signal transduction, interleukin-6 (IL-6), 277*f*
Signal transduction pathway
CCL2, 285*f*
CXCL12 chemokine, 286*f*
interferons (IFNs), 311–312*f*
interleukin-1 (IL-1), 269*f*
tumor necrosis factor (TNF), 323*f*
SIN3 (switchindependent 3), 113, 123–124
Single cancer cell migration, plasticity of, 219–220
Single nucleotide polymorphisms (SNPs), 81–82
chemokine receptor genes, 290–291*t*
genes encoding chemokine ligands, 288–289*t*
genes encoding interleukin, 271–272*t*
Small interfering RNAs (siRNAs), 166
Small nucleolar RNAs (snoRNAs), 166
SNPs. *See* Single nucleotide polymorphisms (SNPs)
Somatic mutation, 74–75
Sphingosine kinase 1 (SK1)/S1P signaling, 124
sRANKL, 54–56
Src family kinases (SFKs), 236–237
Stromal-derived MMPs, 142–143*f*
Suppressor of defective silencing 3 (SUDS3), 118–119
Susceptibility loci, 81–82
metastasis, 83*f*
Systemic metastasis, 211–212
Systems genetics approach, 75–76, 82–99, 83*f*

T

T-cells
on dormancy, 62–63, 62*f*
response, 282*f*
TCGA. *See* The Cancer Genome Atlas (TCGA)
TGFβ, 238–239
The Cancer Genome Atlas (TCGA), 190
Theranostics, 26–29
Therapeutic target, microRNAs as, 197–200
Thrombospondin-1 (TSP-1), 56
TLRs. *See* Toll-like receptors (TLRs)
T lymphocytes, 144–145
ectopic growth, 153–154
invasion and intravasation, 148
survival and extravasation, 150–151
TNF. *See* Tumor necrosis factor (TNF)
TNF receptor-associated factor 6 (TRAF6), 268–270
Toll/IL-1 receptor (TIR) domain, 268–270
Toll-like receptors (TLRs), 312–313
TRAF6. *See* TNF receptor-associated factor 6 (TRAF6)
Transcriptional amplification (TA), 14, 15*f*
Transcriptional enhancers, 13–14
Transfer RNAs (tRNAs), 166
Transgenic mouse model, 83*f*, 84, 99
tRNAs. *See* Transfer RNAs (tRNAs)
TSP-1. *See* Thrombospondin-1 (TSP-1)
Tumor angiogenesis, 282*f*
Tumor cell, 2–5, 210–212
invasion of, 142–143*f*
Tumor cell dormancy
intrinsic and extrinsic signals converge to induce, 51–53
models to study, 48–51, 50*t*
Tumor cell migration program, plasticity of, 219–222
Tumor microenvironment, 47–48
Tumor models, 240
Tumor necrosis factor (TNF), 321–333
RNA and protein expression, 325–326*t*
signal transduction pathway, 323*f*
Tumor necrosis factor-α (TNF-α)
as anticancer therapeutic in clinical trials, 327–328*t*, 331*f*
inhibitors as anticancer therapeutics in clinical trials, 332*t*
pathway in anticancer therapeutics, 331–333
signal transduction and expression patterns, 322–324
single nucleotide polymorphisms, 324*t*
as tumor promoter, 329–331
as tumor suppressor, 324–329

Tumor necrosis factor receptor-associated factor 6 (TRAF6), 269*f*
Tumor necrosis factor-related apoptosis inducing ligand (TRAIL), 154
Tumor promoter, tumor necrosis factor-α as, 329–331
Tumor-specific PEG-3 (progression-elevated gene-3) promoter, 10–11
Tumor suppressor
  microRNAs as, 181–184
  tumor necrosis factor-α (TNF-α) as, 324–329

U

Urokinase/plasminogen activator (uPA), 146

V

Vascular cell adhesion molecule-1 (VCAM-1), 151–152
Vascular endothelial growth factor (VEGF), 59
Vascular endothelial growth factor-A (VEGF-A), 146–147, 149
v-erb-b2 erythroblastic leukemia viral oncogene homolog 2 (HER2-neu), 19–20

W

Warburg effect, 229

X

Xenograft model, 6–7

FSC
www.fsc.org
MIX
Papier | Fördert gute Waldnutzung
FSC® C083411

Kolibri 360 GmbH
Ferdinand-Jühlke-Straße 7
99095 Erfurt, Deutschland
produktsicherheit@kolibri360.de